SSM
（Spring MVC+Spring+MyBatis）源码深入解析与企业项目实战

陈浩翔　厉森彪　石雷◎编著

中国水利水电出版社
www.waterpub.com.cn
· 北京 ·

内 容 提 要

《SSM（Spring MVC+Spring+MyBatis）源码深入解析与企业项目实战》是一本由浅入深，用简单易懂的语言讲解 Spring、Spring MVC、MyBatis 框架魅力的编程宝典。全书共五篇，分别是基础概念篇、Spring MVC 框架篇、MyBatis 框架篇、Spring 框架篇和项目实战篇。内容分为 20 章，从 Spring、Spring MVC、MyBatis 的基础开始，逐步深入至框架、核心应用源码的讲解，以及 SSM 在项目中的应用，让读者能理解框架的设计者为什么如此实现，又是如何实现的。本书不只是让读者学会如何使用框架，还要让读者学会如何实现框架，学习实现技术的方法，授读者以“渔”。

《SSM（Spring MVC+Spring+MyBatis）源码深入解析与企业项目实战》一书语言通俗易懂，案例丰富、实用性强，特别适合 Java 编程爱好者、想深入学习 Spring 源码的读者阅读。另外，本书也适合作为相关培训机构，以及中高等院校、应用型本科的教材使用。

图书在版编目（CIP）数据

SSM（Spring MVC+Spring+MyBatis）源码深入解析与企业项目实战 / 陈浩翔, 厉森彪, 石雷编著. — 北京：中国水利水电出版社，2023.3

ISBN 978-7-5226-1271-3

Ⅰ. ① S… Ⅱ. ①陈… ②厉… ③石… Ⅲ. ①企业—计算机网络 Ⅳ. ① TP393.18

中国国家版本馆 CIP 数据核字 (2023) 第 028609 号

书　　名	SSM（Spring MVC+Spring+MyBatis）源码深入解析与企业项目实战 SSM（Spring MVC+Spring+MyBatis）YUANMA SHENRU JIEXI YU QIYE XIANGMU SHIZHAN
作　　者	陈浩翔　厉森彪　石　雷　编著
出版发行	中国水利水电出版社 （北京市海淀区玉渊潭南路 1 号 D 座 100038） 网址：www.waterpub.com.cn E-mail：zhiboshangshu@163.com 电话：（010）62572966-2205/2266/2201（营销中心）
经　　售	北京科水图书销售有限公司 电话：（010）68545874、63202643 全国各地新华书店和相关出版物销售网点
排　　版	北京智博尚书文化传媒有限公司
印　　刷	北京富博印刷有限公司
规　　格	190mm×235mm　16 开本　37.25 印张　996 千字
版　　次	2023 年 3 月第 1 版　2023 年 3 月第 1 次印刷
印　　数	0001—3000 册
定　　价	129.80 元

在我过往的从业经历中，大多数证券、支付公司、商业银行的系统都会使用Java作为主要编程语言，而在Java企业编程领域，Spring框架一定是所有工程师最优的选择。如何高效、安全地用好Spring框架，是每一个Fintech工程师必须考虑掌握的技能。

本书从Spring源码出发，深入到Spring MVC、MyBatis等组件的原理，结合企业实战项目，深入浅出地带你快速掌握Spring框架，是一本不可多得的好书，十分推荐。

——付呗支付研发负责人　吴建南

此书围绕代码抽丝剥茧地阐述框架原理，把Spring非常实用的生态体系与项目实战结合起来，对于迷茫中需要提升硬核实力的程序员来说，是一个福音。

——阿里巴巴Java开发手册作者 孤尽

前　言

Spring 框架是近 20 年 Java 企业级项目中使用最广泛的框架之一，Spring 生态在 Java 生态中占据了半壁江山。近几年微服务的兴起，SpringBoot、SpringCloud 等基于 Spring 的框架也甚是流行。有些读者甚至只学习 SpringBoot、SpringCloud，跳过了 Spring，当遇到问题时，总要排查半天，甚至找不到问题所在。我的建议是，先学 Spring，再学基于 Spring 的框架（SpringBoot、SpringCloud 等）。有朋友说过："不读 Spring，枉学 Java。"这句话我是赞同的。在写书之初，我也问过一些朋友："如果你的时间只够在 Java 生态中挑选一个框架进行学习，你会选择什么框架？"几乎所有朋友的选择都是"Spring"。

我在学习 Java、学习 Spring 之初，写企业级 Java 项目时遇到过很多困难，当时有一个很简单的 Spring 同名注入问题，排查很久都找不到头绪。如果你学习过 Spring 的源码，便能知道为什么同名无法注入。Spring 就像一个不太智能的"主管"，管理你项目中的"生活起居"：当你需要一个"张三"，你可以找 Spring 要"张三"；但如果项目中有两个"张三"，主管是不清楚你需要哪个"张三"的，解决方案之一是你可以把两个"张三"分别命名为"张三 01""张三 02"。

你平时会遇到 Spring 注入问题、配置问题、初始化失败等问题，我的建议是所有的问题你不需要一一知道答案。本书不教你解决某个问题的方法，每个人在项目中遇到的问题不可能完全相同，但是 Spring 的问题万变不离其宗，解决方案都在 Spring 源码里，都在 Spring 模式上，也都在本书中。

本书将以 Spring 源码解析、小节代码实例、企业级项目实战相结合的方式对 Spring 的相关知识和企业项目进行全面的讲解。

本书内容

本书采用循序渐进的方式，从 Spring、Spring MVC、MyBatis 开始，对每个框架采取"基础知识→核心技术→项目应用→案例实战"的模式进行讲解，以每章的实战案例为辅助，深入浅出地讲解 Spring、Spring MVC、MyBatis 的知识点、解读源码及项目实战。

本书共分为五大篇，整体知识结构如图 0-1 所示。

基础知识
核心技术
项目应用
案例实战
第Ⅰ篇
基础概念
Spring、MyBatis、Spring MVC介绍
Redis介绍与其特性
Java的类加载器（ClassLoader）、反射机制、动态代理详解
单例模式、工厂模式、代理模式、策略模式和模板模式等设计模式详解
第Ⅱ篇
Spring MVC
框架
Spring MVC的架构、体系结构、组件介绍
Spring MVC的项目快速搭建、常用注解源码分析与使用
Spring MVC九大组件深入源码分析
Spring MVC的转发、重定向、文件上传下载、文件缓存、GZIP压缩与主题等基础功能的应用
Spring MVC的全局异常处理、拦截器、过滤器、数据交互与国际化配置等核心功能的应用
第Ⅲ篇
MyBatis
框架
MyBatis组件源码分析
MyBatis的XML配置与映射文件分析
MyBatis中动态SQL技术
MyBatis代码生成器使用与运行
第Ⅳ篇
Spring
框架
深入理解IoC（控制反转）与DI（依赖注入）
Spring中Bean、事件机制、定时器、SpEL等核心机制讲解与源码分析
Spring AOP深入分析与实例演练
Spring中的数据库事务管理
第Ⅴ篇
项目实战
项目之初的思考与设计
项目功能模块划分与数据库设计
项目框架搭建以及集成Spring、Spring MVC、MyBatis、Redis、Log4j等
自动登录、注册、Session共享、日志/异常拦截器、资源发表与下载、用户关注与收藏、用户评论、敏感词过滤、防范XSS攻击、交易等功能的实现
基础知识讲解
示意图、流程图、效果图
小节实例
源码讲解
代码全注释讲解
常用工具类
项目设计、数据设计
企业项目实战

图 0-1　整体知识结构

本书特色

系统的基础知识，丰富的实例案例

本书系统讲解了 Spring 各方面知识和应用，由浅入深、循序渐进，帮助读者奠定坚实的理论基础，通过书中的源码分析与介绍，可掌握源码分析的诀窍。本书的实战案例非常丰富，实战练习是掌握知识点最好的途径，所以在理论与源码分析过程中配合大量的示例、实例和实战案例，透彻详尽地讲述了实际开发中所需的各类知识，让读者做到知其所以然。

完整的企业级项目实战

本书第Ⅴ篇实现了一个企业级资源共享网站的后端功能，涵盖项目设计、项目功能模块划分、数据设计和相关功能的实现等完整企业级项目的开发流程。

GitHub 案例源码上传

本书中所有源码在 GitHub 仓库中均可以找到，下载和使用方便。本书保证仓库中的所有代码均可运行。

在线服务，持续更新

本书提供在线服务，随时随地可交流。技术在迭代，框架也在更新。GitHub 仓库、公众号会持续更新书本知识。另外，提供 QQ 群、读者圈等多渠道面对面交流服务。

本书配套资源及下载方法

为便于读者学习、理解本书内容，本书提供配套资源，读者可以通过以下 4 种方法获取：

（1）关注“程序编程之旅”微信公众号，并在后台回复“Spring 实战”下载所有资源或咨询关于本书的任何问题。

（2）扫描下方二维码加入本书学习交流圈，本书的勘误情况会在此圈中发布。此外，读者可以在此圈中分享读书心得、提出对本书的建议等。

（3）读者可加入 QQ 群：819539788（请注意加群时的提示，并根据提示加入对应的群），编者在线提供本书学习疑难解答等后续服务，让读者无障碍地快速学习本书。

（4）访问 GitHub 仓库下载。下载地址：https://github.com/chenhaoxiang/uifuture-ssm 。

读前须知

（1）本书从 Java 反射（Reflection）等基础出发，通过大量的案例使学习不再枯燥、教条。读者可以边学习边实践，避免学习流于表面、限于理论。

（2）作为源码分析书籍，本书源码较多，但因篇幅限制，分析时又不可能面面俱到，所以本书在很多位置添加了源码阅读的二维码。本书旨在给读者传达一种学习方法（技术学习的关键是方法），在很多实例和源码分析中都体现了学习方法的重要性，读者只要掌握了各种技术的运用方法，在学习更深入的知识时通过历史的学习方法就可大大提高学习效率。

（3）本书提供了大量示例，限于篇幅，部分示例没有提供完整的代码，读者应该将代码补充完整，然后再进行测试练习，或直接参考本书提供的源代码（需下载后使用），边学边练。在这里，强烈建议读者自己先动手编写代码，再参考源代码，这样可能会“踩到一些坑”，但印象会更加深刻，遇到什么问题还可以在读者群中沟通。

（4）本书提供了很多二维码链接以便读者参考学习，一些书中无法详细介绍的问题都可以通过这些链接找到答案。由于这些链接地址会因时间而有所变动或调整，所以在此说明，这些链接地址仅供参考，本书无法保证所有的地址是长期有效的。

本书适用对象

- Java 编程爱好者
- Spring 初学者以及想全面学习 Java 开发技术的人员
- 有一定 Java 基础，想深入学习 Spring 源码的人员
- 中高等院校、应用型本科相关专业学生
- 社会培训班的讲师和学员
- 从业 1~3 年的 Java 工程师

本书编者团队

本书由陈浩翔、厉森彪、石雷编著。陈浩翔承担了本书初稿的编写工作，石雷对初稿进行了第一次的专业知识校对与技术内容的补充，石雷与厉森彪对校对后的初稿再次进行了专业知识校对与技术内容补充。书籍内容偏多，编写耗时两年。在这里，由衷地感谢厉森彪和石雷的支持、指导和帮助。如果没有他们的帮助，本书不可能在短时间内高质量地完成。本书的顺利出版也离不开中国水利水电出版社杨莹莹等编辑的指导和支持，在此一并表示衷心的感谢。

尽管本书经过了编者与出版编辑的精心审读、校对与补充，但限于时间、篇幅，难免有疏漏之处，望各位读者体谅包涵，不吝赐教。本书参考了以下几个链接：

https://spring.io/

https://github.com/spring-projects/spring-framework

https://github.com/mybatis/mybatis-3

编　者
2022 年 9 月

目 录

第Ⅰ篇 基础概念

第Ⅱ篇　Spring MVC框架

第Ⅲ篇　MyBatis框架

第Ⅳ篇 Spring框架

第V篇　项目实战

第 I 篇　基础概念

本篇共分为 3 章，第 1 章主要进行 Spring、Spring MVC、MyBatis、Redis 的相关名词介绍、特性讲解及 Spring、Spring MVC、MyBatis 三者的关系梳理。通过第 1 章，初学者便可轻松入门 Spring。

第 2 章主要介绍 Java 的几个核心知识点。分别是类加载器（ClassLoader）、反射机制、动态代理，理解了这几个知识点，对于后续学习 Spring 有非常大的帮助。

第 3 章着重介绍设计模式，Spring、Spring MVC、MyBatis 中常见的设计模式，都可以在这章学习到。本章通过“理论 + 实例”的方式介绍，不仅能让读者记住设计模式是什么，还能让读者学会在什么场景怎么使用设计模式。

通过第 I 篇的学习，希望读者能够轻松进入 Spring 的大门，知道 Spring、Spring MVC、MyBatis 是什么，它们又有着什么样的联系；可以对反射、动态代理都有一定了解。希望读者对常用的几种设计模式熟记于心（在后续章节的源码分析中，会时刻发现这些设计模式的影子）。

第 1 章 SSM 框架和 Redis

本章要点

1. Spring 简介
2. 如何学习 Spring
3. Spring IoC 和 Spring AOP 的简介与理解
4. MyBatis 简介
5. MyBatis 和 Hibernate 的比较
6. Spring MVC 简介
7. Redis 的特性和优点
8. Spring、Spring MVC、MyBatis 的关系

1.1 了解 Spring 框架

Spring 是一个轻量级的开源企业级 Java 开发框架，主要为了应对复杂的企业项目。学习 Spring，最重要的是理解两个 Spring 的“骨骼架构”，即控制反转（Inversion of Control，IoC）和面向切面编程（Aspect Oriented Programing，AOP）。Spring 之所以在企业中应用如此广泛，完全基于它先进的设计理论：

（1）分层架构，开发者按需引用组件，减少重复性的工作，简化开发与配置过程。对项目侵入性低，在使用 Spring 框架的时候，并不需要继承或者实现框架中的类和接口。

（2）使用 IoC 容器实例化对象及管理对象间的依赖，面向接口编程，降低了模块与模块之间的耦合度，方便扩展。

（3）使用 AOP，让开发者可以更加专注于实现业务本身，而且代码与代码之间没有耦合。

1.1.1 如何学习 Spring

学习 Spring，具有 Java 基础是不可少的。如果对一些设计模式有一定的了解，就会更容易理解 Spring 的设计思想。Spring 主要设计模式有工厂模式（Factory Pattern）、策略模式（Strategy Pattern）、模板模式（Template Pattern）及代理模式（Proxy Pattern），第 3 章会对这几种模式进行详细的理论介绍及实例讲解。

学习 Spring，并不需要强行记住 Spring 的特性，如 IoC 和 AOP 概念。死记硬背是非常痛苦的学习方式，所以如何理解这些概念，并在项目中灵活使用，才是最需要的。

1.1.2 控制反转简介

如果病人没有去医院看病，就只能自行寻找药物服用，还要确定吃药的先后顺序。如果病人去医院看病，那么这些事情交给医生负责就好，病人遵照医嘱吃药即可。病人自己诊断服药，可以类比成传统的 Java 应用程序新建对象的方式。而这里的医生可以理解成 IoC，帮助病人管理药物，病人服用药物即可。

某些药物之间存在依赖关系，比如服用药物 A 之前需要先服用药物 B。在病人没有找医生看病前，这个依赖关系需要病人自行管理，如图 1-1 所示。

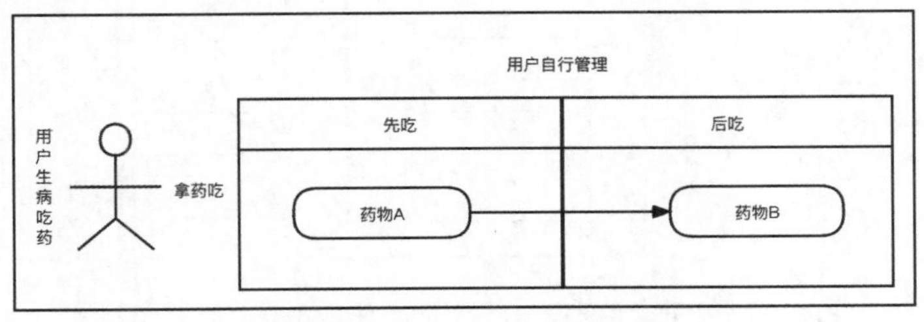

图 1-1　病人自己服药治病的流程

而在病人找医生看病后，医生负责管理病人服用药物的顺序和关系，病人遵照医嘱即可，如图 1-2 所示。

药物之间的依赖可以类比为依赖注入（Dependency Injection，DI），由医生负责管理这种依赖。这里发生了反转（药物服用的顺序开始是由病人管理的），即 IoC。有了 IoC，不需要关心对象的创建、销毁等，只需要在使用时，通过注解 id 或者类名告诉 Spring 需要的对象是什么。

DI 是在"控制反转"之后由 Martin Fowler 于 2004 年年初首次提出的。在配置 Spring 时，Bean 与 Bean 之间通过 ref 相互依赖，动态地为某个对象提供它需要的其他对象，所以称为依赖注入。

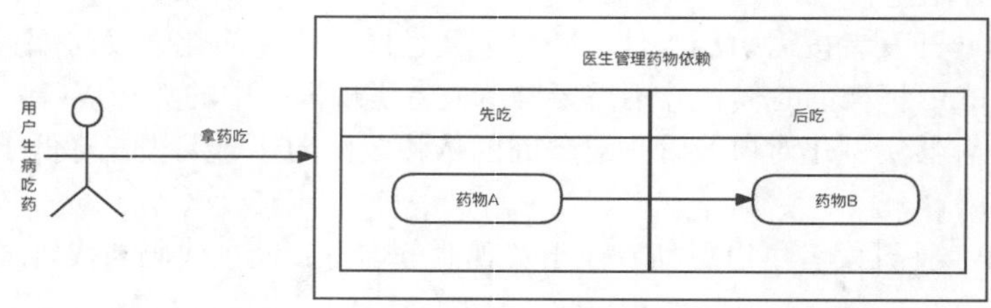

图 1-2　医生负责管理药物

注意： IoC 给编程带来的最大改变是从应用程序主动获取资源，变为应用程序被动接收 IoC 容器为其注入的、需要的资源。

1.1.3　面向切面编程简介

AOP 关注的是程序中的共性功能。在开发时，将共性功能抽取出来制作成独立的功能模块，此时原始功能中将不包含这些被抽取出的共性功能代码，但在项目运行时，不仅具有原始功能，并且具有被抽取的共性功能，对原始代码无侵入性。

下面用一个例子简单讲解 AOP。已上线了某个项目，现在想要全局记录操作 Dao 层的日志，通过 AOP 就可以在不调整原先代码的情况下构造一个切面，通过动态代理技术截取要调用的目标类方法，让调用者优先调用代理类中的方法。

当在执行 Dao 层方法时，会先执行切面，再在切面方法中将调用转发给需要的目标类方法。这样不仅执行了 Dao 层方法，而且可以记录日志的诉求，原始的 Dao 层方法无须改动，如图 1-3 所示。

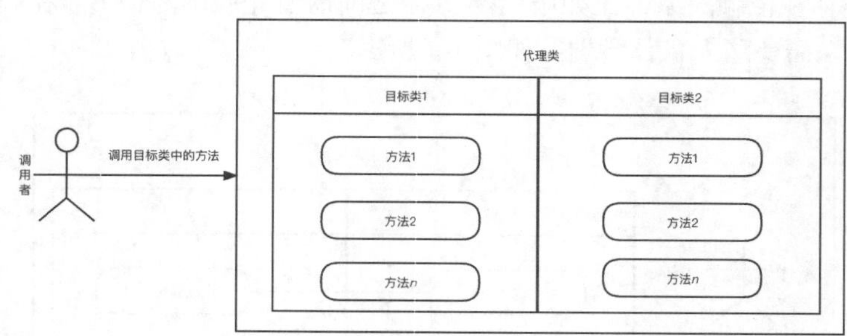

图 1-3　调用者调用目标类中的方法需要先经过代理类

注意： 理解了动态代理之后，对 AOP 的理解会更加深刻，后续章节会详细讲解动态代理。

1.2 了解 MyBatis

MyBatis 是一个基于 Java 的持久层开源框架，主要是将 Java 中的接口方法映射到对应的要执行的 SQL 语句，将普通 Java 对象（Plain Old Java Objects，POJOs）映射成数据库中的记录。对于持久层，可以简单地理解为该层将对象永久保存到数据库中。

1.2.1 MyBatis 简介

MyBatis 原名为 iBatis，在 2010 年由 Apache Software Foundation 迁移到 Google Code，并更名为 MyBatis。其特点如下：

（1）MyBatis 是支持定制化 SQL、存储过程及高级映射的持久层框架。

（2）MyBatis 封装了几乎所有的 JDBC 代码、需手工设置的参数，以及操作结果集的代码。

（3）MyBatis 使用简单的 XML 或注解来配置和映射原生信息，将接口和 Java 的 POJOs 映射成数据库中的记录。

MyBatis 的框架如图 1-4 所示。

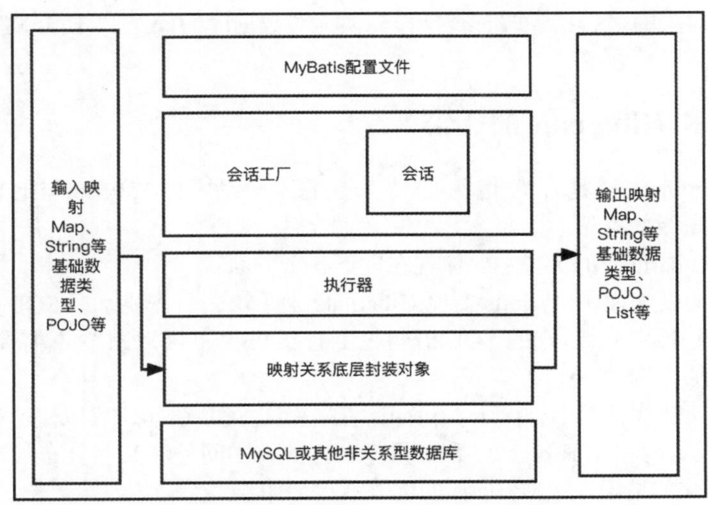

图 1-4 MyBatis 的框架

MyBatis 的框架的说明如下：

（1）MyBatis 的配置文件的名称是自定义的。作为 MyBatis 的全局配置文件，需要在其中配置数据源、事务等 MyBatis 需要的运行环境。此外，需要在全局配置文件中配置自动扫描 *mapper.xml 文件（也就是 SQL 映射文件）的功能，配置接口与文件之间的映射关系。SQL 映射文件中配置的是操作数据库的 SQL 语句。

（2）sqlSessionFaction（会话工厂）负责创建 sqlSession。

（3）sqlSession 是一个接口。开发者通过 sqlSession 操作数据库（增、删、改、查）。

（4）MyBatis 在底层自定义了 Executor 接口来操作数据库。该接口有基本执行器和缓存执行器，每个 SqlSession 对象都有一个 Executor 接口。

（5）Mapped Statement 对象（MyBatis 底层封装对象）包装了一些 MyBatis 的配置信息和 SQL 映射信息等。*mapper.xml 文件中的一个 SQL 对应一个 Mapped Statement 对象，SQL 语句的 id 也就是该对象的 id。Mapped Statement 对象负责对 SQL 执行输入参数和输出参数的定义，包括 Map、POJO、基本数据类型。输入参数映射就是传统 JDBC 编程中对 preparedStatement 设置参数，Executor 通过 Mapped Statement 对象在执行 SQL 语句前将对象映射到 SQL。输出映射相当于以前 JDBC 编程中对结果的解析和封装的处理过程，Executor 执行器通过 Mapped Statement 对象在执行 SQL 语句后将输出的结果映射到对象。

在 MyBatis 之前，绝大部分管理项目及企业级项目在持久层框架上使用的是 Hibernate，Hibernate 也是一个非常优秀的持久层框架（MyBatis 和 Hibernate 也称为 ORM 框架），目前一些老项目还在使用。ORM（Object/Relacation Mapping）也就是对象 / 关系映射，将对数据库的操作转换为对对象的操作。Hibernate 是完全面向 POJO 的开发框架。MyBatis 则不同，更加轻量级和可控，学习成本低。MyBatis 需要自行编写 SQL 语句才能执行，但 MyBatis 有很多优秀的第三方插件，可以支持很多方便的特性，也包括面向对象的开发。

Hibernate 可以在不编写 SQL 语句的情况下，对持久化对象进行操作便可完成对数据库数据的操作（将持久化对象的增、删、改、查操作转换为对数据库的操作），使开发者更加专注业务逻辑代码的实现，提高开发效率。

1.2.2 MyBatis 和 Hibernate 的比较

MyBatis 和 Hibernate 都是非常优秀的持久层框架，在国内 MyBatis 比 Hibernate 更主流，Hibernate 是相对重量级的框架。

MyBatis 相比 Hibernate 的优势主要体现在以下 5 个方面：

（1）从开发者角度来说，MyBatis 要比 Hibernate 上手快，只需要掌握 SQL 语句的编写技巧。

（2）MyBatis 基于 XML 格式的 SQL 语句，更直接可控，可以减少不需要的查询字段，对 SQL 审核和优化更加直接。

（3）在团队快速迭代开发中，由于 MyBatis 直接操作 SQL 语句，可以快速得知问题所在。另外，MyBatis 的高级定制相比 Hibernate 容易很多，比如在中间件和某些分布式环境提供支持需要对 SQL 语句进行优化的情况下，MyBatis 的优势就更加明显。

（4）MyBatis 学习成本低，对技术水平参差不齐的团队来说，MyBatis 用起来更有效。

（5）当在数据量大、表多的情况下进行级联操作时，Hibernate 会使得操作非常复杂，执行效率低。

MyBatis 相比 Hibernate 的不足体现在以下 2 个方面：

（1）从可移植性方面来说，Hibernate 的移植性比 MyBatis 好得多。MyBatis 项目中的所有 SQL 语句都是开发人员自行编写的，依赖于所用的数据库。而对于不同的数据库，SQL 语句的写法略有不同。Hibernate 中使用的 HQL 语言与数据库无关，开发人员只需要关联具体的数据库，移植性好。

（2）从缓存方面来说，Hibernate 和 MyBatis 都可以使用如 Redis 这样的第三方缓存。但是在使用二级缓存方面，由于 Hibernate 对查询对象有着良好的管理机制，用户无须关心 SQL，如果出现脏数据，系统会报出错误并提示。而 MyBatis 需要特别小心，如果不能完全确定数据更新操作的波及范围，应避免 Cache 的盲目使用。否则，脏数据的出现会给系统的正常运行带来很大的隐患。

注意：在日常学习中，建议学习能力强和经验丰富的读者先了解 Hibernate 再学习 MyBatis，这样更加容易理解 MyBatis。但在实际工作中，直接学习 MyBaits 即可。因为现在很少有新的项目用 Hibernate，毕竟想要熟练使用 Hibernate，其学习成本和 MyBaits 相比实在是太高了。

1.3 了解 Spring MVC

Spring MVC 是表现层的框架，属于 Spring 框架 Web Flow 模块的一部分。

Spring MVC 是一个“模型—视图—控制（MVC）”框架，实现了 Model-View-Controller 模式，将数据、业务和展示进行了分离，这种分离使开发分层逻辑更加清晰和可定制。

Spring MVC 中模型、视图、控制的说明如下：

（1）模型：一般用来封装数据，通常由基本的 Java 对象（POJO）组成。

（2）视图：主要用来呈现数据模型，通常是生成浏览器可以解析的 HTML 内容，用户可以浏览查看。

（3）控制：也就是控制器，用来处理用户的请求，构建合适的模型并将其传递到视图中呈现给用户。

Spring MVC 请求流程如图 1-5 所示。

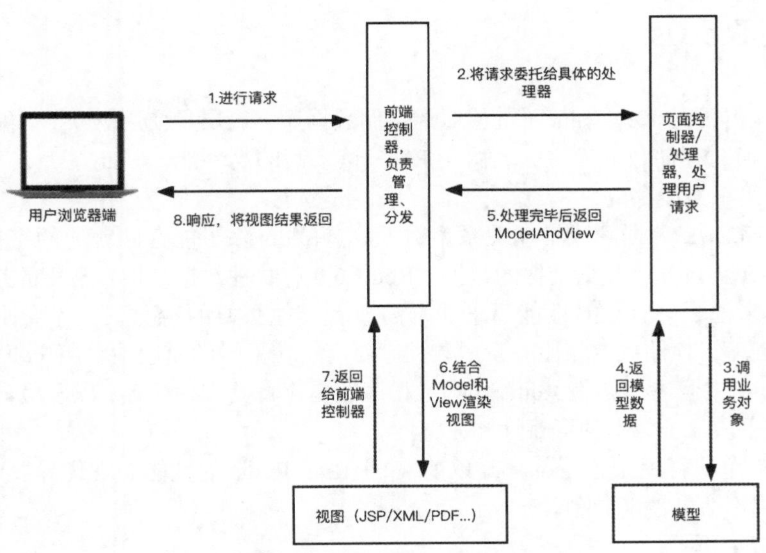

图 1-5 Spring MVC 请求流程

Spring MVC 中前端控制器（DispatcherServlet）和页面控制器（Controller）的说明如下：

（1）DispatcherServlet：Spring 提供的前端控制器，所有的请求都通过它分发给对应的 Handler（根据不同的请求进行分发，如 URL）。在前端控制器将请求分发给 Controller 之前，Controller 先通过 HandlerMapping（处理器映射器）解析请求链接，然后根据请求链接找到执行这个请求的类，也就是 Handler，最后定位到具体 Controller。

（2）Controller：进行功能处理，将收集的参数绑定到一个对象，该对象在 Spring MVC 中称为命令对象，将命令对象委托给业务对象进行业务处理，完成后将返回 ModelAndView（模型数据和逻辑视图）。

注意：DispatcherServlet 根据返回的逻辑视图名选择对应的视图，传入模型数据进行渲染，最后将渲染结果视图返回给用户。

整个运行原理可以概括为以下 6 个步骤：

①用户提交请求到 DispatcherServlet。

②由 DispatcherServlet 根据请求链接查询一个或者多个 HandlerMapping，找到处理请求的 Controller。

③ DispatcherServlet 将请求委托到 Controller。

④ Controller 调用业务对象，处理完逻辑后，返回 ModelAndView。

⑤ DispatcherServlet 根据 Controller 返回的 ModelAndView，找到 ModelAndView 指定的视图。

⑥将视图结果显示到客户端，展示给用户。

了解 Spring MVC 的相关流程是学习 Spring MVC 的第一步，关于 Spring MVC 的更多知识，将在第Ⅱ篇 Spring MVC 框架中详细讲解。

1.4 了解 Redis

Redis 是一个开源的 Key-Value（键值对）存储数据库，底层是使用 ANSI C 编写的。Redis 是目前最流行的键值对存储数据库之一。由于 Redis 读、写的高性能，因此在 Java 项目中，开发者一般使用它作为第三方缓存。

Redis 属于 NoSQL 数据库或者非关系型数据库。Redis 除了能存储普通的字符串，还可以存储 List、Set、Zset、Hash 等数据结构的数据（Redis 6.0 有 9 种数据结构，常用的是这 5 种）。

注意：一般来说，只有在对性能要求非常高或者其他必要的情况下，才会使用 Redis，因为 Redis 是内存数据库，存储成本相比关系型数据库要高。在平时的使用中，用户要根据具体情况选择是否使用 Redis，并且考虑是将 Redis 作为数据库还是仅将其作为缓存，以及应该如何通过复制、持久化和事务等手段保证数据的可靠性和完整性。

Redis 虽然是单线程模型（Redis 6.0 以前），但由于 Redis 的性能非常优异，所以在绝大多数场景下，Redis 都不会是性能瓶颈。

Redis 的优势如下：

（1）Redis 读的速度大约是 11 万次 /s，写的速度大约是 8.1 万次 /s。

（2）Redis 能存储的数据类型丰富，支持 String、List、Hash、Set、Zset 等数据类型。

（3）Redis 的所有操作都具有原子性，所以对于 Redis 来说，操作要么是执行成功，要么就是失败(完全不执行)。Redis 支持事务操作，对于多个操作，可以通过 MULTI 和 EXEX 指令进行包装，然后执行原子性操作。

（4）Redis 具有丰富的特性，支持 publish/subscribe（消息的发布和订阅）、通知、设置 key 过期时间等特性。

（5）Redis 可以作为数据库、队列及缓存系统。

- 当 Redis 作为缓存系统时，可以对每个键值对设置过期时间（强烈建议对每个键值对都设置过期时间，即使设置的过期时间很长，但不要不设置过期时间）。
- Redis 可以通过 LPUSH 和 RPOP 等操作来实现队列，支持阻塞时读取，非常容易实现一个高性能的优先级队列。
- Redis 可以作为数据库直接存储数据，但是这种方法并不实用。Redis 基本上都是作为高速缓存和 session 状态的存储层，搭配其他关系型数据库使用。

Redis 不适合作为数据库存储数据的部分原因如下：

（1）Redis 的数据是占用内存的，在数据量非常大的情况下，每次重启、备份，都需要将数据全部加载进内存，这显然是不合理的，内存的成本比磁盘的高得多。

（2）Redis 在半持久化模式下，数据保存并不是实时的，一旦断电，就会丢失一些数据。

（3）Redis 的查询主要是对 Key 的查询，如果是复杂的数据查询，则非常麻烦且性能低下。

（4）Redis 的事务支持过于简单。

注意：不排除在某些特定场景下使用 Redis 作为数据库更为实用的情况（比如对性能要求非常高、内存足够大、允许一定程度的数据丢失等），但建议还是使用关系型数据库作为存储数据库。

1.5 Spring、Spring MVC 和 MyBatis 的分工与合作

本节将梳理 Spring、Spring MVC 和 MyBatis 这三个框架是如何分工合作的。

Java 程序主要分为视图层、表现层、业务层和持久层，每层的作用如下：

（1）视图层，也就是 View 层。

- 用户通过视图层和应用进行交互。
- 用于展示如 JSP、HTML 等页面。
- 该层与 Controller 层结合开发。

（2）表现层，也就是 Controller 层（Handler 层）。

- 接收用户的输入数据并调用模型和视图去完成用户的需求。
- Controller 层通过调用 Service 层接口来控制具体的业务流程，Service 层的接口通过 Spring 配置进行注入。

（3）业务层，也就是 Service 层。

- 负责业务模块的逻辑处理。
- 先设计接口，再实现接口的实现类，通过 Spring 容器统一管理实现类的对象的生命周期。
- 在 Service 层进行一些逻辑处理后再调用 Dao 层定义的接口实现对数据库的操作。

（4）持久层，也就是 Dao 层。

- 负责和数据库进行交互，用来处理对数据的增、删、改、查操作。
- Dao 层只需要设计接口，通过 MyBatis 的配置来映射到执行的具体的 SQL 语句。

Spring、Spring MVC 和 MyBatis 的分工与合作流程如图 1-6 所示。

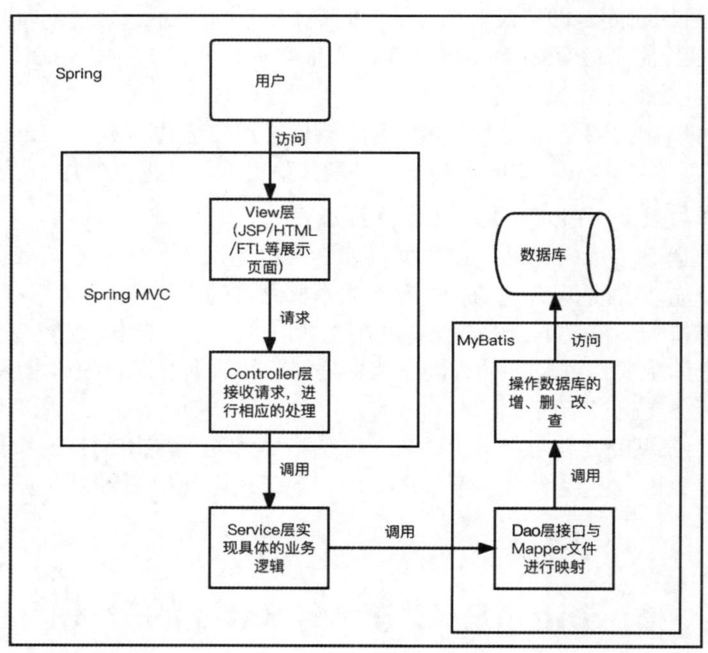

图 1-6　Spring、Spring MVC 和 MyBatis 的分工与合作流程

Spring 负责的是整个项目中 Bean 的管理，可以将其理解为一个容器框架。Spring MVC 是融合在 Spring 中的产品，主要负责 Controller 层，让开发者更容易定制某些技术。例如，使前端展示不局限于 JSP 技术。Spring MVC 将控制器、模型对象及处理程序的对象进行了分离。MyBatis 负责让开发者使用简单的 XML 配置或注解将接口和 POJO 映射成数据库中的数据和操作。

第2章 类加载器、反射和动态代理

本章要点

1. Java 生成对象实例的流程
2. 双亲委派模型
3. 类加载器的工作原理
4. 自定义加载器介绍的实现
5. 反射介绍与理解
6. 类反射示例
7. 获取 Class 对象的三种方式
8. 反射的应用场景与优缺点
9. 理解动态代理
10. JDK 动态代理与 CGLib 动态代理

2.1 类加载器

第 1 章简单介绍了 Spring、Spring MVC 和 MyBatis，以及它们之间的分工与合作。相信读者对 Spring、Spring MVC 和 MyBatis 已经不陌生了。

本节将详细讲解类加载器、反射和动态代理。

关于类与对象：类是一个类型的统称，对象是值（即一个具体的实例）。例如，人是一个统称，具体的某个人就是人的实例。

类加载器（ClassLoader）的作用是加载类文件（或者类）。JVM（Java 虚拟机）可以将本地文件、网络流或者其他来源的字节码文件（class 文件）加载到内存中，并且针对字节码生成对应的 Class 对象，这就是类加载器的功能。

Java 提供的类加载器共有三种：Bootstrap 类加载器（Bootstrap ClassLoader，启动类加载器）、Extension 类加载器（Extension ClassLoader，扩展类加载器）和 System 类加载器（System ClassLoader，也叫 Application 类加载器、应用类加载器、系统类加载器）。

（1）Bootstrap 类加载器：由 C++ 实现，属于 JVM 的一部分，是所有类加载器的父类加载器，作用是加载 {JAVA_HOME}/jre/lib 目录下的文件，并且只加载特定名称的文件（如 rt.jar，仅按照文件名识别），并不是加载该目录下的所有文件。

（2）Extension 类加载器：负责加载 {JAVA_HOME}/jre/lib/ext 目录下或者系统变量 java.ext.dirs 指定目录下的文件。

（3）System 类加载器：负责从 classpath 环境变量中加载与应用相关的 Java 类。

开发者可以直接使用扩展类加载器或者系统类加载器加载类，但是不能直接使用启动类加载器来加载类。开发者可以自定义类加载器，实现类的动态加载、热部署（如 JRebel 插件）等功能。

2.1.1 生成对象的实例的流程

生成一个对象的实例，最初程序会在编译期将 .java 文件使用 Java 编译器编译成对应的 .class 文件。接着 JVM 分配内存空间，当该类被调用时，JVM 自动将 .class 文件加载到虚拟机方法区中（从 JDK 1.8 开始，永久被移除，有了元空间），在加载的过程中，将静态变量和静态方法等静态内容加载到静态区，将非静态内容加载到非静态的区域中（类的加载只会进行一次，下次再创建或者调用对象时，可以直接在方法区获取 class 信息），生成对象的实例的流程如图 2-1 所示。

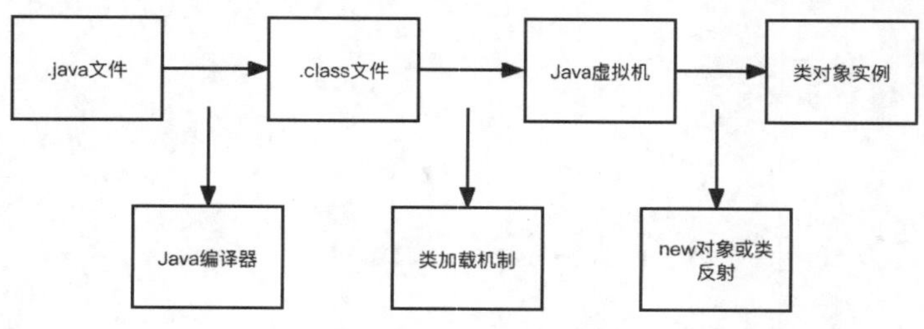

图 2-1 生成对象的实例的流程

由图 2-1 可知，当调用一个对象的静态方法或者创建一个对象时，需要先在 JVM 的方法区中获取对应类的信息，如果方法区中没有该类信息，则需要先加载该类。

类加载机制作用在将 .class 文件加载到 JVM 的过程中，该过程可以细分为以下 3 步：

①通过类的全限定名（如 java.lang.Integer）获取类的字节流（该字节流可以从多方面获取，如网络或者自定义字节流等）。

②将该字节流所代表的静态数据结构转化为方法区运行时的数据结构。

③在内存（堆内存）中生成代表该类的 Class 对象。

2.1.2 双亲委派机制

假设有一个类 ClassLoaderTest.class，首先加载这个类的 System 类加载器，将请求委托给它的父类加载器 Extension 类加载器，然后由 Extension 类加载器委托给 Bootstrap 类加载器。

Bootstrap 类加载器会先查看 rt.jar 中有没有 ClassLoaderTest 类，如果没有，则该请求再由 Extension 类加载器进行加载，Extension 类加载器会查看 {JAVA_HOME}/jre/lib/ext 目录下有没有该类，如果找到了，那么它会被加载，而 System 类加载器就不再加载该类；如果没有找到，那么 System 类加载器从 classpath 环境变量中配置的路径中进行查找。

现在已经不需要自行配置 classpath 了，也不推荐配置 classpath。从 JDK1.5 版本之后，JVM 默认在当前目录下查找 .class 文件，另外 Java 解释器也知道去哪里查找标准类库。

双亲委派机制的工作流程可以简单概述如下。

在一个类加载器在收到了类加载的请求后，并不会直接加载，而是优先将该请求委托给它的父类加载器，每个层次的类加载器都是如此，依次向上（只有 Bootstrap 类加载器是没有父类加载器的，所以 Bootstrap 类加载器是顶层类加载器）。只有父类加载器无法找到所需要的类时，才会让子类加载器尝试加载。

下面通过源码进一步了解。查看 java.lang.ClassLoader 类中的 loadClass(String name, boolean resolve) 方法的源码，如代码清单 2-1 所示。

代码清单 2-1　ClassLoader 类中的 loadClass 方法

```
protected Class<?> loadClass (String name,boolean resolve)throws
ClassNotFoundException {
    //name为类的全限定名
    synchronized (getClassLoadingLock(name)) {
        //通过findLoadedClass方法检查当前类加载器是否已经加载过该类，如果已经加载过了，则直接
返回该Class对象
        Class<?> c = findLoadedClass(name);
        if (c == null) {
            long t0 = System.nanoTime();
            try {
                if (parent != null) {
                    //如果父类加载器不为空，则将加载请求委托给父类加载器
                    c = parent.loadClass(name, false);
                } else {
                    //如果没有父类加载器，则直接委托给启动类加载器
```

```
                        c = findBootstrapClassOrNull(name);
                    }
                } catch (ClassNotFoundException e) {
                    // 如果没有找到类，则出现ClassNotFoundException 异常，这里不进行处理
                }
                if (c == null) {
                    //如果父类加载器都没有完成委托的类加载请求，则使用findClass方法加载该类
                    long t1 = System.nanoTime();
                    c = findClass(name);
                    // 定义类加载器，进行记录统计
                    sun.misc.PerfCounter.getParentDelegationTime().addTime(t1 - t0);
                    sun.misc.PerfCounter.getFindClassTime().addElapsedTimeFrom(t1);
                    sun.misc.PerfCounter.getFindClasses().increment();
                }
            }
        }
    }
    //其他代码省略
```

通过源码可以看出，如果想依照双亲委派机制实现自定义的类加载器，则可以直接继承 ClassLoader 类并重写 findClass 方法；如果不想受双亲委派机制的约束，则可以继承 ClassLoader 类，直接重写 loadClass 方法。

注意：建议还是按照双亲委派机制实现，因为这样可以保证加载出来的类是同一个，而且具有优先级的层次关系，不会引起混乱。例如，用户自行编写了一个名为 java.lang.Object 的类，而且自定义了类加载器，并且不受双亲委派机制的约束，那么系统中将出现两个不同的 Object 类，Java 类型体系的最基础行为无法得到保证，并且程序运行也会一片混乱。

另外，JVM 是根据类的全名及类加载器是否一样来判断两个 Java 类是否相同的，只有两者都一样的情况下才认为两个类是相同的。否则，即使是一样的字节码文件，被不同的类加载器加载后得到的类，对于 JVM 来说，也是不同的类。

出于安全性，双亲委派机制可以防止用户使用自行编写的类动态替换 Java 的一些核心类，如 String、Object 等。

因此，如果要实现自定义的类加载器，则务必遵循双亲委派机制。

2.1.3 类加载器的工作原理

Java 中的类加载器大致可以分为两类，一类是系统提供的，另一类是由开发者编写的（叫作自定义类加载器）。下面根据类加载器的双亲委派机制来理解类加载器的加载流程，如图 2-2 所示。

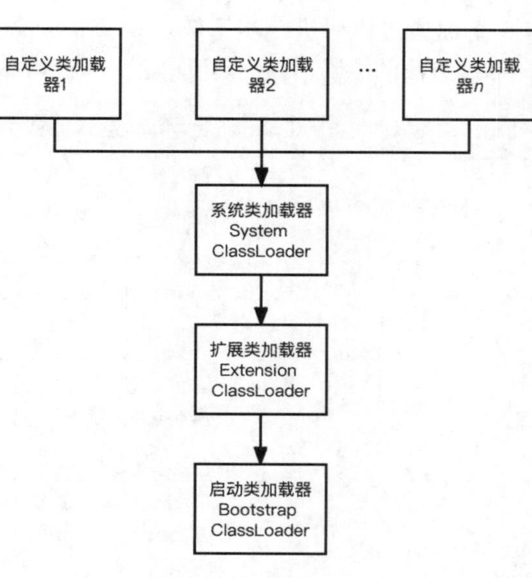

图 2-2 类加载器的加载流程

图 2-2 中的箭头不是代表父类加载器通过继承关系实现，而是代表通过组合关系复用父加载器中的代码实现。

可以通过运行代码清单 2-2 来验证图 2-2 的流程。

代码清单 2-2　ClassLoaderTest 类

```
public class ClassLoaderTest {
    public static void main(String[] args) {
        ClassLoader loader = ClassLoaderTest.class.getClassLoader();
        while (loader != null) {
            System.err.println(loader);
            loader = loader.getParent();}}}
```

运行结果如图 2-3 所示。

```
lassLoaderTest
"C:\Program Files\Java\jdk1.8.0_31\bin\java" ...
sun.misc.Launcher$AppClassLoader@58644d46
sun.misc.Launcher$ExtClassLoader@6d6f6e28
```

图 2-3　运行结果

从结果可以看出，一般情况下是由 AppClassLoader 类加载器加载类的。AppClassLoader 的父类加载器为 ExtClassLoader，而 ExtClassLoader 的父类加载器在运行结果中是看不出来的，这是由于 Bootstrap 类加载器属于 JVM 的一部分，由 C++ 直接实现，在 Java 中无法引入，返回的是空值。

前面的源码中已经很好地说明了双亲委派机制，现在再通过运行程序来验证。将 ClassLoaderTest 打包成 jar 包，包名任意，将该包放到 {JAVA_HOME}/jre/lib/ext 目录下，看一下输出结果是不是只有 ExtClassLoader 类加载器。

在 IDEA 中，基于 Maven 项目，单击 install 选项即可安装当前的 maven project，如图 2-4 所示。

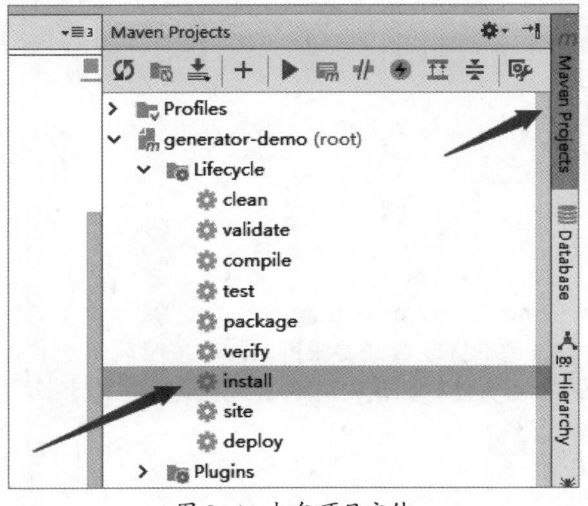

图 2-4　打包项目文件

在传统 Java 项目中，可以通过 jar 命令对项目文件进行打包。

继续运行该类，运行结果如图 2-5 所示。

```
ClassLoaderTest
"C:\Program Files\Java\jdk1.8.0_31\bin\java" ...
sun.misc.Launcher$ExtClassLoader@7ea987ac
```

图 2-5 运行结果

由运行结果可以看出，该类不通过 AppClassLoader 类加载器加载，而是直接通过 ExtClassLoader 类加载器加载的。

如果把该 jar 包放到 Bootstrap 类加载器的搜索路径下，尝试通过顶层类加载器加载该 jar 包的类，那么编译器会无法识别该类，因为启动类除了必须是指定目录下的全限定名，还必须是特定名称的文件才能加载类文件。

2.1.4 自定义类加载器

通过前面的介绍，读者对系统的类加载应有了一定的了解。接下来实现一个自定义类加载器，能更加深入地了解类加载。

首先，介绍自定义类加载器的应用场景。

（1）加密。因为 Java 代码编译后生成了 .class 字节码，反编译软件和 IDE 可以轻易地反编译 class 字节码。因此，如果需要防止 Java 代码被反编译，可以将编译后的 Java 代码使用加密算法进行加密，经过加密后的 .class 字节码不能用类加载器加载。这时，可以利用自定义类加载器在加载类时先解密字节码再加载类。

（2）热部署。例如，在 IDEA 中用到的热部署插件 JRebel 就使用了自定义类加载器。

（3）Java Web 服务器。例如，Tomcat 实现的自定义类加载器，用来解决隔离、共享、JSP 热部署的问题。

然后，先自定义一个实体类，如代码清单 2-3 所示。

代码清单 2-3 User 类

```
public class User {
    private String username;
    private Integer age;
    ...
}
```

注意：代码中忽略了 getter、setter 和 toString 方法的配置。

接下来实现的类加载器是不打破双亲委派机制的。按照前面章节介绍的知识，只需要继承 ClassLoader 类、重写 findClass 方法，如代码清单 2-4 所示。

代码清单 2-4 自定义类加载器

```
//自定义类加载器 遵循双亲委派模型
public class CustomClassLoader extends ClassLoader {
```

```
    private String classPath;
    public CustomClassLoader(String classPath) {
        this.classPath = classPath;
    }
    public CustomClassLoader(ClassLoader parent, String classPath) {
        super(parent);
        this.classPath = classPath;
    }
    public static void main(String[] args) throws ClassNotFoundException, IllegalAccessException,
InstantiationException {
        //注意，这里的是.class文件的路径
        CustomClassLoader customClassLoader = new CustomClassLoader("/Users/chenhx/Desktop/
github/uifuture-ssm/ssm-classLoader/target/classes/");
        Class loadClass = customClassLoader.findClass("com.uifuture.entity.User");
        Object object = loadClass.newInstance();
        System.out.println(object);
        //打印自定义的类加载器
        System.out.println(object.getClass().getClassLoader());
    }
    /**
     * 重写方法
     * @param name 类名
     * @return 类
     * @throws ClassNotFoundException 未找到类抛出异常
     */
    @Override
    public Class<?> findClass(String name) throws ClassNotFoundException {
        //加载类的二进制流
        byte[] classData = getClassBytes(name);
        if (classData == null) { throw new ClassNotFoundException();
        } else { return defineClass(name, classData, 0, classData.length); }
    }
    /**
     * 读取文件的二进制流
     * @param className class类的全限定名
     */
    private byte[] getClassBytes(String className) {
        //拼接路径
        String path = classPath + className.replace('.', '\\') + ".class";
        try {
            //使用字节流获取字节码文件，因为字节码文件是二进制流
            InputStream is = new FileInputStream(path);
            ByteArrayOutputStream stream = new ByteArrayOutputStream();
            byte[] buffer = new byte[2048];
            int num;
            while ((num = is.read(buffer)) != -1) { stream.write(buffer, 0, num); }
            return stream.toByteArray();
        } catch (IOException e) { e.printStackTrace(); }
        return null;
    }
}
```

自定义类加载器的运行结果如图 2-6 所示。

```
CustomClassLoader
"C:\Program Files\Java\jdk1.8.0_31\bin\java" ...
User[username=<null>, age=<null>]
com.uifuture.config.CustomClassLoader@f6f4d33
```

图 2-6　自定义类加载器的运行结果

从运行结果可以看出，类名是 User，类加载器是自定义类加载器 CustomClassLoader。

最后，简单讲解一下热部署的原理。JVM 默认不能热部署类，原因是加载类时 JVM 会调用 findLoadedClass 方法，如果类已经被 JVM 加载过了，就不会再加载。由于不同的类加载器加载同一个字节码生成的类，JVM 会判断为不同的类，所以只需使用 ClassLoader 的不同实例重新加载类来实现热部署。

2.2　反射机制

Java 的反射机制是指在程序运行时，对 JVM 中的任意一个类，都能知道其所有属性和方法，并能任意调用类中的属性和方法。这种能够动态获取类的信息，以及动态调用对象方法和获取对象属性的功能就称为 Java 的反射机制。

反射机制并不是所有语言都具备的，如 C 与 C++，如果要使用反射，则要通过自行开发来实现。

本节主要讲解什么是反射，怎么使用反射，反射的应用场景及反射的优缺点。

2.2.1　什么是反射

一般来说，事物都有两面性，有“反”自然也有“正”。先理解一下什么是“正反射”。“正反射”是通过一般业务逻辑中的代码，new 出对象实例，再获取对象的属性及调用方法等。而反射，简单地说就是不通过对象的实例就可以获取对象的信息。

例如，将 new 对象获取对象的属性和属性值等信息视为按照正规流程查询企业信息，可以在一些公开网站上知道企业名称、统一信用代码等信息。如果使用反射，则需要提前知道对象的身份信息（公司名称、统一信用代码），那么就可以使用非常规方式，直接委托专业人士去调查，将企业的所有信息“暴露”出来，相当于你通过专业人士获取企业信息。

反射允许 Java 程序运行时对自身进行检查（自审），并且能够直接操作对象的内部属性和方法等。类反射的实际应用之一就是 JavaBean 的注入，可以让一些工具通过类反射动态地加载 JavaBean 的属性。Spring、Spring MVC 和 MyBatis 都使用了类反射。

注意： 图 2-7 所示是类的常见组成结构，还有其他组成结构，未全部列出，如 Field、Constructor 和 Method 中都有 Annotation 信息等。

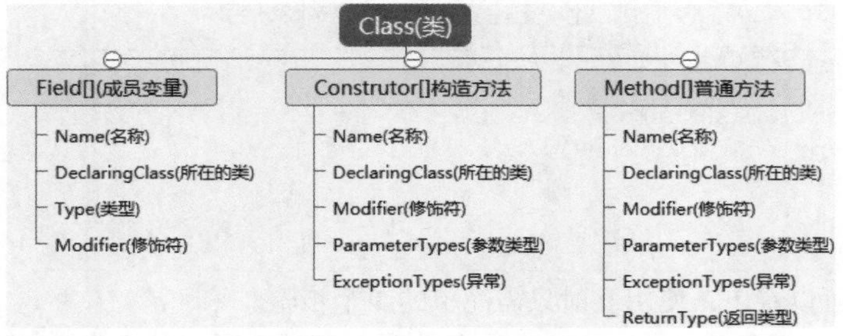

图 2-7　类的常见组成结构

2.2.2　类反射入门示例

下面使用一个简单的示例来介绍类反射。

先创建一个简单的实体类 Person，如代码清单 2-5 所示。

代码清单 2-5　Person 类

```
public class Person {
    Integer age;
    private String name;
    private static Integer sum(Integer age) {return age;}
    public String getName() {return name;}
    public void setName(String name) {this.name = name;}
    public Integer getAge() {return age;}
    public void setAge(Integer age) {this.age = age;}
}
```

Person 就是通过使用反射去获取的类对象，继而获取该类对象进一步的信息的。

再进行 ReflectionHelloWord 类方法的反射，也就是获取 Method[] 信息，如代码清单 2-6 所示。

代码清单 2-6　ReflectionHelloWord 类

```
public class ReflectionHelloWord {
    public static void main(String[] args) {
        try {
            Class c = Class.forName("com.uifuture.helloword.entity.Person");
            System.out.println(MessageFormat.format("类:{0}", c.toString()));
        Method[] methods = c.getDeclaredMethods();
            for (Method m : methods) {
                //获取整个方法
                //包括修饰符、返回类型、方法名字
                 System.out.println(MessageFormat.format("方法名:{0}", m.toString()));
                 System.out.println(MessageFormat.format("修饰符:{0}", m.getModifiers()));
                 System.out.println(MessageFormat.format("返回类型:{0}" , m.getReturnType()));
                 System.out.println(MessageFormat.format("方法名字:{0}", m.getName()));
                Class[] classes = m.getParameterTypes();
                //此处遍历classes数组
                Arrays.asList(classes).stream().forEach(x ->
                        System.out.println(MessageFormat.format("参数:{0}", x)));
                System.out.println(MessageFormat.format("方法所在的类:{0}",
```

```
                    m.getDeclaringClass()));
                System.out.println("------------------------------");
            }
        } catch (ClassNotFoundException e) {
            e.printStackTrace();
        }
    }
}
```

从上面的代码可以看出，使用反射只需简单的 3 个步骤。

①通过调用 Class 对象的 forName 方法，传入类对象的全限定名，获取 Class 对象。

②调用 Class 对象的方法，获取该类中定义的信息。上面示例中调用的是 getDeclaredMethods 方法，用于获取该类定义的所有方法信息列表。

③使用反射 API 中的方法操作这些信息。运行 ReflectionHelloWord 类的 main 方法，可以看到如图 2-8 所示的部分输出结果。

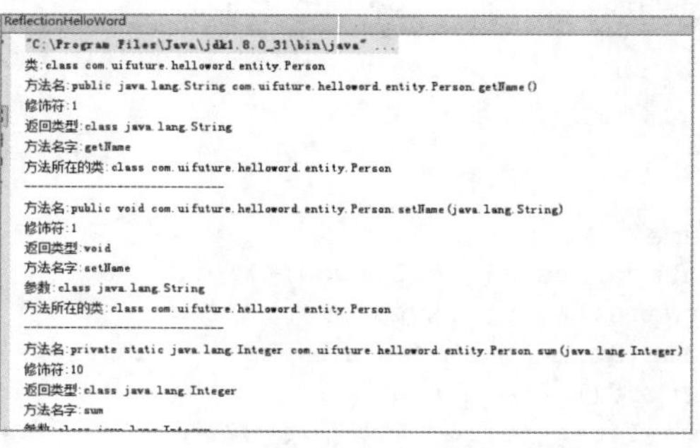

图 2-8　部分输出结果

有一点需要说明，在 JDK 8+ 中，修饰符的输出类型是数字类型。在 java.lang.reflect.Modifier 类中，修饰符实际上是一些标识符常量，如图 2-9 所示。

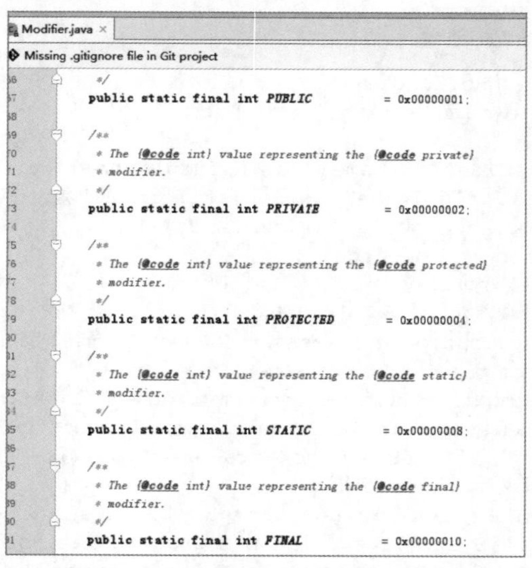

图 2-9　Modifier 类中部分修饰符常量

图 2-8 中显示修饰符为 10，可以通过值相加得出修饰符为 PRIVATE 为 2、STATIC 为 8。图 2-9 所示的 0x 代表数字使用的是十六进制，FINAL 的值为 16。

通过类反射入门示例可以简单地了解反射的使用方法，接下来继续深入了解。

2.2.3 通过类反射实现工厂方法

使用类反射的一个好处就是解耦，方便实现框架给第三方使用。例如，有两个实现类 A 和 B，有时需要使用 A 实现类，有时需要使用 B 实现类。假如不使用类反射，那么一种方法是使用传参（或者读取配置文件中的值），根据参数值 new 不同的类对象；另一种方法就是重启应用，需要哪个类对象，就 new 哪个类对象。

第一种方法的缺点是 A 和 B 两个实现类是固定的，假如要增加新的实现类 C 就必须修改代码，需要将 C 类对象的创建及判断逻辑编入代码。

第二种方法的缺点是每次都需要修改代码。

以上两种方法的代码耦合非常严重，而且基本上只能使用自行实现的 A 实现类和 B 实现类，无法成为工具类。其他人要使用必须修改源码，所以缺陷显而易见。这也是大量 Java 框架中，都使用到了类反射的原因之一。下面使用类反射实现工厂方法，通过配置文件来配置面向接口编程的实例。先定义一个 Worker 接口，如代码清单 2-7 所示。

代码清单 2-7 Worker 接口

```
public interface Worker {
    /** 定义一个接口方法 */
    void work();
}
```

下面定义两个实现类，实际只会使用其中一个，如代码清单 2-8 和代码清单 2-9 所示。

代码清单 2-8 WorkerOneImpl 类

```
public class WorkerOneImpl implements Worker {
    @Override
    public void work() {
        //输出信息以便在控制台识别实际运行的类
        System.out.println("WorkerOneImpl...");}
}
```

代码清单 2-9 WorkerTwoImpl 类

```
public class WorkerTwoImpl implements Worker {
    @Override
    public void work() {System.out.println("WorkerTwoImpl...");}
}
```

因为在实际应用中只会实例化一个实现类，所以要定义一个工厂方法获取实现类。这里要配置需要的那个实现类，然后在工厂方法中实例化对象，如代码清单 2-10 所示。

代码清单 2-10　WorkerFactory 类

```
public class WorkerFactory {
    private static Worker worker = null;
    /**
     * 通过工厂方法获取Worker对象
     * @return Worker对象
     */
    public static Worker getWorker() {
        if (worker != null) { return worker; }
        //配置文件
        Properties p = new Properties();
        FileInputStream in;
        try {
            //读取配置文件
            in = new FileInputStream(WorkerFactory.class.getResource("/").getPath()
              + "worker.properties");
            p.load(in);
            //通过name属性名获取属性值。name 为自定义
            String className = p.getProperty("name");
            //通过name获得name后面=号后面的字符串，这样就可以通过修改配置文件来“new”不同的类
            Class c = Class.forName(className);
            //通过反射获取实例化对象
            worker = (Worker) c.newInstance();
        } catch (Exception e) { e.printStackTrace(); }
        return worker;
    }
}
```

接下来需要一个配置文件，通过类反射入门示例可知，配置文件中只要有类的全限定名就可以通过反射获取对象的实例及信息。所以在配置文件中配置 Worker 实现类的全限定名，如代码清单 2-11 所示。

代码清单 2-11　配置文件

```
# name = com.uifuture.factorydemo.impl.WorkerOneImpl
name=com.uifuture.factorydemo.impl.WorkerTwoImpl
```

通过以上代码可知只需要配置 Worker 实现类的全限定名，不需要配置 Worker 接口类。

最后还需要一个测试类来调用工厂方法。由于不同实例中的输出结果是不同的，所以可以通过输出结果判断实例对象是否与预期一致。在项目开发中，work 方法中的输出结果代表不同的逻辑处理。测试类如代码清单 2-12 所示。

代码清单 2-12　测试类

```
public class TestFactory {
    public static void main(String[] args) {
        Worker worker = WorkerFactory.getWorker();
        worker.work();
    }
}
```

配置 name 为 com.uifuture.factorydemo.impl.WorkerOneImpl 时控制台的输出结果为：

```
WorkerOneImpl...
```

配置 name 为 com.uifuture.factorydemo.impl.WorkerTwoImpl 时控制台的输出结果为：

```
WorkerTwoImpl...
```

修改配置文件中 name 属性的值，再分别运行上面的测试代码，可以看出：输出结果与预期结果一致，也就是配置了哪个实现类，实际运行的就是哪个实现类。

2.2.4 获取 Class 对象的三种方式

首先了解一下 Class 类。Class 对象是 Java 类反射的基础，包含了与类相关的信息。查看 Class 对象的源码，可以发现 Class 对象就是 java.lang.Class<T> 类生成的对象。其中，类型参数 T 表示由该 Class 建模的类的类型。例如，User.class 的类型就是 Class<User>。如果被建模的对象的类型未知，则用"?"号代替，如 Class<?>。

下面的代码是 Class 类源码上的注释。

```
/**
 * Instances of the class {@code Class} represent classes and
 * interfaces in a running Java application.  An enum is a kind of
 * class and an annotation is a kind of interface.  Every array also
 * belongs to a class that is reflected as a {@code Class} object
 * that is shared by all arrays with the same element type and number
 * of dimensions.  The primitive Java types ({@code boolean},
 * {@code byte}, {@code char}, {@code short},
 * {@code int}, {@code long}, {@code float}, and
 * {@code double}), and the keyword {@code void} are also
 * represented as {@code Class} objects.
 *
 * <p> {@code Class} has no public constructor. Instead {@code Class}
 * objects are constructed automatically by the Java Virtual Machine as classes
 * are loaded and by calls to the {@code defineClass} method in the class
 * loader.
```

这里简单地翻译前面两段：Class 类的实例表示正在运行的 Java 应用程序中的类和接口。枚举是一种类，注释是一种接口。每个数组也属于一个被映射为 Class 对象的类，所有具有相同元素类型和维度的数组都共享该 Class 对象。基本的 Java 类型（boolean、byte、char、short、int、long、float 和 double）和关键字 void 也表示为 Class 对象。

Class 没有公共构造方法。Class 对象是在加载类时由 JVM，以及通过调用类加载器中的 defineClass 方法自动构造的。

```
 * <p> The following example uses a {@code Class} object to print the
 * class name of an object:
 *
 * <blockquote><pre>
 *     void printClassName(Object obj) {
```

```
 *          System.out.println("The class of " + obj +" is " + obj.getClass().getName());
 *      }
 * </pre></blockquote>
 *
 * <p> It is also possible to get the {@code Class} object for a named
 * type (or for void) using a class literal.  See Section 15.8.2 of
 * <cite>The Java&trade; Language Specification</cite>.
 * For example:
 *
 * <blockquote>
 *     {@code System.out.println("The name of class Foo is: "+Foo.class.getName());}
 * </blockquote>
 *
 * @param <T> the type of the class modeled by this {@code Class}
 * object.  For example, the type of {@code String.class} is {@code
 * Class<String>}.  Use {@code Class<?>} if the class being modeled is
 * unknown
```

这段注释演示了两个例子。obj.getClass().getName() 表示获取类名，也可以根据命名类型获取类名，如 Foo.class.getName()。通过以上内容可知，每个类都有自己的一个 Class 对象。

下面来获取 Class 对象，有以下 3 种方式：

（1）通过对象的 getClass 方法获取。这种方式需要具体的类和该类的对象，以及调用 getClass 方法。

例如，Class aClass = new String(“test”).getClass();。

（2）任何数据类都有一个静态的属性 class，通过该属性直接获取对应的 Class 对象。这种方式需要使用具体的类，然后调用类中的静态属性 class 完成，不需要调用，性能比第 1 种方式要好一些。

例如，Class aClass =String.class;。

另外，基本数据类型，还可以通过 TYPE 获取 Class 对象。

例如，Class aClass = Integer.TYPE;。

提示：输出为 int 而不是 class java.lang.Integer（这里的 Integer 就是基本数据类型，TYPE 指 Integer.TYPE 这个行为，aClass 就是获取到的 Class 对象）。

（3）通过 Class.forName 方法获取。这种方式仅需使用类名就可以获取该类的 Class 对象，更有利于扩展。也是在一些开源工具项目中常见的用法。

例如，Class aClass =Class.forName("java.lang.String");。

相同的类对象，即使是多次 new 出来的，都是共用同一套类模板的，如代码清单 2-13 所示。

代码清单 2-13　ClassDemo 类

```
public class ClassDemo {
    public static void main(String[] args) throws ClassNotFoundException {
        Person person = new Person();
        Person person1 = new Person();
        System.out.println(person == person1);
```

```
        System.out.println(person.getClass() == person1.getClass());
        System.out.println(Person.class == Person.class);
        System.out.println(Class.forName("com.uifuture.helloword.entity.Person") ==
                Class.forName("com.uifuture.helloword.entity.Person"));
    }
}
```

代码运行后的输出结果分别为 false、true、true、true。可以知道，person 和 person1 对象肯定是不相等的，毕竟都是 new 出来的。而这两个对象的 getClass 是相等的，这是由于同一个类型的对象使用的类模板都是同一个。而 person 3 和 person 4 的输出结果都是 true。

2.2.5 反射的应用场景与优缺点

在平常的业务逻辑中，使用类反射的地方很少。在业务逻辑中用得最多的可能 BeanUtils，也就是传入一个 Map 及 Model 类的 Class，返回该 Model 的实例。

接下来实现这个简单且实用的功能。

设计一个方法 Object getModel(Map map,Class cls)，传入一个包含所有值的 Map，然后传入 Model 类的 class，则会返回 Model 类的实例，这个实例已经包含了所有相关的数据。也就是把 Map 中的数据通过反射设置回 Model 类实例中。

下面用两种方式进行演示，第一种是没有使用泛型的方式，第二种是使用泛型的方式。

（1）没有使用泛型。Users 类如代码清单 2-14 所示。

代码清单 2-14　Users 类

```
@Data
public class Users {
    private String name;
    private Integer age;
    private String address;
    private Date createTime;
}
```

这里使用的 @Data 是 Lombok 工具提供的，在 pom.xml 中引入 Lombok 工具的依赖即可（前提是在 IDEA 中安装 Lombok 插件）。BeanUtilsNoGenerics 类（没有使用泛型）如代码清单 2-15 所示。

代码清单 2-15　BeanUtilsNoGenerics 类（没有使用泛型）

```
public class BeanUtilsNoGenerics {
    public static Object populate(Class cls, Map map) throws ReflectiveOperationException {
        // 用类反射“new”出对象
        Object obj = cls.newInstance();
        // 用类反射为“new”的对象设置属性值(必须遵守Java设置规范)，即通过setter方法设置
        // 遍历所有类属性
        Field[] fields = cls.getDeclaredFields();
        for (Field fld : fields) {
            //获取fld对象的属性名
            String fldName = fld.getName();
```

```
            //根据属性名在Map中读取数据，只有数据非空才需要给该属性设置值
            Object mapValue = map.get(fldName);
            //如果Map中不存在对应的属性数据，则给出提示信息，实际项目中请删除
        if (mapValue == null) {
                System.out.println(fldName + "的数据为空！ ");
            } else {
          //如果Map中存在对应的属性数据，则由属性名得出它的setter方法的名字，注意遵守命名规则
                String setName = "set" + fldName.substring(0, 1).toUpperCase() +
                fldName.substring(1);
                //根据方法名和参数的数据类型(其实就是属性的类型)获得Method对象
                Class[] parameterTypes = new Class[1];
                parameterTypes[0] = fld.getType();
                //最好不要用getMethod，因为这样可能会用到父类的方法
                Method method = cls.getDeclaredMethod(setName, parameterTypes);
                //这里不用担心会访问到私有的方法和属性，因为权限没放开
                //调用该Method对象所代表的方法
                Object[] args = new Object[1];
                args[0] = mapValue;
                method.invoke(obj, args);
            }
        }
        return obj;
    }
    /** 测试类*/
    public static void main(String[] args) throws ReflectiveOperationException {
        Map<String, Object> map = new HashMap<>(16);
        map.put("name", "test");
        map.put("age", 21);
        map.put("address", "浙江杭州");
        map.put("createTime", new Date());
        Users users = (Users) populate(Users.class, map);
        System.out.println(users);
    }
}
```

输出结果如下：

```
Users(name=test, age=21, address=浙江杭州, createTime=Mon Jul 23 21:26:09 CST 2018)
```

（2）使用泛型。

了解泛型的读者应该知道，这里已经将 Class 对象传过来了，就没有必要再经过 Object 到 Model 对象的类型强制转换。BeanUtils 类（使用泛型）如代码清单 2-16 所示。

代码清单 2-16　BeanUtils 类（使用泛型）

```
public class BeanUtilsByGenerics {
    public static <T> T populate(Class<T> cls, Map map) throws
    ReflectiveOperationException {
        // 用类反射“new”出对象
        T t = cls.newInstance();
```

```
        // 用类反射为"new"的对象设置属性值(必须遵守Java设置规范)，即通过setter方法设置
        // 遍历所有类属性
        Field[] flds = cls.getDeclaredFields();
        for (Field fld : flds) {
            //获取该fld对象的属性名
            String fldName = fld.getName();
            Object mapV = map.get(fldName);
            //根据属性名在Map中读取数据，只有数据非空才需要给该属性设置值
            //如果Map中不存在对应的属性数据，则给出提示信息，实际项目中不需要输出，可以灵活处理
        if (mapV == null) {
                System.out.println(fldName + "数据为空！");
            } else {
                //如果Map中存在对应的属性数据，则由属性名得出它的setter方法的名字
                String setName = "set" + fldName.substring(0, 1).toUpperCase() + fldName.
                substring(1);
                //根据方法名和参数的数据类型(其实就是属性的类型)，获得Method对象
                Class[] parameterTypes = new Class[1];
                parameterTypes[0] = fld.getType();
                Method m = cls.getDeclaredMethod(setName, parameterTypes);
                //调用该Method对象所代表的方法
                Object[] args = new Object[1];
                args[0] = mapV;
                m.invoke(t, args);
            }
        }
        return t;
    }
    /** 测试类 */
    public static void main(String[] args) throws ReflectiveOperationException {
        Map<String, Object> map = new HashMap<>(16);
        map.put("name", "test");
        map.put("age", 21);
        map.put("address", "浙江杭州");
        map.put("createTime", new Date());
        Users users = populate(Users.class, map);
        System.out.println(users);
    }
}
```

输出结果与前面是一致的。

注意测试类，已经不再需要强制转换，对用户友好多了。

如果需要访问并调用私有的变量或方法，则需要打开访问权限，然后进行（暴力）访问。也就是让 Field 对象和 Method 对象调用 setAccessible 方法，并设置为 true。该方法也可以用来优化类反射的速度，优点是减少了安全检查的时间。

通过上面的例子可以发现，使用类反射对类进行封装是非常方便的，极大地简化了一些设置和开发流程。另外，JSON/XML 等格式的通用解析也需要用到类反射。

总地来说，框架开发中的一些动态配置、第三方工具的继承、应用的动态扩展、JSON/XML解析成对象、对象的值复制等，都会用到反射。

很多文章或者书上提到了使用反射会使性能变得比较慢，主要是由于反射的调用要增加一些拆箱装箱、参数重排、异常重抛、安全检验等额外操作。其实在 JDK 7+ 以上的 JVM 中，已经对反射进行了很大的优化，关闭安全校验后，性能差距可以控制在 10 倍以内。

另外，如果对反射做了缓存（JDK 7+ 在 Class 类中通过 ReflectionData 内部类缓存类的信息，该缓存会在 JVM 资源紧张时回收），那么性能的 10 倍差距就可以忽略不计了。因为对于计算机的运算力来说，时间粒度是很小的，通常为纳秒级别，纳秒级别的 10 倍差距在业务中可以忽略不计。

注意：①如果使用反射频繁的对象，那么一定要开启缓存，这样也可以自行实现类信息的缓存；②在平时的业务逻辑处理上，尽量不要使用反射，因为代码的可读性较差；但在一些需要使用类反射的情况下，要大胆使用反射。

2.3 动态代理

Spring 中 AOP 的拦截功能就是使用 Java 中的动态代理实现的，也就是在被代理类（方法）的基础上增加切面逻辑，生成代理类（方法）。切面逻辑可以在目标类函数执行之前、之后，或者在目标函数抛出异常的时候执行。本节不过多讲解 Spring 中的 AOP，主要介绍 Java 中的动态代理。

2.3.1 理解动态代理

为了方便读者理解，下面介绍 3 个有关代理的名词。

（1）委托类——被代理的类（也可以叫目标类）。

（2）代理类——进行代理的类。

（3）消费类——调用代理类的类。

在理解动态代理之前，先来了解静态代理。代理类在程序运行之前就已经确定，对该类进行代理的方式称为静态代理。一般情况下，静态代理中的代理类和委托类都会继承相同的父类或者实现相同的接口。

关于静态代理，在有些场景，如需要同时接入支付宝、微信、银联的支付接口，对于一些并没有开发能力的商户来说，是有一定实现难度的。

这时就出现了很多帮助商户直接接入支付宝的第四方支付工具，这里可以将支付宝理解为委托者，第四方封装支付宝的支付接口并作为代理者。商户（消费者）不再需要与支付宝进行对接，而是将接入的需求委托给第四方。

注意：这里讲的是静态代理，也就是第四方必须先知道消费者的需求，并针对不同的消费者（如餐饮行业、游戏行业）进行不同的定制。如果事先不知道消费者是什么行业，第四方就无法进行定制了，除非将所有行业进行全部定制，这会严重浪费资源。或者有的商户需要接入微信等其他支付方式，也会比较麻烦，因为静态代理要求在程序运行前就必须将代理类确定下来。这时就需要动态代理来解决这个问题（静态代理的示例在 ssm-proxy 模块中）。

动态代理是在程序运行时创建的代理方式。也就是说，在程序运行之前，类是不确定的。而是在运行时，根据消费类的不同，运行不同的委托类，不需要针对每个委托类分别实现一个代理类。对代理的理解可以参考图 2-10。

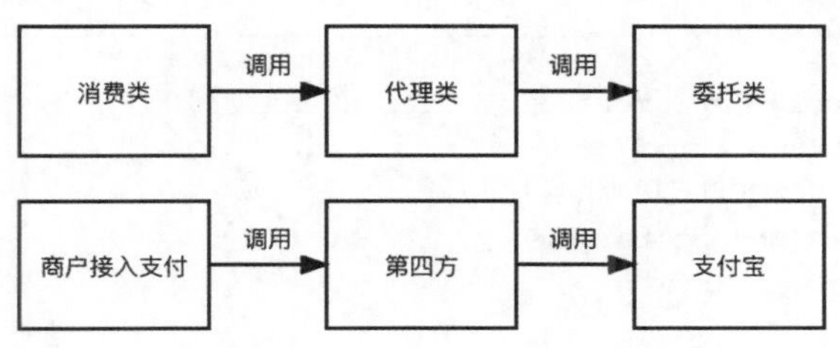

图 2-10　对代理的理解

实现代理后，消费类是不能直接调用委托类的，而是由代理类路由到委托类的。也就是消费类无法直接访问委托类，或者需要对于委托类进行一些特殊的处理。

2.3.2　JDK 动态代理

JDK 动态代理是在 Java 内部使用反射机制实现的。使用 JDK 动态代理的必要条件是委托类实现统一的接口，否则 JDK 动态代理不能应用。虽然有一定的局限性，但是影响不大。java.lang.reflect 包中的 Proxy 类和 InvocationHandler 接口提供了生成动态代理类的能力。所以如果需要代理，那么代理类必须实现 InvocationHandler 接口或者继承 Proxy 类。建议使用 InvocationHandler 接口实现代理类。

继续以前面的支付场景为例，这次用 JDK 动态代理实现。

首先定义一个支付接口和两个实现类，也就是目标类，如代码清单 2-17 所示。

代码清单 2-17　实现类和接口

```
/** 接口——支付的通用接口方法  */
public interface Pay {
    void pay(String operation);
}
/** * 委托类——一种支付方式  */
public class AliPay implements Pay {
    @Override
    public void pay(String operation) {
        System.out.println("进行AliPay支付,操作:" + operation);
    }
}
/** * 委托类——另一种支付方式  */
public class WxPay implements Pay {
    @Override
    public void pay(String operation) {
```

```
            System.out.println("进行WxPay支付,操作:" + operation);
        }
    }
```

接下来是重点，也就是实现代理类 PayProxy，如代码清单 2-18 所示。

代码清单 2-18　代理类 PayProxy

```
/** JDK动态代理类 */
public class PayProxy implements InvocationHandler {
    private Object target;
    /** * 构造方法，需要实现代理的真实对象*/
    public PayProxy(Object target) {
        this.target = target;
    }
    /**
     * 负责处理动态代理类上的方法调用
     * 根据三个参数进行预处理或者将其分派到不同的委托类实例上使用反射执行
     * @param proxy  被代理的对象
     * @param method 要调用的方法
     * @param args   方法调用时所需要的参数
     * @throws Throwable
     */
    @Override
    public Object invoke(Object proxy, Method method, Object[] args) throws Throwable {
        //在执行目标方法前可以进行操作
        System.out.println("调用之前……");
        System.out.println("Method:" + method);
        /*
  method.invoke方法会调用真实对象的方法，会跳转到代理对象关联的handler对象的invoke方法进行调用，invoke内部是通过类反射实现的
  */
        Object result = method.invoke(target, args);
        //执行目标方法后可以进行操作
        System.out.println("调用之后……");
        return result;
    }
}
```

最后，演示如何调用具体的支付实现类，如代码清单 2-19 所示。

代码清单 2-19　演示类

```
//调用类——进行演示
public class Store {
    public static void main(String[] args) {
        Pay aliPay = new AliPay();
        aliPay(aliPay);
        System.out.println("-----------");
        Pay wxPay = new WxPay();
```

```
        aliPay(wxPay);
    }
    public static void aliPay(Pay realPay) {
        /*PayProxy实现了InvocationHandler 接口内部包含指向委托类实例的引用，用于真正执行分派转发过来的方法调用。也就是要代理哪个真实对象，就将该对象传进去，最后通过该真实对象来调用其方法*/
        InvocationHandler handler = new PayProxy(realPay);
        ClassLoader loader = realPay.getClass().getClassLoader();
        Class[] interfaces = realPay.getClass().getInterfaces();
        //该方法用于为指定类装载器、一组接口及调用处理器生成动态代理类实例
        Pay pay = (Pay) Proxy.newProxyInstance(loader, interfaces, handler);
        pay.pay("pay");
    }
}
```

测试结果如下：

```
调用之前……
Method:public abstract void com.uifuture.dynamicproxy.target.Pay.pay(java.lang.String)
进行AliPay支付,操作:pay
调用之后……
------------
调用之前……
Method:public abstract void com.uifuture.dynamicproxy.target.Pay.pay(java.lang.String)
进行WxPay支付,操作:pay
调用之后……
```

通过上面的调用示例可以知道，代理类确实生效了，但是调用时非常麻烦。接下来针对上面的代理类进行一些改造优化，如代码清单 2-20 所示。

代码清单 2-20　优化后的代理类

```
/**
 * 代理类的优化
 */
public class PayProxyOptimize {
    public <T> T create(Class<T> pay, final T t) {
        return (T) Proxy.newProxyInstance(pay.getClassLoader(),
                /* pay.getInterfaces() getInterfaces() 确定此对象所表示类的实现接口，在这里pay是Pay接口类型，而Pay接口没有继承另外的接口，所以getInterfaces方法返回空。报异常Exception in thread "main" java.lang.ClassCastException:com.sun.proxy.$Proxy1 cannot be cast to com.uifuture.dynamicproxy.target.Pay */
                new Class[]{pay}
                //使用匿名内部类，也可以实现InvocationHandler接口
                , new InvocationHandler() {
                    @Override
                        public Object invoke(Object proxy, Method method, Object[] args) throws Throwable {
                        //在执行目标方法前可以进行操作
                        System.out.println("调用之前...");
                        System.out.println("Method:" + method);
```

```
                    /*调用真实对象的方法，会跳转到代理对象关联的handler对象的invoke方法
进行调用，内部通过类反射实现*/
                    Object result = method.invoke(t, args);
                    //在执行目标方法后可以进行操作
                    System.out.println("调用之后...");
                    return result;
                }
            });
    }
}
```

调用的测试类如代码清单 2-21 所示。

代码清单 2-21　测试类

```
public class StoreOptimize {
    public static void main(String[] args) {
        Pay realPay = new AliPay();
        Pay aliPay = new PayProxyOptimize().create(Pay.class, realPay);
        aliPay.pay("测试");
    }
}
```

对比前面的动态代理的调用，可以发现，这个调用简单了很多。因为在代理类的内部进行了优化。

这里有一个需要注意的地方，当调用 public static Object newProxyInstance(ClassLoader loader, Class<?>[] interfaces,InvocationHandler h) 方法传递 Class<?>[] interfaces 参数时，不能直接使用 pay.getInterfaces()。

这里重新改写优化后的代理类，增加一个方法，把 Class<?> [] interfaces 参数直接传递过来，然后对 Class<T> pay 调用 getInterfaces 方法。添加断点，查看 class[] 的值，即可看出差距，如代码清单 2-22 所示。

代码清单 2-22 测试 pay.getInterfaces 与 realPay.getClass().getInterfaces() 的差距的方法

```
/**
 * 在方法中进行断点调试，可以发现classes对象数组为空
 * 所以，在使用泛型的情况下，该方法是无法返回准确的接口类型的
 */
public <T > T create2(Class < T > pay, final T t, Class[] interfaces){
    Class[] classes = pay.getInterfaces();
      return  (T)  Proxy.newProxyInstance(pay.getClassLoader(), interfaces, new
InvocationHandler() {
        @Override
          public Object invoke(Object proxy, Method method, Object[] args) throws
Throwable {
            System.out.println("调用之前...");
            System.out.println("Method:" + method);
            Object result = method.invoke(t, args);
            System.out.println("调用之后...");
```

```
            return result;
        }
    });
}
```

在 return (T)Proxy.newProxyInstance(pay.getClassLoader() 上添加断点，在 StoreOptimize 测试类的 main 方法中增加的测试代码如代码清单 2-23 所示。

代码清单 2-23　测试代码

```
Pay aliPay2 = new PayProxyOptimize().create2(Pay.class, realPay,
        realPay.getClass().getInterfaces()
);
aliPay2.pay("测试2");
```

执行 Debug，出现如图 2-11 所示的结果。

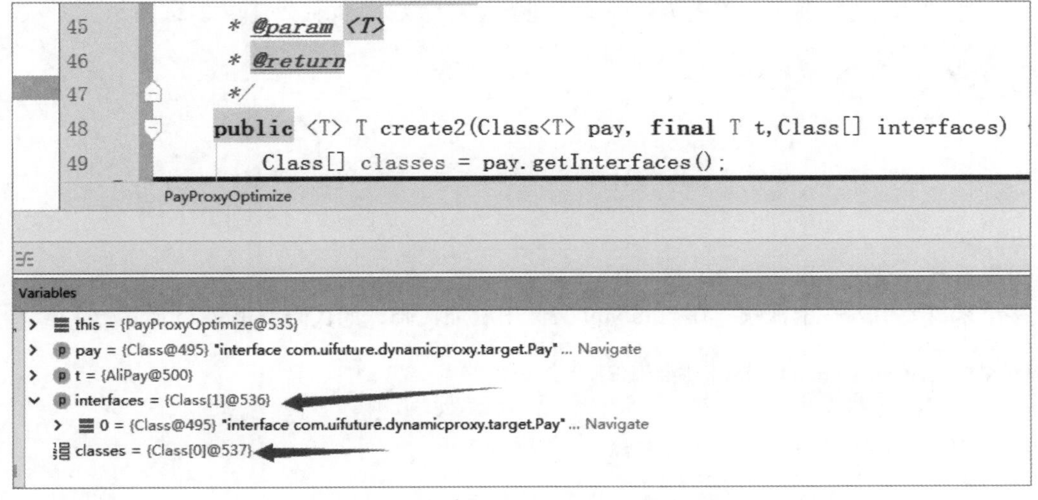

图 2-11　查看 classes 和 interfaces 的不同

从这两个 Class[] 对象的值可以看出，classes 的值为空，这也解释了前面为什么不能使用 getInterfaces，因为会出现类型转换异常。

可以看到 pay 的类型是 Pay 接口，而 Pay 接口是没有继承其他接口的，所以调用 getInterfaces 方法返回的值肯定是空的。

而对于 realPay.getClass().getInterfaces()，由于 realPay 实际上是 new AliPay()，所以 realPay 其实是 AliPay 类型的，继承了 Pay 接口。

如果想查看 getInterfaces 方法反射的结果，则可以在 getInterfaces 方法中添加断点，可以看到 this 对象就是实现类 AliPay。该对象调用 getInterfaces 方法返回的就是 Pay 接口。

JDK 动态代理就是指使用接口类生成新的实现类，新的实现类委托给 InvocationHandler 接口，在 InvocationHandler 接口方法中使用类反射机制调用被代理的类方法。

2.3.3 CGLib 动态代理

前面介绍了 JDK 的动态代理，通过一些示例可以看出，JDK 动态代理是依赖于实现的接口类的。而 CGLib 动态代理弥补了这个缺点，在不需要实现接口的情况下，也可以实现动态代理。JDK 动态代理和 CGLib 动态代理在 Spring AOP 中都有应用，Spring AOP 默认使用 JDK 动态代理来代理接口，但是可以强制使用 CGLib 动态代理。

CGLib 内部使用了 ASM（Java 字节码操控框架）来转换字节码，可以代理没有接口类的类。所以相比 JDK 动态代理，CGLib 更灵活一些。更值得称赞的是，由于 CGLib 通过字节码产生子类来覆盖委托类的非 final 方法从而实现代理，而 JDK 动态代理使用 Java 类反射实现代理，所以 CGLib 动态代理比 JDK 动态代理更快。

注意：CGLib 不能对 final 和私有方法进行代理。

下面通过示例进行演示。首先创建一个委托类（目标类），如代码清单 2-24 所示。

代码清单 2-24　AliPay 委托类

```
public class AliPay {
    public void pay(String operation) {
        System.out.println("进行AliPay支付,操作:" + operation);
    }
}
```

可以看到，这里不再实现 Pay 接口了。

接下来需要引入一个依赖，在 pom.xml 文件中增加代码，如代码清单 2-25 所示。

代码清单 2-25　在 pom.xml 文件增加的代码

```
<dependencies>
    <dependency>
        <groupId>cglib</groupId>
        <artifactId>cglib</artifactId>
        <version>3.1</version>
    </dependency>
</dependencies>
```

然后编写 CGLib 的代理类，如代码清单 2-26 所示。

代码清单 2-26　CGLib 动态代理的代理类

```
/**
 * 使用CGLib动态代理的代理类
 * 实现一个MethodInterceptor，方法调用会被转发到该类的intercept方法
 */
public class PayProxy<T> implements MethodInterceptor {
    private Enhancer enhancer = new Enhancer();
    public T getInstance(Class clazz) {
        enhancer.setSuperclass(clazz);
        enhancer.setCallback(this);
```

```
        return (T) enhancer.create();
    }
    @Override
     public Object intercept(Object o, Method method, Object[] objects, MethodProxy
methodProxy) throws Throwable {
        System.out.println("代理开始");
        //调用真实对象的方法
        Object result = methodProxy.invokeSuper(o, objects);
        System.out.println("代理结束");
        return result;
    }
}
```

最后编写测试类，查看实际的运行效果，如代码清单 2-27 所示。

代码清单 2-27　测试类

```
/** 进行CGLib代理的测试  */
public class TestProxy {

    public static void main(String[] args) {
        PayProxy<AliPay> payProxy = new PayProxy();
        AliPay aliPay = payProxy.getInstance(AliPay.class);
        aliPay.pay("测试cglib动态代理");
    }
}
```

输出结果如下：

```
代理开始
进行AliPay支付,操作:测试cglib动态代理
代理结束
```

CGLib 动态代理和 JDK 动态代理的主要区别是，JDK 动态代理生成的代理类实现了和委托类相同的接口类，是面向接口的，JDK 动态代理使用反射机制调用委托类的方法。

而 CGLib 动态代理通过委托类生成字节码，生成的代理类是委托类的子类，所以通过 CGLibB 代理的类不能处理 final 和私有方法，而且 CGLib 没有采用反射机制，而是直接调用了委托类方法，所以相对来说，CGLib 动态代理的性能要高一些。

第3章 设计模式

本章要点

1. 懒汉式单例模式
2. 饿汉式单例模式
3. 枚举单例模式
4. 简单工厂模式
5. 工厂方法模式
6. 抽象工厂模式
7. 代理模式
8. 策略模式介绍与实例
9. 策略模式与模板模式
10. MVC 模式与组件

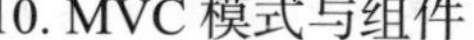

设计模式是对面向对象设计中反复出现的问题的解决方案，使用设计模式是为了可重用代码，让代码更容易被他人理解，保证代码的可靠性和程序的重用性。本章针对六个常用的设计模式进行深入的讲解。

3.1 单例模式

单例模式（Singleton Pattern）指项目全局的一个单例类只能有一个实例，是 Java 中最容易理解的设计模式之一。单例模式提供的是对象创建的方式，确保对象只会被创建一次，并提供唯一获取该对象的方法。单例具有以下 3 个特点：

（1）单例类必须只能有一个实例。

（2）单例类必须自行进行唯一实例化。

（3）单例类必须提供一个公共方法给全局访问。

接下来讲解几种创建单例类的方式，以及如何保证线程安全。

3.1.1 懒汉式单例模式

懒汉式单例模式可以理解为单例对象比较懒，在没有用到之前，该对象一直在“睡觉”。也就是在没有用到单例对象之前，对象不实例化，是种懒加载（LazyLoad）模式。懒汉式单例模式是单例模式最基本的实现方式。下面基于懒汉式单例模式演示线程安全和线程不安全的实现方式。

先实现线程不安全的懒汉式单例类，如代码清单 3-1 所示。

代码清单 3-1　线程不安全的懒汉式单例模式

```
public class ThreadUnsafeSingleton {
    /** 类的内部自行定义一个对象，注意是私有对象 */
    private static ThreadUnsafeSingleton threadUnsafeSingleton;
    /** 将构造方法设置为私有，不让外部访问 */
    private ThreadUnsawwfeSingleton() {
    }
    /** 接下来实现一个全局可以访问的方法，通过该方法获取该单例对象 */
    public static ThreadUnsafeSingleton getInstance() {
        if (threadUnsafeSingleton == null) {
        threadUnsafeSingleton = new ThreadUnsafeSingleton(); }
        return threadUnsafeSingleton;
    }
}
```

从代码中可以看出，getInstance 方法是判断 threadUnsafeSingleton 对象是否为 null 并为其赋值。在这两行代码的执行过程中，存在多线程并发问题，因为它们不是原子操作，为 threadUnsafeSingleton 对象赋值存在多次赋值的可能性。因此，严格来讲，在多线程的情况下，该实例不能算单例模式。

接下来介绍线程安全的懒汉式单例类的编写，如代码清单 3-2 所示。

代码清单 3-2　线程安全的懒汉式单例模式

```
public class ThreadSafeSingleton {
    /** 类的内部自行定义一个对象，注意是私有对象 */
    private static ThreadSafeSingleton threadSafeSingleton;
    /** 将构造方法设置为私有，不让外部访问 */
    private ThreadSafeSingleton() { }
    /**
     * 接下来依旧实现一个全局可以访问的方法，通过该方法获取该单例对象
     * 但是这里通过synchronized关键字对这个方法加锁
     */
    public static synchronized ThreadSafeSingleton getInstance() {
            if (threadSafeSingleton == null) { threadSafeSingleton = new
ThreadSafeSingleton(); }
        return threadSafeSingleton;
    }
}
```

这种方式在多线程下也能得到单例类，但是每次调用 getInstance 方法都需要加锁，造成了同步开销，而这个开销在 99% 的情况下都是不需要的，所以不建议使用这种方式。

接下来看一下双重校验模式（Double-Checked-Locking，DCL），如代码清单 3-3 所示。

代码清单 3-3　双重校验模式

```
// 双重校验模式，一种更加高效的多线程下安全的懒汉式单例模式
public class DCLThreadSafeSingleton {
    /**
     * 类的内部自己定义一个对象，注意是私有对象
     * 使用volatile关键字，可以理解为轻量级的synchronized
     * 保证了变量的可见性，在JDK1.5以后保证指令重排，JDK1.5+在此例中不会出现DCL失效的情况
     */
    private static volatile DCLThreadSafeSingleton dclThreadSafeSingleton;
    // 将构造方法设置为私有，不让外部访问
    private DCLThreadSafeSingleton() {}
    // 接下来依旧实现一个全局可以访问的方法，通过该方法获取该单例对象
    public static DCLThreadSafeSingleton getInstance() {
        /* 第一次判断空是为了减少synchronized不必要的同步开销，因为只有在第一次调用该方法时才需
要同步，以后实例不为null，是不需要同步的，提高了性能 */
        if (dclThreadSafeSingleton == null) {
            synchronized (DCLThreadSafeSingleton.class) {
                //第二次判断空是为了在dclThreadSafeSingleton为null的情况下才创建实例
                if (dclThreadSafeSingleton == null) {
                    /* 该句的执行并不是原子操作，所以可能出现指令重排的情况，在这种情况下，
可能出现dclThreadSafeSingleton被赋值，但是构造方法还未执行的情况。此时另外一个线程调用
getInstance方法时，dclThreadSafeSingleton已经不是null的情况，为了避免这种情况，使用
volatile关键字，防止指令重排 */
                    dclThreadSafeSingleton = new DCLThreadSafeSingleton();
                }
            }
        }
```

```
            return dclThreadSafeSingleton;
        }
    }
```

截至本书写作，已经发布 JDK 16 了，相信使用 JDK 5 以下的读者基本没有了，所以该实例在绝大多数场景下都能够满足单例要求。缺点是第 1 次加载时比较慢，毕竟有同步开销。而且 volatile 也会稍微影响一些性能，但是为了线程安全，这点性能的影响是可以接受的。

另外，可以使用懒汉式静态内部类实现单例模式。这种方式利用了 Java 类加载器机制，保证类的加载过程是线程安全的。静态内部类只有在被引用后才会被装载到内存中，所以也是懒加载模式。在懒加载模式中，类的加载与其是不是内部类没有关系，类的加载都是在有了对象的引用之后才会被加载到内存中的。懒汉式静态内部类的实现方式如代码清单 3-4 所示。

代码清单 3-4　懒汉式静态内部类实现方式

```
//懒汉式静态内部类实现方式
public class StaticThreadSafeSingleton {
    // 将构造方法设置为私有，不让外部访问
    private StaticThreadSafeSingleton() { }
    //实现一个全局可以访问的方法，通过该方法获取该单例对象
    public static StaticThreadSafeSingleton getInstance() {
        return HolderClass.staticThreadSafeSingleton;
    }
    //类的内部自己定义一个类，注意是私有类
    private static class HolderClass {
        /* 静态内部类中定义一个私有变量，通过直接实例化StaticThreadSafeSingleton这个类生成对
象，赋值给私有变量，HolderClass类得到了单例 */
        private final static StaticThreadSafeSingleton staticThreadSafeSingleton =
new StaticThreadSafeSingleton();
    }
}
```

上面的代码非常简洁，既实现了延迟加载，又保证了线程安全，并且不影响系统的性能，推荐这种写法。

3.1.2　饿汉式单例模式

饿汉式单例模式是指在对象引用之前，单例类已经初始化并加载好了。相比于加锁，饿汉式单例模式的优点是提高了效率，以“空间换时间”；缺点是类在加载时就进行了初始化，比较浪费内存（即使没有用到类，类也会在内存中）。与前面的静态内部类懒汉式单例模式类似，饿汉式单例模式也是利用了 Java 的类加载机制，避免了多线程下的同步问题。饿汉式单例模式的实现方式如代码清单 3-5 所示。

代码清单 3-5　饿汉式单例模式的实现方式

```
/** 饿汉式单例模式实现方式  */
public class EagerSingleton {
    /** 静态的内部变量 */
```

```
    private static EagerSingleton eagerSingleton = new EagerSingleton();
    /** 将构造方法设置为私有，不让外部访问 */
    private EagerSingleton() { }
    /** 实现一个全局可以访问的方法，通过该方法获取该单例对象 */
    public static EagerSingleton getInstance() { return eagerSingleton; }
}
```

可以看到，饿汉式单例模式的实现方法非常简单，这也是比较常用的设计模式，缺点是浪费了一些内存，在内存够大的情况下，这个缺点是可以接受的。

3.1.3 枚举单例模式

JDK 5 新增的 enum 关键字用于定义枚举类，如果枚举类中只有一个成员，那么该成员作为单例对象，是单例模式的一种实现。枚举类中没有可以访问的构造器，所以枚举单例类是 *Effective Java* 一书推荐的实现方式。枚举单例模式的实现方式如代码清单 3-6 所示。

代码清单 3-6　枚举单例模式

```
/** 枚举单例类实现方式  */
public enum EnumSingleton {
    /** 定义一个枚举元素，它代表了单例模式的一种实现 */
    uniqueInstance;
    // 添加任意方法
}
```

通过 Java 自带的关键字实现单例模式非常简单，另外 JDK 底层为了防止反射获取对象，做了特殊处理，如果尝试获取则会抛出以下异常：

Exception in thread "main" java.lang.IllegalArgumentException: Cannot reflectively create enum objects at java.lang.reflect.Constructor.newInstance(Constructor.java:417)。

3.1.4 注意事项

Java 对象的创建方式有 new、克隆、序列化、反射 4 种，懒汉式单例模式和饿汉式单例模式只是解决了不能直接通过“new”来创建对象的问题，后面 3 种方式可以破坏单例模式，下面简单介绍这 3 种方式。

1. 克隆方式

使用克隆方式创建对象的原理是直接从内存中 copy 内存区域，生成新的实例，该方式不通过构造方法创建对象。通过比对克隆出来的对象值和从单例获取的对象值，就可以发现两者的 Hash 值是不同的，说明是两个不同的对象。防止克隆破坏单例模式的方法两种：

（1）单例类不要实现 Cloneable 接口。

（2）如果实现了 Cloneable 接口，则重写 clone 方法，内部实现直接返回已有实例。

2. 序列化方式

一个类实现 Serializable 接口，就可以存储该类对象（保存到文件中或者通过网络传输到其他地方），该方式通过反序列化技术读取到内存中并把它组装成跟原来值一样的对象（值一样，但引用不一样的对象），这样就破坏了单例模式。防止序列化破坏单例模式的方法有两种：

（1）单例类不要实现 Serializable 接口。

（2）如果实现了 Serializable 接口，则重写 readResolve 方法，在内部实现直接返回已有实例。

3. 反射方式

通过反射可以获取类的构造方法，只要再加一行 setAccessible(true) 代码来解决访问权限，接下来就可以调用私有的构造方法创建对象。防止反射破坏单例模式的方法：当第二次调用构造函数时抛出异常。

以饿汉式单例模式为例，防止反射破坏单例模式的实现方式如代码清单 3-7 所示。

代码清单 3-7　防止反射破坏单例模式的方法

```
// 饿汉式实现方式
public class EagerSingleton {
    // 静态的内部变量
    private static EagerSingleton eagerSingleton = new EagerSingleton();
    private static volatile boolean flag = true;
    // 将构造方法设置为私有，不让外部访问
    private EagerSingleton() {
        if (flag) {
            flag = false;
            return;
        }
        throw new RuntimeException("单例模式险些被破坏，第二个对象未创建成功");
    }
    // 实现一个全局可以访问的方法，通过该方法获取该单例对象
    public static EagerSingleton getInstance() {
        return eagerSingleton;
    }
}
```

注意：枚举类就算实现了 Cloneable 接口和 Serializable 接口，也是无法被破坏的。枚举类无法克隆，没有构造方法。对于反序列化，Java 仅仅是将枚举对象的 name 属性输出到结果中，反序列化时则通过 java.lang.Enum 的 valueOf 方法根据名字查找枚举对象。由于编译器不允许定制序列化机制，所示禁用了 writeObject、readObject、readObjectNoData、writeReplace 和 readResolve 等方法。因此，枚举才是实现单例模式的最好方式。

3.2　工厂模式

工厂模式（Factory Pattern）是一种常用的设计模式，在基于 Java 语言的系统中随处可见。工厂模式分为简单工厂模式、工厂方法模式和抽象工厂模式。通常意义上的工厂模式是指工厂方法模式。在工厂模式中，通过一个共同的接口可以返回不同的新建对象。接下来分别讲解三种工厂模式。

3.2.1　简单工厂模式

简单工厂模式（Simple Factory Pattern）又称为静态工厂方法模式，属于类创建型模式。在简单工

厂模式中，可以根据工厂方法的参数值返回类实例。一般创建的实例会有共同的父类，或要实现相同的接口。简单工厂模式适用于要创建的对象较少的场景，如果要创建的对象太多，其中的业务逻辑就很复杂。另外，调用方关心的不是对象的创建过程，而是返回的对象，调用方只需知道对应的参数。

下面实现一个具体的实例。编写一个可以获取不同的支付对象的方法，如获取微信对象和支付宝对象。

支付宝和微信都可以用于支付，所以先为其定义一个接口或者抽象类，作为微信对象和支付宝对象的公共父类，并且在公共父类中声明公共的 pay 方法。这里建议使用接口，因为 Java 支持多个接口被一个类实现，也便于以后的扩展。

代码清单 3-8　支付的公共接口

```
/** 支付方式的公共接口  */
public interface Pay {
    /**公共接口的方法 */
    void pay();
    //还有退款、关闭订单、查询等公共方法，此处省略
}
```

编写支付宝和微信支付的实现类，如代码清单 3-9 所示。

代码清单 3-9　支付宝和微信支付方式的实现类

```
/** 支付宝支付方式  */
public class AliPayImpl implements Pay {
    @Override
    public void pay() {
        System.out.println("使用支付宝支付");
    }
}
/** 微信支付方式  */
public class WxPayImpl implements Pay {
    @Override
    public void pay() {
        System.out.println("使用微信支付");
    }
}
```

每个类代表不同的支付渠道，并实现了支付的方法（这里简化了支付功能的实现，只进行了简单输出）。

编写支付渠道的工厂类，如代码清单 3-10 所示。

代码清单 3-10　工厂类

```
/** 支付渠道的工厂方法  */
public class PayFactory {
    /** 根据传入不同的参数返回不同的实例对象 */
    public static Pay getPay(String type) {
        Pay pay = null;
        if (type.equals("ali")) { pay = new AliPayImpl();
```

```
            } else if (type.equals("wx")) { pay = new WxPayImpl();
            }
            return pay;
        }
    }
```

在工厂方法中，传入不同的 type 值就可以返回不同的实例对象（此处为支付渠道的对象），其为 Pay 类型。对简单工厂模式进行简单的测试，如代码清单 3-11 所示。

代码清单 3-11　测试类

```
    /** 简单工厂模式的测试  */
    public class TestPayFactory {
        public static void main(String[] args) {
            //传入不同的type值，运行测试类查看输出结果是否与预期的一致
            Pay aliPay = PayFactory.getPay("ali");
            aliPay.pay();
            Pay wxPay = PayFactory.getPay("wx");
            wxPay.pay();
        }
    }
```

运行测试类，查看测试结果，如图 3-1 所示。

```
TestPayFactory ×
"C:\Program Files\Java\jdk1.8.0_31\bin\java.exe" ...
进行支付宝支付
进行微信支付

Process finished with exit code 0
```

图 3-1　测试结果

可以看到，测试结果与预期的一样，传入 ali 值，返回的是 AliPayImpl 对象；传入 wx 值，返回的是 WxPayImpl 对象。

注意：这里的 type 值是可以配置的，为了演示方便，代码清单 3-11 中没有通过读取配置的方式来实现。

拓展：简单工厂模式分为 3 个角色。

（1）工厂角色，也就是实例中的 PayFactory，负责创建实例。

（2）抽象产品角色，如 Pay 接口，也就是需要创建的对象的父类，负责描述所有实例的公共接口。

（3）具体的产品角色，也就是实际需要创建的对象，如 AliPayImpl、WxPayImpl 等。

简单工厂模式的最大特点在于不需要知道具体的实现过程，只要知道类型对应的参数是什么就能获取正确的对象类型，无须关心细节。

简单工厂模式的缺点也是显而易见的，如果在上面的支付渠道中，再增加京东支付、百度支付等，则需要修改工厂类的创建逻辑，随着产品数量增加，逻辑就会越来越复杂。

3.2.2　工厂方法模式

工厂方法模式（Factory Method Pattern）也就是工厂模式，还可以称为虚拟构造器模式（Virtual Constructor Pattern）、多态工厂模式（Polymorphic Factory Pattern），也属于类创建型模式。工厂方

法模式是对简单工厂模式的优化。在工厂方法模式中，不再通过一个工厂类创建所有的对象，而是针对不同的对象提供不同的工厂。也就是说，针对每一个对象，都有与之对应的工厂类。

在工厂方法模式中，工厂父类只负责定义创建产品对象的公共接口，产品类的实例化延迟到工厂子类中进行。通过工厂子类确定具体实例化哪个产品类。

对代码清单 3-10 进行修改，不再由一个支付工厂类负责所有支付渠道的创建，而是引进工厂子类，将之前的工厂类修改为工厂接口，如代码清单 3-12 所示。

代码清单 3-12　公共工厂接口

```
/** 公共的工厂接口 */
public interface PayFactory {
    /** 声明公共的工厂方法 */
    Pay getPay();
}
```

为每个支付渠道编写一个工厂类，如代码清单 3-13 和代码清单 3-14 所示。

代码清单 3-13　AliPayImplFactory 工厂类

```
/** 支付宝的具体工厂方法 */
public class AliPayImplFactory implements PayFactory {
    /** 返回具体的实例对象 */
    @Override
    public Pay getPay() {
        return new AliPayImpl();
    }
}
```

代码清单 3-14　WxPayImplFactory 工厂类

```
/** 微信的具体工厂类 */
public class WxPayImplFactory implements PayFactory {
    @Override
    public Pay getPay() {
        return new WxPayImpl();
    }
}
```

从代码中可以看出每个具体的工厂类都返回一个具体的实例对象。

对工厂方法模式进行测试，如代码清单 3-15 所示。

代码清单 3-15　工厂方法模式测试类

```
/** 工厂方法模式的测试 */
public class TestPayFactory {
    public static void main(String[] args) {
        //获取支付宝支付的渠道
        PayFactory payFactory = new AliPayImplFactory();
        Pay pay = payFactory.getPay();
        pay.pay();
```

```
        //获取微信支付的渠道
        PayFactory payFactory2 = new WxPayImplFactory();
        Pay pay2 = payFactory2.getPay();
        pay2.pay();
    }
}
```

工厂方法模式和简单工厂模式的区别在于，简单工厂模式通过不同实例类型对应的参数获取不同的实例，而工厂方法模式通过不同的工厂类获取不同的对象实例。代码清单 3-15 的测试代码获取了不同的支付渠道，调用了相应的支付方法。测试结果如图 3-2 所示。

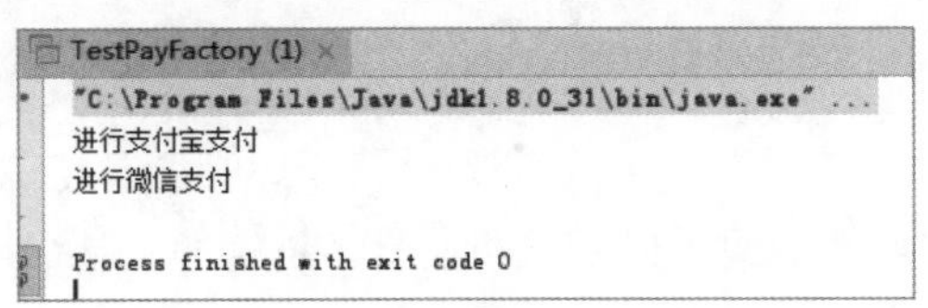

图 3-2　测试结果

通过上面的实例，可以将工厂方法模式和简单工厂模式进行对比。两者根本的区别在于，简单工厂模式只有一个工厂类，而工厂方法模式针对每一个不同的实例对象都提供了一个工厂类，这些工厂类都实现了一个工厂父类，也就是代码清单 3-12 中的 PayFactory 接口。

工厂方法模式由于使用了面向对象的多态性，所以允许在不修改现有项目代码的情况下，引进新的实例产品。但工厂方法模式也有缺点，就是每个产品实例都需要一个实例工厂与之对应，所以项目中的类是成对增加的，在一定程度上增加了项目的复杂性。另外，工厂方法模式可能增加项目的实现难度（如为了便于扩展，在调用方代码中使用了反射技术）。

工厂方法模式的使用场景：调用方不需要知道所创建对象的类。在本实例中，调用方不需要知道具体的支付渠道是什么，知道创建它的工厂名即可。另外，调用方可以通过子类指定创建哪个对象。在本例中，由于 Java 的多态性，在程序运行时，子类对象可以覆盖父类对象，所以可以由工厂子类确定具体的创建对象。

拓展：工厂方法模式比简单工厂模式多了一个具体工厂的角色。

3.2.3　抽象工厂模式

抽象工厂模式（Abstract Factory Pattern）是工厂模式中最难理解的一种模式，该模式是对工厂方法模式的进一步优化。与工厂方法模式相比，抽象工厂模式的工厂类不是创建一个对象，而是创建一系列对象，有效弥补了工厂方法模式中每个工厂只能创建一个对象的缺点。

在抽象工厂模式中，每个具体工厂都提供了多个工厂方法用于获取不同类型的对象。与工厂方法模式一样，抽象工厂模式也可以分为 4 个角色：

（1）AbstractFactory：抽象工厂，与工厂方法模式最大的不同是声明了多个用于创建对象的方法。

（2）ConcreteFactory：具体工厂，也就是抽象工厂的实现类，提供具体的对象实现方法。

（3）AbstractProduct：抽象产品，为每种对象声明接口，在接口中声明对象具有的业务方法。

（4）Product：具体产品，也就是具体工厂中负责实现的具体对象。

对 3.2.2 小节的支付实例进行扩展，实现跨平台支付，具体的支付渠道就是所需的具体产品。

优先声明支付宝支付和微信支付的抽象产品接口，如代码清单 3-16 所示。

代码清单 3-16　支付抽象产品接口

```
/** 为支付宝支付对象声明接口 */
public interface AliPay {
    /** App支付方式 */
    void appPay();
}
/** 为微信支付声明接口 */
public interface WxPay {
    /** App支付方式 */
    void appPpay();
}
```

上面创建了两个支付的抽象产品接口。下面针对支付产品，在不同的平台实现相关接口，如代码清单 3-17 和代码清单 3-18 所示。

代码清单 3-17　Android 平台下的具体产品实现

```
/** Android平台下支付宝支付的具体产品，由工厂负责生产的具体对象 */
public class AndroidAliPay implements AliPay {
    @Override
    public void appPay() {
        System.out.println("Android平台下的支付宝App支付");
    }
}
/** Android平台下微信支付的具体产品，由工厂负责生产的具体对象 */
public class AndroidWxPay implements WxPay {
    @Override
    public void appPpay() {
        System.out.println("Android平台下的微信App支付");
    }
}
```

代码清单 3-18　iOS 平台下的具体产品实现

```
/** iOS平台下支付宝支付的具体产品，由工厂负责生产的具体对象 */
public class IOSAliPay implements AliPay {
    @Override
    public void appPay() {
        System.out.println("iOS平台下的支付宝App支付");
    }
}
/** iOS平台下微信支付的具体产品，由工厂负责生产的具体对象 */
public class IOSWxPay implements WxPay {
    @Override
    public void appPpay() {
        System.out.println("iOS平台下的微信App支付");
    }
}
```

上面针对不同的平台创建了不同具体产品的支付类。下面继续定义一个抽象工厂，该工厂可以创建支付宝支付和微信支付的对象，如代码清单 3-19 所示。

代码清单 3-19　抽象工厂

```
/**
 * 抽象工厂，负责创建支付渠道的对象，在这里是创建AliPay和WxPay
 * 注意：是创建多个支付渠道的对象。这是与工厂方法模式最大的区别
 */
public interface PayFactory {
    /** 创建支付宝支付的对象 */
    public AliPay createAliPay();
    /** 创建微信支付的对象 */
    public WxPay createWxPay();
}
```

在不同的平台分别实现该抽象工厂接口，如代码清单 3-20 和代码清单 3-21 所示。

代码清单 3-20　Android 平台下抽象工厂的实现

```
/** Android平台下支付的具体支付工厂*/
public class AndroidPayFactory implements PayFactory {
    /** 创建Android平台下支付宝支付的对象 */
    @Override
    public AliPay createAliPay() { return new AndroidAliPay(); }
    /** 创建Android平台下微信支付的对象 */
    @Override
    public WxPay createWxPay() { return new AndroidWxPay(); }
}
```

代码清单 3-21　iOS 平台下抽象工厂的实现

```
/** iOS平台下支付的具体工厂*/
public class IOSFactory implements PayFactory {
    /** 创建iOS平台下支付宝支付的对象 */
    @Override
    public AliPay createAliPay() {
        return new IOSAliPay();
    }
    /** 创建iOS平台下微信支付的对象 */
    @Override
    public WxPay createWxPay() {
        return new IOSWxPay();
    }
}
```

对抽象工厂模式进行测试，如代码清单 3-22 所示。

代码清单 3-22　抽象工厂模式测试

```
// 抽象工厂模式测试
public class TestAbstractFactory {
```

```
    public static void main(String[] args) {
        PayFactory payFactory;
        AliPay aliPay;
        WxPay wxPay;
        // 在Android平台下
        payFactory = new AndroidPayFactory();
        aliPay = payFactory.createAliPay();
        wxPay = payFactory.createWxPay();
        aliPay.appPay();
        wxPay.appPpay();
        // 在iOS平台下，仅需要改抽象工厂的创建过程
        payFactory = new IOSFactory();
        aliPay = payFactory.createAliPay();
        wxPay = payFactory.createWxPay();
        aliPay.appPay();
        wxPay.appPpay();
    }
}
```

测试结果如图 3-3 所示。

图 3-3　测试结果

上面是将支付对象作为具体产品，针对不同的平台获取不同的支付对象。

拓展：相比工厂方法模式，一个工厂方法能返回多个对象是抽象工厂的优势，但是也由于这个原因，一旦增加支付的渠道对象，就需要修改原有代码（不符合开闭原则，即对扩展开放、对修改封闭）。但是对于增加平台的需求，如增加 Windows 平台，在 Windows 平台上实现支付宝和微信的支付功能，则是符合开闭原则的——只需增加一个新的具体工厂，对已有的代码无须做任何的修改。

使用哪种工厂模式，需要根据具体的业务场景来选择，通过一些项目的历练才能不断地积累经验，这个过程并不是一蹴而就的。所以平时应该在设计模式的理解上多下功夫，多使用、多总结，只有这样，才知道在什么情况下使用哪种模式会更好。

3.3　代理模式

代理模式（Proxy Pattern）是指给一个对象或类提供一个代理（理解为第三方）。例如，小明想吃煮熟的蔬菜，他可以通过电话联系厨师，委托厨师煮菜，而不是直接面向“蔬菜”自己动手。通过厨师这个第三方，小明可以吃到煮熟的蔬菜。

这里就出现了两个角色，一个是真实对象角色（煮熟的蔬菜），另一个是代理角色（厨师）。而对于 JDK 动态代理，需要面向接口编程，所以引入了另一个角色——抽象角色。此处可以将抽象角色理解为电话，想要吃煮熟的蔬菜的信息由小明通过电话联系厨师发布，而小明需要通过厨师来获得煮熟的蔬菜，如图 3-4 所示。

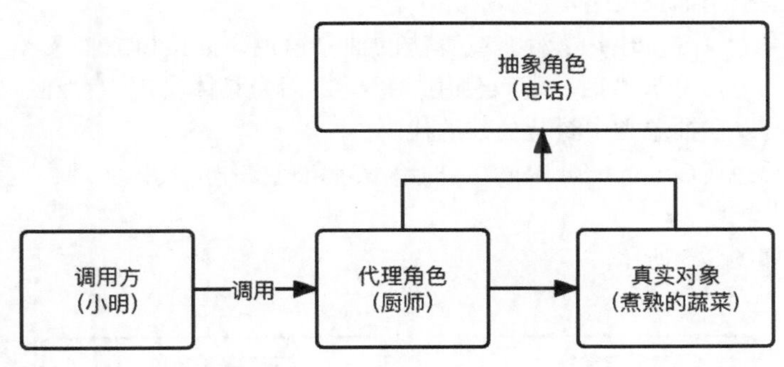

图 3-4 代理模式结构

另外需要记住关键的类和接口，也就是 InvocationHandler 接口和 Proxy 类。一般来说，使用最多的是 JDK 动态代理。对于没有实现接口的类，可以使用 CGLIB 动态代理，关键点是记住 MethodInterceptor 类。

代理模式一般用于在不改变原来接口的情况下，对对象进行访问控制的场景。

3.4 策略模式

不同的问题有不同的解决方案，而不同解决方案的合集就是策略。在实际开发中，某个功能有多种实现方式，使用策略模式选择一种最适合实际场景的方式，可以更加灵活地解决问题。策略即方法或方式，将实现的方法或方式（算法）解耦，并在另外的接口中实现。

3.4.1 策略模式的定义

Design Patterns (*GOF*) 一书中对策略模式的定义是 Define a family of algorithms, encapsulate each one, and make them interchangeable，翻译后就是定义一组算法，并且将每个算法封装到具有公共接口的一系列类中，从而让每个算法的变化不会影响客户端。

3.4.2 策略模式的作用

策略模式（Strategy Pattern）的作用如下：

（1）算法本身（实现方）和调用方解耦，由客户端选择使用哪种算法。

（2）避免了客户端的多重条件判断，扩展性好。

3.4.3 策略模式的结构

策略模式的底层逻辑是使用面向对象的继承和多态机制。在策略模式中共有 3 种角色：

（1）上下文角色（Context）：对应不同场景下的角色，对策略进行了二次封装，让调用方不直接调用策略，而是引用抽象策略角色。

（2）抽象策略角色（Stratrgy）：策略或算法的抽象角色，如接口或抽象类。当要声明每个策略或算法必须具有的方法和属性时，建议使用抽象类，因为具体实现中存在一些相同的逻辑，为了避免代码重复，可以在抽象类中封装公共的代码。

（3）具体策略角色（ConcreteStrategy）：抽象策略角色操作的具体实现，也就是算法的具体实现，一般是多个。

策略模式结构如图 3-5 所示。

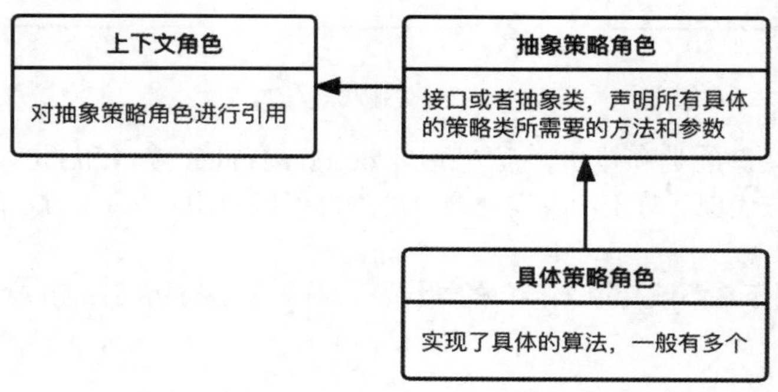

图 3-5　策略模式的结构

3.4.4 策略模式的应用

1. 应用场景

想象下面一个场景，网吧在某些时间段生意不好，需要举办一些促销活动，如在生意不好的时间段对上网费用进行打折促销，吸引用户。

梳理该场景的背景、问题和解决方案。

- 背景：吸引用户在某些时间段在网吧消费。
- 问题：早上（6:00—8:00）、中午（12:00—14:00）、晚上（18:00—20:00）的生意不好。
- 解决方案：在不同的时间段举办不同的打折促销活动。

2. 策略模式的实现

首先，定义抽象策略角色（Strategy），也就是网吧所有时间段的打折活动接口，如代码清单 3-23 所示。

代码清单 3-23　抽象策略角色类

```
/** 抽象策略角色  */
public abstract class Strategy {
    /**
     * 打折活动
     * @param price 实际价格
     * @return 通过打折后的价格
     */
    public abstract BigDecimal discount(BigDecimal price);
}
```

其次，实现具体策略角色（Concrete Strategy），针对不同的时间段进行不同的打折活动，如代码清单 3-24 ~ 代码清单 3-26 所示（这里仅作策略模式展示）。

代码清单 3-24　早上时间段的具体策略角色

```
/** 早上时间段的具体策略角色 */
public class MorningStrategyImpl implements Strategy {
    /**
     * 早上进行促销打折活动。在实际应用中，还会有其他操作
     * @param price 实际价格
     */
    @Override
    public BigDecimal discount(BigDecimal price) {
        System.out.println("实际价格:"+price.doubleValue());
        //直接返回打折后的价格
        return price.multiply(new BigDecimal(0.8));
    }
}
```

代码清单 3-25　中午时间段的具体策略角色

```
/** 中午时间段的具体策略角色 */
public class NooningStrategyImpl extends Strategy{
    /**
     * 中午进行促销打折活动
     * @param price 实际价格
     */
    @Override
    public BigDecimal discount(BigDecimal price) {
        System.out.println("实际价格:" + price.doubleValue());
        //直接返回打折后的价格
        return price.multiply(new BigDecimal(0.6));
    }
}
```

代码清单 3-26　晚上时间段的具体策略角色

```
/**  晚上时间段的具体策略角色  */
public class EveningStrategyImpl extends Strategy{
```

```
    /**
     * 晚上进行促销打折活动
     * @param price，实际价格
     * @return
     */
    @Override
    public BigDecimal discount(BigDecimal price) {
        System.out.println("实际价格:" + price.doubleValue());
        //直接返回打折后的价格
        return price.multiply(new BigDecimal(0.5));
    }
}
```

以上 3 个具体策略角色都有 discount 方法，返回打折后的具体价格，只是每个方法实现的过程不同。

然后，实现上下文角色（Context）。将 Context 传入不同的时间段来计算具体价格，并将其返回，如代码清单 3-27 所示。

代码清单 3-27　上下文角色

```
/** 根据时间选择不同的具体策略角色类 */
public class Context {
    private Strategy strategy;
    /** 构造函数，根据不同的具体策略给strategy 赋值*/
    public Context(Strategy strategy) {
        this.strategy = strategy;
    }
    /** 构造函数，根据不同的参数，实现不同的具体策略 */
    public Context(String timeName) {
        //不建议使用魔法值，请在实际开发中定义一个枚举类
        if("evening".equals(timeName)){
            this.strategy = new EveningStrategyImpl();
        }else if("nooning".equals(timeName)){
            this.strategy = new NooningStrategyImpl();
        }else if("morning".equals(timeName)){
            this.strategy = new MorningStrategyImpl();
        }
    }
    /**促销活动的方法 */
    public BigDecimal execute(BigDecimal price){
        return strategy.discount(price);
    }
}
```

Context 类中的 Context(String timeName) 构造方法将代理的判断逻辑交给了传入的 timeName 参数，根据 timeName 参数，创建不同的策略对象；Context(Strategy strategy) 方法将选择权交给了客户端，更加便于扩展。

拓展：在该 Context 上下文角色中，主要是根据条件选择不同的策略。

最后，实现测试类，如代码清单 3-28 所示。

代码清单 3-28　测试类

```java
/** 客户端测试类 */
public class TestClient {
    public static void main(String[] args) {
        BigDecimal price = new BigDecimal(20);
        Context context;
        System.out.println("----执行早上的策略----");
        context = new Context("morning");
        System.out.println("折扣后价格:"+context.execute(price).doubleValue());

        System.out.println("----执行中午的策略----");
        context = new Context("nooning");
        System.out.println("折扣后价格:"+context.execute(price).doubleValue());

        System.out.println("----执行晚上的策略----");
        context = new Context("evening");
        System.out.println("折扣后价格:"+context.execute(price).doubleValue());

        System.out.println("----执行早上的策略----");
        //只演示早上和中午的
        context = new Context(new MorningStrategyImpl());
        System.out.println("折扣后价格:"+context.execute(price).doubleValue());

        System.out.println("----执行中午的策略----");
        //只演示早上和中午的
        context = new Context(new NooningStrategyImpl());
        System.out.println("折扣后价格:"+context.execute(price).doubleValue());
    }
}
```

测试结果如图 3-6 所示。

```
TestClient ×
"C:\Program Files\Java\jdk1.8.0_31\bin\java.exe" ...
----执行早上的策略----
实际价格:20.0
折扣后价格: 16.0
----执行中午的策略----
实际价格:20.0
折扣后价格: 12.0
----执行晚上的策略----
实际价格:20.0
折扣后价格: 10.0
----执行早上的策略----
实际价格:20.0
折扣后价格: 16.0
----执行中午的策略----
实际价格:20.0
折扣后价格: 12.0
```

图 3-6　测试结果

策略模式是一种简单、常用的模式，一般不会单独使用，与模板模式、工厂模式等混合使用的情况比较多。策略模式有两个缺点：第一，维护各个策略类会给开发带来额外开销；第二，客户端必须知道具体策略是什么。也就是说，策略类如果不全部暴露给客户端，客户端是不知道该实现哪个策略的。例如，冒泡排序、桶排序、快速排序等实现的功能都是排序，但是在客户端调用策略之前，必须知道策略的适用情况，如快速排序、桶排序在哪种情况下效率高等，否则可能引起效率低、资源浪费的后果。也就是说，客户端必须知道算法及具体的行为才能使用策略模式。

策略模式的核心不在于如何实现算法，如前面的排序算法、折扣算法等，而是通过上下文对一系列具体的算法进行选择、调用。也就是说，这一系列的算法都是平等的，所有的策略算法在实现上是相互独立的，相互之间并没有依赖。

3.5 模板模式

模板模式（Template Pattern）又称为模板方法模式。当需要实现一些步骤，而这些步骤都有一些公共逻辑时，就可以使用模板方法，将公共逻辑抽取出来并放到抽象父类中形成公共方法。在公共方法中可以定义逻辑的骨架，也就是需要子类实现的业务逻辑的顺序；在子类中实现不同的逻辑，可以提高代码的复用度，并且不影响项目架构的逻辑（不会被子类破坏）。

前面提到，策略模式一般与模板模式或者工厂模式等混用，本节就讲解策略模式与模板模式的混用。如果要在抽象策略类中实现具体策略类的公共逻辑，则需要定义一些必要的抽象方法，这时便用到了模板模式，也就是混合应用了策略模式和模板模式。

3.5.1 策略模式与模板模式

策略模式与模板模式有较大区别，策略模式是基于多态实现里氏替换原则的体现（也可以说依赖倒置原则的简单体现），而模板模式是基于继承进行代码复用的技术。

策略模式的具体实现过程可以理解为：接口 A 中有一个抽象方法 a，另一组类使用不同的算法实现了 a 方法。根据上下文，当客户端调用 a 方法时也要调用 a 方法的具体实现方法，通过引用具体的实现类来判断调用哪个类的 a 方法。

而模板模式的一个抽象类中有多个方法，其中，一个方法 a 是 final 的抽象方法，另外的是一些不同的业务方法。不同的子类继承抽象类并实现不同的业务方法。抽象类需要在方法 a 中给出其他业务方法的调用顺序，形成逻辑的骨架。最后，在客户端中构造具体的子类并使用抽象类调用。

注意：可以将策略模式理解为面向接口编程的一个例子，策略模式体现了多态性。可以将模板模式理解为代码的复用。

3.5.2 模板模式的应用

模板模式主要有抽象模板角色（Abstract Template Role）和具体模板角色（Concrete Template Role）两个角色。

（1）抽象模板角色：定义的一个或者多个抽象方法。

抽象方法是一些基本方法，由子类实现。另外还需要实现一个抽象模板方法，这个模板方法一般被声明为 final，在该方法中会按照顺序调用基本方法，从而实现不会让子类破坏项目的逻辑顺序。

（2）具体模板角色：实现父类定义的基本抽象方法。

抽象模板角色可以对应任意多个具体模板角色，具体模板角色可以实现基本方法，这使得模板方法的实现逻辑也各不相同。

模板模式中的角色如图 3-7 所示。

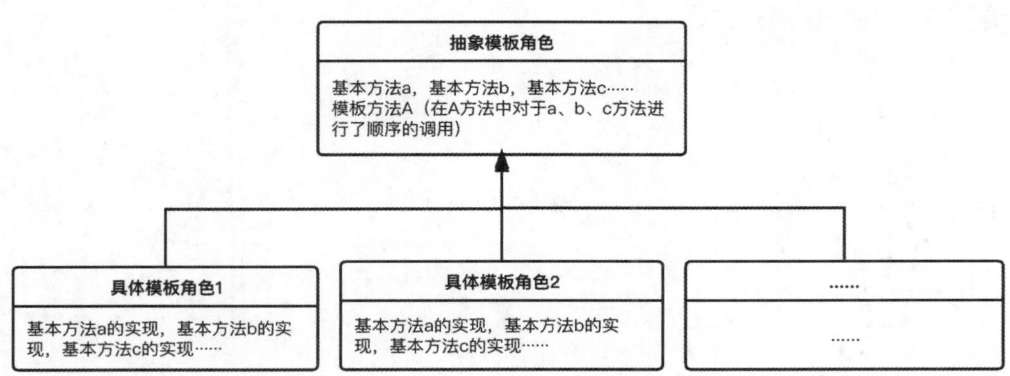

图 3-7　模板模式中的角色

1. 应用场景

现在有这样一个场景：我们需要实现喝茶这个过程的代码。这个过程包含 3 个步骤：首先烧开水，然后选择茶叶和杯子，最后喝茶。在代码中如何保证这个顺序不会被破坏呢？

2. 模板模式的实现

首先，定义一个抽象模板类，如代码清单 3-29 所示。

代码清单 3-29　抽象模板类

```
//抽象模板角色，现在模拟的是喝茶的场景
public abstract class AbstractTemplateRole {
    // 基本抽象方法，由子类实现，可能有柴火烧水、电烧水等方式
    protected abstract void boil();
    /* 基本抽象方法，由子类实现泡茶，可能用圆杯泡红茶、方杯泡绿茶，等等 */
    protected abstract void makeTea();
    /* 基本方法，空方法，子类可以选择是否覆盖实现默认的钩子方法。配料的选择，不是必需的 */
    protected void burdening(){};
    //模板方法，喝茶
    public final void drinkTea(){
        //按照顺序调用基本方法
        System.out.println("开始泡茶");
```

```
        boil();
        makeTea();
        burdening();
        System.out.println("可以喝茶了");
    }
}
```

其次，实现具体模板类，这里实现两个。

模板类一：选择用柴火烧水、圆杯泡绿茶，并且不放入其他配料，如代码清单 3-30 所示。

代码清单 3-30　GreenTeaImpl 类

```
/**实现具体模板类*/
public class GreenTeaImpl extends AbstractTemplateRole {
    /** 选择烧火方式*/
    @Override
    protected void boil() {
        System.out.println("使用柴火烧水……");
    }
    /** 选择茶叶和杯子*/
    @Override
    protected void makeTea() {
        System.out.println("使用圆杯泡绿茶");
    }
}
```

模板类二：选择用电烧水、方杯泡红茶，并且放点糖进去，如代码清单 3-31 所示。

代码清单 3-31　BlackTeaImpl 类

```
/** 实现具体模板类*/
public class BlackTeaImpl extends AbstractTemplateRole {
    /** 选择烧火方式*/
    @Override
    protected void boil() {
        System.out.println("使用电烧水……");
    }
    /** 选择茶叶和杯子*/
    @Override
    protected void makeTea() {
        System.out.println("使用方杯泡红茶");
    }
    /** 加配料*/
    @Override
    protected void burdening() {
        System.out.println("放点糖进去");
    }
}
```

最后，实现测试类，如代码清单 3-32 所示。AbstractTemplateRole 类中用 final 修饰了 drinkTea 方法，该方法不能被覆盖。子类可以实现或者覆盖父类的可变部分，但是逻辑的骨架是在父类模板方法中实现的，子类不能破坏。

代码清单 3-32　测试类调用

```
/** 实现测试类 */
public class TestClient {
    public static void main(String[] args) {
        AbstractTemplateRole templateRole = new GreenTeaImpl();
        templateRole.drinkTea();
        System.out.println("============");
        //接下来调用另外一种方式
        templateRole = new BlackTeaImpl();
        templateRole.drinkTea();
    }
}
```

测试结果如图 3-8 所示。

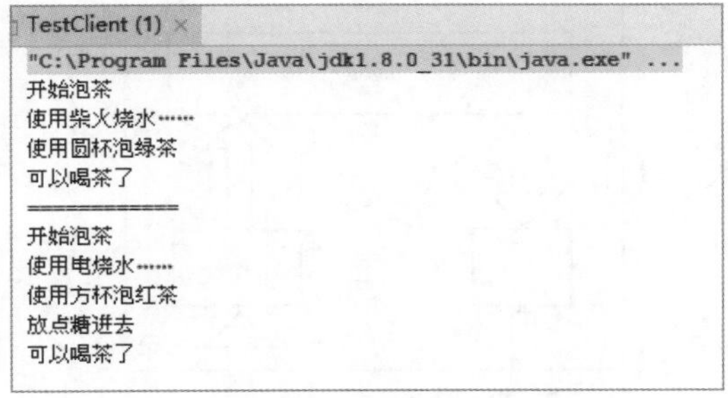

图 3-8　测试结果

拓展：模板模式的缺点是，每种实现方式都需要定义一个子类，这会导致子类的数量增加，使项目更加庞大，提高了系统的复杂度。但是，这样更加符合单一职责原则和开闭原则，使用反向控制，通过子类的实现扩展不同的行为。而这个缺点，从另外的角度来看，其实是优点。

模板模式的优点：一是提高了代码的复用度，也就是前面所说的将相同逻辑的代码放到抽象的父类方法中实现；二是提高了系统的扩展性，对于单个步骤实现方式不同的问题，在子类实现了就能解决，通过扩展子类即可增加新的方式。前面的抽象模板角色实现了一个空的方法（称为钩子方法），通过选择子类是否覆盖父类的钩子方法来决定该步骤是否执行，也就是实现了反向控制。

这些设计模式需要读者平时在项目及不同的场景中多使用才可以加深理解。

3.6　MVC 模式

MVC 即 Model-View-Controller（模型—视图—控制器），MVC 模式用在系统的分层开发中，是一种系统的架构模式，最早由 Trygve Reenskaug 在 1978 年提出。使用 MVC 模式是为了实现一种动态的程序设计，让后续程序的修改和扩展更加简化，并且让程序的部分代码能够得到重复利用。需要注意的是，MVC 模式不能说是一种技术，而是一种思想、一种理念。

3.6.1　MVC 模式的三大组件

MVC 模式的三大组件分别为模型（Model）、视图（View）、控制器（Controller）。

（1）模型：封装了一些数据，以及与这些数据有关的操作，在一个组件中，模型往往是表示组件的状态和操作状态的方法。

（2）视图：封装了模型，是对模型的一种展示。一个模型可以由多个视图展示，一个视图也可以和多个模型关联。

（3）控制器：封装外界（如用户）对模型的操作。这些操作最后都会作用于模型，调用模型中的一个或者多个方法。控制器一般在模型和视图之间起沟通作用，如用户在视图上输入文本，由控制器转发给模型。这样控制器负责连接两方，模型和视图之间就松耦合了。

组件间的关系如图 3-9 所示。

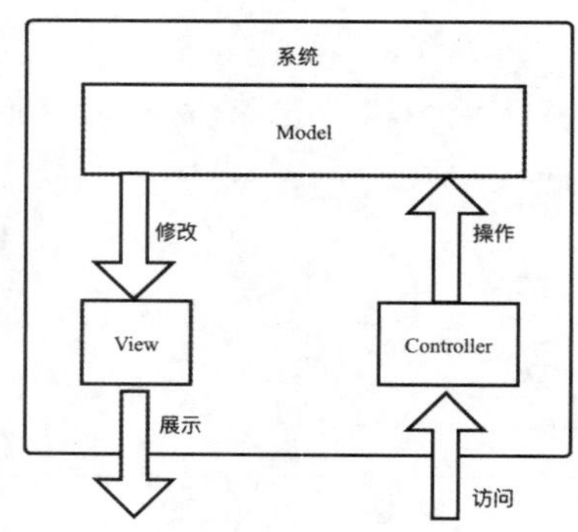

图 3-9　组件间的关系

在 MVC 模式下，当用户访问视图时，控制器处理请求，并且将请求转发给模型，从而操作模型中的方法，将结果返回给视图，展示给用户。

3.6.2　MVC 模式的优点

在使用 MVC 模式之前，查询数据和处理表现层展示代码全部写在了 JSP 页面里。页面越多，冗余的代码就越多，系统的维护成本也越高。MVC 模式强调职责分离，将业务逻辑和表现层分离，从而复用很多功能代码，大大减少了代码冗余，并降低了各模块之间的耦合度。

另外，控制器提高了项目的灵活性和可配置性。控制器可以连接不同的模型、渲染不同的视图，根据用户的需求，选择更合适的视图展示方式。

3.6.3　MVC 模式的应用

下面使用 Java 代码演示模型、视图和控制器之间的相互协作。其中，User 作为模型；UserView 作为视图，将 User 信息输出到控制台；UserController 作为控制器，负责模拟用户输入数据，将其存储到 User 对象中，更新视图并展示到控制台。

创建模型，如代码清单 3-33 所示。

代码清单 3-33　User 模型

```
/** 模型对象 */
public class User {
    /** 唯一标识*/
    private Integer id;
    /**用户名*/
    private String username;
    /**密码*/
    private String password;
    /**此处省略了setter方法和getter方法，GitHub中的demo代码是完整的*/
    @Override
    public String toString() {
        final StringBuilder sb = new StringBuilder("User{");
        sb.append("id=").append(id);
        sb.append(", username='").append(username).append('\'');
        sb.append(", password='").append(password).append('\'');
        sb.append('}');
        return sb.toString();
    }
}
```

创建视图，如代码清单 3-34 所示。

代码清单 3-34　UserView 视图

```
/**视图*/
public class UserView {
    public void show(User user){
        System.out.println("展示数据:"+user);
    }
}
```

创建控制器，如代码清单 3-35 所示。

代码清单 3-35　控制器

```
/**控制器*/
public class UserController {
    private User user;
    private UserView userView;
    public UserController(User user, UserView userView) {
        this.user = user;
        this.userView = userView;
    }
    /**更新模型数据*/
    public void setUserPassword(String password) {
        user.setPassword(password);
    }
```

```
    /**更新视图*/
    public void updateView(){
        userView.show(user);
    }
}
```

最后，创建一个演示类来模拟用户的操作、访问控制器，演示 MVC 设计模式，如代码清单 3-36 所示。

代码清单 3-36　MVC 模式演示

```
/**MVC设计模式演示*/
public class MVCDemo {
    public static void main(String[] args) {
        //获取输入的数据
        User user=new User();
        user.setId(1);
        user.setUsername("chenhx");
        user.setPassword("1234");
        //创建视图，将信息输出到控制台
        UserView userView=new UserView();
        //使用控制器更新数据
        UserController userController=new UserController(user,userView);
        userController.updateView();
        //更新模型数据
        userController.setUserPassword("6666");
        //展示模型数据的视图
        userController.updateView();
    }
}
```

查看演示结果，如图 3-10 所示。

```
"C:\Program Files\Java\jdk1.8.0_31\bin\java.exe" ...
展示数据：User{id=1, username='chenhx', password='1234'}
展示数据：User{id=1, username='chenhx', password='6666'}

Process finished with exit code 0
```

图 3-10　演示结果

上面只是简单地演示了 MVC 设计模式，并没有进行功能扩展。现在的 Java Web 项目基本上都用到了 MVC 模式。也可以简单地将 MVC 模式理解为限制开发者不能像使用 JSP/PHP 等一样直接将数据层和表现层全部放在一起的开发模式。尽管优秀的开发者会对项目业务实现结构分层，但是 MVC 模式可以从根本上强制开发者对项目业务实现结构分层。

注意：记住一点，设计模式的最终目的都是实现（方便系统拓展）代码复用、功能复用、松耦合，解决冲突等。使用合适的设计模式，开始可能觉得比较麻烦，但是从长远角度看，随着项目越来越庞大，设计模式的好处就越来越明显。

第Ⅱ篇　Spring MVC 框架

本篇内容为了解 Spring MVC、Spring MVC 基础、深入理解 Spring MVC 九大组件、Spring MVC 基础应用和 Spring MVC 核心应用。重点讲解了转发与重定向、GZIP 压缩、文件上传与下载、主题的配置、全局异常处理、过滤器与拦截器、JSON 数据交互、在国际化等项目中的实际应用。

通过本篇的学习，希望读者能够搭建好 Spring MVC 项目，并能与前端页面进行数据交互、上传与下载文件等。在 Spring MVC 九大组件的学习中，希望读者能够掌握源码的学习技巧，记住源码阅读的精髓——学代码初始化流程可学细节，读框架结构与模块设计要懂全貌。

第4章

了解 Spring MVC

本章要点

1. Spring MVC 架构
2. Spring MVC 体系结构、运行流程
3. Spring MVC 组件 DispatcherServlet（前端控制器）
4. Spring MVC 组件 HandlerMapping（映射处理器）
5. Spring MVC 组件 Handler（处理器）
6. Spring MVC 组件 HandlerAdapter（适配处理器）
7. Spring MVC 组件 ViewResolver（视图解析器）
8. Spring MVC 组件 View（视图）

4.1 Spring MVC 特点

Spring MVC 是一款非常优秀的 MVC 框架，也是目前在 Java 开发中被应用最多的 MVC 框架。一般情况下，Spring MVC 会和 Spring、MyBatis 结合使用。本节主要讲解 Spring MVC 是什么及 Spring MVC 能做什么。

1. Spring MVC 是什么

Spring MVC 是一个使用 Java 语言实现 Web MVC 设计模式的轻量级 Web 应用框架。

Spring MVC 基于 MVC 模式的思想理念，将 Web 层进行解耦，并按职责分层，简化了开发过程，方便以后的扩展。

2. Spring MVC 能做什么

（1）使 Web 层开发更加简洁。

（2）可重用业务代码，方便以后的扩展。

（3）与 Spring 框架集成非常简单、方便。

（4）更方便的国际化处理和主题的切换。

（5）灵活的 URL 页面映射。

（6）提供了表单数据校验、格式化、数据绑定等功能。

（7）提供了 JSP 标签库，使 JSP 的开发更加简单。

（8）能够进行 Web 层的单元测试。

（9）异常处理更加方便。

（10）提供了静态资源支持。

（11）对功能模块进行划分，实现层级之间的解耦。

4.2 Spring MVC 体系结构

Spring MVC 将前端控制器作为整个框架的中枢。下面对前端控制器及 Spring MVC 的处理流程进行讲解。

4.2.1 Spring MVC 运行流程

Spring MVC 有一个入口类，类名为 DispatcherServlet，该类的继承关系如图 4-1 所示。

在 HttpServletBean 类中，重写了 GenericServlet 类中的 init 方法，该方法是 HttpServletBean 初始化的入口。

GenericServlet 类中的 init 方法为空方法，HttpServletBean 类中 init 方法源码如代码清单 4-1 所示。

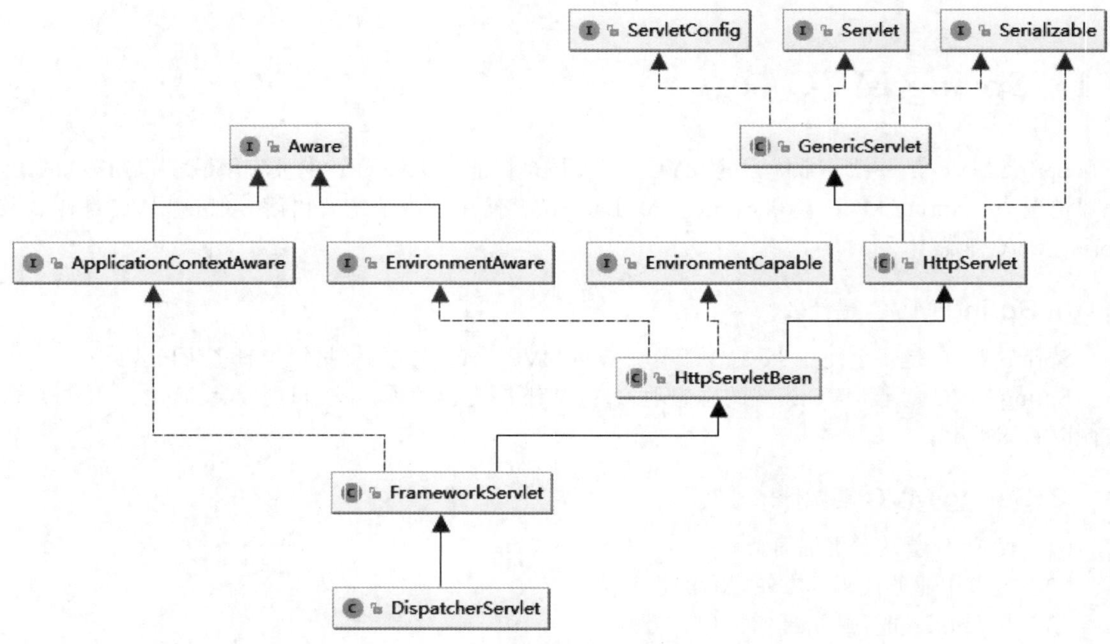

图 4-1　DispatcherServlet 类的继承关系

代码清单 4-1　HttpServletBean 类中的 init 方法

```
public final void init() throws ServletException {
    if (this.logger.isDebugEnabled()) {
        this.logger.debug("Initializing servlet '" + this.getServletName() + "'");
    }
    try {
        PropertyValues pvs = new
        HttpServletBean.ServletConfigPropertyValues(this.getServletConfig(), this.
requiredProperties);// ①
        BeanWrapper bw = PropertyAccessorFactory.forBeanPropertyAccess(this);// ②
        ResourceLoader resourceLoader =
                new ServletContextResourceLoader(this.getServletContext());// ③
        bw.registerCustomEditor(Resource.class,
                new ResourceEditor(resourceLoader, this.getEnvironment()));// ④
        this.initBeanWrapper(bw);// ⑤
        bw.setPropertyValues(pvs, true);// ⑥
    } catch (BeansException var4) {
          this.logger.error("Failed to set bean properties on servlet '" + this.
getServletName() + "'", var4);
        throw var4;
    }
    this.initServletBean();// ⑦
    if (this.logger.isDebugEnabled()) {
```

```
            this.logger.debug("Servlet '" + this.getServletName() + "' configured
    successfully");
        }
    }
```

该方法已被声明为 final，说明子类无法再重写。首先看看该方法实现的功能。

①创建 PropertyValues 对象。

PropertyValues 中保存了 PropertyValue 数组，PropertyValue 对象包含 name、value 等属性，new HttpServletBean.ServletConfigPropertyValues 方法实际上创建了 ServletConfigPropertyValues 对象，该对象的构造方法主要实现的是将 PropertyValue 对象加进 PropertyValue 数组。创建 PropertyValues 对象，实际上就是将 Spring MVC 的关于 Servlet 的名称和值添加进 List<Property Value> 中的过程。

②创建 BeanWrapper。

BeanWrapper 接口是对 JavaBean 的封装，并且提供了对 JavaBean 的操作，可以获取属性值及属性的描述符等。通常并不会直接使用该对象，而是通过 BeanFactory 或者 DataBinder 进行操作。

在 PropertyAccessorFactory.forBeanPropertyAccess(this) 方法中，实际上创建的是 BeanWrapper Impl 对象，BeanWrapperImpl 继承了 AbstractNestablePropertyAccessor 抽象类并实现了 BeanWrapper 接口，BeanWrapperImpl 对象提供了一些默认的属性编辑器，以及不同类型之间的转换、设置或获取 Bean 的属性等操作。这里的 Bean 对应的是 this，也就是 DispatcherServlet 对象。

简单地理解该过程，就是创建了 BeanWrapper，并将 DispatcherServlet 对象进行封装。

③创建 ResourceLoader。

ResourceLoader 也就是资源的加载器。实际上创建的是 ServletContextResourceLoader 资源加载器，该加载器通过路径加载上下文资源。

④将 ResourceLoader 添加到 BeanWrapper 中。

首先将 ResourceLoader 和 ConfigurableEnvironment 封装到 ResourceEditor 对象中。然后将 ResourceEditor 对象注册到 BeanWrapper 中。

⑤初始化 initBeanWrapper 方法，该方法为空方法，留给用户进行扩展。

⑥将 PropertyValues 添加到 BeanWrapper 中。

将 PropertyValues 对象也注册到 BeanWrapper 中。

⑦再次调用了一个空方法 initServletBean。该方法在其子类 FrameworkServlet 中实现。在 FrameworkServlet 类的 initServletBean 方法中，也预留了 initFrameworkServlet 空方法。

通过以上分析可知，HttpServletBean 初始化就是运行 init 方法，init 方法的主要作用是创建 BeanWrapper。通过 BeanWrapper 对象可以访问 Servlet 的所有参数，以及添加进去的资源、DispatcherServlet 的属性和其他被添加进去的 Bean。

接下来，通过图 4-2 讲解 Spring MVC 的运行流程：用户发起请求→进入 DispatcherServlet（前端控制器）→进入 HandlerMapping（处理器映射器）→进入 Controller（或 Handler，页面控制器）→进入 ModelAndView（模型和视图）→进入 ViewResolver（视图解析器）→返回 View（视图）→返回 Response（响应）→用户查看结果。

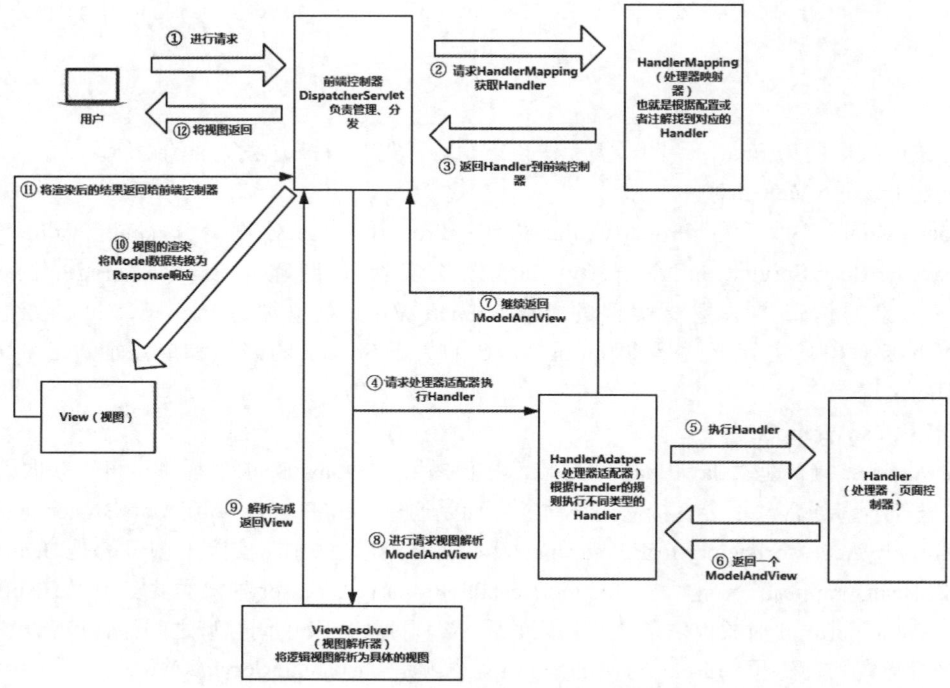

图 4-2　Spring MVC 运行流程

4.2.2　Handler 和 Controller 的联系

HandlerMapping 接口实现从 URL 请求映射到对应的请求处理 Bean，支持映射到 Bean 实例和 Bean names。

Base Controller 接口是接收和处理 HttpServletRequest 和 HttpServletResponse 实例的组件，就像 HttpServlet 一样，但能够参与 MVC 工作流程。Controller 可以与 Struts Action 相媲美。

Controller 接口的任何实现都应该是可重用的线程安全类，能够在应用程序的整个生命周期中处理多个 HTTP 请求。为了轻松配置控制器，建议使用 Controller 实现 JavaBeans。

Controller 接口的任何直接实现都只是处理 HttpServletRequests，并且应该返回一个 ModelAndView，以便由 DispatcherServlet 进一步解析。

Handler 和 Controller 接口在这里可以认为是一个概念，但是 Handler 并不是一个类或者接口，只是一个名词术语。Controller 是控制器，Handler 是控制器的处理方法。

4.3　Spring MVC 组件说明

Spring MVC 框架涉及的组件较多，本节简单地介绍部分组件。

1.DispatcherServlet（前端控制器）

DispatcherServlet 是 Spring MVC 框架的流程控制中心，是最核心的类，负责控制其他组件协调执行、统一调度，降低组件的耦合度，方便组件的扩展。DispatcherServlet 并不实现某个具体的功能，仅充当中央处理器和转发器。

2.HandlerMapping（映射处理器）

HandlerMapping 的主要作用就是接收前端控制器转发过来的请求，根据请求的信息（URL 或者请求参数等）按照某种映射的规则和机制找到对应的处理器（Handler），将请求的处理工作委托给具体的处理器。

其中某种映射的规则和机制由开发者实现，也就是通过配置文件、接口、注解等方式实现不同的映射操作。

3.Handler（处理器）

Handler 是页面控制器，对应 MVC 中的 C（也就是 Controller 层），可以对具体的请求进行处理。也可以将 Handler 理解为 Controller 中的具体方法。该组件需要开发者根据实际的业务进行开发。Handler 完成业务逻辑后会返回 ModelAndView（包含逻辑视图名和模型对象信息），这里的逻辑视图名还不是真正的视图，还需要交给前端控制器进行处理成为真正的视图。

4.HandlerAdapter（适配处理器）

HandlerAdapter 用于执行具体的 Handler。通过该组件调用 Handler 并执行，这里用到了适配器模式，通过扩展处理器可以执行更多类型的 Handler。AnnotationHandlerAdapter 类（Spring 3.2 以后已被弃用，建议使用 RequestMappingHandlerAdapter）、RequestMappingHandlerMappingAdapter 类、HttpRequestHandlerAdapter 类等都是对 HandlerAdapter 接口的不同实现。

5.ViewResolver（视图解析器）

通过 ResolveViewName 方法解析视图，将逻辑视图名解析成真正的视图。

ViewResolver 接口通过 ResolveViewName 方法，根据逻辑视图名和视图类型，返回 View。

6.View（视图）

View 是真正展示给用户看的一个接口，其实现类有很多，分别支持不同的 View 类型，如 JSP、Freemarker、PDF、Excel 等。

第5章 Spring MVC 基础

本章要点

1. 了解 Spring MVC 的使用
2. 使用 IDEA 快速搭建一个 Spring MVC 项目
3. Spring MVC 项目配置与运行
4. Spring MVC 常用注解讲解
5. Java 对象与 JSON 数据之间的转换
6. Java 对象与 XML 数据之间的转换
7. 编程式静态资源映射
8. XML 配置静态资源映射
9. ControllerAdvice 注解讲解

本章结合示例对 Spring MVC 的知识点进行讲解，包括 Spring MVC 中常用的注解，Bean 对象与 JOSN/XML 之间的转换，静态资源在 Spring MVC 中的映射，以及 ControllerAdvice 注解和其他加快配置的方法等。

5.1 快速搭建 Spring MVC 项目

Spring MVC 提供了 DispatcherServlet 来开发 Web 应用。Servlet 3.0+ 环境下，在一个类中实现 WebApplicationInitialzer 接口时无须配置 web.xml，从而简化了 Web 应用的开发。

下面使用无 web.xml 的方式开发 Web 应用，开发工具是 IntelliJ IDEA。

5.1.1 创建 Maven 项目

搭建项目环境所需的软件和版本如下：

- IntelliJ IDEA Version 2018.1.6。
- JDK 1.8。
- Tomcat 9。
- Maven 3.3.9。
- Spring 的版本为 5.0.8（写书时的最新版本）。

通过 IntelliJ IDEA（简称 IDEA）创建 Maven 项目，由于是在父项目中创建子项目，所以在父项目上选择 New → Module 命令，会出现图 5-1 所示的界面。在该界面，按图 5-1 所示箭头所指步骤执行。

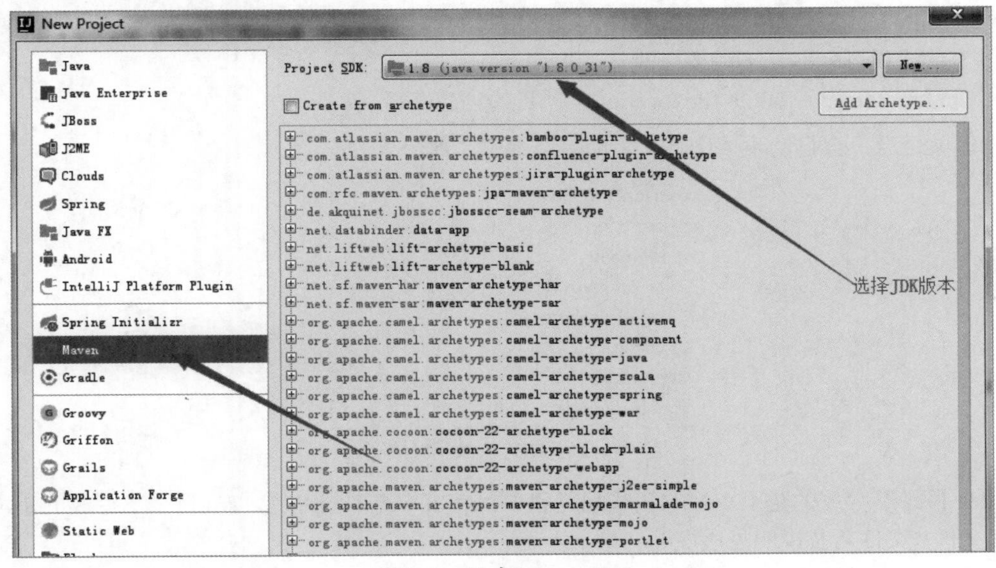

图 5-1　创建 Maven 项目

单击 Next 按钮执行下一步操作，如图 5-2 所示。

- GroupId：项目的唯一标识，最好分为多个段。一般按照域名命名，笔者的域名是 uifuture.com，所以选择 com.uifuture。
- ArtifactId：对应项目的实际名称，也就是项目根目录的名称。

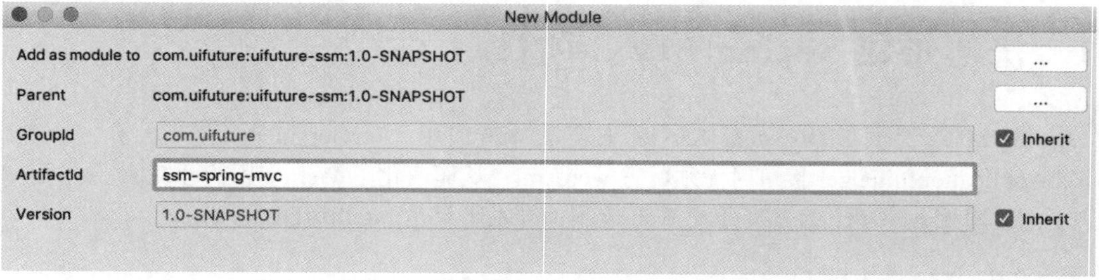

New Module

Add as module to com.uifuture:uifuture-ssm:1.0-SNAPSHOT

Parent com.uifuture:uifuture-ssm:1.0-SNAPSHOT

GroupId com.uifuture ☑ Inherit

ArtifactId ssm-spring-mvc

Version 1.0-SNAPSHOT ☑ Inherit

图 5-2 设置项目唯一的标识 ID

GroupId 和 ArtifactId 的作用是保证项目的唯一性。如果要把项目发布到 Maven 本地仓库 / 中央仓库，那么这两个 ID 在仓库中就是“坐标”，根据这两个 ID 在仓库中就可以找到对应的 jar 包。

填写完成后，单击 Next 按钮执行下一步操作，进入图 5-3 所示界面。

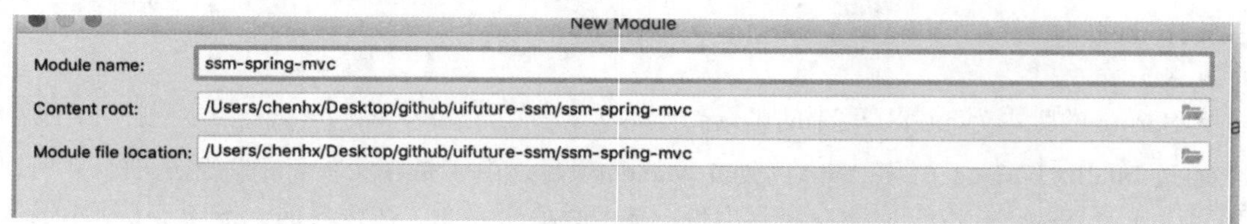

图 5-3 项目名和存储路径

- Module name：项目名，默认 ArtifactId 填写的名称，建议不要修改。其中，ssm-spring-mvc 为默认项目名称，不需要修改。
- Content root、Module file location：项目的创建路径。

创建项目后，会自动生成图 5-4 所示的目录树（部分）。

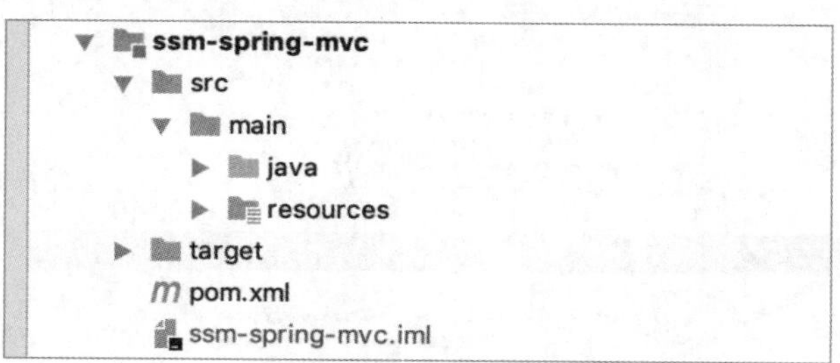

图 5-4 项目的目录树（部分）

.idea 目录下存放的是 IDEA 的一些配置和项目在 IDEA 中的配置。src 目录下存放的是项目的源码。文件 / 文件名及说明见表 5-1。

表 5-1 文件 / 文件名及说明

文件 / 文件名	说 明
main/java	存放项目的 Java 代码
main/resources	存放项目配置文件
test/java	存放单元测试代码
pom.xml	项目依赖的配置信息
ssm-spring-mvc.iml	IDEA 的项目配置文件，由 IDEA 自动生成
External Libraries	项目中引入的所有 jar 包文件

5.1.2 配置项目依赖

pom.xml 文件用于配置项目需要依赖的 jar 包，如代码清单 5-1 所示。

代码清单 5-1 pom.xml 配置文件

```
<?xml version="1.0" encoding="UTF-8"?>
<project xmlns="http://maven.apache.org/POM/4.0.0"
         xmlns:xsi="http://www.w3.org/2001/XMLSchema-instance"
             xsi:schemaLocation="http://maven.apache.org/POM/4.0.0 http://maven.
apache.org/xsd/maven-4.0.0.xsd">
    <!--指定了当前POM模型的版本，为了兼容Maven2和Maven3，modelVersion值固定设置为4.0.0-->
    <modelVersion>4.0.0</modelVersion>
    <!-- 项目的唯一标识，groupId和artifactId -->
    <artifactId>ssm-spring-mvc</artifactId>
    <!--groupId定义项目属于哪个组-->
    <groupId>com.uifuture</groupId>
    <!--项目的版本号-->
    <version>1.0-SNAPSHOT</version>
    <!-- 通用配置，方便管理依赖包的版本号及其他常量 -->
    <properties>
        <!--JDK版本-->
        <java.version>1.8</java.version>
        <!-- 编码 -->
        <project.build.encoding>UTF-8</project.build.encoding>
        <!-- Spring版本号 -->
        <spring-framework.version>5.0.8.RELEASE</spring-framework.version>
        <!--Logging版本 -->
        <logback.version>1.2.3</logback.version>
        <slf4j.version>1.7.25</slf4j.version>
    </properties>
    <!-- 此项目所依赖的jar包-->
    <dependencies>
        <!-- Spring MVC 依赖的jar包 -->
        <dependency>
            <groupId>org.springframework</groupId>
            <artifactId>spring-webmvc</artifactId>
            <version>${spring-framework.version}</version>
```

```
        </dependency>
          <!-- logback日志依赖的jar包 https://mvnrepository.com/artifact/org.slf4j/
slf4j-api -->
        <dependency>
            <groupId>org.slf4j</groupId>
            <artifactId>slf4j-api</artifactId>
            <version>${slf4j.version}</version>
        </dependency>
        <dependency>
            <groupId>ch.qos.logback</groupId>
            <artifactId>logback-core</artifactId>
            <version>${logback.version}</version>
        </dependency>
        <!-- https://mvnrepository.com/artifact/ch.qos.logback/logback-classic -->
        <dependency>
            <groupId>ch.qos.logback</groupId>
            <artifactId>logback-classic</artifactId>
            <version>${logback.version}</version>
        </dependency>
        <!-- Web依赖 -->
        <dependency>
            <groupId>javax.servlet</groupId>
            <artifactId>javax.servlet-api</artifactId>
            <version>3.0.1</version>
        </dependency>
        <!--JSTL 解析JSP标签依赖-->
        <dependency>
            <groupId>javax.servlet</groupId>
            <artifactId>jstl</artifactId>
            <version>1.2</version>
        </dependency>
    </dependencies>
    <build>
        <plugins>
            <plugin>
                <!--指定JDK版本和编码,防止编译问题和编码错误。如果不指定,那么Maven 3+会默认使用
JDK 1.5-->
                <groupId>org.apache.maven.plugins</groupId>
                <artifactId>maven-compiler-plugin</artifactId>
                <version>2.3.2</version>
                <configuration>
                    <source>${java.version}</source>
                    <target>${java.version}</target>
                    <encoding>${project.build.encoding}</encoding>
                </configuration>
            </plugin>
        </plugins>
    </build>
</project>
```

5.1.3 项目日志配置

在 src/main/resources 目录下，新建 logback.xml，用来配置日志格式、级别等，如代码清单 5-2 所示。

代码清单 5-2 logback.xml 配置文件

```
<?xml version="1.0" encoding="UTF-8"?>
<configuration debug="false" scan="true" scanPeriod="1 seconds">
    <contexListener class="ch.qos.logback.classic.jul.LevelChangePropagator">
        <resetJUL>true</resetJUL>
    </contexListener>
    <jmxConfigurator/>
    <appender name="console" class="ch.qos.logback.core.ConsoleAppender">
        <encoder>
            <!-- 配置日志输出格式 -->
            <pattern>logbak: %d{HH:mm:ss.SSS} %logger{36} - %msg%n</pattern>
        </encoder>
    </appender>
    <logger name="org.springframework.web" level="DEBUG"></logger>
    <root level="info">
        <appender-ref ref="console"></appender-ref>
    </root>
</configuration>
```

日志配置文件的基本结构：文件以 <configuration> 开头，后面有 0 或者多个 <appender> 标记、0 或者多个 <logger> 标记，最多只能有 1 个 <root> 标记。

（1）<configuration> 标记的属性如下：

- debug 属性的值有 true 或者 false，默认值为 false。当值为 true 时，可以实时查看 logback 的运行状态，还可以打印出 logback 内部的日志信息。
- scan 属性的值也是 true 或者 false，默认值为 true。当值为 true 时，在配置文件改变的情况下，Web 容器会重新加载配置文件。
- scanPeriod 属性用于设置监听配置文件是否变化的时间间隔，单位默认是 ms，可以自行给出时间单位。当 scan 属性为 true 时，此属性生效。

（2）<appender> 标记是负责写日志的组件，它是根标记 <configuration> 的子标记，可以有多个。<appender> 标记既可以定义日志的输出格式，也可以指定日志内容记录到哪个文件等操作。

（3）<logger> 标记也是 <configuration> 的子标记，可以有多个，用来设置某一个包或者具体到某一个类的日志打印级别，DEBUG、INFO、WARN、ERROR 是常用的日志级别。<logger> 标记中的 name 属性用来指定受到该 logger 标记约束的包或者类。level 属性用来设置打印的级别，级别设置是忽略字符串大小写的。

（4）<root> 标记是标记的配置根 logger，只能有一个。该标记只有一个 level 属性，没有 name 属性。该标记的默认日志级别是 INFO。在 <root> 标记中可以配置子节点，<appender-ref> 子标记可以用来配置使用哪个 appender。

注意：在 logback.xml 的 <logger> 标记中，将 org.springframwork.web 包下的类的日志级别

设置为 DEBUG 级别，这是因为在使用 Spring MVC 开发的过程中，不可避免地会出现和参数类型有关系的以 4 开头的错误码（4XX），设置此 logger 可以看到更加详细的错误信息，方便调试与改错。

5.1.4 演示页面

需要注意页面的创建位置。在 src/main/resources 下建立 views 目录，并在此目录下新建 index.jsp，如代码清单 5-3 所示。

代码清单 5-3 index.jsp

```
<%@ page contentType="text/html;charset=UTF-8" language="java" %>
<html>
<head>
    <title>MVC Title</title>
</head>
<body>
    <pre>Welcome to Spring MVC</pre>
    <br/>欢迎!
</body>
</html>
```

这里的 JSP 页面不是放在 Maven 标准的 src/main/webapp/WEB-INF 目录下，而是遵循 Spring Boot 规范放在 src/main/resources 目录下，因为在实际项目中 Spring Boot 的应用更加广泛。

5.1.5 Spring MVC 配置类

下面是一个简单的 Spring MVC 配置类。这里配置了一个 JSP 的 ViewResolver，用来映射路径和实际页面的位置，如代码清单 5-4 所示。

代码清单 5-4 Spring MVC 配置类

```
import org.springframework.context.annotation.Bean;
import org.springframework.context.annotation.ComponentScan;
import org.springframework.context.annotation.Configuration;
import org.springframework.web.servlet.config.annotation.EnableWebMvc;
import org.springframework.web.servlet.view.InternalResourceViewResolver;
import org.springframework.web.servlet.view.JstlView;
/** Spring MVC配置类 */
@Configuration
@EnableWebMvc
@ComponentScan("com.uifuture.basics")
public class MvcConfig {
    @Bean
    public InternalResourceViewResolver viewResolver(){
        InternalResourceViewResolver viewResolver = new InternalResourceViewResolver();
        //映射路径：运行时代码会将页面自动编译到/WEB-INF/classes/views/下
        viewResolver.setPrefix("/WEB-INF/classes/views/");
        //实际页面后缀
```

```
        viewResolver.setSuffix(".jsp");
        viewResolver.setViewClass(JstlView.class);
        return viewResolver;
    }
}
```

需要注意的注解知识点如下。

1.Configuration 注解

Configuration 并不是 Spring MVC 中的注解，而是 Spring 中的一个定义配置类的注解。在这个注解配置的类中，会有一个或者多个 Bean 注解的方法，这些方法对应的是 XML 配置文件中实例化的 Bean，有 Configuration 注解配置的类会被 AnnotationConfigApplicationContext 或者 AnnotationConfigWebApplicationContext 类扫描，Spring 会对 Bean 注解的方法的返回值 Bean 进行统一管理，后续项目中需要使用这些 Bean 时，会通过 Spring 容器获取。

2.EnableWebMvc 注解

EnableWebMvc 注解会开启一些默认的配置，如 ViewResolver 或者 MessageConverter 等。EnableWebMvc 注解源码如代码清单 5-5 所示。

代码清单 5-5　EnableWebMvc 注解源码

```
@Retention(RetentionPolicy.RUNTIME)
@Target({ElementType.TYPE})
@Documented
@Import({DelegatingWebMvcConfiguration.class})
public @interface EnableWebMvc {
}
```

可以看到源码中使用 Import 注解导入了 DelegatingWebMvcConfiguration 类，该类继承了 WebMvc ConfigurationSupport 类，WebMvcConfigurationSupport 类是提供 MVC 背后的配置的主类。

3.ComponentScan 注解

ComponentScan 注解的功能是告知 Spring 框架在启动时要扫描哪个类所在的包，并通过 value 属性指定具体的 Bean。

注意：如果类没有添加 Component 注解，则不会自动注入。

Spring MVC 中有一个接口叫作 ViewResolver，实现这个接口要重写 resolverViewName 方法，这个方法的返回值是接口 View，而 View 的职责就是使用 model、request、response 对象，并将渲染的视图（不一定是 HTML 数据，可能是 JSON、XML、PDF 数据）返回给浏览器。

读者可能会对将映射路径前缀配置为 /WEB-INF/classes/views/ 而感到奇怪，因为这不是存放页面的路径。实际上用户看到的是运行时的页面效果而不是开发时的代码。页面通过编译器会自动编译到 /WEB-INF/classes/views/ 下，编译后的路径结构如图 5-5 所示。

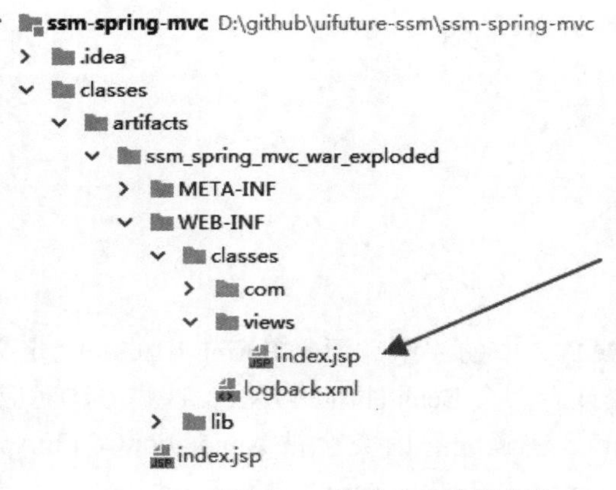

图 5-5　编译后的路径结构

5.1.6　Web 配置类

如果说上面的 MvcConfig 类的作用是替代 Spring 和 Spring MVC 的配置文件，那么下面这个 WebInitializer 类的作用就是为了替代 web.xml 配置文件。WebInitializer 配置类如代码清单 5-6 所示。

代码清单 5-6　WebInitializer 配置类

```
/** Web配置 */
public class WebInitializer implements WebApplicationInitializer {
    @Override
    public void onStartup(ServletContext servletContext) throws ServletException {
        AnnotationConfigWebApplicationContext context =
                new AnnotationConfigWebApplicationContext();
        context.register(MvcConfig.class);
        //新建WebApplicationContext注册配置类，并和当前servletContext关联
        context.setServletContext(servletContext);
        //注册并映射调度程序servlet
        ServletRegistration.Dynamic servlet =
                servletContext.addServlet("dispatcher",new Dispatcher Servlet (context));
        //注册Spring MVC的DispatcherServlet
        servlet.addMapping("/");
        servlet.setLoadOnStartup(1);
    }
}
```

WebApplicationInitializer 接口是 Spring 提供的，Servlet 3.0+ 环境下才能使用，只需要实现 onStartup 方法就可以替代 web.xml 配置文件的访问路径。可以在实现类方法中添加 servlet、listener 等，当启动项目时会加载这个接口的实现类，从而起到与 web.xml 配置文件一样的作用。

WebApplicationInitializer 接口源码如下：

```
public interface WebApplicationInitializer {
    void onStartup(ServletContext var1) throws ServletException;
}
```

用 IDEA 下载 WebApplicationInitializer 接口源码（单击在 IDEA 工具界面右上角出现的 Download Sources 链接），在接口上有很多注释，如代码清单 5-7 所示。

代码清单 5-7　WebApplicationInitializer 接口部分注释及源码

```
/**
 * Interface to be implemented in Servlet 3.0+ environments in order to configure
the
 * {@link ServletContext} programmatically -- as opposed to (or possibly in
conjunction
 * with) the traditional {@code web.xml}-based approach.
 * * <p>Implementations of this SPI will be detected automatically by {@link
 * SpringServletContainerInitializer}, which itself is bootstrapped automatically
 * by any Servlet 3.0 container. See {@linkplain SpringServletContainerInitializer its
 * Javadoc} for details on this bootstrapping mechanism.
 *
 * <h2>Example</h2>
 * <h3>The traditional, XML-based approach</h3>
 * Most Spring users building a web application will need to register Spring's {@
code
 * DispatcherServlet}. For reference, in WEB-INF/web.xml, this would typically be
done as
 * follows:
 * <pre class="code">
 ...
 * {@code
....
</pre>
....
 **/
```

从上面这些注释可以看出，WebApplicationInitializer 是在 Servlet 3.0+ 环境中实现的接口，便于进行编程式配置，其与传统的基于 XML 的方法不同，当然也可以与 XML 的方式结合使用。

SPI 机制（Service Provider Interface，JDK 内置的一种服务提供发现机制）的实现由 SpringServletContainerInitializer 类自动检测，该类本身可以由任意 Servlet 3.0+ 容器自动引导，类似于以前的 WEB-INF/web.xml 中的 XML 配置，如代码清单 5-8 所示。

代码清单 5-8　类似的 XML 配置

```
<servlet>
    <servlet-name>dispatcher</servlet-name>
    <servlet-class>
        org.springframework.web.servlet.DispatcherServlet
    </servlet-class>
    <init-param>
        <param-name>contextConfigLocation</param-name>
        <param-value>/WEB-INF/spring/dispatcher-config.xml</param-value>
    </init-param>
    <load-on-startup>1</load-on-startup>
```

```
    </servlet>
    <servlet-mapping>
        <servlet-name>dispatcher</servlet-name>
        <url-pattern>/</url-pattern>
    </servlet-mapping>
```

SpringServletContainerInitializer 类源码如代码清单 5-9 所示。

代码清单 5-9　SpringServletContainerInitializer 类源码

```
@HandlesTypes(WebApplicationInitializer.class)
public class SpringServletContainerInitializer implements ServletContainerInitializer {
    @Override
     public void onStartup(Set<Class<?>> webAppInitializerClasses, ServletContext
servletContext) throws ServletException {
        List<WebApplicationInitializer> initializers = new LinkedList<WebApplicatio
nInitializer>();
        if (webAppInitializerClasses != null) {
            for (Class<?> waiClass : webAppInitializerClasses) {
        if (!waiClass.isInterface() && !Modifier.isAbstract(waiClass.getModifiers())
&& WebApplicationInitializer.class.isAssignableFrom(waiClass)) {
                    try {
                        initializers.add((WebApplicationInitializer) waiClass.
newInstance());
                    } catch (Throwable ex) {
                        throw new ServletException("Failed to instantiate
WebApplicationInitializer class", ex);
                    }
                }
            }
        }
        if (initializers.isEmpty()) {
            servletContext.log("No Spring WebApplicationInitializer types detected
on classpath");
            return;
        }
        AnnotationAwareOrderComparator.sort(initializers);
           servletContext.log("Spring WebApplicationInitializers detected on
classpath: " + initializers);
        for (WebApplicationInitializer initializer : initializers) {
            initializer.onStartup(servletContext);
        }
    }
}
```

这段代码首先判断一个 Set 类型是否为 null，如果不为 null，则遍历这个 Set 类型，找到这个 Set 类型中既不是接口也不是抽象类，并且实现了 WebApplicationInitializer 接口的类（执行判断逻辑的代码：!wai Class.isInterface()&&!Modifier.isAbstract(waiClass.getModifiers())&&WebApplication-Initializer.class.isAssignableFrom(waiClass)），将这些类存到 list 变量中。遍历完之后，如果 list 变

量为空，就返回。如果 list 变量不为空，则通过 sort 方法对 list 变量中的元素排序，再使用增强 for 循环遍历 initializers 中的元素，然后按顺序调用 list 变量中元素的 onStartup 方法。

SpringServletContainerInitializer 实现了 ServletContainerInitializer 接口，而 ServletContainer Initializer 接口是 javax.servlet 包下的。官方有一个解释：为了支持不使用 web.xml 配置文件实现零配置提供了 ServletContainerInitializer 接口。该接口可以应用 SPI 机制，当启动 Web 项目时，Web 容器会自动到 META-INF/services 目录下查找以 ServletContainerInitializer 命名的文件，如图 5-6 所示。SpringServletContainerInitializer 的内容为 ServletContainerInitializer 实现类的全路径，如图 5-7 所示。

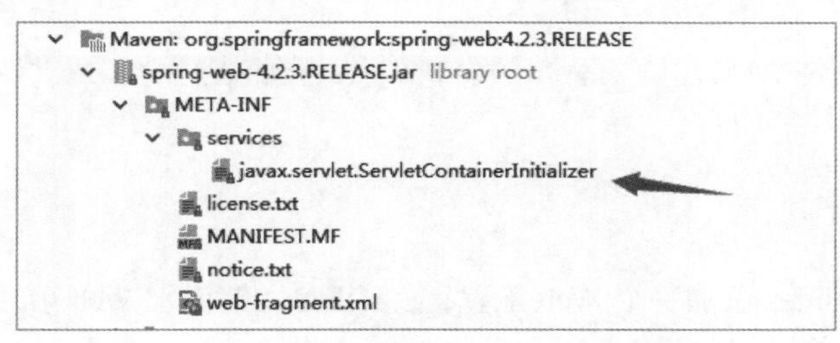

图 5-6　META-INF/services 目录下的文件

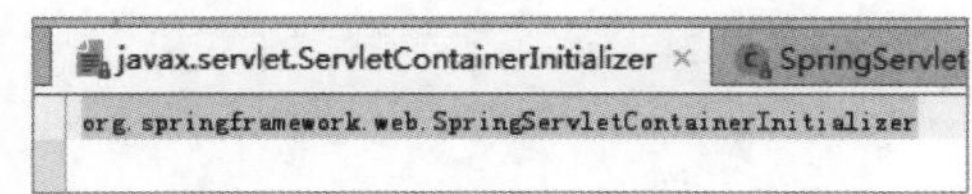

图 5-7　ServletContainerInitializer 文件中的内容

由此可知 SpringServletContainerInitializer 类中的方法为什么会随着项目的启动而调用。只需要使用 HandlesTypes 注解和实现 ServletContainerInitializer 接口，应用 SPI 机制，通过一个实例即可实现类似的功能。

5.1.7　Controller 层开发

下面开发 Controller 层的代码，这里演示如何跳转到 index.jsp 页面和直接返回文本，如代码清单 5-10 所示。

代码清单 5-10　IndexController 代码

```
import org.slf4j.Logger;
import org.slf4j.LoggerFactory;
import org.springframework.stereotype.Controller;
import org.springframework.web.bind.annotation.RequestMapping;
import org.springframework.web.bind.annotation.ResponseBody;
/** Controller层演示 */
@Controller
public class IndexController {
    private static Logger logger = LoggerFactory.getLogger(IndexController.class);
    /** 返回值中内容通过RequestMapping会映射到对应的页面  */
    @RequestMapping({"/index",""})
    public String index () {
        logger.info("访问首页");
```

```
        return "/index";
    }
    /** 使用ResponseBody注解，返回字符串 */
    @RequestMapping("/info")
    @ResponseBody
    public String info () {
        logger.info("获取信息");
        return "username:chenhx";
    }
}
```

在 IDEA 中运行 Web 项目需要 war 包或者进行 Web 编译，接下来讲解如何设置 Artifacts 和 Tomcat 以启动项目。

5.1.8 配置 Tomcat

首先给项目的 Modules 添加一个 Web 配置，让 IDEA 能够识别 Web 项目。选择 File → Project Structure 命令，接下来的操作如图 5-8 所示。

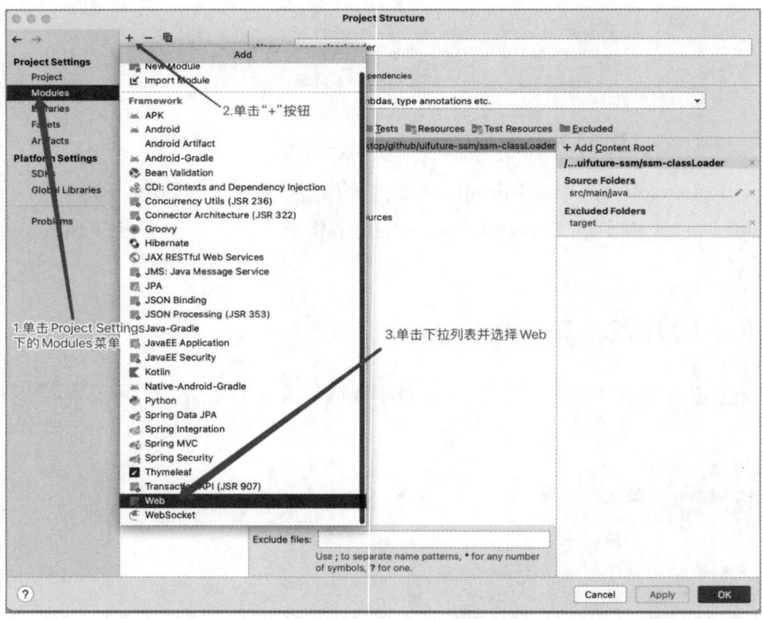

图 5-8　添加一个 Web 配置

按照正常流程，接下来要配置 web.xml 文件和 Web 项目启动的根路径，因为这里演示的是无 web.xml 文件形式，所以无须配置。后面将通过在代码中配置一些特定的注解来实现 web.xml 文件中的功能，如图 5-9 所示。

这里直接将上下两块的内容（Deployment Descriptors 和 Web Resource Directories）保存后，会发现有一个警告的感叹号，提示 Web 资源没有被包含在 Artifacts 中。单击 create Artifact 按钮即可自动创建 Artifact，跳转到如图 5-10 所示的页面。

如果没有自动创建 Artifact，则可以手动创建 Artifact，如图 5-11 所示。

单击“From Modules...”选项，在弹出的页面中选择模块，这里只有一个项目，单击 OK 按钮即可。然后来配置 Tomcat。

在 IDEA 左上角，单击下拉菜单弹出如图 5-12 所示的界面。

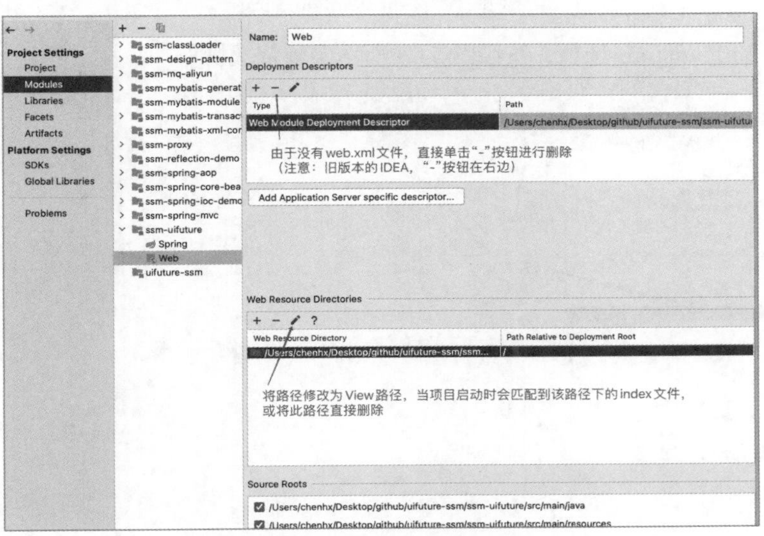

图 5-9　Web 的配置

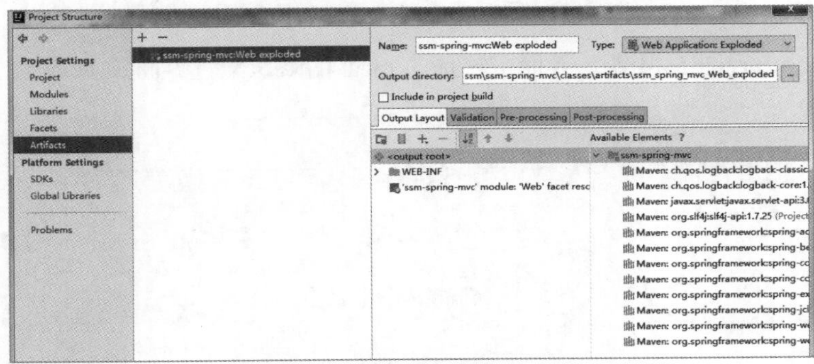

图 5-10　自动创建的 Artifact

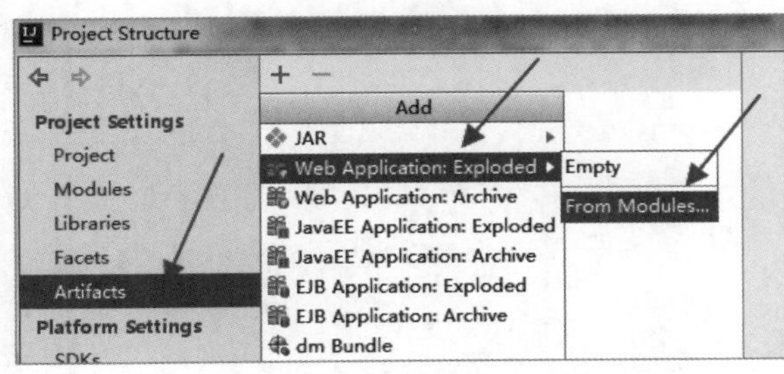

图 5-11　手动创建 Artifacts

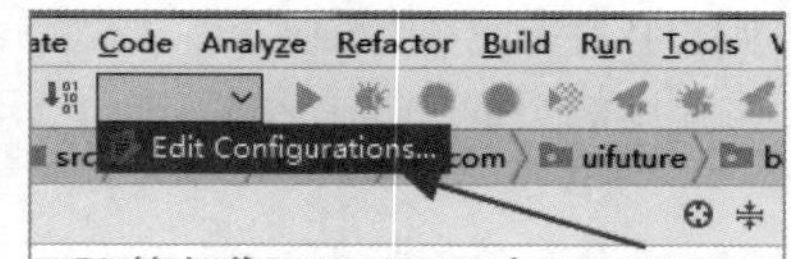

图 5-12　Edit Configurations

单击 Edit Configurations 选项，按照图 5-13 所示步骤依次操作。

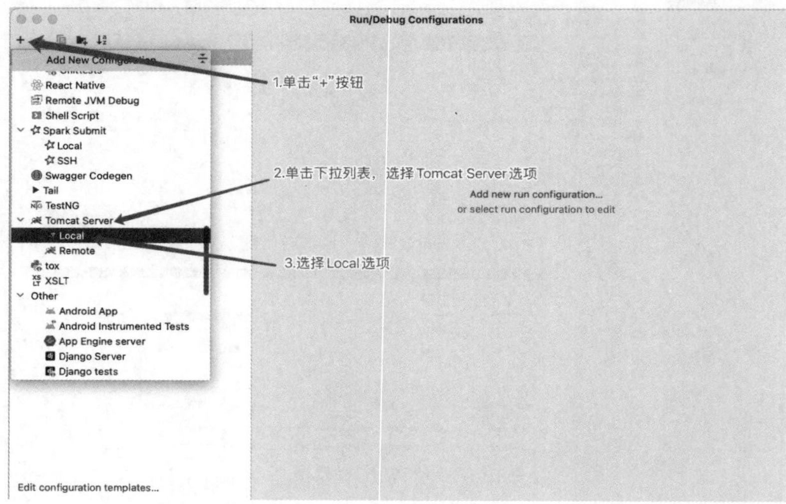

图 5-13　配置 Tomcat

操作完成后会出现图 5-14 所示界面，这是 Tomcat 的默认配置界面。

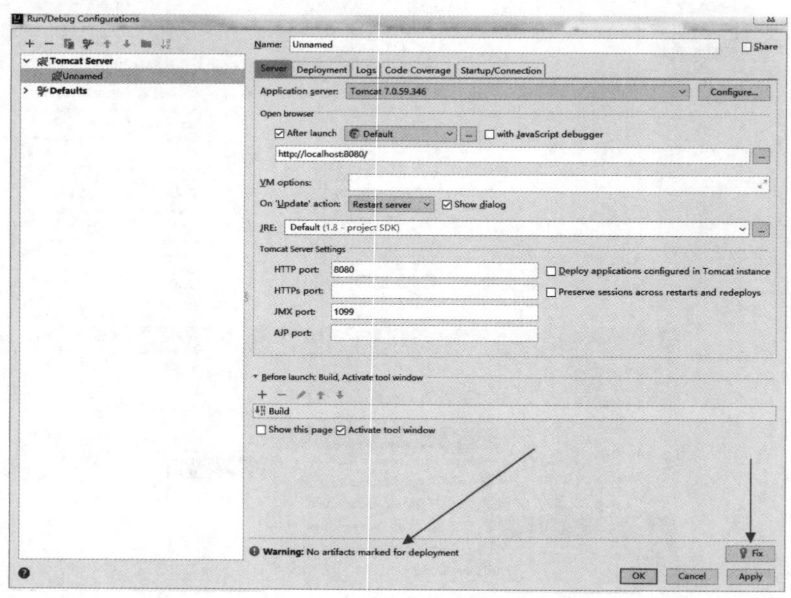

图 5-14　Tomcat 默认配置界面

可以明显看到 Warning 警告，提示没有配置 Artifacts。单击右下角的 Fix 按钮即可快速完成 Tomcat 的配置。

如果没有使用上述快捷方式，则可以参考图 5-15 进行手动配置。

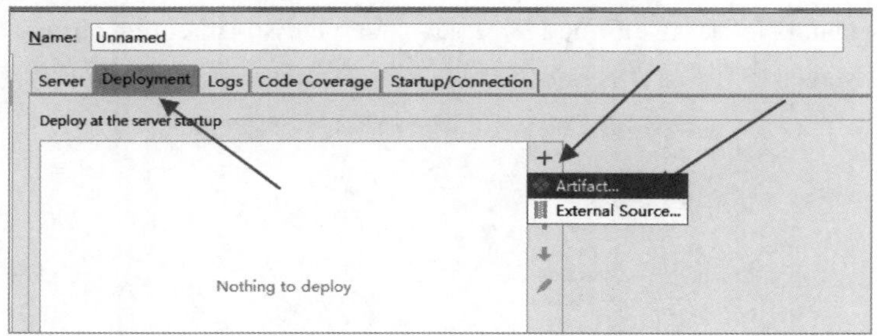

图 5-15　为 Tomcat 手动配置 Artifacts

在 Tomcat 的配置页面，单击 Deployment 标签，再单击右边的 + 按钮，然后选择 Artifact 选项即可。建议直接单击 Fix 警告按钮，一次配置到位。

最后修改配置名，这里将其修改为 tomcat 7。单击 Apply 按钮应用配置，再单击 OK 按钮退出该配置页面，如图 5-16 所示。

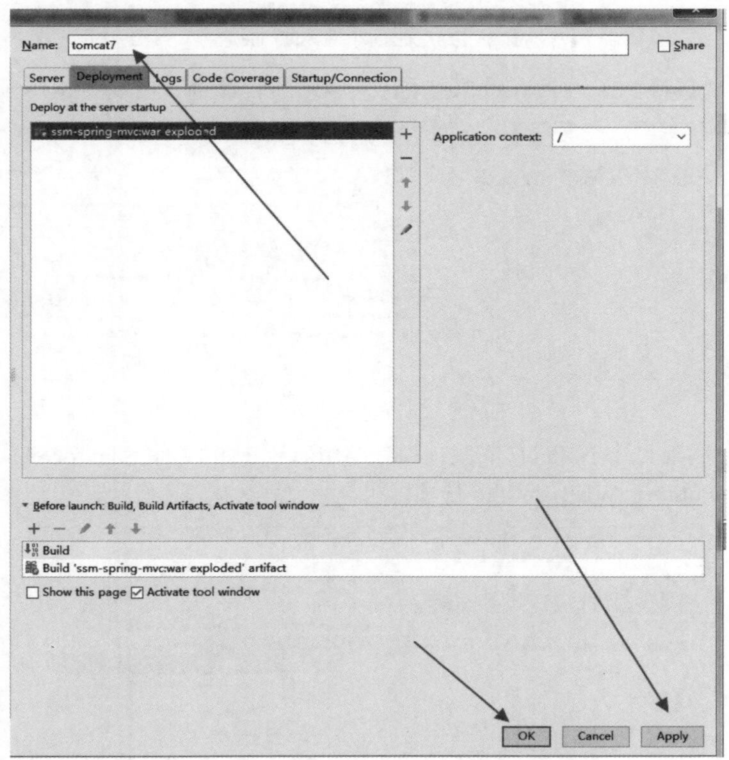

图 5-16　修改配置名并退出

现在可以看到之前空白的地方出现了配置的 tomcat 7。单击运行按钮即可运行项目。

5.1.9 NoClassDefFoundError-jstl 异常处理

如果项目启动后提示以下异常信息，则说明项目中没有引入 JSTL 包：

Caused by: java.lang.NoClassDefFoundError: javax/servlet/jsp/jstl/core/Config。

原因是虽然在 Maven 中导入了 JSTL 标签，但在实际的 classes 中并没有此依赖包，如图 5-17 所示。

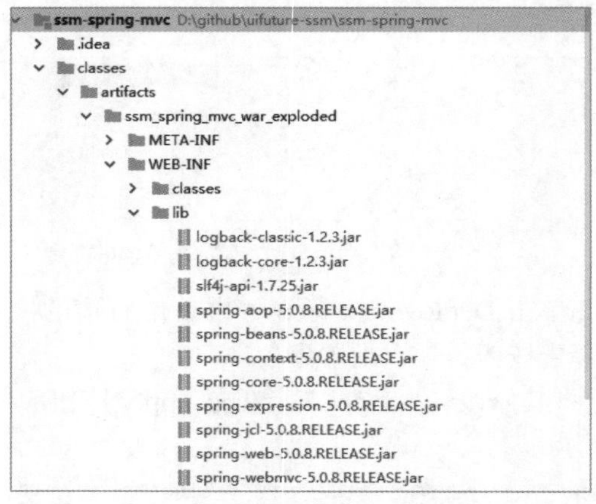

图 5-17　classe 中导入的 jar 包

解决方案：依次选择 File → Project Structure → Artifacts 命令，出现图 5-18 所示界面。

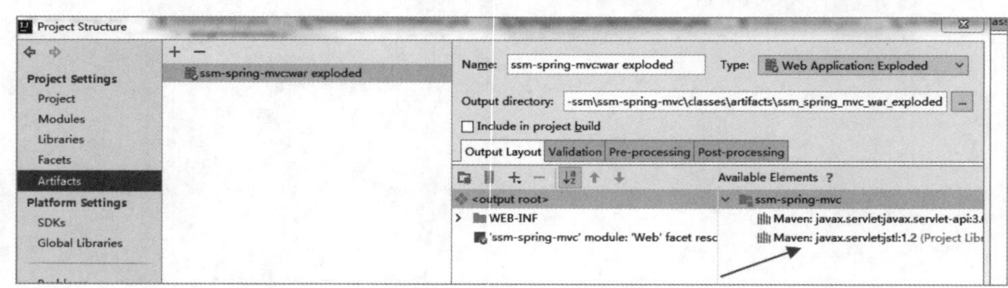

图 5-18　Artifacts 界面

可以看到，虽然项目中有 JSTL 依赖，但在 Artifacts 中并没有。右击 JSTL 包，出现图 5-19 所示的界面，选择 Put into /WEB-INF/lib 选项即可。

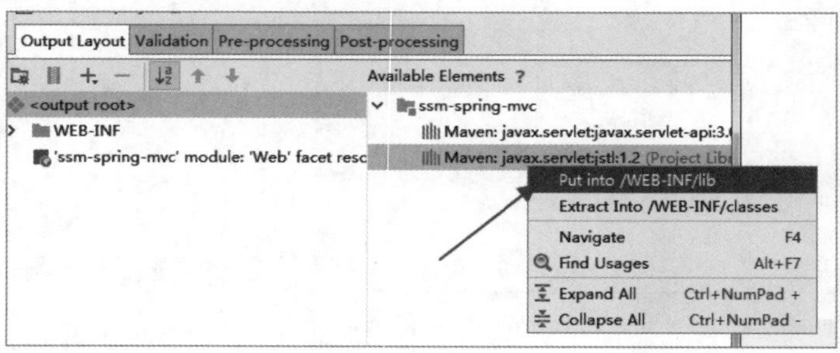

图 5-19　将 JSTL 包加入发布路径

单击 Apply 按钮，然后单击 OK 按钮退出即可。

5.1.10 页面演示

运行项目后，可访问网址：http://localhost:8080，返回的页面效果如图 5-20 所示。

再访问网址 http://localhost:8080/info，直接返回文本的页面效果如图 5-21 所示。

图 5-20 首页效果

图 5-21 直接返回文本的页面效果

以上就是最简单的 Spring MVC 项目搭建过程，后面会深入讲解 Spring MVC 中的核心技术。

5.2 Spring MVC 中的常用注解

在进行更加深入的开发之前，必须先了解一些 Spring MVC 中的常用注解。

5.2.1 Controller 注解

Controller 注解的源码如代码清单 5-11 所示。

代码清单 5-11 Controller 注解的源码

```
@Target({ElementType.TYPE})
@Retention(RetentionPolicy.RUNTIME)
@Documented
@Component
public @interface Controller {
    @AliasFor(
            annotation = Component.class
    )
    String value() default "";
}
```

由 @Target({ElementType.TYPE})（元注解）可知，Controller 注解通常配置在类、接口（包括注解类型）或枚举声明上。一般将 Controller 注解配置在 Controller 层的类上，表明该类是 Spring MVC 中的 Controller，带该注解的类是 Web 控制器。从源码中可以看到，Controller 上又组合了 Component 注解，将被注解的类声明为一个 Bean，DispatcherServlet 会自动扫描注解了 @Component 的类，并且将 Web 请求映射到注解了 @RequestMapping 的方法上。

当声明普通 JavaBean 时，使用的无论是 @Component、@Service、@Repository，还是 @Controller，作用都是相同的，因为 @Service、@Repository 和 @Controller 都包含了 Component 注解。但是当

在 Spring MVC 中声明控制器 Bean 时，只能使用 Controller 注解。

Controller 注解中有一个 value 值，用来指定逻辑组件名称，也就是在其他地方按名称注入 Bean 时需要使用该值，一般不会用到。如果不指定 value 值，默认将 Controller 注解标识的类的名称作为 Bean 的名称。

5.2.2 RequestMapping 注解

RequestMapping 注解的源码如代码清单 5-12 所示。

代码清单 5-12 RequestMapping 注解的源码

```
@Target({ElementType.METHOD, ElementType.TYPE})
@Retention(RetentionPolicy.RUNTIME)
@Documented
@Mapping
public @interface RequestMapping {
    String name() default "";
    @AliasFor("path")
    String[] value() default {};
    @AliasFor("value")
    String[] path() default {};
    RequestMethod[] method() default {};
    String[] params() default {};
    String[] headers() default {};
    String[] consumes() default {};
    String[] produces() default {};
}
```

该注解的主要作用是将 Web 请求（请求路径和参数）映射到请求处理类中的具体方法。由 @Target({ElementType.METHOD, ElementType.TYPE}) 可知，该注解是作用在类、接口、枚举或者方法上的。接下来介绍该注解中的 6 个属性。

（1）value/path：value 其实就是 path，命名不同而已。在 RequestMapping 注解中使用最频繁的就是 value，用来指定请求的 URL 路径。当使用 RequestMapping 注解时，value 属性可以省略，省略后默认是 value。而且可以使用通配符“*”和“**”。value 属性值的演示如代码清单 5-13 所示。

代码清单 5-13 value 属性值的演示

```
/** SpringMVC中常用注解的演示 */
@Controller
@RequestMapping("annotation")
public class AnnotationController {
    /**
     * 可以通过 http://localhost:8080/annotation/aaa/test 访问
     * 无法通过http://localhost:8080/annotation/test 访问
     */
    @RequestMapping("*/test")
    @ResponseBody
    public String test() {
```

```
        System.out.println( "-----AnnotationController.test-----" );
        return "test" ;
    }
    /**
     * 可以通过 http://localhost:8080/annotation/test2 访问
     * 也可以通过http://localhost:8080/annotation/aaa/test2 访问
     */
    @RequestMapping({"**/test2","test3"})
    @ResponseBody
    public String test2() {
        System.out.println( "-----AnnotationController.test2-----" );
        return "test2" ;
    }
}
```

在类上的 URL 参数会最先匹配，方法上的 RequestMapping 注解中的 value 值是在类注解 value 值之后匹配的。value 中的“annotation”和“/annotation”值的效果是一致的。声明多个值（匹配多个 URL 路径）需要使用大括号，如上面的 {"**/test2","test3"}。

“**”和“*”的主要区别是，使用“**”时，中间可以没有匹配路径，即任意或者为空。而“*”的作用是匹配任意的字符，但是不能为空。**提示**：这里的字符可以是多个。

（2）method：指定请求的类型，也就是 GET、HEAD、POST、PUT、PATCH、DELETE、OPTIONS、TRACE 类型。如果指定了这个属性值，则建议使用对应的 GetMapping、PostMapping、PutMapping、DeleteMapping 等注解替换 RequestMapping。

（3）params：指定 request 请求中必须包含某些参数的值。

（4）headers：指定 request 请求中必须包含某些 header 的值。

（5）consumes：映射请求的提交类型，对应请求中的 Content-Type 值，如 consumes="text/plain" 或者指定多个值 consumes={"text/plain", "application/*"}，最终匹配其中一个即可。

（6）produces：映射请求的可生成媒体类型，缩小主映射，也就是指定返回的内容类型。produces 对应 request 请求头中的 Accept 类型，需要该类型包含 Accept 中指定类型才会返回数据。produces 可以有多个，如 produces="text/plain"、produces={"text/plain", "application/*"}。另外，还可以使用 MediaType 类中的静态属性指定媒体文件。例如，produces=MediaType.APPLICATION_JSON_UTF8_VALUE。**提示**：可以使用“!”反向指定除了该类型的所有类型。

5.2.3 ResponseBody 注解

ResponseBody 注解的作用是将 Controller 类的方法返回的对象，通过适当的转换器转换为指定的格式之后，写入 Response 对象的 Body 区，通常用来返回 JSON 数据或者 XML 数据。

从 Spring 4.0 开始，该注解可以直接添加在类级别上，即该类的方法不需要再添加注解。

5.2.4 RequestBody 注解

RequestBody 注解主要用来接收前端传递给后端的数据（请求体中的数据）。因为 GET 方式无请求体，所以当使用 @RequestBody 接收数据时，前端不能以 GET 方式提交数据，而是要以

POST 或者 PUT 方式提交数据。在后端的同一个接收方法中，@RequestBody 与 @RequestParam() 可以同时使用，但 @RequestBody 最多只能有一个，而 @RequestParam() 可以有多个。

RequestBody 注解是可选的，如果前端传递过来的是 JSON/XML 格式的数据，在没有使用该注解的情况下，在 Java 中收到的值是空的。另外，MappingJackson2HttpMessageConverter 转换器需要导入包（默认使用 Jackson 处理 JSON 数据）。代码清单 5-14 主要用来说明当请求参数为 JSON 数据时，使用 Spring MVC 依然能够进行类型转换。

代码清单 5-14　配置 JSON 格式传输数据解析

```
//第1步，在AnnotationController 类中增加一个方法
/**
 * produces = MediaType.APPLICATION_JSON_UTF8_VALUE - 防止返回数据乱码
 * ResponseBody注解返回的默认编码为ISO-8859-1 */
@PostMapping(value = "/testRequestBody",produces = MediaType.APPLICATION_JSON_
UTF8_VALUE)
@ResponseBody
public User testRequestBody(@RequestBody User user) {
    System.out.println( "-----user="+user);
    return user;
}
//END
//第2步，在pom.xml中增加两个依赖
<!--JSON转换器依赖包-->
<dependency>
    <groupId>com.fasterxml.jackson.core</groupId>
    <artifactId>jackson-core</artifactId>
    <version>2.9.6</version>
</dependency>
<dependency>
    <groupId>com.fasterxml.jackson.core</groupId>
    <artifactId>jackson-databind</artifactId>
    <version>2.9.6</version>
</dependency>
```

完成上面的配置后，启动 Tomcat 项目可能会遇到包不存在的问题，这是由于项目中虽然导入了包，但还没有将包“put”到 Artifacts 组件中。可以参考 5.1.9 小节中的解决方案解决该问题，将三个 JSON 包“put”到 Artifacts 组件中即可。请求的配置如图 5-22 和图 5-23 所示。

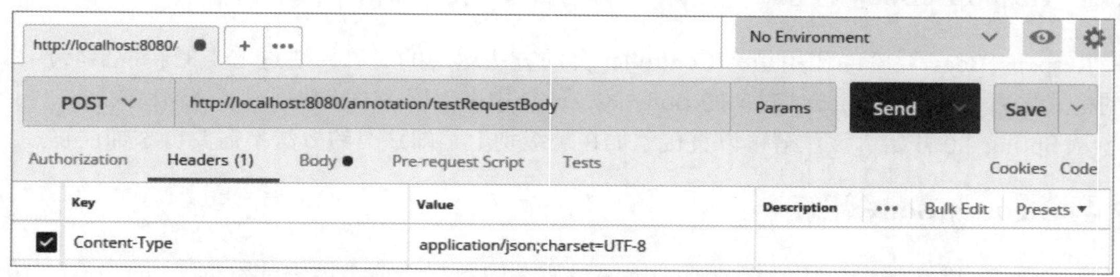

图 5-22　设置 Content-Type 的值为 application/json;charset=UTF-8

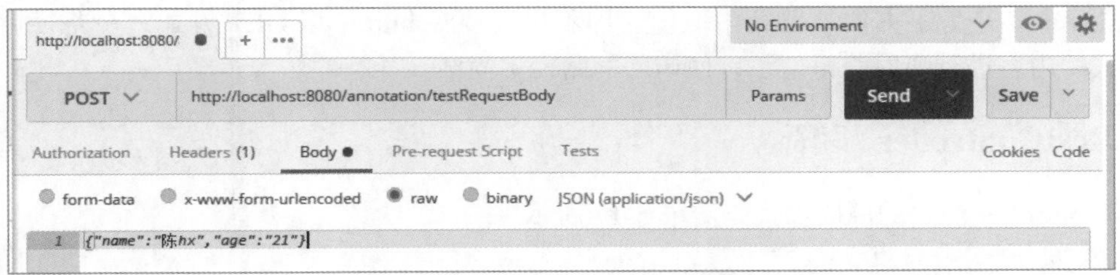

图 5-23 设置数据类型和值

填好请求链接，在 Body 中选中 raw 发送文本，选择数据类型为 JSON (application/json)，填写 JSON 数据。最后单击 Send 按钮，返回结果如图 5-24 所示。

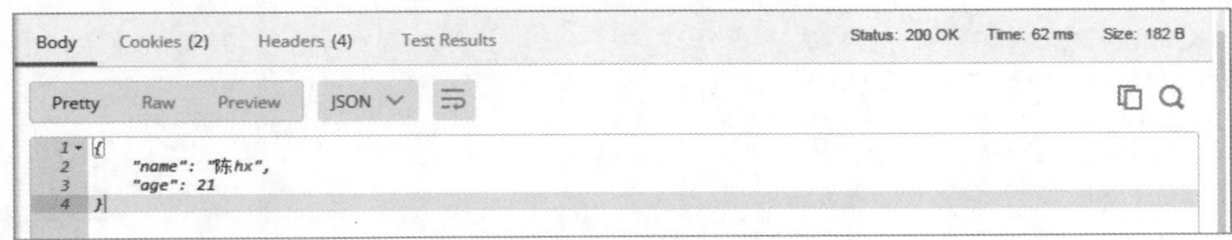

图 5-24 测试结果

5.2.5 PathVariable 注解

PathVariable 注解在方法参数前使用，用来接收路径中的值并作为参数。例如，在 AnnotationController 类中增加 testPathVariable 方法，如代码清单 5-15 所示。

代码清单 5-15 PathVariable 注解包含 name 演示

```
/** 演示PathVariable注解 */
@RequestMapping(value = "/{str}")
public @ResponseBody String testPathVariable(@PathVariable("str") String name){
    return name;
}
```

访问 http://localhost:8080/annotation/testRequestBody，如果是 POST 方式，则会优先匹配前面的 testRequestBody 方法。也就是说，Spring MVC 会依次匹配链接，优先匹配最接近的链接，只有全部都匹配不到，才会返回 404。

对于路径中的 {str}，使用 PathVariable 注解中的 name 值转换成对应的 Java 对象，如果在 PathVariable 注解中的值不和 RequestMapping 的 value 名称进行映射，则请求会查找方法参数中的 str 属性名，如果没有，则无法赋值，如代码清单 5-16 所示。

代码清单 5-16 PathVariable 注解不包含 name 演示

```
/** 演示PathVariable注解 */
@RequestMapping(value = "/{str}")
public @ResponseBody String testPathVariable(@PathVariable String str){
    return str;
}
```

该注解主要是为了方便从路径中取值，即不再需要在 http 中以 GET 方式，在路径后带 key-value 形式的内容进行映射获得值了，使访问的路径更简洁。

5.2.6 RestController 注解

RestController 注解的源码如代码清单 5-17 所示。

代码清单 5-17　RestController 注解源码

```
@Target(ElementType.TYPE)
@Retention(RetentionPolicy.RUNTIME)
@Documented
@Controller
@ResponseBody
public @interface RestController {
    @AliasFor(annotation = Controller.class)
    String value() default "";
}
```

从源码中可以看到，该注解是一个组合注解，组合了 @Controller 和 @ResponseBody，如果只是开发一个交互数据的控制器，就可以使用该注解。

5.2.7 CookieValue 注解

CookieValue 注解用在方法参数上，使用该注解可以将请求头中关于 Cookie 的值绑定到方法的参数上，如 jsessionId 的 Cookie 值，如代码清单 5-18 所示。

代码清单 5-18　CookieValue 注解演示

```
@RequestMapping("/testCookieValue")
@ResponseBody
public String testCookieValue(@CookieValue("JSESSIONID") String jsessionId)  {
    System.out.println("jsessionId="+jsessionId);
    return jsessionId;
}
```

代码清单 5-18 演示了如何获取 CookieValue 中的 JSESSIONID 值，并绑定到了 jsessionId 参数上。

5.2.8 RequestParam 注解

RequestParam 注解用在 Controller 类中的方法参数前，用来标识请求参数的名称。例如，以前在 Controller 类中使用 request.getParameter ("参数名") 获取参数值，现在可以使用 RequestParam 注解直接绑定参数来得到值。RequestParam 注解源码如代码清单 5-19 所示。

代码清单 5-19　RequestParam 注解源码

```
@Target(ElementType.PARAMETER)
@Retention(RetentionPolicy.RUNTIME)
@Documented
```

```
public @interface RequestParam {
    @AliasFor("name")
    String value() default "";
    @AliasFor("value")
    String name() default "";
    boolean required() default true;
    String defaultValue() default ValueConstants.DEFAULT_NONE;
}
```

该注解有 4 个常用参数，在 Spring MVC 4.2 以前只有 3 个参数，参数 name 在 Spring MVC 4.2 以后添加，与 value 等价。

（1）value/name：参数的名字，也就是前端请求数据时，请求中参数的名字。可以不指定，默认是方法中的参数名。

（2）required：参数是否必填，默认为 true，表示请求中必须要有该参数，否则会报找不到请求的异常。

（3）defaultValue：参数的默认值，也就是当在请求中没有匹配的参数名时，方法参数中的默认值。当设置了 defaultValue，没有设置 required 时，默认的 required 将变成 false，也就是参数不是必填的了。默认值可以设置为 SpEL（Spring Expression Language）表达式。

在 Controller 类中增加如代码清单 5-20 的代码，运行 Tomcat，分别访问 http://localhost:8080/annotation/testRequestParam 和 http://localhost:8080/annotation/testRequestParam?name=1234。访问前面的链接，返回的结果是 springmvc；访问后面链接，返回的结果是 1234。

代码清单 5-20　testRequestParam 方法

```
@RequestMapping("/testRequestParam")
@ResponseBody
public String testRequestParam(@RequestParam(value = "name",defaultValue =
"springmvc")
                                    String username) {
    System.out.println("username=" + username);
    return username;
}
```

5.2.9　InitBinder 注解

InitBinder 注解用于标注 Controller 类中的方法，可以在标注的方法中对 Spring 的 WebDataBinder 对象（DataBinder 的子类）进行初始化，只对当前的 Controller 类有效。WebDataBinder 对象主要用在表单中的日期字符串和 JavaBean 中 Date 类型的转换上，Spring MVC 默认不支持该格式的转换，需要手动配置，所以使用 InitBinder 注解可以解决数据绑定的问题。另外，还可以使用该注解对 JSP 页面的参数进行绑定（很少使用）。

下面的例子演示了日期类型的转换，如代码清单 5-21 所示。

代码清单 5-21　InitBinder 注解演示

```
/**
```

```
 * 该参数会在InitBinder注解的方法中进行转换
 * 对于string类型，例如2018-08-05 20:19:20 会解析成Date对象
 */
@GetMapping("/testDate")
@ResponseBody
public String testDate(@RequestParam("data") Date date){
    System.out.println("--------------data="+date);
    return "data:"+date;
}
/**
 * InitBinder注解用于Date类型的转换
 * ServletRequestDataBinder类是WebDataBinder的子类
 */
@InitBinder
public void initBinder(ServletRequestDataBinder binder){
    SimpleDateFormat dateFormat = new SimpleDateFormat("yyyy-MM-dd HH:mm:ss");
    //指定日期/时间解析规则是否宽松
    dateFormat.setLenient(false);
     binder.registerCustomEditor(Date.class, new CustomDateEditor(dateFormat,
true));
}
```

访问链接 http://localhost:8080/annotation/testDate?data=2018-8-5 22:12:52 即可进入 testDate 方法，Spring MVC 会自动将符合解析规则的日期字符串转换为 Date 对象。

5.3 对象与 JSON 或 XML 数据之间的转换

前面在讲解 RequestBody 注解时，解释了如何返回 JSON 数据和传入 JSON 数据。本节主要讲解对象与 JSON 或 XML 数据之间的转换。

5.3.1 对象与 JSON 数据之间的转换

在 AnnotationController 类中增加如代码清单 5-22 所示的代码，运行项目，使用 Postman 工具进行测试。请求链接和请求参数如图 5-25 所示。JSON 数据的返回结果如图 5-26 所示。

代码清单 5-22　将对象转换成 JSON 数据并返回

```
/** 将对象转换成JSON格式数据并返回到前端  */
@PostMapping(value = "/testUserToJson" , produces = MediaType.APPLICATION_JSON_
UTF8_VALUE)
@ResponseBody
public User testUserToJson(@RequestBody User user) {
    System.out.println("-----user=" + user);
    return user;
}
```

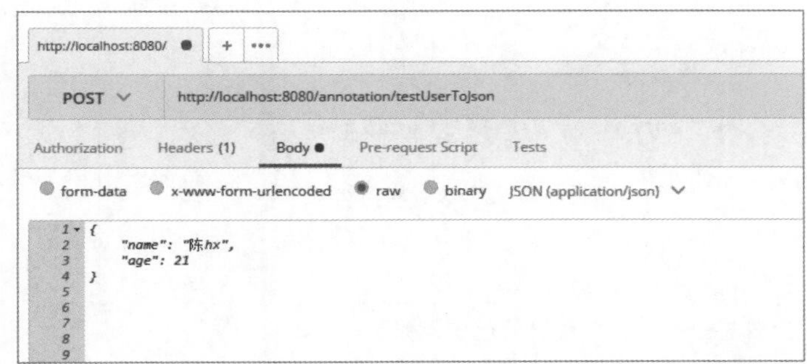

图 5-25 请求链接和请求参数

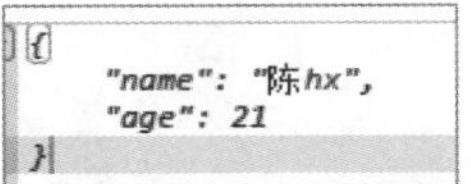

图 5-26 JSON 数据的返回结果

5.3.2 对象与 XML 数据之间的转换

如果 Controller 类中的方法需要将对象转换成 XML 数据并返回，则必须在对象上增加 XmlRootElement 注解；如果只在 PostMapping 注解中编写了 produces=MediaType.APPLICATION_XML_VALUE，会无法返回正确的 XML 格式数据；如果并没有编写返回数据类型，则默认的返回数据类型就是 JSON 格式。因此，要记得给需要使用 XML 格式返回的对象类上加上 XmlRootElement 注解。

在本项目中，在 User 类上增加 XmlRootElement 注解后（该代码清单省略，请自行加上），再在 AnnotationController 类中增加代码清单 5-23 所示的代码。

代码清单 5-23 将对象转换为 XML 数据

```
/** 将对象转换成XML格式数据返回到前端 */
@PostMapping(value = "/testUserToXml"
        , produces = MediaType.APPLICATION_XML_VALUE)
@ResponseBody
public User testUserToXml(@RequestBody User user) {
    System.out.println("-----user=" + user);
    return user;
}
```

接下来使用 Postman 工具进行测试。测试链接和参数如图 5-27 所示。

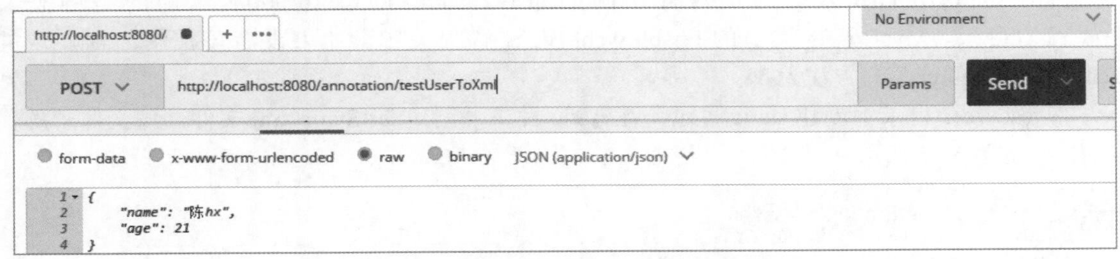

图 5-27 演示 XML 返回数据请求参数

XML 数据的返回结果如图 5-28 所示。

```
<?xml version="1.0" encoding="UTF-8" standalone="yes"?>
<user>
    <age>21</age>
    <name>陈hx</name>
</user>
```

图 5-28　XML 数据的返回结果

5.4　静态资源映射

下面先讲解如何通过编程式配置静态资源映射，再讲解如何通过 XML 文件配置静态资源的映射。

5.4.1　通过编程式配置静态资源映射

使用 MvcConfig 类实现 WebMvcConfigurer 接口，然后在 MvcConfig 类中增加 add Resource Handlers 方法，如代码清单 5-24 所示。

注意：在 Spring 5.0 之前是可以继承 WebMvcConfigurerAdapter 类的，Spring 5.0 之后弃用了。

代码清单 5-24　添加静态资源映射方法

```
@Override
public void addResourceHandlers(ResourceHandlerRegistry registry) {
    //配置URL中的路径
    registry.addResourceHandler("/static/**")
            //配置静态文件的路径
            .addResourceLocations("classpath:/static/");
}
```

addResourceHandler 添加的是对外访问的路径，addResourceLocations 添加的是静态文件放置的目录。MvcConfig 类上的 EnableWebMvc 注解很关键，如果没有该注解，则实现的 WebMvcConfigurer 接口方法无效。

在 resources 目录下创建 static 目录，在 static 目录下创建 testStatic.xml 文件，该文件中的代码如代码清单 5-25 所示。

代码清单 5-25　testStatic.xml 文件

```
<!DOCTYPE html>
<html lang="en">
<head>
<meta charset="UTF-8">
```

```
<title>SpringMVC</title> </head>
<body> 静态资源访问测试 </body>
</html>
```

运行项目，访问 http://localhost:8080/static/testStatic.html，静态资源演示效果如图 5-29 所示。

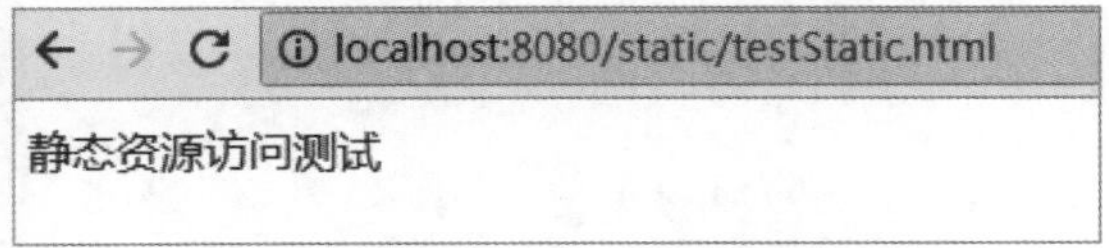

图 5-29　静态资源演示效果

5.4.2　通过 XML 文件配置静态资源映射

本小节讲解如何使用 XML 文件进行静态资源的映射配置，将在之前项目的基础上继续演示。首先将 MvcConfig 类中代码全部注释（这是因为只能注册一个 DispatcherServlet），然后改造 WebInitializer 类，如代码清单 5-26 所示。

代码清单 5-26　WebInitializer 类

```
//Web配置
public class WebInitializer implements WebApplicationInitializer {
    @Override
    public void onStartup(ServletContext servletContext){
        //一、与XML文件组合使用
        XmlWebApplicationContext appContext = new XmlWebApplicationContext();
        appContext.setConfigLocation("/WEB-INF/classes/spring-mvc-servlet.xml");
        /* 二、使用编程式方式配置静态资源的映射
        AnnotationConfigWebApplicationContext appContext = new AnnotationConfigWebAp
plicationContext(); appContext.register(MvcConfig.class);新建WebApplicationContext，
注册配置类，并将其和当前servletContext关联appContext.setServletContext(servletContext)
; 注册并映射调度程序servlet */
            ServletRegistration.Dynamic servlet = servletContext.addServlet
("dispatcher", new DispatcherServlet(appContext));
        //注册Spring MVC的DispatcherServlet，替代web.xml
        servlet.addMapping("/");
        servlet.setLoadOnStartup(1);
    }
}
```

使用 XmlWebApplicationContext 注册 DispatcherServlet，而不是使用编程式的 AnnotationConfigWebApplicationContext 类注册 DispatcherServlet。配置路径为 /WEB-INF/classes/ 的原因是项目编译后，文件路径配置在 /WEB-INF/classes/ 目录下，如图 5-30 所示。

接下来配置 Spring MVC 的配置文件。在 resources 目录下新建 XML 文件，命名为 spring-mvc-servlet.xml，如代码清单 5-27 所示。

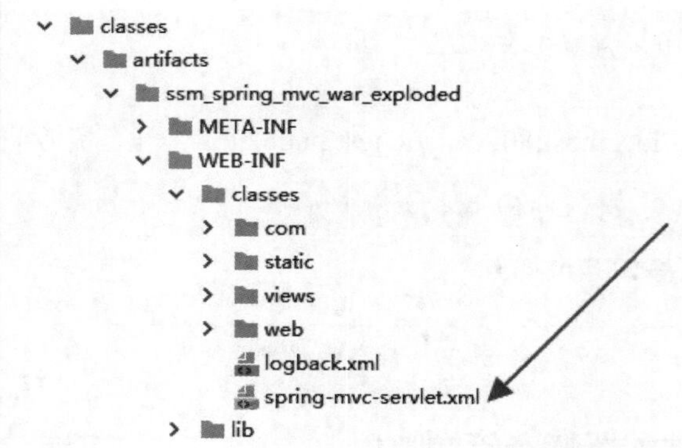

图 5-30　Spring MVC 配置的文件目录

代码清单 5-27　Spring MVC 配置文件

```
<?xml version="1.0" encoding="UTF-8"?>
<beans xmlns="http://www.springframework.org/schema/beans"
       xmlns:mvc="http://www.springframework.org/schema/mvc"
       xmlns:xsi="http://www.w3.org/2001/XMLSchema-instance"
       xmlns:context="http://www.springframework.org/schema/context"
       xsi:schemaLocation="http://www.springframework.org/schema/beans
http://www.springframework.org/schema/beans/spring-beans.xsd
http://www.springframework.org/schema/context
http://www.springframework.org/schema/context/spring-context.xsd
http://www.springframework.org/schema/mvc
http://www.springframework.org/schema/mvc/spring-mvc.xsd">
<!--加入注解驱动，保证控制器不会被影响-->
<mvc:annotation-driven/>
<!--配置静态页面扫描-->
<mvc:resources mapping="/static/**" location="classpath:/static/"/>
<!--扫描Controller-->
<context:component-scan base-package="com.uifuture.basics.controller">
    <context:include-filter type="annotation"
                            expression="org.springframework.stereotype.Controller"
/>
</context:component-scan>
<!--配置Controller层的映射关系-->
<bean class="org.springframework.web.servlet.view.InternalResourceViewResolver">
    <property name="prefix" value="/WEB-INF/classes/views/" />
    <property name="suffix" value=".jsp" />
</bean>
</beans>
```

在配置文件中添加了自动扫描 com.uifuture.basics.controller 包下的 Controller 注解。另外执行了静态资源的映射和 Controller 层的映射。接下来是运行 Tomcat 项目进行演示。演示结果不再展示，结果和预期的一致。

5.5 ControllerAdvice 注解

在讲解该注解前，先把代码还原到 5.4.2 小节之前，简单的操作就是注释 WebInitializer 类中 XML 注册的前端控制器，改为使用编程式方式进行配置，将 MvcConfig 类中的注释全部放开。

ControllerAdvice 注解是在 Spring 3.2 中才引入的。首先看一下该注解的源码，如代码清单 5-28 所示。

代码清单 5-28 ControllerAdvice 注解的源码

```
@Target(ElementType.TYPE)
@Retention(RetentionPolicy.RUNTIME)
@Documented
@Component
public @interface ControllerAdvice {
    @AliasFor("basePackages")
    String[] value() default {};
    @AliasFor("value")
    String[] basePackages() default {};
    Class<?>[] basePackageClasses() default {};
    Class<?>[] assignableTypes() default {};
    Class<? extends Annotation>[] annotations() default {};
}
```

ControllerAdvice 注解的实现过程组合了 Component 注解，所以被 ControllerAdvice 注解的类也能被自动扫描注入。ControllerAdvice 注解的官方注释如代码清单 5-29 所示。

代码清单 5-29 ControllerAdvice 官方注释

```
/**
 * Specialization of {@link Component @Component} for classes that declare
 * {@link ExceptionHandler @ExceptionHandler}, {@link InitBinder @InitBinder}, or
 * {@link ModelAttribute @ModelAttribute} methods to be shared across
 * multiple {@code @Controller} classes.
 *
 * <p>Classes with {@code @ControllerAdvice} can be declared explicitly as Spring
 * beans or auto-detected via classpath scanning. All such beans are sorted via
 * {@link org.springframework.core.annotation.AnnotationAwareOrderComparator
 * AnnotationAwareOrderComparator}, i.e. based on
 * {@link org.springframework.core.annotation.Order @Order} and
 * {@link org.springframework.core.Ordered Ordered}, and applied in that order
 * at runtime. For handling exceptions, an {@code @ExceptionHandler} will be
 * picked on the first advice with a matching exception handler method. For
 * model attributes and {@code InitBinder} initialization, {@code @ModelAttribute}
 * and {@code @InitBinder} methods will also follow {@code @ControllerAdvice}
order.
 *
 * <p>Note: For {@code @ExceptionHandler} methods, a root exception match will be
 * preferred to just matching a cause of the current exception, among the handler
```

```
 * methods of a particular advice bean. However, a cause match on a higher-priority
 * advice will still be preferred to a any match (whether root or cause level)
 * on a lower-priority advice bean. As a consequence, please declare your primary
 * root exception mappings on a prioritized advice bean with a corresponding order!
 * <p>By default the methods in an {@code @ControllerAdvice} apply globally to
 * all Controllers. Use selectors {@link #annotations()},
 * {@link #basePackageClasses()}, and {@link #basePackages()} (or its alias
 * {@link #value()}) to define a more narrow subset of targeted Controllers.
 * If multiple selectors are declared, OR logic is applied, meaning selected
 * Controllers should match at least one selector. Note that selector checks
 * are performed at runtime and so adding many selectors may negatively impact
 * performance and add complexity
 */
```

（1）ControllerAdvice 注解是对 Component 注解的增强，用来声明 @ExceptionHandler、@InitBinder、@ModelAttribute 的方法类，实现在多个 Controller 类中共享配置（默认是全局所有的控制器）。

（2）使用 ControllerAdvice 注解可以显式声明该类为 Spring Bean，通过类路径自动扫描检测，所有的 Bean 都基于 @Order 和 Ordered 进行排序，在运行时按照优先级顺序执行。对于异常的处理，将使用第 1 个匹配的异常处理程序方法（使用了 @ExceptionHandler）进行通知。

（3）当一个异常被多个 @ExceptionHandler 标注的方法处理时，只有优先级高的处理器会去处理异常内容。**提示**：优先级高的通知会先于优先级低的通知匹配，所以当定义多个异常处理方法时，应该注意顺序。

（4）在 Spring 4.2 以前，只能使用 ControllerAdvice 注解统一处理所有的控制器，在 Spring 4.2 以后，可以使用一些属性定义目标控制器的子集，主要使用 annotations()、basePackageClasses()、basePackages() 这几个选择器配置控制器子集。项目运行时动态确认执行具体的控制器，所以添加过多的选择器会对性能造成一定的影响，并且增加了代码的复杂度。

虽然 ExceptionHandler 注解可以和 @ExceptionHandler、@InitBinder、@ModelAttribute 配合使用，但是最多的场景是与 ControllerAdvice 注解配合使用。下面使用 ControllerAdvice 注解配合 ExceptionHandler 注解处理全局异常，如代码清单 5-30 所示。

代码清单 5-30　使用 ControllerAdvice 注解定义全局异常处理类

```
/** 全局异常类 */
@ControllerAdvice
public class ExceptionAdvice {
    /**
     * 设置需要捕获的异常为Exception异常，也就是所有的异常
     * 这里返回的是视图，也可以返回JSON等数据类型的数据，可将其当作Controller进行处理
     */
    @ExceptionHandler(value = Exception.class)
    public ModelAndView handException(Exception e) {
        e.printStackTrace();
        ModelAndView modelAndView = new ModelAndView("error/error");
        Map map = new HashMap(4);
        map.put("code", 404);
```

```
            map.put("message", e.getMessage());
            modelAndView.addAllObjects(map);
            return modelAndView;
        }
    }
```

这里捕获了所有的异常，如果配置了自定义的异常，则可以在该方法中进行参数 e 的类型判断；如果是自定义的异常，直接抛出异常即可，然后在自定义的异常类中捕获该异常。

在 AnnotationController 类中增加的代码如代码清单 5-31 所示。

代码清单 5-31　异常处理演示

```
    /** 测试异常处理 */
    @RequestMapping(value = "/errorDemo")
    @ResponseBody
    public String errorDemo() {
        System.out.println("异常处理演示");
        throw new RuntimeException("测试运行时异常处理");
    }
```

最后增加定制的异常显示页面，在 views 目录下新建 error 目录，再在 error 目录下新建 error.jsp 文件，文件代码如代码清单 5-32 所示。

代码清单 5-32　异常显示页面的代码

```
<%@ page contentType="text/html;charset=UTF-8" language="java" %>
<html>
<head>
<title>错误页面</title>
</head>
<body>
<pre>定制的错误页面</pre>
<br/>${code},${message}
</body>
</html>
```

接下来运行项目，在浏览器中输入 http://localhost:8080/annotation/errorDemo 并访问，异常处理页面显示结果如图 5-31 所示。

图 5-31　异常处理页面的显示结果

通过该注解，再配合 @InitBinder、@ModelAttribute 还可以定制全局所有控制器的初始化数据绑定。

通过本章的学习，读者应该对 Spring MVC 中一些常用的注解有了比较深刻的理解，后面会对 Spring MVC 的组件及源码进行讲解。

第6章 深入理解 Spring MVC 的九大组件

本章要点

深入理解 Spring MVC 的九大组件：

1. HandlerMapping：映射处理器
2. HandlerAdapter：适配处理器
3. HandlerExceptionResolver：异常处理器
4. ViewResolver：视图解析器
5. RequestToViewNameTranslator：视图名称转换器
6. LocaleResolver：国际化解析器
7. ThemeResolver：主题样式解析器
8. MultipartResolver：文件上传解析器
9. FlashMapManager：重定向管理器

Spring MVC 中最核心的部分就是 DispatcherServlet 类，所有由 Spring MVC 处理的请求都会经过 DispatcherServlet。由 DispatcherServlet 负责初始化的九大组件，就是进行请求的实际处理。下面将结合源码分析和讲解这九大组件。

6.1 HandlerMapping

6.1.1 HandlerMapping 的作用

在 DispatcherServlet 收到客户端请求之后，在 HandlerMapping（映射处理器）接口类中通过映射规则找到对应的 Controller。HandlerMapping 接口类负责路由，找到对应请求的 Controller。

6.1.2 HandlerMapping 源码

HandlerMapping 的源码如代码清单 6-1 所示。

代码清单 6-1 HandlerMapping 的源码

```
public interface HandlerMapping {
    String PATH_WITHIN_HANDLER_MAPPING_ATTRIBUTE = HandlerMapping.class.getName() +
            ".pathWithinHandlerMapping";
    String BEST_MATCHING_PATTERN_ATTRIBUTE = HandlerMapping.class.getName() +
            ".bestMatchingPattern";
    String INTROSPECT_TYPE_LEVEL_MAPPING = HandlerMapping.class.getName() +
            ".introspectTypeLevelMapping";
    String URI_TEMPLATE_VARIABLES_ATTRIBUTE = HandlerMapping.class.getName() +
            ".uriTemplateVariables";
    String MATRIX_VARIABLES_ATTRIBUTE = HandlerMapping.class.getName() +
            ".matrixVariables";
    String PRODUCIBLE_MEDIA_TYPES_ATTRIBUTE = HandlerMapping.class.getName() +
            ".producibleMediaTypes";
    @Nullable
    HandlerExecutionChain getHandler(HttpServletRequest var1) throws Exception;
}
```

根据源码可以看出 getHandler 方法中根据请求获取了 HandlerExecutionChain 处理执行链。

每个实现类获取 HandlerExecutionChain 对象的处理规则都是不同的，主要的实现类有 SimpleUrlHandlerMapping（需要手动配置，通过配置的请求路径与 Controller 建立映射关系）、BeanNameUrlHandlerMapping（通过定义 Controller 的 beanName 匹配请求对应的 Controller）、RequestMappingHandlerMapping（通过 RequestMapping 注解查找请求对应的 Controller，Spring 3.1 以前是 DefaultAnnotationHandlerMapping 注解）。

6.1.3 HandlerMapping 的初始化

首先了解 Web 容器初始化的顺序：HttpServletBean.init（Web 容器启动时调用）→ Framework

Servlet.initServletBean → DispatcherServlet.onRefresh → DispatcherServlet.initStrategies。

因为 HandlerMapping 是由 DispatcherServlet 负责管理的，所以可以进入 DispatcherServlet 类的源码查看其相关的方法，如代码清单 6-2 所示。

代码清单 6-2　DispatcherServlet 类中的 initStrategies 方法

```
/**
 * Initialize the strategy objects that this servlet uses.
 * <p>May be overridden in subclasses in order to initialize further strategy
objects.
 */
protected void initStrategies(ApplicationContext context) {
    initMultipartResolver(context);
    initLocaleResolver(context);
    initThemeResolver(context);
    initHandlerMappings(context);
    initHandlerAdapters(context);
    initHandlerExceptionResolvers(context);
    initRequestToViewNameTranslator(context);
    initViewResolvers(context);
    initFlashMapManager(context);
}
```

在该方法中初始化 DispatcherServlet 时使用了一些策略对象，用来执行整个 Spring MVC 框架的初始化，该方法可以在子类中被重写，以便初始化更多的策略对象。

initHandlerMappings(context) 方法用来初始化 HandlerMapping，initHandlerMappings 方法的源码如代码清单 6-3 所示。

代码清单 6-3　initHandlerMappings 方法的源码

```
private void initHandlerMappings(ApplicationContext context) {
    this.handlerMappings = null;
    if (this.detectAllHandlerMappings) {                              //        ①
         // Find all HandlerMappings in the ApplicationContext, including ancestor
contexts
        Map<String, HandlerMapping> matchingBeans =
                BeanFactoryUtils.beansOfTypeIncludingAncestors(context,
                        HandlerMapping.class, true, false);           //        ②
        if (!matchingBeans.isEmpty()) {                               //        ③
            this.handlerMappings = new ArrayList<>(matchingBeans.values()); // ④
            // We keep handlerMappings in sorted order
            AnnotationAwareOrderComparator.sort(this.handlerMappings);      // ⑤
        }
    }
    else {
        try {
```

```
                HandlerMapping hm = context.getBean(HANDLER_MAPPING_BEAN_NAME,
HandlerMapping.class);                                                    // ⑥
            this.handlerMappings = Collections.singletonList(hm);          // ⑦
        }
        catch (NoSuchBeanDefinitionException ex) {
            // Ignore, we'll add a default HandlerMapping later
        }
    }
    if (this.handlerMappings == null) {                                    // ⑧
            this.handlerMappings = getDefaultStrategies(context, HandlerMapping.
class);// ⑨
        if (logger.isDebugEnabled()) {
            logger.debug("No HandlerMappings found in servlet '" +
                    getServletName() + "': using default");
        }
    }
}
```

初始化此类使用的是HandlerMapping，如果BeanFactory中没有为此命名空间定义的HandlerMapping Bean，则默认使用BeanNameUrlHandlerMapping。下面分析带序号的代码，下面序号对应代码中注释的序号。

①判断detectAllHandlerMappings是否为true(该值默认为true)，如果为ture,则检测并导入所有的HandlerMapping Bean（在当前DispatcherServlet的IoC容器中或者是父类/接口的上下文中），如果在代码或者XML配置中设置该值为false，那么只需一个名称为handlerMapping的HandlerMapping（可参照标记为⑥的代码）。

②在ApplicationContext中查找所有HandlerMapping，包括父类/接口的上下文。

③判断matchingBeans是否为空值。

④将matchingBeans中的HandlerMappings复制到this.handlerMappings中。

⑤按照order属性将HandlerMappings集合排序，order越小，优先级越高。

⑥ detectAllHandlerMappings被设置为false，在ApplicationContext中获取一个名为hm的HandlerMapping实例。

⑦返回一个只包含HandlerMapping的不可变列表。

⑧判断handlerMappings是否为null。

⑨如果没有找到其他映射，则注册DispatcherServlet.properties中默认的HandlerMapping，确保至少有一个HandlerMapping。

6.1.4 DispatcherServlet.properties文件

通过DispatcherServlet.properties文件可以知道Spring MVC中一些默认的注册组件：

```
# Default implementation classes for DispatcherServlet's strategy interfaces.
# Used as fallback when no matching beans are found in the DispatcherServlet
context.
# Not meant to be customized by application developers.
```

```
org.springframework.web.servlet.LocaleResolver=org.springframework.web.servlet.
i18n.AcceptHeaderLocaleResolver
org.springframework.web.servlet.ThemeResolver=org.springframework.web.servlet.
theme.FixedThemeResolver

org.springframework.web.servlet.HandlerMapping=org.springframework.web.servlet.
handler.BeanNameUrlHandlerMapping,\
org.springframework.web.servlet.mvc.method.annotation.RequestMappingHandlerMapping
org.springframework.web.servlet.HandlerAdapter=org.springframework.web.servlet.
mvc.HttpRequestHandlerAdapter,\
org.springframework.web.servlet.mvc.SimpleControllerHandlerAdapter,\
org.springframework.web.servlet.mvc.method.annotation.RequestMappingHandlerAdapter
org.springframework.web.servlet.HandlerExceptionResolver=org.springframework.web.
servlet.mvc.method.annotation.ExceptionHandlerExceptionResolver,\
org.springframework.web.servlet.mvc.annotation.ResponseStatusExceptionResolver,\
org.springframework.web.servlet.mvc.support.DefaultHandlerExceptionResolver
org.springframework.web.servlet.RequestToViewNameTranslator=org.springframework.
web.servlet.view.DefaultRequestToViewNameTranslator
org.springframework.web.servlet.ViewResolver=org.springframework.web.servlet.
view.InternalResourceViewResolver
org.springframework.web.servlet.FlashMapManager=org.springframework.web.servlet.
support.SessionFlashMapManager
```

在这个文件中对 DispatcherServlet 策略接口进行了一些默认的实现。

6.1.5 RequestMappingHandlerMapping 分析

Spring MVC 中默认的映射处理器是 BeanNameUrlHandlerMapping，为什么不讲解它呢？主要原因是BeanNameUrlHandlerMapping 中一个Controller 只能对应一个URL 请求，所以在平时开发中，一般使用注解来映射页面。

首先看一下 RequestMappingHandlerMapping 的继承关系，如图 6-1 所示。

RequestMappingHandlerMapping 映射处理器会处理所有带 @Controller 的控制器类的方法，通过@RequestMapping 将一个 URL 映射到这个具体的方法中。

RequestMappingHandlerMapping 重写了 InitalizingBean 接口的 afterPropertiesSet 方法，Spring 容器在启动阶段会执行 InitalizingBean.afterPropertiesSet 方法，而 RequestMappingHandlerMapping 对其进行了重写，可以理解为 afterPropertiesSet 方法是 RequestMappingHandlerMapping 初始化的入口。RequestMappingHandlerMapping 中的 afterPropertiesSet 方法如代码清单 6-4 所示。

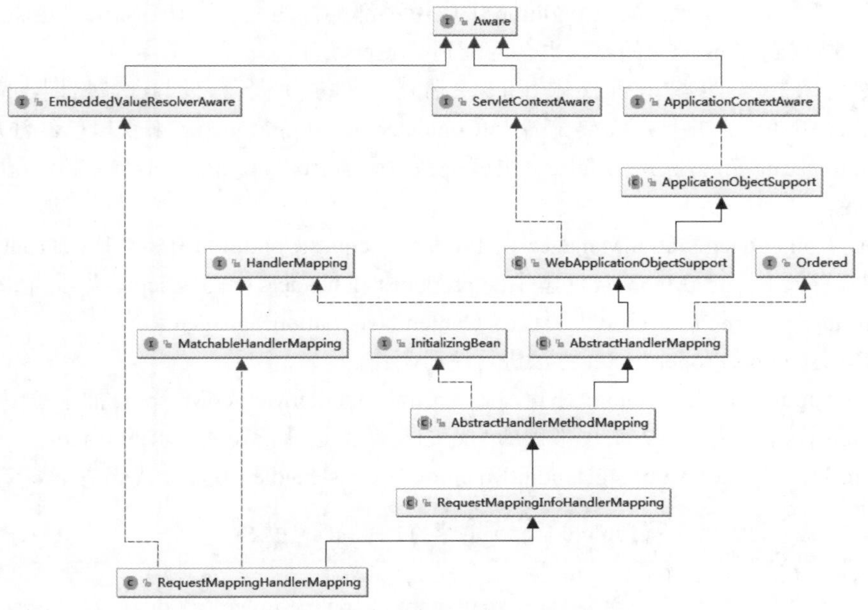

图 6-1　RequestMappingHandlerMapping 的继承关系

代码清单 6-4　RequestMappingHandlerMapping 中的 afterPropertiesSet 方法

```
@Override
public void afterPropertiesSet() {
this.config = new RequestMappingInfo.BuilderConfiguration();                    //①
this.config.setUrlPathHelper(getUrlPathHelper());                               //②
this.config.setPathMatcher(getPathMatcher());                                   //③
this.config.setSuffixPatternMatch(this.useSuffixPatternMatch);                  //④
this.config.setTrailingSlashMatch(this.useTrailingSlashMatch);                  //⑤
this.config.setRegisteredSuffixPatternMatch(this.useRegisteredSuffixPatternMat
ch);                                                                            //⑥
this.config.setContentNegotiationManager(getContentNegotiationManager());        //⑦
super.afterPropertiesSet();                                                     //⑧
}
```

下面讲解其中的重要代码，下面的序号对应代码中注释的序号。

①创建了 RequestMappingInfo.BuilderConfiguration 类，该类是用于存放匹配请求映射的配置选项的容器。接下来的配置是创建 RequestMappingInfo 实例所必需的，通常适用于所有 RequestMappingInfo 实例，也就是为创建 RequestMappingInfo 对象做准备，是 Spring 4.2 版本之后才有的代码。

②为 PatternsRequestCondition 设置 UrlPathHelper，默认是 UrlPathHelper，用来解析和查找路径。

③为 PatternsRequestCondition 设置一个 URL 路径匹配器，默认是 AntPathMatcher。

④设置是否在 PatternsRequestCondition 中应用后缀匹配模式，默认设置为 true。也就是说，在设置为 ture 的情况下，映射到“/user”的路径也会匹配到“/user.*”。

⑤设置是否在 PatternsRequestCondition 中应用尾部斜杠匹配，默认为 true。也就是说，在启用的情况下，映射到 "/users" 的路径也会匹配到 "/users/"。

⑥设置后缀模式匹配是否仅限于已注册的文件扩展名。设置此属性同时还会设置 suffixPatternMatch=true，并且还需要配置 setContentNegotiationManager 来获取已经注册的文件扩展名（如 .json、.xml 等），默认为 false。也就是在设置 suffixPatternMatch=true 的情况下，可以匹配所有的扩展名。

⑦设置 ContentNegotiationManager 在 ProducesRequestCondition 中使用。Produces Request Condition 用于存储 RequestMapping 注解中的 produces 和 headers 中的 accept。当设置 set Registered SuffixPatternMatch 为 true 后，注意也要设置 ContentNegotiationManager。

⑧调用父类的 afterPropertiesSet 方法进行注册工作。

RequestMappingHandlerMapping 还重写了 AbstractHandlerMethodMapping 抽象类中的 isHandler (Class<?> beanType) 方法。该方法用来判断给定的类是不是具备处理请求的 Handler，也就是判断是不是 Controller。RequestMappingHandlerMapping 中的 isHandler 方法如代码清单 6-5 所示。

代码清单 6-5　RequestMappingHandlerMapping 中的 isHandler 方法

```
protected boolean isHandler(Class<?> beanType) {
    return (AnnotatedElementUtils.hasAnnotation(beanType, Controller.class) ||
            AnnotatedElementUtils.hasAnnotation(beanType, RequestMapping.class));
}
```

从源码可以知道，当匹配 @Controller 和 @RequestMapping 时，优先匹配的是 @Controller。

在 RequestMappingHandlerMapping 中会创建 RequestMappingInfo 对象，RequestMappingHandler Mapping 中的 createRequestMappingInfo 方法如代码清单 6-6 所示。

代码清单 6-6　RequestMappingHandlerMapping 中的 createRequestMappingInfo 方法

```
@Nullable
private RequestMappingInfo createRequestMappingInfo(AnnotatedElement element) {
    RequestMapping requestMapping =
              AnnotatedElementUtils.findMergedAnnotation(element, RequestMapping.
class);
    RequestCondition<?> condition = (element instanceof Class ?getCustomTypeConditi
on((Class<?>) element) : getCustomMethodCondition((Method) element));
    return (requestMapping != null ?
            createRequestMappingInfo(requestMapping, condition) : null);
}
protected RequestMappingInfo createRequestMappingInfo(
        RequestMapping requestMapping,
        @Nullable RequestCondition<?> customCondition) {
    RequestMappingInfo.Builder builder = RequestMappingInfo
            .paths(resolveEmbeddedValuesInPatterns(requestMapping.path()))
            .methods(requestMapping.method())
            .params(requestMapping.params())
            .headers(requestMapping.headers())
            .consumes(requestMapping.consumes())
```

```
            .produces(requestMapping.produces())
            .mappingName(requestMapping.name());
    if (customCondition != null) {
        builder.customCondition(customCondition);
    }
    return builder.options(this.config).build();
}
```

AnnotatedElementUtils.findMergedAnnotation(element, RequestMapping.class) 方法的作用就是在给定的 element 对象结构中获取第 1 个返回的 RequestMapping 注解对象。该注解并不是只匹配一个单独的 RequestMapping 注解，也适用于组合注解。

createRequestMappingInfo(RequestMappingrequestMapping,@Nullable RequestCondition<?> customCondition) 方法是从提供的 RequestMapping 注解中创建 RequestMappingInfo，RequestMapping 注解可以是直接声明的注解，也可以是其他组合了 RequestMapping 的注解（如 PostMapping 等）。

createRequestMappingInfo 方法是在 AbstractHandlerMethodMapping 抽象类中的 detectHandlerMethods 方法中调用的。调用顺序：afterPropertiesSet → initHandlerMethods → detectHandlerMethods → getMappingForMethod → createRequestMappingInfo。

Spring 中的很多方法都是 protected 修饰的，这意味着开发者通过继承可以更灵活地定制控件。

6.2 HandlerAdapter

如果将 HandlerMapping 比作一个工具，那么可以将 HandlerAdapter（适配处理器）理解为使用工具的人。HandlerAdapter 类是典型的适配器模式。基于该模式，可以使接口不兼容的类协同工作，也就是将自己的接口类包含在另一个已经存在的类中。

6.2.1 HandlerAdapter 的源码

HandlerAdapter 的源码如代码清单 6-7 所示。

代码清单 6-7　HandlerAdapter 的源码

```
public interface HandlerAdapter {
    boolean supports(Object handler);
    @Nullable
    ModelAndView handle(HttpServletRequest request, HttpServletResponse response,
                        Object handler) throws Exception;
    long getLastModified(HttpServletRequest request, Object handler);
}
```

HandlerAdapter 接口只有 3 个方法：

（1）supports 方法：传入一个 Handler 对象并检查，判断 HandlerAdapter 是否支持该 Handler。

（2）handle 方法：使用给定的 Handler 处理请求并返回 ModelAndView，这是最关键的方法。

（3）getLastModified 方法：与 HttpServlet 的 getLastModified 方法相同，如果处理程序类中不支持 getLastModified 方法，则可以简单地返回 –1。

6.2.2 HandlerAdapter 的初始化

如代码清单 6-2 所示，在 DispatcherServlet 类的 initStrategies 方法中进行了 HandlerAdapter 的初始化。HandlerAdapter 初始化的源码如代码清单 6-8 所示。

代码清单 6–8 HandlerAdapter 初始化的源码

```
private void initHandlerAdapters(ApplicationContext context) {
    this.handlerAdapters = null;
    if (this.detectAllHandlerAdapters) {
        //在ApplicationContext中找到所有的HandlerAdapters，包括所有的上下文
        Map<String, HandlerAdapter> matchingBeans =
                BeanFactoryUtils.beansOfTypeIncludingAncestors(context,
                        HandlerAdapter.class, true, false);
        if (!matchingBeans.isEmpty()) {
            this.handlerAdapters = new ArrayList<>(matchingBeans.values());
            // 保持HandlerAdapters 按照顺序排序
            AnnotationAwareOrderComparator.sort(this.handlerAdapters);
        }
    }
    else {
        try {
                HandlerAdapter ha = context.getBean(HANDLER_ADAPTER_BEAN_NAME,
HandlerAdapter.class);
            this.handlerAdapters = Collections.singletonList(ha);
        }
        catch (NoSuchBeanDefinitionException ex) {
            // 忽略，后续会添加一个默认的 HandlerAdapter
        }
    }
        // 如果一直没有找到 HandlerAdapters,则注册一个默认的 HandlerAdapters，确保有一个
HandlerAdapters
    if (this.handlerAdapters == null) {
            this.handlerAdapters = getDefaultStrategies(context, HandlerAdapter.
class);
        if (logger.isDebugEnabled()) {
            logger.debug("No HandlerAdapters found in servlet '"+getServletName()
+ "': using default");
        }
    }
}
```

在上面初始化的方法中，与 HandlerMapping 相似。首先根据 detectAllHandlerAdapters 的值（默认为 true）查找所有的 HandlerAdapters 类，或者只查找名字为 handlerAdapter 的 Bean。最后创建

HandlerAdapter 对象并保存在 DispatcherServlet 的 handlerAdapters 中。

如果开发者没有定义 HandlerAdapter，那么 Spring MVC 会加载默认的 HandlerAdapter。

6.2.3 对 RequestMappingHandlerAdapter 的分析

RequestMappingHandlerAdapter 的继承关系如图 6-2 所示。

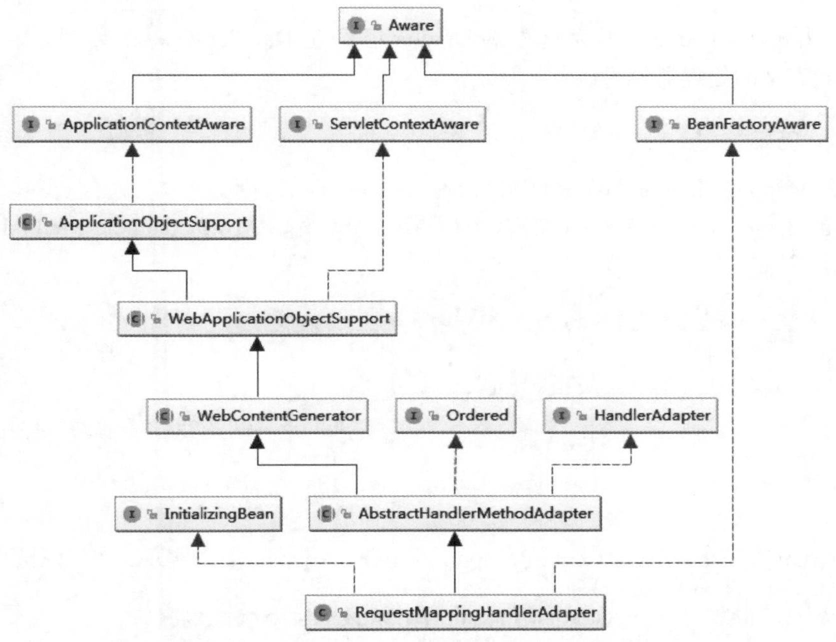

图 6-2 RequestMappingHandlerAdapter 的继承关系

RequestMappingHandlerAdapter 继承了 AbstractHandlerMethodAdapter，而 AbstractHandlerMethodAdapter 是一个抽象类，基本没有具体的实现逻辑代码，所以 RequestMappingHandlerAdapter 真正意义上实现了 HandlerAdapter 接口的功能。

下面讲解 HandlerAdapter 接口的 3 个方法。

1.supports 方法

实际上，supports 方法在 AbstractHandlerMethodAdapter 类中已经进行了 final 实现，保证 supports 传进来的参数类型为 HandlerMethod 类型。AbstractHandlerMethodAdapter 类中 supports 方法的实现如代码清单 6-9 所示。

代码清单 6-9 AbstractHandlerMethodAdapter 中 supports 方法的源码

```
public final boolean supports(Object handler) {
    return (handler instanceof HandlerMethod
            && supportsInternal((HandlerMethod) handler));
}
```

该方法进行了类型判断，调用了 supportsInternal 方法，该方法的具体实现在 RequestMappingHandlerAdapter 类中，如代码清单 6-10 所示。

代码清单 6-10　RequestMappingHandlerAdapter 中 supportsInternal 方法的源码

```
protected boolean supportsInternal(HandlerMethod handlerMethod) {
    return true;
}
```

该方法直接返回了 true，因为这里的 handler 已经是 HandlerMethod 类型了，所以只要是 HandlerMethod 类型的处理器就可以。

2.handle 方法

在 RequestMappingHandlerAdapter 中也没有 handle 方法，因为该方法已经在其父类 AbstractHandlerMethodAdapter 中进行了 final 实现，只是换了个名称。AbstractHandlerMethodAdapter 中的 handle 方法的源码如代码清单 6-11 所示。

代码清单 6-11　AbstractHandlerMethodAdapter 中 handle 方法的源码

```
@Nullable
public final ModelAndView handle(HttpServletRequest request,
                                HttpServletResponse response, Object handler)throws
Exception {
    return handleInternal(request, response, (HandlerMethod) handler);
}
```

handleInternal 方法的实际在 RequestMappingHandlerAdapter 类中实现，如代码清单 6-12 所示。

代码清单 6-12　RequestMappingHandlerAdapter 中 handleInternal 方法的源码

```
    protected ModelAndView handleInternal(HttpServletRequest request,
HttpServletResponse response, HandlerMethod handlerMethod) throws Exception {
    ModelAndView mav;
    checkRequest(request);                                              // ①
    if (this.synchronizeOnSession) {                                    // ②
        HttpSession session = request.getSession(false);
        if (session != null) {
            Object mutex = WebUtils.getSessionMutex(session);
            synchronized (mutex) {
                mav = invokeHandlerMethod(request, response, handlerMethod); // ③
            }
        }
        else {
            mav = invokeHandlerMethod(request, response, handlerMethod);
        }
    }
    else {
        mav = invokeHandlerMethod(request, response, handlerMethod);
    }
```

```
        if (!response.containsHeader(HEADER_CACHE_CONTROL)) {                    // ④
            if (getSessionAttributesHandler(handlerMethod).hasSessionAttributes()) {
// ⑤
                applyCacheSeconds(response,
                        this.cacheSecondsForSessionAttributeHandlers);            // ⑥
            }
            else {
                prepareResponse(response);                                         // ⑦
            }
        }
        return mav;
    }
```

下面讲解其中的重要代码，下面的序号对应代码中注释的序号。

①检查请求中的 method 和 session 是否被支持。

② synchronizeOnSession 值默认为 false。如果为 true，则表示在同步块中执行 invokeHandlerMethod 方法。HttpSession 引用本身是安全的，因为总是一个对象的引用，但不保证不同 servlet 容器中的安全，只能使用会话互斥锁来保证不同 servlet 容器中的安全。

③利用动态代理，调用 RequestMapping 注解的处理方法，准备生成 ModelAndView。

④判断响应头中是否包含 Cache-Control 字段，即控制缓存行为的字段。

⑤根据 handlerMethod 获取 SessionAttributesHandler，并判断 SessionAttributesHandler 控制器是否有 SessionAttributes 注解。

⑥进行缓存设置。

⑦准备 HTTP 响应。

简单分析后，得出以下步骤：

（1）校验请求。

（2）调用处理器，得到 ModelAndView。

（3）判断响应头中是否有 Cache-Control 字段，然后判断 Controller 是否有 SessionAttributes 注解，如果有，则设置缓存，如果没有，则准备 HTTP 响应。

（4）返回 ModelAndView。

3. getLastModified 方法

getLastModified 方法在 RequestMappingHandlerAdapter 类中，直接返回 -1。

6.3 HandlerExceptionResolver

HandlerExceptionResolver（异常处理器）该组件是 Spring MVC 中专门负责处理异常的组件，HandlerExceptionResolver 的作用就是根据异常来设置返回的 ModelAndView。实现该接口的类一般会在 Spring 的上下文中提前通过注解注册。

6.3.1 HandlerExceptionResolver 的源码

代码清单 6-13 HandlerExceptionResolver 的源码

```
public interface HandlerExceptionResolver {
    @Nullable
    ModelAndView resolveException(
            HttpServletRequest request, HttpServletResponse response,
            @Nullable Object handler, Exception ex);
}
```

HandlerExceptionResolver 只有一个方法，这个方法用来处理程序运行期间抛出的异常，返回到指定的 ModelAndView。可以定义多个异常返回视图，在实际实现中，可以维护一个 Map，异常作为 key，View 作为 value，直接根据异常 key 返回 ModelAndView 即可。

HandlerExceptionResolver 中的 resolveException 是如何被调用的呢？调用顺序：DispatcherServlet.doService（每次的 Web 请求都会进来）→ DispatcherServlet.doDispatch（该方法中的异常交给 processDispatchResult 方法处理）→ DispatcherServlet.processDispatchResult（在该方法中对异常进行判断，如果异常不为 null，且异常不是继承自 ModelAndViewDefiningException）→ DispatcherServlet.processHandlerException → handlerExceptionResolver.resolveException（遍历 DispatcherServlet 中的 handlerExceptionResolvers 集合，调用第 1 个 handlerExceptionResolver 执行 resolveException 方法）。

6.3.2 HandlerExceptionResolver 的初始化

根据代码清单 6-2 可知，在 DispatcherServlet 类的 initStrategies 方法中调用了 initHandlerExceptionResolvers 方法来初始化 handlerExceptionResolvers。

代码清单 6-14 DispatcherServlet 中的 initHandlerExceptionResolvers 方法

```
private void initHandlerExceptionResolvers (ApplicationContext context){
    this.handlerExceptionResolvers = null;
    // detectAllHandlerExceptionResolvers值默认为true
    if (this.detectAllHandlerExceptionResolvers) {
        // 从ApplicationContext中找出所有HandlerExceptionResolver对象
        Map<String, HandlerExceptionResolver> matchingBeans
                = BeanFactoryUtils.beansOfTypeIncludingAncestors(context,
HandlerExceptionResolver.class, true, false);
        if (!matchingBeans.isEmpty()) {
            //将matchingBeans中的值赋值到handlerExceptionResolvers 中
                this.handlerExceptionResolvers = new ArrayList<>(matchingBeans.
values());
            // 根据order的值对handlerExceptionResolvers排序
            AnnotationAwareOrderComparator.sort(this.handlerExceptionResolvers);
        }
    } else {
```

```
        try {
            /* 开发者设置detectAllHandlerExceptionResolvers值为false后将会运行这里。获取
名称为handlerExceptionResolver的Bane */
            HandlerExceptionResolver her =
                    context.getBean(HANDLER_EXCEPTION_RESOLVER_BEAN_NAME,
                            HandlerExceptionResolver.class);
            this.handlerExceptionResolvers = Collections.singletonList(her);
        } catch (NoSuchBeanDefinitionException ex) {
            //忽略，这里没有找到用户自定义的HandlerExceptionResolver
        }
    }
    //如果 HandlerExceptionResolvers为空，那么注册默认的handlerExceptionResolvers
    if (this.handlerExceptionResolvers == null) {
            /*创建默认策略对象列表。默认实现使用“DispatcherServlet.properties”文件(在与
DispatcherServlet类相同的包中)来确定类名。通过上下文的BeanFactory实例化策略对象*/
            this.handlerExceptionResolvers = getDefaultStrategies(context,
HandlerExceptionResolver.class);
        if (logger.isDebugEnabled()) {
                logger.debug("No HandlerExceptionResolvers found in servlet '" +
getServletName() + "': using default");
        }
    }
}
```

（1）通过判断 detectAllHandlerExceptionResolvers 的值来确定使用所有的 handler Exception Resolvers 还是使用自定义的名为 handlerExceptionResolver 的 Bean。

（2）如果没有找到 handlerExceptionResolvers，那么注册默认的 HandlerExceptionResolver，Spring MVC 在 DispatcherServlet.properties 文件中配置了默认的注册信息。

6.4 ViewResolver

ViewResolver（视图解析器）的作用就是将 String 类型的逻辑视图名解析为 View 类型的视图。由于在程序运行期间视图的状态不会被改变，所以缓存了视图。

6.4.1 ViewResolver 的源码

ViewResolver 的源码如代码清单 6-15 所示。

代码清单 6-15 ViewResolver 的源码

```
public interface ViewResolver {
    @Nullable
    View resolveViewName(String viewName, Locale locale) throws Exception;
}
```

ViewResolver 的源码非常简单，只有一个方法 resolveViewName。该方法的作用就是按照名

称解析并返回视图。resolveViewName 方法有 viewName 和 locale 两个参数，一般情况下只会使用视图名，根据视图名找到视图，然后渲染即可。如果需要国际化，才会用到 local 参数。视图的实际作用就是用来渲染页面，也就是将程序方法返回的结果封装到具体的模板中，生成具体的视图文件（JSP、FTL、HTML 等）。

需要什么类型的视图，就需要对应的视图解析器来返回视图。例如，如果需要解析出 JSP 页面（InternalResourceView 视图），则要用到 InternalResourceViewResolver 视图解析器；如果使用了 GroovyMarkupViewResolver 视图解析器，则会解析出 GroovyMarkupView 视图。还有其他视图解析器，有的只支持一种类型的视图解析，有的则支持多种类型的视图解析。

6.4.2 ViewResolver 的初始化

从代码清单 6-2 可以知道，该组件的初始化也是在 DispatcherServlet 类的 initStrategies 方法中完成的。查看 DispatcherServlet 类中的 initViewResolvers(context) 方法，如代码清单 6-16 所示。

代码清单 6–16 DispatcherServlet 类中的 initViewResolvers 方法

```
private void initViewResolvers (ApplicationContext context){
    this.viewResolvers = null;
    //判断detectAllViewResolvers，默认为true
    if (this.detectAllViewResolvers) {
        //在ApplicationContext中查找所有的ViewResolver，包括父类的上下文
        Map<String, ViewResolver> matchingBeans =
                       BeanFactoryUtils.beansOfTypeIncludingAncestors(context,
ViewResolver.class, true, false);
        if (!matchingBeans.isEmpty()) {
            this.viewResolvers = new ArrayList<>(matchingBeans.values());
            //根据order排序。order越小的越先调用
            AnnotationAwareOrderComparator.sort(this.viewResolvers);
        }
    } else {
        //自定义detectAllViewResolvers的值为false会调用到这里
        try {
            //获取名称为viewResolver的Bean
                  ViewResolver vr = context.getBean(VIEW_RESOLVER_BEAN_NAME,
ViewResolver.class);
            //返回一个包含指定对象的集合。返回的集合是可序列化的
            this.viewResolvers = Collections.singletonList(vr);
        } catch (NoSuchBeanDefinitionException ex) {
            // 忽略，后面代码会添加一个默认的ViewResolver
        }
    }
     /* 如果没有找到ViewResolver，则到“DispatcherServlet.properties”配置文件中获取默认的
ViewResolver，保证至少有一个ViewResolver */
    if (this.viewResolvers == null) {
        this.viewResolvers = getDefaultStrategies(context, ViewResolver.class);
        if (logger.isDebugEnabled()) {
            logger.debug("No ViewResolvers found in servlet '" + getServletName()
```

```
                + "': using default");
            }
        }
    }
```

6.4.3 resolveViewName 方法被调用的过程

ViewResolver 接口的 resolveViewName 方法是从 DispatcherServlet 类的 render 方法中调用过来的。对于每一个请求，都会运行到 DispatcherServlet.doService 方法中，调用顺序如下：

DispatcherServlet.doService → DispatcherServlet.doDispatch → DispatcherServlet.processDispatchResult → DispatcherServlet.render → DispatcherServlet.resolveViewName → ViewResolver.resolveViewName

在第 4 步之前，resolveViewName 方法运行的步骤与前面的 HandlerExceptionResolver.resolveException 方法的运行步骤是一致的。而在调用 render 方法之前，异常的 ModelAndView 已经被设置了出来，图 6-3 所示为 processDispatchResult 方法的部分源码。但是 render 方法尚未执行，也就是 View 还没有被渲染。当使用 HandlerExceptionResolver 进行全局异常捕获时，如果是在渲染的过程中出现了异常，则该异常是无法被 HandlerExceptionResolver 处理的。

```
if (exception != null) {
    if (exception instanceof ModelAndViewDefiningException) {
        logger.debug( o: "ModelAndViewDefiningException encountered", exception);
        mv = ((ModelAndViewDefiningException) exception).getModelAndView();
    }
    else {
        Object handler = (mappedHandler != null ? mappedHandler.getHandler() : null);
        mv = processHandlerException(request, response, handler, exception);
        errorView = (mv != null);
    }
}

// Did the handler return a view to render?
if (mv != null && !mv.wasCleared()) {
    render(mv, request, response);
    if (errorView) {
        WebUtils.clearErrorRequestAttributes(request);
    }
}
```

图 6-3 processDispatchResult 方法的部分源码

6.4.4 对 ViewResolverRegistry 的分析

ViewResolverRegistry 类是配置一个视图解析器的链，内部使用 List 对 ViewResolver 进行维护，可以支持不同的模板机制。另外可以根据请求的类型配置默认的 View 进行渲染，如 JSON、XML 等，还可以在 ViewResolverRegistry 类内部注册 ViewResolver 视图解析器。例如，ViewResolverRegistry 类中的内部类 FreeMarkerRegistration 注册了 FreeMarkerViewResolver，如代

码清单 6-17 所示。

代码清单 6-17　FreeMarkerRegistration 类

```
private static class FreeMarkerRegistration extends UrlBasedViewResolverRegistration {
    public FreeMarkerRegistration() {
        super(new FreeMarkerViewResolver());
        getViewResolver().setSuffix(".ftl");
    }
}
```

通过该源码可以知道，为什么 FreeMarker 文件需要以 .ftl 后缀结尾了。另外，其他的视图解析器也在该类中有静态内部类。

通过实现 WebMvcConfigurer 接口，重写 configureViewResolvers(ViewResolverRegistry registry) 方法，还可以对该类实现一些定制的功能。

6.5 RequestToViewNameTranslator

RequestToViewNameTranslator（视图名称转换器，请求到视图名称的转换）的使用场景是在 Handler 方法中未显式提供视图名称，在方法返回时将传入的 HttpServletRequest 转换为逻辑视图名称。

还有一种场景，即如果 Spring MVC 在处理业务的过程中出现了异常，且没有返回视图名 ModelAndView，在这种情况下，也是需要 RequestToViewNameTranslator 来进行默认的处理，返回默认的视图。

6.5.1 RequestToViewNameTranslator 被调用的情况

第 1 种被调用的情况是 6.3 节中提到的异常处理，如果异常并不是 ModelAndViewDefiningException 的异常，则会运行 processHandlerException 方法。processHandlerException 方法如代码清单 6-18 所示。

代码清单 6-18　processHandlerException 方法

```
@Nullable
protected ModelAndView processHandlerException(HttpServletRequest request,
HttpServletResponse response,@Nullable Object handler, Exception ex) throws
Exception {
    ModelAndView exMv = null;
    if (this.handlerExceptionResolvers != null) {
            for (HandlerExceptionResolver handlerExceptionResolver : this.
handlerExceptionResolvers) {
                exMv = handlerExceptionResolver.resolveException(request, response,
handler, ex);
            if (exMv != null) {
                break;
```

```
            }
        }
    }
    if (exMv != null) {
        if (exMv.isEmpty()) {
            request.setAttribute(EXCEPTION_ATTRIBUTE, ex);
            return null;
        }
        //如果没有视图，则获取默认的视图名称，并设置视图名称
        if (!exMv.hasView()) {
            String defaultViewName = getDefaultViewName(request);
            if (defaultViewName != null) {
                exMv.setViewName(defaultViewName);
            }
        }
        if (logger.isDebugEnabled()) {
              logger.debug("Handler execution resulted in exception-forwarding to
resolved error view: "
                    + exMv, ex);
        }
        WebUtils.exposeErrorRequestAttributes(request, ex, getServletName());
        return exMv;
    }
    throw ex;
}
```

在该方法中，如果没有提供视图名称（也就是没有返回异常的代理视图），则会调用 getDefaultViewName 获取默认的视图名称。

代码清单 6-19　getDefaultViewName 方法的实现

```
@Nullable
protected String getDefaultViewName(HttpServletRequest request) throws Exception {
    return (this.viewNameTranslator != null ?
            this.viewNameTranslator.getViewName(request) : null);
}
```

可以看到代码清单 6-19 中调用了 viewNameTranslator 的 getViewName 方法，回到了本节的主题 RequestToViewNameTranslator 接口。

第 2 种情况是，如果在自定义的方法中，对 Handler 方法没有提供视图返回，如代码清单 6-20 所示的情况，则会根据请求链接来获取默认的视图。

代码清单 6-20　不返回视图

```
@RequestMapping("/testRequestToViewNameTranslator")
public void testRequestToViewNameTranslator() {
    logger.info("testRequestToViewNameTranslator");
}
```

在 DispatcherServlet 的 doDispatch 方法中调用了 applyDefaultViewName 方法，applyDefault-

ViewName 方法如代码清单 6-21 所示。

代码清单 6-21　applyDefaultViewName 方法

```
private void applyDefaultViewName(HttpServletRequest request, @Nullable
ModelAndView mv) throws Exception {
    if (mv != null && !mv.hasView()) {
        String defaultViewName = getDefaultViewName(request);
        if (defaultViewName != null) {
            mv.setViewName(defaultViewName);
        }
    }
}
```

一般情况下，方法是有视图名称的，但是在没有视图名称的情况下，会调用 getDefaultViewName 方法，然后调用 viewNameTranslator 中的 getViewName 方法。

6.5.2 RequestToViewNameTranslator 的初始化

该组件的初始化也是在 DispatcherServlet 的 initStrategies 方法中完成的，initStrategies 调用了 initRequestToViewNameTranslator 方法，initRequestToViewNameTranslator 方法的源码如代码清单 6-22 所示。

代码清单 6-22　initRequestToViewNameTranslator 方法的源码

```
private void initRequestToViewNameTranslator(ApplicationContext context) {
    try {
        this.viewNameTranslator =
                context.getBean(REQUEST_TO_VIEW_NAME_TRANSLATOR_BEAN_NAME
                        , RequestToViewNameTranslator.class);
        if (logger.isDebugEnabled()) {
            logger.debug("Using RequestToViewNameTranslator ["
                    + this.viewNameTranslator + "]");
        }
    }
    catch (NoSuchBeanDefinitionException ex) {
        // 获取默认的RequestToViewNameTranslator
            this.viewNameTranslator = getDefaultStrategy(context,
RequestToViewNameTranslator.class);
        if (logger.isDebugEnabled()) {
             logger.debug("Unable to locate RequestToViewNameTranslator with name
'"
                    +REQUEST_TO_VIEW_NAME_TRANSLATOR_BEAN_NAME
                    + "': using default [" + this.viewNameTranslator +"]");
        }
    }
}
```

这里只获取了名称为 viewNameTranslator 的 Bean，如果没有获取到 Bean，则会进入 getDefault-

Strategy 方法，继续调用 getDefaultStrategies 方法，通过 DispatcherServlet.properties 配置文件获取默认的 RequestToViewNameTranslator，并且默认的配置只会是一个 RequestToViewNameTranslator。

6.5.3 对 DefaultRequestToViewNameTranslator 的分析

DefaultRequestToViewNameTranslator 类是 RequestToViewNameTranslator 接口在 Spring MVC 中唯一的实现类，如果开发者没有实现，则进入 DefaultRequestToViewNameTranslator 类进行处理。

RequestToViewNameTranslator 接口的源码如代码清单 6-23 所示。

代码清单 6-23　RequestToViewNameTranslator 接口的源码

```
public interface RequestToViewNameTranslator {
    @Nullable
    String getViewName(HttpServletRequest request) throws Exception;
}
```

RequestToViewNameTranslator 接口只有一个 getViewName 方法，即 getViewName 方法根据 HttpServletRequest 获取到视图名称。getViewName 方法的实现如代码清单 6-24 所示。

代码清单 6-24　getViewName 方法的实现

```
public String getViewName(HttpServletRequest request) {
    String lookupPath = this.urlPathHelper.getLookupPathForRequest(request);
    return (this.prefix + transformPath(lookupPath) + this.suffix);
}
```

getLookupPathForRequest 方法就是根据请求返回映射路径，可以添加断点进行调试，访问 http://127.0.0.1:8080/testRequestToViewNameTranslator，getViewName 方法中的参数值如图 6-4 所示。

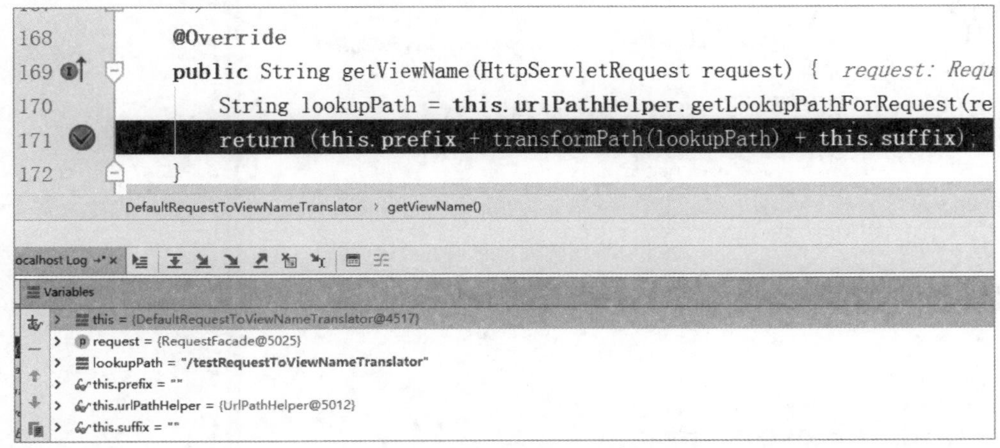

图 6-4　getViewName 方法中的参数值

从图 6-4 可以看到，lookupPath 的值为“/testRequestToViewNameTranslator”，也就是请求的路径，这里的前缀和后缀都为空。transformPath 方法会将“/testRequestToViewNameTranslator”最前面的“/”去掉，所以该方法最后返回的值是 testRequestToViewNameTranslator。

最后，在 MvcConfig 有以下配置，如代码清单 6-25 所示。

代码清单 6-25　MvcConfig 中视图解析器的配置

```
@Bean
public InternalResourceViewResolver viewResolver() {
...//省略部分代码
    viewResolver.setPrefix("/WEB-INF/classes/views/");
...//省略部分代码
    viewResolver.setSuffix(".jsp");
...//省略部分代码
}
```

由前缀和后缀组成完整的路径映射，最后获取的实际视图为“/WEB-INF/classes/views/testRequestToViewNameTranslator.jsp”。由图 6-5 可知，返回值与期望的一致。

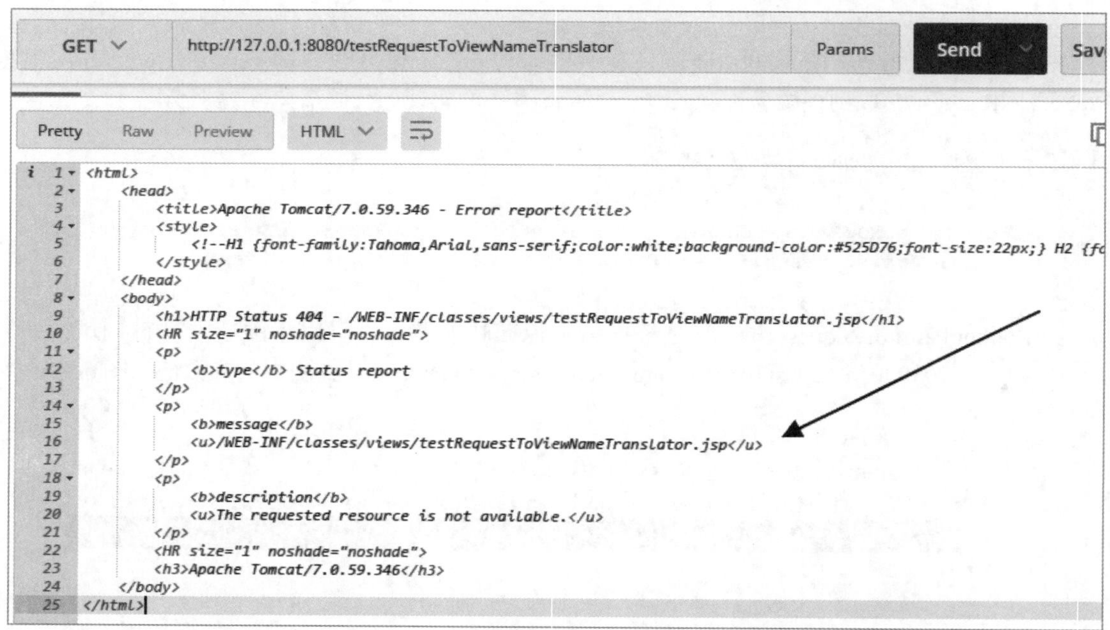

图 6–5　访问实际视图后的返回值

实际视图的路径就是映射到了该视图的文件，但是项目中是没有该文件的，所以返回 404。

6.6　LocaleResolver

在讲解 ViewResolver 视图解析器时，在 ViewResolver 接口的 resolveViewName 方法中可以看到 locale 参数（区域），该参数类型便与本节所讲的 LocaleResolver（国际化解析器）相关。

6.6.1 LocaleResolver 的初始化

LocaleResolver 组件与前面的几个组件一样，也是在 DispatcherServlet 的 initStrategies 方法中完成初始化的，调用了 initLocaleResolver(context) 方法，initLocaleResolver 方法如代码清单 6-26 所示。

代码清单 6-26 initLocaleResolver 方法

```
private void initLocaleResolver(ApplicationContext context) {
    try {
            this.localeResolver = context.getBean(LOCALE_RESOLVER_BEAN_NAME ,
LocaleResolver.class);
        if (logger.isDebugEnabled()) {
            logger.debug("Using LocaleResolver [" + this.localeResolver + "]");
        }
    }
    catch (NoSuchBeanDefinitionException ex) {
        //配置默认的LocaleResolver
        this.localeResolver = getDefaultStrategy(context, LocaleResolver.class);
        if (logger.isDebugEnabled()) {
            logger.debug("Unable to locate LocaleResolver with name '"
                    + LOCALE_RESOLVER_BEAN_NAME
                    +"': using default [" + this.localeResolver + "]");
        }
    }
}
```

由代码清单 6-26 可知，首先获取名称为 localeResolver 的 Bean，如果没有找到，则会出现 NoSuchBeanDefinitionException 异常；然后进入 getDefaultStrategy 方法，在该方法中获取配置文件配置的 Bean，接着调用 getDefaultStrategies 方法，在 Spring 5.0.8 中默认配置的 LocaleResolver 为 org.springframework.web.servlet.i18n.AcceptHeaderLocaleResolver。

6.6.2 对 AcceptHeaderLocaleResolver 的分析

本节分析默认的国际化组件的实现。LocaleResolver 的源码如代码清单 6-27 所示。

代码清单 6-27 LocaleResolver 的源码

```
public interface LocaleResolver {
    Locale resolveLocale(HttpServletRequest request);
      void setLocale(HttpServletRequest request, @Nullable HttpServletResponse
response, @Nullable Locale locale);
}
```

该接口中只有两个方法，setLocale 方法用于设置当前的环境区域，resolveLocale 方法是通过给定的请求解析出当前的语言环境，无论在什么情况下，都可以返回默认的语言环境。

setLocale 在 AcceptHeaderLocaleResolver 中的实现如代码清单 6-28 所示。

代码清单 6-28　setLocale 在 AcceptHeaderLocaleResolver 中的实现

```
public void setLocale(HttpServletRequest request, @Nullable HttpServletResponse
response , @Nullable Locale locale) { throw new UnsupportedOperationException("Cann
ot change HTTP accept header - use a different locale resolution strategy");
}
```

该方法直接抛出了 UnsupportedOperationException 异常，也就是不让开发者通过该类修改 Locale 语言区域。

接下来查看 resolveLocale 方法在 AcceptHeaderLocaleResolver 中的实现，如代码清单 6-29 所示。

代码清单 6-29　resolveLocale 在 AcceptHeaderLocaleResolver 中的实现

```
public Locale resolveLocale (HttpServletRequest request){
    //获取默认的语言环境，可能会没有配置语言环境
    Locale defaultLocale = getDefaultLocale();
    //如果默认语言环境不为空，且请求头中Accept-Language为空，则直接返回defaultLocale
    if (defaultLocale != null && request.getHeader("Accept-Language") == null) {
        return defaultLocale;
    }
    //获取请求头中的Accept-Language标志的语言环境
    Locale requestLocale = request.getLocale();
    //返回已配置的受支持的语言环境列表
    List<Locale> supportedLocales = getSupportedLocales();
     /*如果没有配置受支持的语言环境，也就是supportedLocales列表为空，或者supportedLocales不
为空，且supportedLocales包含请求的语言环境，则返回请求的语言环境*/
    if (supportedLocales.isEmpty() || supportedLocales.contains(requestLocale)) {
        return requestLocale;
    }
     /* 通过在配置的语言环境列表中查找语言环境，匹配请求中带的任一语言环境，该匹配有匹配顺序的，会
优先匹配首选的语言环境 */
    Locale supportedLocale = findSupportedLocale(request, supportedLocales);
    if (supportedLocale != null) {
        return supportedLocale;
    }
    //如果默认的语言环境不为空，则直接返回默认的语言环境，否则返回请求中首选的语言环境
    return (defaultLocale != null ? defaultLocale : requestLocale);
}
```

基于这两个方法的源码可以知道，AcceptHeaderLocaleResolver 只是简单地使用 HTTP 请求头中的 Accept-Language 来指定 Locale 对象（即客户端浏览器发送的语言环境），这也就理解了为什么不让使用 setLocale 方法，因为需要根据客户端的区域来设置 Locale，如果需要更改 Locale，那么只能更改请求头中的 Accept-Language。

6.6.3　对 SessionLocaleResolver 的分析

通常情况下，如果开发者进行国际化开发，需要让用户自己选择语言环境，则需要配置 SessionLocaleResolver，也就是设置名称为 localeResolver 的 Bean 实例为 org.springframework.web.servlet.i18n.SessionLocaleResolver（这就是为什么使用 XML 和编程式代码配置都可以，只需要保

证 Bean 的名称为 Spring MVC 中注入的 Bean 名称），然后进行一些定制化的配置。

SessionLocaleResolver 的结构如图 6-6 所示。

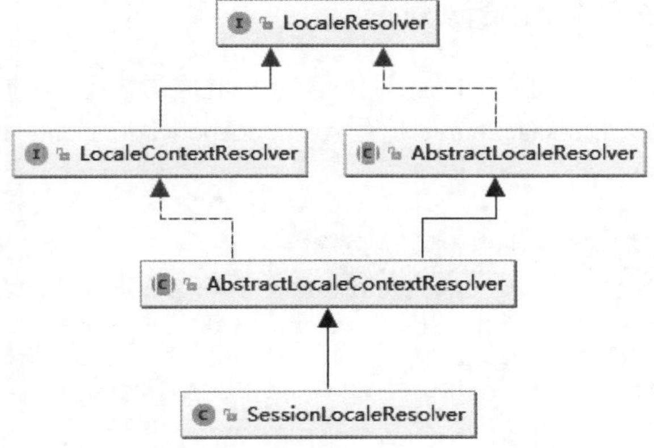

图 6-6　SessionLocaleResolver 的结构

在 AbstractLocaleContextResolver 抽象类中实现了 setLocale 方法，在该方法中调用了抽象方法 setLocaleContext，如代码清单 6-30 所示。

代码清单 6-30　AbstractLocaleContextResolver 类中的 setLocale 方法

```
public void setLocale(HttpServletRequest request, @Nullable HttpServletResponse
response , @Nullable Locale locale) {
    setLocaleContext(request, response, (locale != null ? new SimpleLocaleContext(locale)
: null));
}
```

在 SessionLocaleResolver 类中实现了 setLocaleContext 方法，如代码清单 6-31 所示。

代码清单 6-31　SessionLocaleResolver 类中的 setLocaleContext 方法

```
public void setLocaleContext (HttpServletRequest request
        , @Nullable HttpServletResponse response, @Nullable LocaleContext
localeContext){
    Locale locale = null;
    TimeZone timeZone = null;
    if (localeContext != null) {
        locale = localeContext.getLocale();
        if (localeContext instanceof TimeZoneAwareLocaleContext) {
            timeZone = ((TimeZoneAwareLocaleContext) localeContext).getTimeZone();
        }
    }
    //存储语言环境(地区信息)
    WebUtils.setSessionAttribute(request, this.localeAttributeName, locale);
    //存储时区信息
    WebUtils.setSessionAttribute(request, this.timeZoneAttributeName, timeZone);
}
```

该方法将请求中携带的 locale 对象信息存储到了 Session 中，然后在调用 resolveLocale 方法获取 Locale 对象时，再通过 Session 获取之前存储的 locale 对象。

6.6.4 MessageSource 国际化资源分析

虽然 MessageSource 实例不是在 DispatcherServlet 中完成初始化的，但该类也是和国际化相关的，负责加载国际化资源。MessageSource 是在 AbstractApplicationContext 类中完成初始化的，如代码清单 6-32 所示。

代码清单 6-32 AbstractApplicationContext 类部分源码

```
public abstract class AbstractApplicationContext extends DefaultResourceLoader
        implements ConfigurableApplicationContext {
    //注册的MessageSource Bean名称
    public static final String MESSAGE_SOURCE_BEAN_NAME = "messageSource";
    //...省略部分代码
    //实际的消息解析对象
    @Nullable
    private MessageSource messageSource; //...省略部分代码

    public void refresh() throws BeansException, IllegalStateException {
        //...省略部分代码
        // 初始化ApplicationContext的消息源
        initMessageSource(); //...省略部分代码
    }//...省略部分代码

    //初始化MessageSource 对象
    protected void initMessageSource() {
        //获取Bean工厂
        ConfigurableListableBeanFactory beanFactory = getBeanFactory();
        //先判断名称为“messageSource”的Bean是否已经实例化了(是否已经注册)
        if (beanFactory.containsLocalBean(MESSAGE_SOURCE_BEAN_NAME)) {
            //获取Bean名称为“messageSource”的实例
            this.messageSource = beanFactory.getBean(MESSAGE_SOURCE_BEAN_NAME, MessageSource.
class);
            // 判断messageSource是否有父类消息资源
            if (this.parent != null && this.messageSource instanceof Hierarchical
MessageSource) {
                HierarchicalMessageSource hms = (HierarchicalMessageSource) this.
messageSource;
                if (hms.getParentMessageSource() == null) {
                    /* 如果父messageSource尚未注册, 则设置messageSource的
parentMessageSource为父上下文消息源, 或者父上下文本身 */
                    hms.setParentMessageSource(getInternalParentMessageSource());
                }
            }//...省略部分代码
        } else {
            //没有对应的Bean实例, 则“new”一个默认的DelegatingMessageSource
```

```
            DelegatingMessageSource dms = new DelegatingMessageSource();
            //设置父消息源
            dms.setParentMessageSource(getInternalParentMessageSource());
            this.messageSource = dms;
            //以单例模式注册到工厂中
            beanFactory.registerSingleton(MESSAGE_SOURCE_BEAN_NAME, this.messageSource); //...
        }
    } //...省略部分代码
    //获取父上下文的消息源
    protected MessageSource getInternalParentMessageSource() {
        //如果实现AbstractApplicationContext接口，则返回父上下文的消息源，否则返回父上下文本身
        return (getParent() instanceof AbstractApplicationContext ? ((AbstractApp
licationContext) getParent()).messageSource : getParent());
    }
}
```

代码清单 6-32 只列出了和 MessageSource 相关的代码，这里完成了国际化资源信息的初始化，主要是将 messageSource 这个 Bean 定义的信息资源加载为 Spring 容器级的国际化信息资源。

MessageSource 接口的实例有很多，从代码清单 6-32 中可以知道 Spring 默认使用的是 DelegatingMessageSource 消息源。DelegatingMessageSource 的功能比较简单，只是将字符串和参数数组格式化为一个消息字符串，并且将所有调用委托给父 MessageSource，如果没有可用的父实例，则该类不会解析任何消息。DelegatingMessageSource 实例只是在 AbstractApplicationContext 中被用作占位符，如果开发者在上下文中没有定义自己的 MessageSource，则 DelegatingMessageSource 不会在应用程序中被直接使用。

也就是说，如果要进行国际化处理，则开发者要自行实例化 MessageSource，一般使用 Spring 的 ResourceBundleMessageSource 和 ReloadableResourceBundleMessageSource，将其简单地注入 Spring 容器后可以直接使用。

- ResourceBundleMessageSource：修改资源后，需要重启项目。依赖于 JDK 的 ResourceBundle 加载资源，支持配置绝对路径和项目相对路径，只支持 properties 文件。
- ReloadableResourceBundleMessageSource：使用指定的名称访问资源，参与 ApplicationContext 的资源加载，ReloadableResourceBundleMessageSource 使用 java.util.Properties 实例作为消息的自定义数据结构，通过 PropertiesPersister 加载资源文件。另外还可以使用 Spring 的 Resource 加载 XML 文件，加载文件时遵循基本的 ResourceBundle 规则。相比 ResourceBundleMessageSource，ReloadableResourceBundleMessageSource 的功能强大很多。

如果需要在 Spring MVC 项目业务中进行国际化处理，那么只需要在 Spring 容器的上下文中注入名称为 localeResolver 和 messageSource 的两个 Bean 实例。

6.7 ThemeResolver

主题（Theme）就是整个系统的样式和风格。Spring MVC 提供的主题可以设置应用的整体样

式风格，也就是设置一些静态资源的集合（CSS 样式、图片等），用来控制系统的视觉效果。开发者只需准备资源，新建“*.properties”文件，将资源配置进去，放在 classes 或者其他自定义路径下，便可以使用相关的主题。

6.7.1 ThemeResolver 的初始化

在 DispatcherServlet 类的 initStrategies 方法中初始化 ThemeResolver（主题样式解析器）、运行 initThemeResolver 方法。DispatcherServlet 类中的 initThemeResolver 方法如代码清单 6-33 所示。

代码清单 6-33 DispatcherServlet 类中的 initThemeResolver 方法

```
private void initThemeResolver (ApplicationContext context){
    try {
        // 获取上下文中配置的名称为“themeResolver”的Bean
        this.themeResolver = context.getBean(THEME_RESOLVER_BEAN_NAME
                , ThemeResolver.class);
        if (logger.isDebugEnabled()) {
            logger.debug("Using ThemeResolver [" + this.themeResolver + "]");
        }
    } catch (NoSuchBeanDefinitionException ex) {
        // 没有找到名称为“themeResolver”的Bean，则注册一个默认的ThemeResolver
        this.themeResolver = getDefaultStrategy(context, ThemeResolver.class);
        if (logger.isDebugEnabled()) {
                logger.debug("Unable to locate ThemeResolver with name '" + THEME_
RESOLVER_BEAN_NAME + "': using default [" + this.themeResolver + "]");
        }
    }
}
```

从上面的源码和注释可知，同前面几个组件的初始化类似，initThemeResolver 方法也是先获取上下文中配置的 Bean，如果没有找到，那么再获取默认的 ThemeResolver。Spring MVC 中默认的 ThemeResolver 为 org.springframework.web.servlet.theme.FixedThemeResolver。

6.7.2 ThemeResolver 的源码

接下来分析 ThemeResolver 源码，如代码清单 6-34 所示。

代码清单 6-34 ThemeResolver 源码

```
public interface ThemeResolver {
    String resolveThemeName(HttpServletRequest request);
    void setThemeName(HttpServletRequest request
            , @Nullable HttpServletResponse response, @Nullable String themeName);
}
```

ThemeResolver 接口中只有 2 个方法。

（1）resolveThemeName：根据请求解析当前的主题名称。在任何情况下，该方法都会返回一个默认的主题。

（2）setThemeName：将当前的主题名称设置为给定的主题名称。如果该方法的实现抛出 UnsupportedOperationException 异常，则说明不支持主题的动态修改。

6.7.3 ThemeResolver 的实现类

ThemeResolver 接口的实现类一共有 4 个，其中 AbstractThemeResolver 为抽象类，如图 6-7 所示。

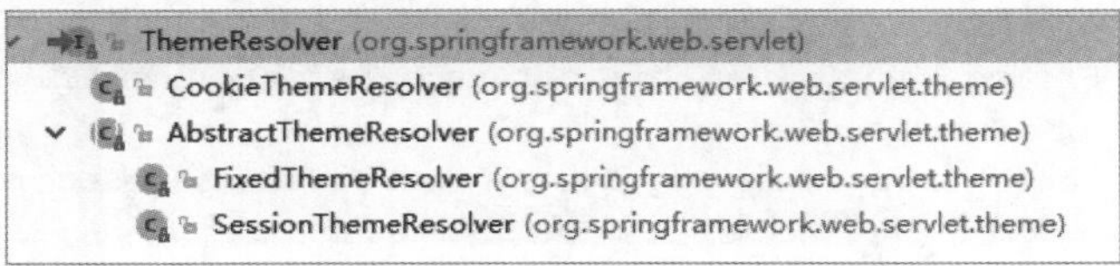

图 6-7 ThemeResolver 接口的实现类

（1）CookieThemeResolver：通过客户端浏览器的 Cookie 来存储主题。

（2）FixedThemeResolver：使用默认的主题。可以通过 defaultThemeName 设置主题名，但不支持动态设置。

（3）SessionThemeResolver：通过会话（Session）来保存主题，每个会话只需要设置一次，该会话中的所有请求都共享主题，多个 Session 之间无法共享。

根据需要，开发者可以自定义 Bean 名称为 ThemeResolver 的 CookieThemeResolver 或 SessionThemeResolver。

6.7.4 对 ThemeSource 主题资源的分析

前面只是介绍了主题解析器，同 LocaleResolver 国际化解析器一样，LocaleResolver 需要配合国际化资源使用。在使用主题解析器时，也需要对应的主题资源。

为了使 Web 项目能够使用主题，需要设置 ThemeSource，通过 ThemeSource 接口来实现 Theme 资源存放。

ThemeSource 接口的源码如代码清单 6-35 所示。

代码清单 6-35 ThemeSource 接口的源码

```
public interface ThemeSource {
    @Nullable
    Theme getTheme(String themeName);
}
```

在该接口中只有一个方法 getTheme，也就是通过主题名称来获取主题资源。虽然 Theme Source 的实现类有多个，但实际上真正实现 getTheme 方法的只有 ResourceBundleThemeSource 类，ResourceBundleThemeSource 类中的 getTheme 实现方法如代码清单 6-36 所示。

代码清单 6-36 ResourceBundleThemeSource 类中 getTheme 实现

```
@Nullable
public Theme getTheme(String themeName){
    //首先判断缓存中是否有该主题，主题第一次被使用时肯定是没有的
```

6 深入理解Spring MVC的九大组件

```
    Theme theme = this.themeCache.get(themeName);
    if (theme == null) {
        //同步使用
        synchronized (this.themeCache) {
            //可能出现并发，再获取一次
            theme = this.themeCache.get(themeName);
            if (theme == null) {
                //根据basenamePrefix和themeName拼接资源名称
                String basename = this.basenamePrefix + themeName;
                //创建MessageSource
                MessageSource messageSource = createMessageSource(basename);
                //创建一个简单的主题，包含主题名和MessageSource
                theme = new SimpleTheme(themeName, messageSource);
                //使用来自此ThemeSource的父级中的一个给定主题初始化MessageSource
                initParent(theme);
                //将主题添加进缓存
                this.themeCache.put(themeName, theme);
                //...省略部分代码
            }
        }
    }
    return theme;
}
```

首先根据主题名创建 MessageSource，然后创建主题，最后返回主题。ResourceBundleThemeSource 类默认在 classpath 的根部（/WEB-INF/classes 目录下）寻找合适的配置文件完成配置。

如果需要将主题资源配置在其他路径下，可以选择定制 ResourceBundleThemeSource 类中的 basenamePrefix。例如，basenamePrefix 配置为"theme."，代表 classes/theme/ 目录下的主题资源，basenamePrefix 中的"."代表路径。或者可以选择自己实现 TemeSource 接口，建议直接使用 ResourceBundleThemeSource 即可。

注意：Bean 的名称需要是 themeSource，这是由于 UiApplicationContextUtils 在 initThemeSource 方法中进行初始化时是使用 themeSource 从上下文中获取 ThemeSource 的。

6.8 MultipartResolver

MultipartResolver（文件上传解析器）是处理文件上传的接口类，主要作用是将一个普通的请求（HttpServletRequest）包装成 MultipartHttpServletRequest 进行实现。在 Servlet 3.0 以前，没有使用 Spring MVC，上传文件是比较麻烦的（需要手动解析请求）。现在 Spring MVC 对其进行了包装，可以很方便地处理文件的上传。

6.8.1 MultipartResolver 接口的源码

MultipartResolver 接口的源码如代码清单 6-37 所示。

代码清单 6-37　MultipartResolver 接口的源码

```
public interface MultipartResolver {
    boolean isMultipart(HttpServletRequest request);
      MultipartHttpServletRequest  resolveMultipart(HttpServletRequest  request)
throws MultipartException;
    void cleanupMultipart(MultipartHttpServletRequest request);
}
```

该接口中有 3 个方法：

（1）isMultipart：判断给定的请求是否包含文件内容，通常情况下是检查请求中的 multipart/form-data，当客户端选择使用 AJAX 请求时，需要特别注意请求类型 type，实际应用中可接受的请求取决于 MultipartResolver 解析器的实现。

（2）resolveMultipart：将给定的 HttpServletRequest 类型请求转换为 MultipartHttpServlet-Request 类型请求（MultipartHttpServletRequest 继承 HttpServletRequest 接口）。

（3）cleanupMultipart：清理 MultipartHttpServletRequest 请求中的资源。

6.8.2　MultipartResolver 的初始化

与前面的组件一样，DispatcherServlet 类在完成初始化的时候，会在 initStrategies 方法中调用 initMultipartResolver 方法初始化 MultipartResolver。DispatcherServlet 类中 initMultipartResolver 方法如代码清单 6-38 所示。

代码清单 6-38　DispatcherServlet 类中的 initMultipartResolver 方法

```
private void initMultipartResolver (ApplicationContext context){
    try {
        //获取上下文中名称为“multipartResolver”的Bean
          this.multipartResolver = context.getBean(MULTIPART_RESOLVER_BEAN_NAME ,
MultipartResolver.class);
        //...省略部分代码
    } catch (NoSuchBeanDefinitionException ex) {
        //如果没有进行配置multipartResolver实现，则直接设置multipartResolver为null
        this.multipartResolver = null;
        //...省略部分代码
    }
}
```

通过上面的源码可以知道，如果开发者需要处理文件上传，则一定要配置名称为 multipart Resolver 且类型为 MultipartResolver 的 Bean。

6.8.3　对 CommonsMultipartResolver 的分析

CommonsMultipartResolver 是 MultipartResolver 的一个实现类，CommonsMultipartResolver 类中的 resolveMultipart 方法如代码清单 6-39 所示。

代码清单 6-39　CommonsMultipartResolver 类中 resolveMultipart 方法

```
public MultipartHttpServletRequest resolveMultipart( final HttpServletRequest
request) throws MultipartException {
    Assert.notNull(request, "Request must not be null");
    //resolveLazily判断是否要延迟解析文件，默认为false。开发者可以配置
    if (this.resolveLazily) {
          //new DefaultMultipartHttpServletRequest(request)，并没有将请求中的数据封装到
MultipartParsingResult 中
        return new DefaultMultipartHttpServletRequest(request) {
              //只有当getMultipartFiles方法被调用，且还没有解析请求数据时，才调用
initializeMultipart方法封装数据
            @Override
            protected void initializeMultipart() {
                MultipartParsingResult parsingResult = parseRequest(request);
                setMultipartFiles(parsingResult.getMultipartFiles());
                setMultipartParameters(parsingResult.getMultipartParameters());
                setMultipartParameterContentTypes(parsingResult
                        .getMultipartParameterContentTypes());
            }
        };
    } else {
        //解析请求的数据，并且将解析结果封装到DefaultMultipartHttpServletRequest对象中
        MultipartParsingResult parsingResult = parseRequest(request);
        return new DefaultMultipartHttpServletRequest(request
                , parsingResult.getMultipartFiles()
                , parsingResult.getMultipartParameters()
                , parsingResult.getMultipartParameterContentTypes());
    }
}
```

从 MultipartParsingResult parsingResult=parseRequest(request) 可以看出，数据是在 parseRequest 方法中解析的。

CommonsMultipartResolver 类中的 parseRequest 方法的源码，如代码清单 6-40 所示。

代码清单 6-40　CommonsMultipartResolver 类中 parseRequest 方法的源码

```
protected MultipartParsingResult parseRequest(HttpServletRequest request) throws
MultipartException {
    //获取请求的编码
    String encoding = determineEncoding(request);
    //根据编码匹配FileUpload 对象
    FileUpload fileUpload = prepareFileUpload(encoding);
    try {
        //通过FileUpload实例解析请求数据，获取文件的列表
            List<FileItem> fileItems = ((ServletFileUpload) fileUpload).
parseRequest(request);
        /*将给定的文件项列表解析为一个MultipartParsingResult，MultipartParsingResult中包
含MultipartFile实例和multipart参数映射*/
        return parseFileItems(fileItems, encoding);
    }
```

```
    //...省略部分捕获异常代码
}
```

parseFileItems 方法在 CommonsMultipartResolver 的父类 CommonsFileUploadSupport 中被实现。CommonsFileUploadSupport 类的部分源码如代码清单 6-41 所示。

代码清单 6-41　CommonsFileUploadSupport 类的部分源码

```
public abstract class CommonsFileUploadSupport {
    ...//
    private boolean preserveFilename = false;
    /** maxUploadSize设置能够上传的文件的最大值(单位：字节)*/
    public void setMaxUploadSize(long maxUploadSize) {
        this.fileUpload.setSizeMax(maxUploadSize);
    }
    /** maxUploadSizePerFile在Spring 4.2以后使用，与maxUploadSize用处一致*/
    public void setMaxUploadSizePerFile(long maxUploadSizePerFile) {
        this.fileUpload.setFileSizeMax(maxUploadSizePerFile);
    }
    /**
     * maxInMemorySize设定文件上传到磁盘之前写入内存的最大值(单位：字节)，超过这个参数将生成临
时文件，默认为10240
     */
    public void setMaxInMemorySize(int maxInMemorySize) {
        this.fileItemFactory.setSizeThreshold(maxInMemorySize);
    }
    /** 设置文件的编码，默认为ISO-8859-1 */
    public void setDefaultEncoding(String defaultEncoding) {
        this.fileUpload.setHeaderEncoding(defaultEncoding);
    }
    /** uploadTempDir设置文件上传的临时路径
     */
    public void setUploadTempDir(Resource uploadTempDir) throws IOException {
        if (!uploadTempDir.exists() && !uploadTempDir.getFile().mkdirs()) {
            throw new IllegalArgumentException("Given uploadTempDir ["
                    + uploadTempDir + "] could not be created");
        }
        this.fileItemFactory.setRepository(uploadTempDir.getFile());
        this.uploadTempDirSpecified = true;
    }
    ...
    /**
     * preserveFilename是Spring 4.3.5之后的参数
     * 设置是否保留客户端发送过来的文件名，默认为false。可以通过CommonsMultipartFile.
getOriginalFilename()获取路径信息
     */
    public void setPreserveFilename(boolean preserveFilename) {
        this.preserveFilename = preserveFilename;
    }
}
```

由代码清单 6-41 可以得知，开发者可以在注入 CommonsMultipartResolver 进行文件上传解析时，设置一些文件上传的限制。

另外还有一个 MultipartResolver 接口的实现类 StandardServletMultipartResolver，因为篇幅有限，留待读者自行研究。

6.9 FlashMapManager

FlashMapManager（重定向管理器）是用于检索和保存 FlashMap 实例的处理接口。Spring 3.1 之后引入了 Flash Attribute 功能，主要解决表单重复提交数据的问题，应用 POST/Redirect/GET（PRG）模式来防止重复提交数据（表单通过 HTTP POST 请求提交之后，用户在服务器端返回之前刷新响应的页面，会导致原始的表单内容重复提交，以及出现一些难以预料的结果），所以采用重定向请求到成功页面的方式。这样用户刷新页面后不会重复提交表单，而是加载新的 GET 请求。但是重定向会引入无法传递请求参数和属性的问题，而 Spring 的 Flash Attribute 就是为了解决重定向之前请求临时存储的问题。

FlashMap 为请求提供了方法，该方法用于获取和存储在另一个请求中使用的数据属性。当从一个 URL 重定向到另一个 URL 时，FlashMap 会在重定向之前将请求数据保存（通常保存在 Session 中），并且在重定向时将数据写入请求，在重定向后立即将原来保存在 Session 中的请求数据删除，而 FlashMapManager 便是用来管理 FlashMap 的。

6.9.1 FlashMapManager 的初始化

FlashMapManager 的初始化和前面的组件一样，也是在 DispatcherServlet 类的 initStrategies 方法中调用 initFlashMapManager 方法完成的。DispatcherServlet 类中的 initFlashMapManager 方法如代码清单 6-42 所示。

代码清单 6-42　DispatcherServlet 类中的 initFlashMapManager 方法

```
private void initFlashMapManager(ApplicationContext context) {
    try {
        //获取名称为flashMapManager的Bean实例
            this.flashMapManager = context.getBean(FLASH_MAP_MANAGER_BEAN_NAME ,
FlashMapManager.class);
        //...省略部分代码
    }
    catch (NoSuchBeanDefinitionException ex) {
            /* 没有获取到名称为“flashMapManager”的Bean，则取默认的FlashMapManager，默认的
FlashMapManager为SessionFlashMapManager */
        this.flashMapManager = getDefaultStrategy(context, FlashMapManager.class);
        //...省略部分代码
    }
}
```

默认的 FlashMapManager 实例的实现是 SessionFlashMapManager。下面看一下 FlashMapManager 的源码。

6.9.2 FlashMapManager 的源码

FlashMapManager 接口在 Spring 中只有一个实现，也就是 AbstractFlashMapManager 抽象类，而 AbstractFlashMapManager 的子类也只有一个，即 SessionFlashMapManager 类。FlashMapManager 接口只有两个方法，其源码如代码清单 6-43 所示。

代码清单 6-43　FlashMapManager 接口的源码

```
public interface FlashMapManager {
    @Nullable
    FlashMap retrieveAndUpdate(HttpServletRequest request
            , HttpServletResponse response);
    void saveOutputFlashMap(FlashMap flashMap, HttpServletRequest request
            , HttpServletResponse response);
}
```

（1）retrieveAndUpdate 方法：获取由先前请求保存的 FlashMap，当该请求参数与之前请求（保存的）的参数匹配时，则将其从底层存储中删除，并且删除其他过期的 FlashMap 实例。该方法在每个请求的方法中均被调用。

（2）saveOutputFlashMap 方法：将给定的 FlashMap 保存到某个底层存储中，并设置其有效期。该方法只有在重定向之前才会被调用。

（3）FlashMapManager 接口：正如前面所说，FlashMapManager 就是用来存储、获取和管理 FlashMap 实例的，保证重定向过程中数据的传输。

retrieveAndUpdate 方法被调用的地方为 DispatchServlet 类的 doService 的方法，如代码清单 6-44 所示。

代码清单 6-44　DispatchServlet 类的 doService 的方法

```
protected void doService(HttpServletRequest request, HttpServletResponse response)
throws Exception {
    ...
    if (this.flashMapManager != null) {
        // 获取FlashMap对象
        FlashMap inputFlashMap = this.flashMapManager.retrieveAndUpdate(request
                , response);
        // 设置FlashMap信息到请求中，方便后面的saveOutputFlashMap获取
        if (inputFlashMap != null) {
            request.setAttribute(INPUT_FLASH_MAP_ATTRIBUTE
                    , Collections.unmodifiableMap(inputFlashMap));
        }
        /*将值保存到request中，就不需要通过客户端浏览器的请求跳转来组装链接并传递参数了*/
        request.setAttribute(OUTPUT_FLASH_MAP_ATTRIBUTE, new FlashMap());
        request.setAttribute(FLASH_MAP_MANAGER_ATTRIBUTE, this.flashMapManager);
    }
```

```
    try {
        // 分发请求
        doDispatch(request, response);
    }
    finally {
        ...
    }
}
```

每次请求都会调用 DispatcherServlet 类中的 doService 方法，如果 flashMapManager!=null，则会调用 retrieveAndUpdate 方法。

具体在什么情况下会调用 saveOutputFlashMap 方法，可参考代码清单 6-45 和代码清单 6-46。

代码清单 6-45　RequestContextUtils 类中的 saveOutputFlashMap 方法

```
public static void saveOutputFlashMap(String location, HttpServletRequest request
, HttpServletResponse response) {
     //获取前面的request.setAttribute(OUTPUT_FLASH_MAP_ATTRIBUTE,FlashMap)设置的
FlashMap
    FlashMap flashMap = getOutputFlashMap(request);
    //...省略部分代码;获取URI组件的不可变集合
     UriComponents uriComponents = UriComponentsBuilder.fromUriString(location).
build();
    //设置URL路径以帮助识别此FlashMap的目标请求
    flashMap.setTargetRequestPath(uriComponents.getPath());
    //添加标识此FlashMap请求的请求参数
    flashMap.addTargetRequestParams(uriComponents.getQueryParams());
     /* 返回FlashMapManager实例保存Flash属性。也就是在request.setAttribute(FLASH_MAP_
MANAGER_ATTRIBUTE, this.flashMapManager)中设置的flashMapManager  */
    FlashMapManager manager = getFlashMapManager(request);
    //...省略部分代码;调用saveOutputFlashMap方法
    manager.saveOutputFlashMap(flashMap, request, response);
}
```

RequestContextUtils 类中的 saveOutputFlashMap 方法是在 Spring 5.0 以后才提供的，而 Request ContextUtils 类中的 saveOutputFlashMap 方法是从 RedirectView 视图中 renderMerged OutputModel 方法调用过来的。

代码清单 6-46　RedirectView 类中的 renderMergedOutputModel 方法

```
protected void renderMergedOutputModel(Map<String, Object> model,
HttpServletRequest request,
        HttpServletResponse response) throws IOException {
    String targetUrl = createTargetUrl(model, request);
    //获取目标URL
    targetUrl = updateTargetUrl(targetUrl, model, request, response);
    //保存flash attributes
    RequestContextUtils.saveOutputFlashMap(targetUrl, request, response);
    //重定向
    sendRedirect(request, response, targetUrl, this.http10Compatible);
}
```

在 Spring 5.0 以前，其实也只是直接在 renderMergedOutputModel 方法中执行了 Request-ContextUtils 类中的 saveOutputFlashMap 方法的逻辑，前后版本差别不大。

6.9.3 对 SessionFlashMapManager 的分析

retrieveAndUpdate 方法和 saveOutputFlashMap 方法都是在 SessionFlashMapManager 的父类 AbstractFlashMapManager 中实现的。saveOutputFlashMap 方法的实现如代码清单 6-47 所示。

代码清单 6-47 saveOutputFlashMap 方法的实现

```
public final void saveOutputFlashMap(FlashMap flashMap, HttpServletRequest request
        , HttpServletResponse response) {
    //...省略部分代码
     String path = decodeAndNormalizePath(flashMap.getTargetRequestPath(),
request);
    //设置目标地址
    flashMap.setTargetRequestPath(path);
    //...省略部分代码；启动此实例的到期时间，默认是180秒
    flashMap.startExpirationPeriod(getFlashMapTimeout());
    //获取用于修改FlashMap列表的互斥锁
    Object mutex = getFlashMapsMutex(request);
    if (mutex != null) {
        //加锁
        synchronized (mutex) {
            List<FlashMap> allFlashMaps = retrieveFlashMaps(request);
            allFlashMaps = (allFlashMaps != null ? allFlashMaps
                    : new CopyOnWriteArrayList<>());
            allFlashMaps.add(flashMap);
            //调用子类的updateFlashMaps实现
            updateFlashMaps(allFlashMaps, request, response);
        }
    }
    else {
        List<FlashMap> allFlashMaps = retrieveFlashMaps(request);
        allFlashMaps = (allFlashMaps != null ? allFlashMaps : new LinkedList<>());
        allFlashMaps.add(flashMap);
        //调用子类的updateFlashMaps实现
        updateFlashMaps(allFlashMaps, request, response);
    }
}
```

该方法保存了 FlashMap，并设置了过期时间。这里为了防止不同重定向请求的数据相互影响，使用了锁进行处理控制。

最后调用了子类的 updateFlashMaps 方法，如代码清单 6-48 所示。

代码清单 6-48 updateFlashMaps 方法

```
protected void updateFlashMaps(List<FlashMap> flashMaps, HttpServletRequest
request
```

```
        , HttpServletResponse response) {
    /*在Session中设置名称为 SessionFlashMapManager.class.getName() + ".FLASH_MAPS"的
flashMaps*/
    WebUtils.setSessionAttribute(request, FLASH_MAPS_SESSION_ATTRIBUTE
            , (!flashMaps.isEmpty() ? flashMaps : null));
}
```

updateFlashMaps 方法的实现就是将数据保存到请求的 Session 中，这样跳转后的请求可以从 Session 中获取数据。

retrieveAndUpdate 方法实现如代码清单 6-49 所示。

代码清单 6-49　retrieveAndUpdate 方法的实现

```
public final FlashMap retrieveAndUpdate(HttpServletRequest request
        , HttpServletResponse response) {
    //从Session中获取List<FlashMap>
    List<FlashMap> allFlashMaps = retrieveFlashMaps(request);
    //...省略部分代码;返回给定列表中包含的过期FlashMap实例列表
    List<FlashMap> mapsToRemove = getExpiredFlashMaps(allFlashMaps);
    //返回在给定列表中与请求匹配的FlashMap
    FlashMap match = getMatchingFlashMap(allFlashMaps, request);
    if (match != null) {
        mapsToRemove.add(match);
    }
    if (!mapsToRemove.isEmpty()) {
        //...省略部分代码;获取互斥锁
        Object mutex = getFlashMapsMutex(request);
        if (mutex != null) {
            //加锁
            synchronized (mutex) {
                //再次从Session中获取List<FlashMap>
                allFlashMaps = retrieveFlashMaps(request);
                if (allFlashMaps != null) {
                    //移除过期的FlashMap
                    allFlashMaps.removeAll(mapsToRemove);
                    //更新Session中的flashMaps
                    updateFlashMaps(allFlashMaps, request, response);
                }
            }
        }
        else {
            //移除过期的FlashMap
            allFlashMaps.removeAll(mapsToRemove);
            //更新Session中的flashMaps
            updateFlashMaps(allFlashMaps, request, response);
        }
    }
    return match;
}
```

retrieveAndUpdate 方法就是获取之前请求中的 FlashMap（FlashAttribute），并将一些过期的 FlashMap 从 Session 中移除，最后更新 Session 中的 FlashMap 列表。

SessionFlashMapManager 类中 retrieveFlashMaps 方法如代码清单 6-50 所示。

代码清单 6-50　SessionFlashMapManager 类中的 retrieveFlashMaps 方法

```
@Nullable
protected List<FlashMap> retrieveFlashMaps(HttpServletRequest request) {
    HttpSession session = request.getSession(false);
     return (session != null ? (List<FlashMap>) session.getAttribute(FLASH_MAPS_
SESSION_ATTRIBUTE) : null);
}
```

retrieveFlashMaps 方法从 Session 中获取之前保存的 FlashMap 列表并返回。

本节主要介绍了 FlashMapManager 中两个方法的调用与实现，而 FlashMap 本质上就是存储的 key-value。但要注意，缓存的时间只有 3min，这里不建议使用 PRG 模式处理表单重复提交的问题，此功能局限性比较大，性能也一般。目前的互联网产品一般都是集群部署的，所以服务端以基于网关、第三方的分布式锁作为避免前端重复提交的处理方案，是更通用的一种做法。

第7章

Spring MVC 基础应用

本章要点

1. Spring MVC 中三种请求转发方式
2. Spring MVC 中的请求重定向
3. 转发与重定向中的绝对路径
4. 静态资源缓存的配置
5. 通过 GzipResourceResolver 压缩静态资源
6. 文件的上传与下载
7. 大文件的下载，避免内容溢出
8. Spring MVC 中多种主题的使用与切换

7.1 转发与重定向

在日常开发中，经常会遇到转发与重定向的情况，如用户登录或者注册成功后的操作，服务器端直接转发到登录或注册成功的页面或者重定向到另外的网站，又或者是错误页面等。

7.1.1 Spring MVC 中的转发方式

转发是服务器的行为，也就是在服务器内部直接将请求转发到相应的 URL。转发的路径必须是在同一个 Web 容器下的 URL，用户是无法感知服务器的转发的。

在 Spring MVC 中，可以使用 String 映射到 View（通过 ViewResolver 解析），也可以通过 ModelAndView 对象进行转发，还可以使用 HttpServletRequest 进行转发。接下来针对以上三种方式进行实例的讲解。

由于第 5 章已经使用 Spring MVC 搭建了一个项目，这里不再重复搭建，代码部分参考第 5 章的项目。

7.1.2 通过 ViewResolver 请求转发

首先在项目的 resources/views/forward 目录下创建 login.jsp 文件，用来作为展示页面，如代码清单 7-1 所示。

代码清单 7-1 forward 目录下 login.jsp 文件内容

```
<%@ page contentType="text/html;charset=UTF-8" language="java" %>
<html>
<head>    <title>MVC Title</title></head>
<body>
<pre>演示转发</pre><br/>
        演示转发,username=${username}
</body>
</html>
```

这里的 username 用来演示转发请求中参数的传递。

下面在项目的 controller 包下创建 StringForwardController 类，如代码清单 7-2 所示。

代码清单 7-2 StringForwardController 类

```
/** 演示通过ViewResolver进行请求转发 */
@Controller
@RequestMapping("stringForward")
public class StringForwardController {
      private Logger logger = LoggerFactory.getLogger(StringForwardController.
class);
    /** 返回ModelAndView，映射到forward/login页面 */
    @RequestMapping(value="/login")
```

```
    public ModelAndView login(String username){
        logger.info("login...username={}",username);
        ModelAndView modelAndView = new ModelAndView("forward/login");
        modelAndView.addObject("username",username);
        return modelAndView;
    }
    /** 转发到login方法，/代表使用绝对路径访问，在同一个Web容器中 */
    @RequestMapping(value="/backslashForward")
    public  String backslashForward(){
        logger.info("backslashForward...");
        return "forward:/stringForward/login";
    }
    /**
     * 转发到login方法
     * 由于在一个Controller中，可以不使用“/”代表相对路径访问
     * 转发会自动携带请求中的参数
     * 也就是同一个request在服务器内部进行转发
     */
    @RequestMapping(value="/parameterForward")
    public  String parameterForward(String username){
        logger.info("parameterForward...username={}",username);
        return "forward:login";
    }
}
```

StringForwardController 类中实现了 3 个方法：

（1）login 方法只是将 URL 映射到对应的 View。

（2）backslashForward 方法在转发中使用了“forward:/stringForward/login”，“/”表示的是项目中的绝对访问路径。

（3）parameterForward 方法省略了该 Controller 的前缀 URL，只能跳转到 StringForwardController 类。在 parameterForward 方法中演示带参数的转发。

演示的过程可以通过访问 http://localhost:8080/stringForward/backslashForward 和 http://localhost:8080/stringForward/parameterForward?username=test 查看，结合日志可以看出转发的过程。由于转发是直接在 Web 服务器内部将请求（request 和 response）的处理权转交给另外的一个 Controller，所以参数属性都可以带过去，传输的数据不会丢失。由于是在 Web 服务器内部的转发，所以转发是不能跨 Web 服务器进行的。

7.1.3　通过 ModelAndView 请求转发

下面使用 ModelAndView 对象演示请求的转发。创建 ModelAndViewForwardController 类，如代码清单 7-3 所示。

代码清单 7-3　ModelAndViewForwardController 类

```
/** 演示通过ModelAndView进行请求转发 */
@Controller
```

```
@RequestMapping("forward")
public class ModelAndViewForwardController {
    private Logger logger = LoggerFactory.getLogger(ModelAndViewForwardController.
class);
    /** 返回ModelAndView */
    @RequestMapping(value = "/login")
     public ModelAndView login2(HttpServletRequest request, @RequestParam String
username) {
        logger.info("login...request.getAttribute(\"username\")={}," +
                        "username={},request.getParameter(\"username\")={}",
                request.getAttribute("username"), username,
                request.getParameter("username"));
        ModelAndView modelAndView = new ModelAndView("/forward/login");
        modelAndView.addObject("username", request.getParameter("username"));
        return modelAndView;
    }
    /**
     * 转发到login方法，"/"代表使用绝对路径访问，在同一个Web容器中
     * 使用HttpServletRequest设置值进行传递
     */
    @RequestMapping(value = "/testForward")
    public ModelAndView testForward(HttpServletRequest request, String username) {
            logger.info("testForward...username={},request.
getAttribute(\"username\")={}," +
                        "request.getParameter(\"username\")={}",
                username, request.getAttribute("username"),
                request.getParameter("username"));
            /*使用此种方式虽然能够传递值到下一个方法中，但是注意，如果使用了转发，那么这里
HttpServletRequest设置的参数是无法通过RequestParam注解获取该方法中设置的Attribute的。同样，
在下一个方法中需要使用HttpServletRequest的getAttribute方法获取该方法中设置的Attribute*/
        request.setAttribute("username", "springmvc");
        ModelAndView model = new ModelAndView("forward:/forward/login");
        return model;
    }
    /** 使用ModelAndView设置值进行传递 */
    @RequestMapping(value = "/testForward2")
    public ModelAndView testForward2(String username) {
        logger.info("testForward2...username={}", username);
        ModelAndView model = new ModelAndView("forward:login");
        /*使用此种方式也是可以将值传递到下一个请求方法中。实际上就是Spring MVC内部使用ModelMap
(本质上就是HashMap)进行维护key-value。简单地理解为request.setAttribute即可。建议使用该种方
式*/
        model.addObject("username", username);
        return model;
    }
}
```

接下来访问 http://localhost:8080/forward/testForward?username=test，控制台输出的日志如图 7-1 所示。

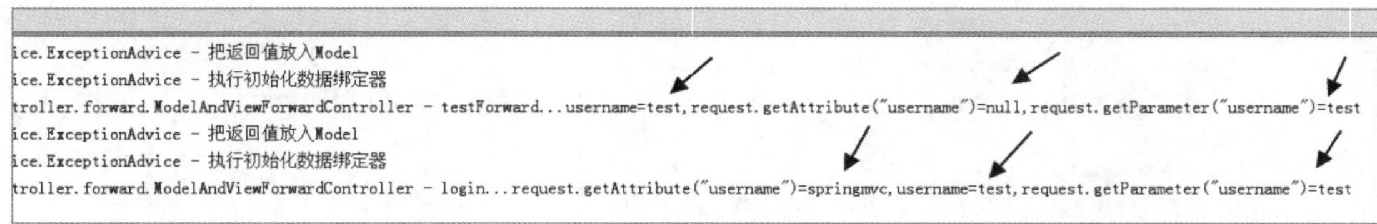
ice.ExceptionAdvice - 把返回值放入Model
ice.ExceptionAdvice - 执行初始化数据绑定器
troller.forward.ModelAndViewForwardController - testForward...username=test,request.getAttribute("username")=null,request.getParameter("username")=test
ice.ExceptionAdvice - 把返回值放入Model
ice.ExceptionAdvice - 执行初始化数据绑定器
troller.forward.ModelAndViewForwardController - login...request.getAttribute("username")=springmvc,username=test,request.getParameter("username")=test

图 7-1　控制台输出的日志

从输出结果可知，从路径中传递的值是 username=test，在 testForward 方法中使用 getAttribute("username") 是无法获取到值的。只能通过 getParameter("username") 获取 POST/GET 等请求方式传过来的参数值。接下来使用 request.setAttribute("username","springmvc")，在 login 方法中，方法参数 username 的值为 test，getParameter("username") 的值也为 test，request.getAttribute("username") 获取的值是 springmvc。

在使用 getAttribute 获取值之前，必须由容器内的对象（HttpServletRequest、HttpSession 等）调用 setAttribute 方法设置值。request.setAttribute 方法和 request.getAttribute 方法传递的数据只会存在于 Web 容器内部，并且只会在同一个 request 中共享。另外，通过 session.setAttribute 方法设置的值，能够在 session 所有的请求中共享。

7.1.4　通过 HttpServletRequest 请求转发

创建 RequestForwardController 类，如代码清单 7-4 所示。

代码清单 7-4　RequestForwardController 类

```
/** 使用HttpServletRequest进行转发 */
@Controller
@RequestMapping("requestForward")
public class RequestForwardController {
    private Logger logger = LoggerFactory.getLogger(RequestForwardController.
class);
   /** 返回ModelAndView */
   @RequestMapping(value = "/login")
    public ModelAndView login(HttpServletRequest request, @RequestParam String
username) {
          logger.info("RequestForwardController.login...request.getAttribute(\
"username\")={},username={},request.getParameter(\"username\")={}", request.
getAttribute("username"), username, request.getParameter("username"));
       ModelAndView modelAndView = new ModelAndView("/forward/login");
       modelAndView.addObject("username", request.getAttribute("username"));
       return modelAndView;
   }
   /**
     * 使用HttpServletRequest方式进行转发，需要HttpServletRequest类和HttpServlet-
Response类
    * 使用绝对路径进行转发,可以转发到不同路径的Controller类
    */
   @RequestMapping(value = "/testForward")
```

```
    public void testForward(HttpServletRequest request, HttpServletResponse
response, String username) throws ServletException, IOException {
        logger.info("testForward...username={}", username);
        request.getRequestDispatcher("/forward/login").forward(request, response);
    }
    /** 使用相对路径进行转发，前缀匹配类上的路径 */
    @RequestMapping(value = "/testForward1")
    public void testForward4(HttpServletRequest request, HttpServletResponse
response, String username) throws ServletException, IOException {
        logger.info("testForward1...username={}", username);
        request.getRequestDispatcher("login").forward(request, response);
    }
}
```

与前面的知识点基本一致，可以访问 http://localhost:8080/requestForward/test Forward1?username=test 和 http://localhost:8080/requestForward/testForward?username=test 进行测试。

以上共介绍了三种转发方式，推荐读者使用 ModelAndView 对象的转发方式，并且建议使用 ModelAndView 中的 addObject 方法传递参数值。

7.1.5 Servlet 中的重定向方式

常用的重定向方式有两种，分别通过 sendRedirect 方法和字符串映射实现。sendRedirect 方法通过 HttpServletResponse 对象调用 sendRedirect 方法传递目标 URL 参数。例如，response.sendRedirect("/otherRedirect/login")。通过字符串映射实现重定向的方式是直接返回由 "redirect:" 开头的映射路径。例如，return "redirect:/otherRedirect/login"。

7.1.6 Spring MVC 通过 ModelAndView 实现重定向

下面的示例是通过 ModelAndView 对象实现重定向，先在 resources/redirect 目录下创建一个演示页面，如代码清单 7-5 所示。

代码清单 7-5 创建 resources/redirect/login.jsp 演示页面

```
<%@ page contentType="text/html;charset=UTF-8" language="java" %>
<html>
<head><title>MVC Title</title></head>
<body>
<pre>演示重定向</pre><br/>
        演示重定向,username=${username},password=${password}
</body>
</html>
```

接下来创建 ModelAndViewRedirectController 类，如代码清单 7-6 所示。

代码清单 7-6 使用 ModelAndView 实现重定向

```
/** 使用ModelAndView实现重定向 */
@Controller
```

```
@RequestMapping("modelAndViewRedirect")
public class ModelAndViewRedirectController {
     private Logger logger = LoggerFactory.getLogger(ModelAndViewRedirectControll
er.class);
    /** 返回ModelAndView */
    @RequestMapping(value = "/login")
    public ModelAndView login(String username) {
        logger.info("login...username={}", username);
        ModelAndView modelAndView = new ModelAndView("/redirect/login");
        modelAndView.addObject("username", username);
        return modelAndView;
    }
    /** 使用绝对路径 */
    @RequestMapping(value = "/testRedirect")
    public ModelAndView testRedirect(String username) {
        logger.info("testRedirect...username={}", username);
             return new ModelAndView("redirect:/modelAndViewRedirect/
login?username="+username);
    }
    /** 不带"/",也就是前面的路径匹配类路径 */
    @RequestMapping(value = "/testRedirect2")
    public ModelAndView testRedirect2(String username) {
        logger.info("testRedirect2...username={}", username);
        ModelAndView modelAndView = new ModelAndView("redirect:login");
        /* 重定向中使用该种方式传值时，相当于在URL链接后拼接参数和值进行传递
     不建议使用拼接的方式进行传值，这样可能会出现字符值乱码问题，以及数据泄露的问题 */
        modelAndView.addObject("username", "testRedirect");
        return modelAndView;
    }
    /**
     * 演示重定向传递参数
     * ModelAttribute注解中的name(参数名称)必须填写，负责获取重定向传递的参数
     */
    @RequestMapping(value = "/login3")
    public ModelAndView login3(@ModelAttribute("username") String name,
                               @ModelAttribute("password") String password) {
        logger.info("login3...username={},password={}", name, password);
        ModelAndView modelAndView = new ModelAndView("/redirect/login");
        modelAndView.addObject("username", name);
        modelAndView.addObject("password", password);
        return modelAndView;
    }
    /**
     * 这里使用ModelAttribute注解接收参数，可使用addAttribute和addFlashAttribute传递参数，
但方式略有不同
     */
    @RequestMapping(value = "/testRedirect3")
```

```
    public ModelAndView testRedirect3(RedirectAttributes attributes, String
username) {
        logger.info("testRedirect3...username={}", username);
        // 使用addAttribute方式传值，也是在URL后面拼接参数
        attributes.addAttribute("username", username);
        /* 使用addFlashAttribute，也就是将参数存储到FlashMap中
          参数先保存在Session中，等待下一次请求访问，缓存时间为180s，并且在访问后会删除添加到
attributes的参数*/
        attributes.addFlashAttribute("password", "1234");
        return new ModelAndView("redirect:/modelAndViewRedirect/login3");
    }
    /** 不在方法参数上绑定参数，使用RequestContextUtils类的getInputFlashMap方法传递
request并获取传递的参数*/
    @RequestMapping(value = "/login4")
    public ModelAndView login4(HttpServletRequest request) {
        Map<String, ?> map = RequestContextUtils.getInputFlashMap(request);
        logger.info("login4...map={}", map);
        ModelAndView modelAndView = new ModelAndView("/redirect/login");
        modelAndView.addObject("username", map.get("username"));
        return modelAndView;
    }
    /** 使用RequestContextUtils获取传递过去的参数值  */
    @RequestMapping(value = "/testRedirect4")
    public ModelAndView testRedirect4(RedirectAttributes attributes, String
username) {
        logger.info("testRedirect4...username={}", username);
        attributes.addFlashAttribute("username", username);
        return new ModelAndView("redirect:/modelAndViewRedirect/login4");
    }
}
```

在浏览器中访问 http://127.0.0.1:8080/modelAndViewRedirect/testRedirect?username=springmvc 和 http://127.0.0.1:8080/modelAndViewRedirect/testRedirect2?username=springmvc。URL 请求参数是通过在 URL 后面拼接参数将值传递过去的，这样传递参数的缺点显而易见——容易出现乱码，明文展示在链接中，无法直接传递对象。

接下来访问 http://127.0.0.1:8080/modelAndViewRedirect/testRedirect3?username=springmvc 和 http://127.0.0.1:8080/modelAndViewRedirect/testRedirect4?username=springmvc，并且使用两种不同方式接收重定向传递过来的参数。可以发现，使用 addFlashAttribute 方法传递的参数不会暴露在 URL 中，这是因为参数是存在于 Session 中的。login4 方法中接收参数的 Map 其实就是 FlashMap 类型的 Map。

7.1.7 通过 RedirectView 对象实现重定向

使用 RedirectView 对象实现重定向，也就是不用再写重定向的字符串“redirect:”来标志重定向，如代码清单 7-7 所示。

代码清单 7-7　通过 RedirectView 对象实现重定向

```
/** 通过RedirectView对象实现重定向 */
@Controller
@RequestMapping("redirectViewRedirect")
public class RedirectViewRedirectController {
     private Logger logger = LoggerFactory.getLogger(RedirectViewRedirectControll
er.class);
    /** 仅作为展示的返回页面 */
    @RequestMapping(value = "/login")
    public ModelAndView login() {
        logger.info("login...");
        return new ModelAndView("/redirect/login");
    }
    /** 通过RedirectView实现重定向，使用绝对路径进行重定向 */
    @RequestMapping(value = "/redirect")
    public RedirectView redirect() {
        logger.info("redirect...");
        return new RedirectView("/redirectViewRedirect/login");
    }
    /** 使用相对路径进行重定向 */
    @RequestMapping(value = "/redirect1")
    public RedirectView redirect1() {
        logger.info("redirect1...");
        return new RedirectView("login");
    }
}
```

接下来访问 http://localhost:8080/redirectViewRedirect/redirect 和 http://localhost:8080/redirectViewRedirect/redirect1 查看重定向结果。

7.1.8　转发与重定向中的绝对路径

绝对路径是指以“/”开头的路径信息。如果在两个项目路径中转发请求，当使用绝对路径时，则需要加上项目路径；如果在同一个项目中转发请求，则可以不加项目名。

由于重定向时服务器不确定该请求为项目内部还是项目外部的请求，所以在重定向时，绝对路径是从项目名（上下文路径）开始的。

这里使用 RedirectView 时对重定向进行了增强，重定向也可以省略上下文路径，是因为有了 contextRelative 参数，RedirectView 自动增加了上下文路径，如 return new RedirectView("/redirectViewRedirect/login",true)。在本项目中，由于上下文路径为空，所以使用不使用上下文路径都是一样的。

下面简单分析 RedirectView 如何增加上下文路径。首先找到 Spring MVC 渲染 View 的地方，当访问页面时，请求调用的步骤如图 7-2 所示。

由于这里使用的 View 是 RedirectView 类，所以会调用 RedirectView 类的 renderMergedOutputModel 方法，如代码清单 7-8 所示。

```
renderMergedOutputModel:314, RedirectView (org.springframework.web.servlet.view)
render:314, AbstractView (org.springframework.web.servlet.view)
render:1325, DispatcherServlet (org.springframework.web.servlet)
processDispatchResult:1069, DispatcherServlet (org.springframework.web.servlet)
doDispatch:1008, DispatcherServlet (org.springframework.web.servlet)
doService:925, DispatcherServlet (org.springframework.web.servlet)
processRequest:974, FrameworkServlet (org.springframework.web.servlet)
doGet:866, FrameworkServlet (org.springframework.web.servlet)
service:620, HttpServlet (javax.servlet.http)
```

图 7-2 请求调用的步骤

代码清单 7-8 RedirectView 类的 renderMergedOutputModel 方法

```
protected void renderMergedOutputModel(Map<String, Object> model,
HttpServletRequest request,HttpServletResponse response) throws IOException {
    // 构造目标路径
    String targetUrl = createTargetUrl(model, request);
    targetUrl = updateTargetUrl(targetUrl, model, request, response);
    // 保存Flash Attributes，用来重定向传递参数
    RequestContextUtils.saveOutputFlashMap(targetUrl, request, response);
    // 重定向
    sendRedirect(request, response, targetUrl, this.http10Compatible);
}
```

下面分析构造目标路径。RedirectView 类中的 createTargetUrl 方法如代码清单 7-9 所示。

代码清单 7-9 RedirectView 类中的 createTargetUrl 方法

```
protected final String createTargetUrl(Map<String, Object> model,
HttpServletRequest request) throws UnsupportedEncodingException {
    // 准备目标路径
    StringBuilder targetUrl = new StringBuilder();
    String url = getUrl();
    Assert.state(url != null, "'url' not set");
    /*在这里增加上下文的路径。例如，项目的上下文为“demo”，访问根路径为“http://127.0.0.1:8080/
demo”，如果设置contextRelative为true且使用相对路径“/redirectViewRedirect/login”，那么路
径会变为“/demo/redirectViewRedirect/login”。在本实例中，由于上下文路径为空，所以看不出效果*/
    if (this.contextRelative && getUrl().startsWith("/")) {
        // 加上项目的上下文路径
        targetUrl.append(getContextPath(request));}
    targetUrl.append(getUrl());
    String enc = this.encodingScheme;
//...省略部分代码
    return targetUrl.toString();
}
```

contextRelative 变量用来判断是否将给定的 URL 解释为相对于当前 ServletContext 的 URL（即是否使用的是上下文的相对路径，默认是 false）。

将 contextRelative 设置为 true，这样的好处是，开发者在同一个项目中进行重定向时，不用

再写项目的上下文名称了（不建议直接在路径中写上下文名称，因为路径可能会变，可以使用 request.getContextPath() 代替）。

7.2 静态资源缓存与加载 GZIP 资源

静态资源（CSS 文件、JS 文件、HTML 文件等）一般不会经常改动，完全可以缓存起来，在没有更改之前都可以使用缓存文件，另外还可以进行压缩，优化了静态资源的访问，对于项目的访问速度提升有一些帮助。如果静态资源很多，那么优化效果会更加明显。本节讲解如何在 Spring MVC 中配置静态资源的缓存及压缩。

由于项目已经配置了静态资源的映射，以及创建了 static 目录，所以可以直接在 static 下新建 bootstrap 路径，引入一些静态资源，并引入 GZIP 压缩文件。访问 https://github.com/chenhaoxiang/uifuture-ssm/tree/master/ssm-spring-mvc/src/main/resources/static 下载测试的资源文件，如图 7-3 所示。

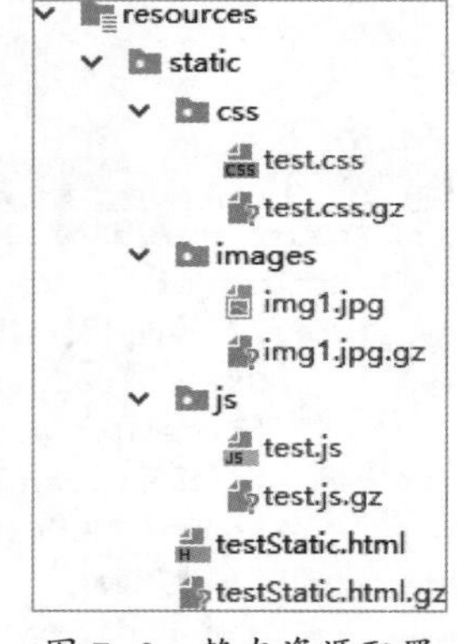

图 7-3　静态资源配置

7.2.1 配置静态资源的缓存

由于使用的是编程式配置，使用缓存非常简单。在 MvcConfig 类的 addResourceHandlers 方法中设置缓存时间即可（Spring 4.2 以后），如代码清单 7-10 所示。

代码清单 7-10　配置静态资源缓存

```
public void addResourceHandlers(ResourceHandlerRegistry registry) {
    // 配置URL中的路径
    registry.addResourceHandler("/static/**")
            // 配置静态文件的路径
            .addResourceLocations("classpath:/static/")
            // 设置静态资源的缓存时间，单位为秒
} .setCachePeriod(60*60*24*365);
```

使用 setCachePeriod 设置时间即可。接下来创建一个动态页面加载静态资源。在 views 目录下创建 cache/index.jsp 文件，并且配置一个 Controller 跳转至该页面，如代码清单 7-11 和代码清单 7-12 所示。

代码清单 7-11　创建 index.jsp 文件

```
<%@ page contentType="text/html;charset=UTF-8" language="java" %>
<%@taglib prefix="c" uri="http://java.sun.com/jsp/jstl/core" %>
<html>
<head>
<title>Title</title>
<link rel="stylesheet"
```

```
        href="<c:url value='/static/css/test.css'/>" />
</head>
<body>
<h1 class="title">演示静态资源的缓存</h1>
<img src="<c:url value='/static/images/img1.jpg'/>">
<script src="<c:url value='/static/js/test.js'/>"></script>
</body>
</html>
```

代码清单 7-12　创建 StaticCacheController 类

```
@Controller
@RequestMapping("cache")
public class StaticCacheController {
    /**跳转到JSP页面，里面加载了静态资源 */
    @RequestMapping("index")
    public ModelAndView cache(){
        return new ModelAndView("/cache/index");
    }
}
```

由于跳转到 index.jsp 页面的 Controller 没有任何的业务处理逻辑，所以可以使用 addViewController 方法进行简单的页面跳转。在 MvcConfig 类中添加如代码清单 7-13 所示的代码。

代码清单 7-13　添加 addViewControllers 方法

```
/**
 * 配置直接映射、定义ParameterizableViewController的简单方式
 * 当该控制器被访问的时候没有任何操作，如果只是进行相应的请求映射，则可以使用该方法
 */
@Override
public void addViewControllers(ViewControllerRegistry registry) {
    // 添加访问路径
    registry.addViewController("/cacheController/index")
            // 设置视图名称——对应文件路径
            .setViewName("/cache/index");
}
```

进行以上设置后，访问 /cacheController/index 时会被映射到 /cache/index.jsp 文件。

下面访问 http://localhost:8080/cache/index，在浏览器中打开开发者模式，可以看到资源的加载时间等信息，如图 7-4 所示。

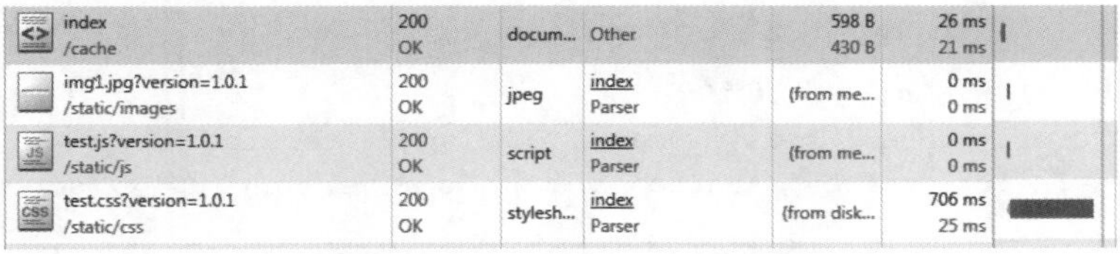

index /cache	200 OK	docum...	Other	598 B 430 B	26 ms 21 ms
img1.jpg?version=1.0.1 /static/images	200 OK	jpeg	index Parser	(from me...	0 ms 0 ms
test.js?version=1.0.1 /static/js	200 OK	script	index Parser	(from me...	0 ms 0 ms
test.css?version=1.0.1 /static/css	200 OK	stylesh...	index Parser	(from disk...	706 ms 25 ms

图 7-4　静态资源缓存后的加载时间

可以看出，经过缓存之后，访问时间大大减少。但是应该注意，开启缓存后当服务器资源被

修改，用户会请求不到最新的资源。可以使用版本号解决这个问题，在每次修改静态文件后修改版本号即可获取最新资源。

新建一个静态资源版本号管理类 StaticResourcesVersion，在构造方法中使用 ServletContext 接口设置属性，如代码清单 7-14 所示。

代码清单 7-14　StaticResourcesVersion 类

```
/** 配置静态资源的版本号 */
@Component
public class StaticResourcesVersion {
    /** 版本号，实际项目中设置在配置文件中 */
    private static String version = "1.0.1";
    /** 全局版本号名称，实际项目中设置在配置文件中 */
    private static String name="version";
    /** 设置全局变量，使用EL表达式即可获取版本号 */
    public StaticResourcesVersion(ServletContext servletContext) {
        servletContext.setAttribute(name,name+"="+version);
    }
    public static String getVersion() {
        return version;
    }
    public static String getName() {
        return name;
    }
}
```

可以在 Spring MVC 初始化时设置 ServletContext 对象的值，也可以使用 AOP 拦截请求设置版本号（不建议）。

既然设置了全局的模板变量，那么在前端页面使用时就需要通过 EL 表达式来获取版本，版本控制的静态页面引入如代码清单 7-15 所示。

代码清单 7-15　版本控制的静态页面引入

```
<%@ page contentType="text/html;charset=UTF-8" language="java" %>
<%@taglib prefix="c" uri="http://java.sun.com/jsp/jstl/core" %>
<html>
<head>
<title>Title</title>
<link rel="stylesheet"
        href="<c:url value='/static/css/test.css?${version}'/>" />
</head>
<body>
<h1 class="title">演示静态资源的缓存</h1>
<img src="<c:url value='/static/images/img1.jpg?${version}'/>">
<script src="<c:url value='/static/js/test.js?${version}'/>"></script>
</body>
</html>
```

使用 ${version} 引入设置的版本控制属性即可，如果更改了静态资源，那么修改版本号即可。

7.2.2 通过 GzipResourceResolver 压缩静态资源

对于数据量大的静态资源，第 1 次加载文件需要花费很长时间。为了降低传输的数据量，就需要对资源进行压缩。现代浏览器基本上都支持 GZIP 协议、支持客户端解压处理，主要目的还是降低传输量，提升速度，如将前面的 bootstrap.css 压缩为 bootstrap.css.gz。

GZIP 压缩是 Linux 中常用的压缩方式。在 Linux 系统下，可以使用命令一键递归目录中的文件并分别压缩目录中的每个文件。“gzip -r /java/static/” 命令的作用是遍历压缩 /java/static/ 目录下的所有文件。**提示**：源文件会被删除。

在文件被压缩的情况下，如果直接加载 bootstrap.css，那么 Spring MVC 肯定是无法正常处理的。若要在请求 bootstrap.css 时能正确返回 bootstrap.css.gz，就需要添加一个处理压缩的资源解析器（GzipResourceResolver）。

代码清单 7-16 所示为 MvcConfig 类中的 addResourceHandlers 方法，可在该方法中设置 resourceChain 为 true（当用户请求过来时，会优先匹配 CachingResourceResolver 缓存资源解析器），然后添加 GzipResourceResolver 资源链即可。

代码清单 7-16　MvcConfig 类中的 addResourceHandlers 方法

```
public void addResourceHandlers(ResourceHandlerRegistry registry) {
    // 配置url中的路径
    registry.addResourceHandler("/static/**")
            // 配置静态文件的路径，最后需要以“/”结尾，否则访问不到静态资源
            .addResourceLocations("classpath:/static/")
            // 设置静态资源的缓存时间，单位为s
            .setCachePeriod(60*60*24*365)
            .resourceChain(true)
            .addResolver(new GzipResourceResolver())
            // 如果是资源处理链，则将默认的PathResourceResolver路径资源解析器放到最后（Spring推荐）
            .addResolver(new PathResourceResolver())
    ;
}
```

注意：并不是 GzipResourceResolver 将静态资源压缩，而是需要开发人员添加压缩文件将文件压缩，GzipResourceResolver 只是负责映射到对应的 GZIP 压缩文件（以 .gz 为后缀的压缩文件，注意与 ZIP 压缩文件区分），并且正确解析。所以开发人员需要到对应的路径下添加 GZIP 文件（适用一些比较大的资源文件）。如果未找到压缩文件，那么还是会加载压缩当前的源文件。另外，也可以多次调用 registry.addResourceHandler，分别配置资源处理器。

先查询 test.js 的源文件大小。文件大小和对应的压缩文件大小如图 7-5 所示。

在开发者模式下访问 http://127.0.0.1:8080/cache/index。在第 1 次加载的情况下，可以看到加载的文件大小，如图 7-6 所示。

名称	修改日期	类型	大小
test.js	2018/8/19 下午 ...	JScript Script 文件	54 KB
test.js.gz	2018/8/19 下午 ...	WinRAR 压缩文件	1 KB

图 7-5　test.js 文件大小和压缩文件大小

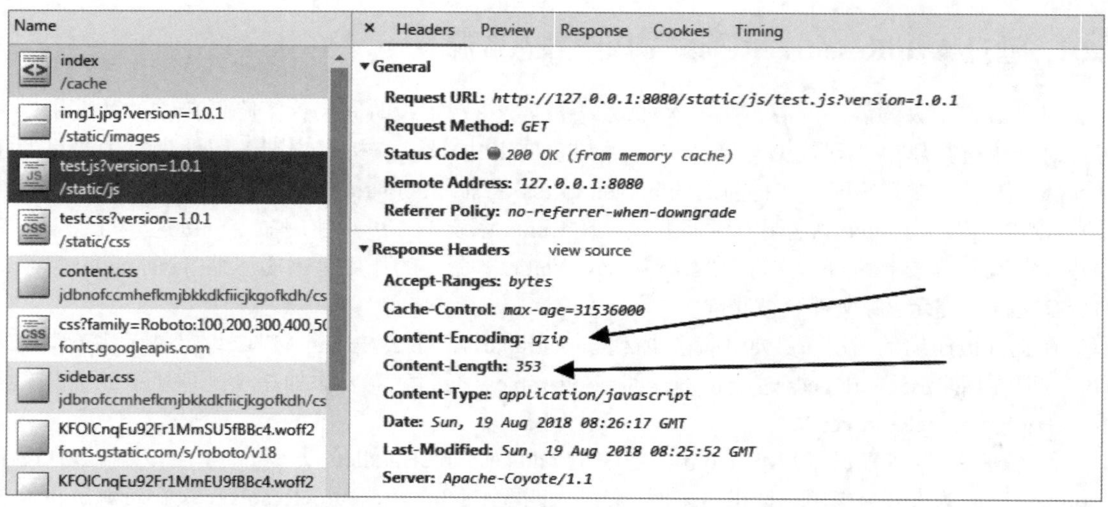

图 7-6　访问后接收 test.js 的编码类型和数据大小

从 Response Headers 中可以看到 Content-Encoding 为 gzip，且 Content-Length 为 353Bytes，数据传输量大大减少。推荐使用该种方法压缩大文件，将压缩文件放在源文件的同目录下即可。

7.2.3 GZIP 一键压缩工具

为了方便压缩静态文件，下面编写一个 GZIP 的压缩工具类，能够一键递归压缩“resources/static”目录下的所有文件并保留源文件。首先创建一个文件工具类，如代码清单 7-17 所示。

代码清单 7-17　File 工具类

```
/** 文件工具类 ——辅助GZIP压缩 */
public class FileUtils {
    /** 保存字节数组到文件 */
    public static void saveFile(String filePath,byte[] data) throws IOException {
        if(data != null){
            File file  = new File(filePath);
            if(file.exists()){
                // 原文件若存在，就先删除原文件
                file.delete();
            }
            FileOutputStream fos = null;
            fos = new FileOutputStream(file);
            fos.write(data,0,data.length);
            fos.flush();
            fos.close();
        }
    }
    /** 从文件名中获取文件的字节数组 */
    public static byte[] readFileName(String filename) throws IOException {
        return readInputStream(new FileInputStream(filename));
```

```
    }
    /** 从输入流中获取字节数组 */
     public static byte[] readInputStream(InputStream inputStream) throws
IOException {
        byte[] buffer = new byte[1024];
        int len = 0;
        ByteArrayOutputStream bos = new ByteArrayOutputStream();
        while ((len = inputStream.read(buffer)) != -1) {
            bos.write(buffer, 0, len);
        }
        bos.close();
        return bos.toByteArray();
    }
    /**
     * 递归取到当前目录所有文件
     * @return 返回的是文件的绝对路径
     */
    public static List<String> getFilesNames(String dir) {
        List<String> lstFiles = new ArrayList<>();
        File[] files = new File(dir).listFiles();
        if(files==null || files.length==0){
            return lstFiles;
        }
        for (File f : files) {
            // 如果是目录，则继续递归
            if (f.isDirectory()) {
                lstFiles.addAll(getFilesNames(f.getAbsolutePath()));
            } else {
                //否则加入list
                lstFiles.add(f.getAbsolutePath());
            }
        }
        return lstFiles;
    }
}
```

代码清单 7-18 所示为 GZIP 的压缩工具类。

代码清单 7-18　GZIP 压缩工具类

```
/** GZIP压缩工具类 */
public class GZIPUtils {
    /** GZIP压缩文件后缀 */
public static final String GZIP_FILE_SUFFIX = ".gz";
    /** 静态文件在项目下的路径 */
    private static final String STATIC_PATH = "/src/main/resources/static/";
    /** 运行该方法即可将静态资源遍历压缩，并且过滤不想压缩的文件 */
    public static void main(String[] args) throws IOException {
        String contestPath = new File("").getCanonicalPath();
        contestPath = contestPath+STATIC_PATH;
```

```
        System.out.println(contestPath);
        gzipPath(contestPath);
    }
    /** 排除 GZIP_FILE_SUFFIX 后缀的文件压缩 */
    public static void gzipPath(String path) throws IOException {
        Set<String> stringSet = new HashSet<>();
        stringSet.add(GZIP_FILE_SUFFIX);
        gzipPath(path,stringSet);
    }
    /**
     * 递归压缩目录下的所有文件，并且在文件同目录下创建压缩文件
     * @param excludeSuffix  过滤文件后缀名包含在Set中的文件
     */
     public static void gzipPath(String path, Set<String> excludeSuffix) throws
IOException {
        if(path==null || path.trim().length()==0){
            return;
        }
        // 遍历所有的文件
        List<String> stringList = FileUtils.getFilesNames(path);
        for(String filename:stringList){
            // 如果文件的后缀名为需要排除的后缀，则跳过
             String suffix = filename.substring(filename.lastIndexOf("."),filename.
length());
            if(excludeSuffix.contains(suffix)){
                continue;
            }
            //获取文件的字节数组
            byte[] fileBytes = FileUtils.readFileName(filename);
            //压缩
            byte[] gzipBytes = gzip(fileBytes);
            //存储文件
            FileUtils.saveFile(filename+GZIP_FILE_SUFFIX,gzipBytes);
        }
    }
    /** 将字节数组进行GZIP压缩  */
    public static byte[] gzip(byte[] fileBytes) throws IOException {
        if (fileBytes == null || fileBytes.length == 0) {
            return null;
    }
        ByteArrayOutputStream out = new ByteArrayOutputStream();
        GZIPOutputStream gzip;
        gzip = new GZIPOutputStream(out);
        gzip.write(fileBytes);
        gzip.close();
        return out.toByteArray();
    }
    /** 解压 */
    public static byte[] ungzip(byte[] bytes) throws IOException {
```

```
        if (bytes == null || bytes.length == 0) {
            return null;
        }
        ByteArrayOutputStream out = new ByteArrayOutputStream();
        ByteArrayInputStream in = new ByteArrayInputStream(bytes);
        GZIPInputStream ungzip = new GZIPInputStream(in);
        byte[] buffer = new byte[512];
        int n;
        while ((n = ungzip.read(buffer)) >= 0) {
            out.write(buffer, 0, n);
        }
        return out.toByteArray();
    }
}
```

这样，只需定义好静态文件，在项目打包前，将该类运行一遍就可以将静态资源进行 GZIP 压缩。更好的方法就是在单元测试方法中添加该压缩静态资源的测试方法，这样在使用 Maven 打包时，只要不关闭单元测试，就不用再运行该类的 main 方法了。

7.3 文件上传与下载

从 6.8 节可以知道，如果要上传文件，则需要文件上传的解析器，也就是需要配置名称为 multipartResolver 类型的 Bean。

7.3.1 文件的上传与注册拦截器

首先配置名称为 multipartResolver 类型的 Bean，使用 CommonsMultipartResolver 注入，并在 MvcConfig 中添加 multipartResolver 方法，如代码清单 7-19 所示。

代码清单 7-19 创建 CommonsMultipartResolver 方法

```
/** 文件上传解析器 */
@Bean
public MultipartResolver multipartResolver(){
        CommonsMultipartResolver commonsMultipartResolver = new
CommonsMultipartResolver();
    // 文件上传最大值为50MB
    commonsMultipartResolver.setMaxUploadSize(1024*1024*50);
    // 设置编码为UTF-8
    commonsMultipartResolver.setDefaultEncoding("UIF-8");
    return commonsMultipartResolver;
}
```

虽然创建了 MultipartResolver，但是还需要一个 .jar 包作为依赖。在 pom.xml 中加入 FileItemFactory 依赖，如代码清单 7-20 所示。

代码清单 7-20　FileItemFactory 依赖

```
<!--文件上传依赖的包-->
<dependency>
    <groupId>commons-fileupload</groupId>
    <artifactId>commons-fileupload</artifactId>
    <version>1.3.2</version>
</dependency>
```

因为 CommonsMultipartResolver 父类（CommonsMultipartResolver 类）的构造函数中使用了 newFileItemFactory 方法，CommonsMultipartResolver 的构造函数直接调用父类构造函数，所以需要该类的依赖（Spring MVC 对文件的上传提供了封装，是由 MultipartResolver 文件解析器实现的。而 Spring MVC 使用的 CommonsMultipartResolver 类是使用了 Apache CommonsFileUpload 技术实现的，所以才依赖了 Apache CommonsFileUpload 的组件），如代码清单 7-21 所示。

代码清单 7-21　CommonsMultipartResolver 构造函数

```
public CommonsFileUploadSupport() {
    this.fileItemFactory = newFileItemFactory();
    this.fileUpload = newFileUpload(getFileItemFactory());
}
```

有时加入依赖后可能会出现 java.lang.NoClassDefFoundError: org/apache/commons/fileupload/FileItemFactory 异常（解决方法在 5.1 节中讲过），这是由于发布包中 lib 下并没有该包，需要在 IDEA 中依次单击 File → Project Structure → Artifacts，如图 7-7 所示，将 commons-fileupload 包添加到 <output root>（/WEB-INF/lib）中，同时将下面的 commons-io 包也 Put 进去。

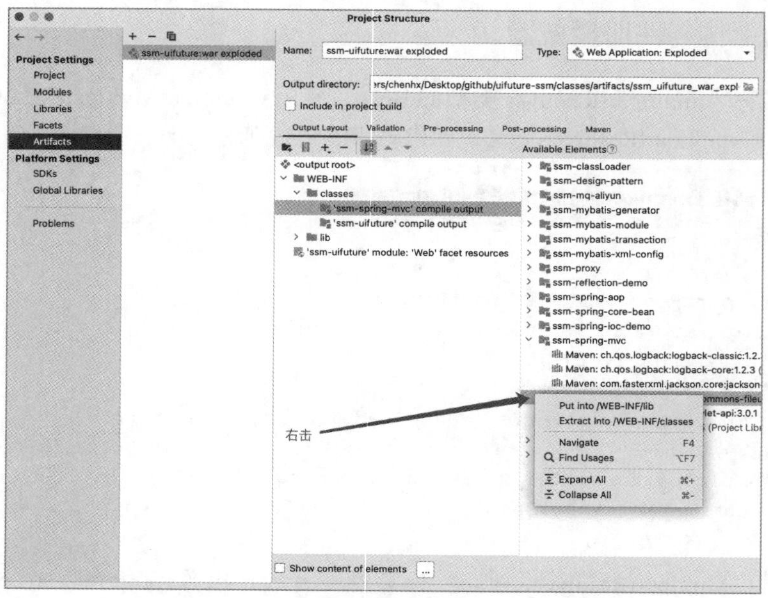

图 7-7　将 .jar 包添加到 WEB-INF/lib 下

接下来创建文件上传页面。利用表单进行文件的上传，表单中携带用户信息，并将文件保存在 WEB-INF/file/images/ 目录下。

在文件上传过程中，必须把表单的 method 设置为 POST，并且将类型 enctype 设置为 multipart/form-data，只有这样，浏览器才会将用户选择的文件以二进制流的方式提交给服务器。接下来在 views/upload 目录下创建 index.jsp 文件，如代码清单 7-22 所示。

代码清单 7-22　文件上传演示页面

```
<%@ page contentType="text/html;charset=UTF-8" language="java" %>
<%@taglib prefix="c" uri="http://java.sun.com/jsp/jstl/core" %>
<html>
<head>
<title>文件上传</title>
</head>
<body>
<h1>文件上传演示</h1>
<%--表单的method必须为post, enctype为multipart/form-data--%>
<form action="<c:url value= '/file/upload'/>" enctype="multipart/form-data"
method="post">
<table>
<tr>
<td><input type="text" placeholder="用户名" name="name"></td>
</tr>
<tr>
<td><input type="number" placeholder="年龄" name="age"></td>
</tr>
<tr>
<td>请上传头像:</td>
<td><input type="file" name="image"></td>
</tr>
<tr>
<td><input type="submit" value="提交"></td>
</tr>
</table>
</form></body></html>
```

在同一目录下创建消息提醒文件 message.jsp，如代码清单 7-23 所示。

代码清单 7-23　message.jsp 文件

```
<%@ page contentType="text/html;charset=UTF-8" language="java" %>
<html>
<head>    <title>文件上传提示</title></head>
<body>
<h1>文件上传提示</h1>
        ${message}
</body></html>
```

接下来在 MvcConfig 类的 addViewControllers 方法中添加页面映射代码，如代码清单 7-24 所示。

代码清单 7-24　添加映射代码

```
//映射文件上传的页面
registry.addViewController("/upload/index").setViewName("/upload/index");
```

由于 User 实体类已经有了，这里不再创建。一般不会在数据库中存储图片二进制文件，而是存储图片的路径。这里只对图片进行存储，方便演示。

另外，由于使用了 enctype="multipart/form-data" 方式上传表单，需要将请求的编码设置为 UTF-8，可以使用 Filter 将全局的请求和响应的编码统一设置为 UTF-8。由于使用的是零 XML 配置，所以不使用通过 web.xml 配置拦截器的方式，在 Spring MVC 初始化方法中注册拦截器即可。在 WebInitializer 类的 onStartup 方法中注册拦截器、设置全局编码，如代码清单 7-25 所示。

代码清单 7-25　注册拦截器设置全局编码

```
// 注册Filter字符编码的过滤器
FilterRegistration.Dynamic dynamic=servletContext.addFilter("characterEncodingFilter",
new CharacterEncodingFilter("UTF-8",true,true));
// 映射Filter
dynamic.addMappingForUrlPatterns(EnumSet.of(DispatcherType.REQUEST), true, "/*");
```

在 web.xml 中注册的类均可以在 WebInitializer 的 onStartup 方法中注册。

接下来创建 Controller，用于控制表单的上传，如代码清单 7-26 所示。

代码清单 7-26　文件上传的 Controller

```
/** 文件上传和下载的Controller */
@Controller
@RequestMapping("file")
public class FileController {
    private Logger logger = LoggerFactory.getLogger(FileController.class);

    @PostMapping(value = "/upload")
    public ModelAndView register(HttpServletRequest request,
                               User user, MultipartFile image) throws Exception {
        ModelAndView modelAndView = new ModelAndView("/upload/message");
        if (user == null || image == null) {
            modelAndView.addObject("message","参数错误");
            return modelAndView;
        }
            logger.info("参  数，user={},imageName={}", user, image.getOrigina
lFilename());
/* 上传的绝对路径。例如，该项目路径为***/项目名/classes/artifacts/ssm_spring_mvc_war_
exploded/WEB-INF/file/images/，***为省略的项目路径。getRealPath方法返回服务器filesystem
上的绝对路径。getServletContext().getRealPath返回上下文所在根路径的绝对路径*/
        String path = request.getServletContext().getRealPath("/WEB-INF/file/images/");
        // 获取上传文件名
        String filename = image.getOriginalFilename();
        File filepath = new File(path, filename);
        logger.info("路径:"+filepath);
```

```
            // 判断路径是否存在，若不存在则先创建一个
            if (!filepath.getParentFile().exists()) {
                filepath.getParentFile().mkdirs();
            }
            // 将文件保存到目标文件中，File.separator为文件分隔符
            image.transferTo(new File(path + File.separator + filename));
            modelAndView.addObject("message","上传成功");
            return modelAndView;
        }
    }
```

启动项目，访问 http://127.0.0.1:8080/upload/index，如图 7-8 所示。填好表单，选择图片（或其他文件）之后单击“提交”按钮。

单击“提交”按钮之后，可以看到返回了“上传成功”的页面，如图 7-9 所示。

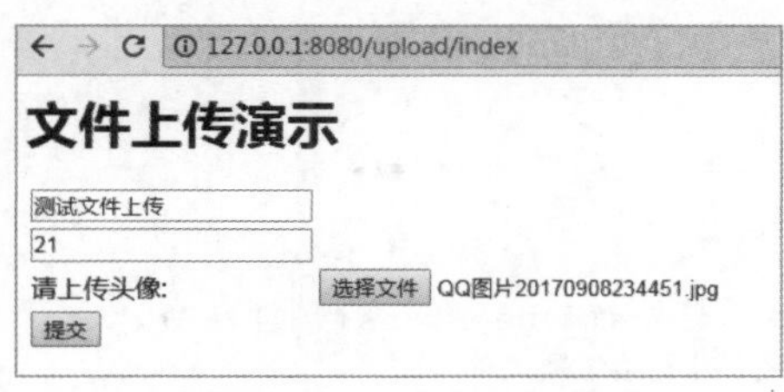

图 7-8　文件上传表单

图 7-9　成功上传返回页面

还可以在对应的目录下看到该文件。路径为项目根路径 \classes\artifacts\ssm_spring_mvc_war_exploded\WEB-INF\file\images。request.getServletContext().getRealPath("/") 获取的路径就是 Output directory 的路径（容器的发布路径），如图 7-10 所示。

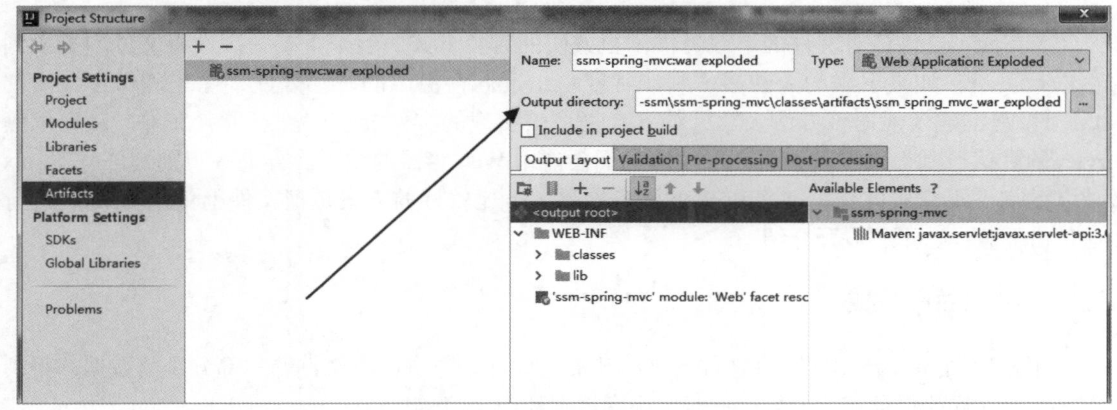

图 7-10　Output directory 路径

7.3.2　通过 ResponseEntity 下载文件

7.3.1 小节演示了文件和对象的传输，接下来演示 Spring MVC 中的文件下载。文件的下载比较简单，在页面中访问文件的超链接即可。如果文件名为中文名，在某些早期的浏览器中则可能

会出现乱码，从而导致下载失败。

Spring MVC 提供了 ResponseEntity 类型，可以很方便地定义返回头、返回文件名和状态码等信息。Controller 类提供的下载文件的方法如代码清单 7-27 所示。

代码清单 7–27　Controller 类的下载方法

```
/**
 * 如果出现大文件，会提示OOM（OutOfMemory Error）信息，会出现java.lang.OutOfMemoryError:
Java heap space
 */
@RequestMapping(value = "/download")
public ResponseEntity<byte[]> download(HttpServletRequest request,
                                       String filename) throws Exception {
    logger.info("filename={}",filename);
    // 下载文件路径
      String path = request.getServletContext().getRealPath("/WEB-INF/file/
images/");
    File file = new File(path + File.separator + filename);
    HttpHeaders headers = new HttpHeaders();
    // 下载显示的文件名，解决中文名称乱码问题
    String downloadFielName = new String(filename.getBytes("UTF-8"),"iso-8859-1");
    // 通知浏览器以attachment下载方式，那么响应时直接下载文件
    headers.setContentDispositionFormData("attachment", downloadFielName);
    //application/octet-stream：二进制流数据（最常见的文件下载）
    headers.setContentType(MediaType.APPLICATION_OCTET_STREAM);
     return new ResponseEntity<>(FileUtils.readFileToByteArray(file),     headers,
HttpStatus.CREATED);
}
```

虽然 Spring MVC 对文件的下载提供了很方便的封装方式，但是直接一次性将文件内容读到内存进行发送，如果下载的人数多，则非常容易出现内存溢出问题。因此，除非确保不会出现 OOM，否则不建议使用该方法。常见的方式是将被下载的文件配置成静态资源文件，从而依赖 Servlet 的默认机制提供下载，但高并发情况下会占用 Web 容器的线程。再进一步则是依赖 Nginx 做独立的文件下载服务器，彻底和 Web 服务器隔离。比较好的方案是把文件上传到 CDN，进行多地备份下载，这里不展开介绍。

7.3.3　大文件的下载

接下来使用 byte 数组实现大文件的下载功能。原理是每次读取文件的一部分字节数据并写到输入流中，这样能保证在读取大文件时，稳定地消耗内存。

在 FileController 中增加一个方法来进行大文件的下载，如代码清单 7-28 所示。

代码清单 7–28　大文件的下载

```
/** 大文件下载  */
@RequestMapping("downloadLargeFile")
```

```
public void downloadLargeFile(String filename, HttpServletResponse response,
HttpServletRequest request) throws UnsupportedEncodingException {
    //设置响应头和客户端保存文件名
    response.setCharacterEncoding("utf-8");
    response.setContentType("multipart/form-data");
    //设置response内容的类型，二进制流
    response.setContentType("application/octet-stream");
    String downloadFielName = new String(filename.getBytes("UTF-8"), "iso-8859-1");
    logger.info("filename={},downloadFielName={}", filename, downloadFielName);
    //设置头部信息
    response.setHeader("Content-Disposition", "attachment;fileName=" + downloadFielName);
    //下载文件路径
    String path = request.getServletContext().getRealPath("/WEB-INF/file/images/");
    File file = new File(path + File.separator + filename);
    //文件不存在
    if (!file.exists()) {
        return;
    }
    //设置文件的大小
    response.setContentLength((int) file.length());
     /* 利用BufferedInputStream从文件流读取数据。因为在读取大文件时，BufferedInputStream的
速度比InputStream快很多 */
    BufferedInputStream inputStream = null;
    OutputStream os = null;
    try {
        //打开本地文件流
        inputStream = new BufferedInputStream(new FileInputStream(file));
        //激活下载操作
        os = new BufferedOutputStream(response.getOutputStream());
        //循环写入输出流 每次读取10MB，防止内存溢出
        byte[] b = new byte[1024 * 1024 * 10];
        int length;
        //遍历读取输入流与写到response的输出流中
        while ((length = inputStream.read(b)) > 0) {
            os.write(b, 0, length);
        }
        inputStream.close();
        // 这里注意关闭。一定要调用flush()方法
        os.flush();
        os.close();
    } catch (SocketException e) {
        logger.warn("用户取消了下载:" + e.getMessage());
    } catch (Exception e) {
        logger.error("下载文件出现异常:" + e.getMessage());
    } finally {
        try {
            if (os != null) {os.close(); }
            if (inputStream != null) { inputStream.close(); }
```

```
            } catch (IOException e) {
            logger.error("关闭下载流出现异常:" + e.getMessage());
        }
        }
    }
```

其实文件下载很简单，直接访问 http://localhost:8080/file/downloadLargeFile?filename=*** 或者 http://localhost:8080/file/download?filename=***（文件名自定义）就可以下载文件了。

这里做一个静态页面测试文件的下载，如代码清单 7-29 所示。

代码清单 7-29　测试下载的静态页面

```
<%@ page contentType="text/html;charset=UTF-8" language="java" %>
<%@taglib prefix="c" uri="http://java.sun.com/jsp/jstl/core" %>
<html>
<head>
<title>MVC Title</title>
</head>
<body>
<pre>文件下载演示</pre>
<br/>
<h1>下载小文件</h1>
<form action="<c:url value='/file/download'/>" method="get">
<input type="text" name="filename" placeholder="文件名"/>
<button type="submit">下载小文件</button>
</form>
<h1>下载大文件</h1>
<form action="<c:url value='/file/downloadLargeFile'/>" method="get">
<input type="text" name="filename" placeholder="文件名"/>
<button type="submit">下载大文件</button>
</form>
</body></html>
```

一个是 Spring MVC 方式的文件下载，另一个是表单式大文件的下载。

另外，还需要将此下载页面映射到路径，由于没有业务逻辑，所以在实现了 WebMvcConfigurer 接口的 MvcConfig 类的 addViewController 方法中增加一行代码，如代码清单 7-30 所示。

代码清单 7-30　addViewControllers 方法

```
//映射下载的页面
registry.addViewController("/download").setViewName("/download");
```

测试时，需要保证项目路径“\ssm-spring-mvc\classes\artifacts\ssm_spring_mvc_war_exploded\WEB-INF\file\images\”下有提到的该文件，或将大文件复制到该目录下，演示出现 OOM 的情况。当使用 Spring MVC 的方式下载大文件时，只需要多单击几次下载按钮就会出现 OOM（“文件大小 × 下载次数”超过 Java 内存即会出现 OOM）。而使用 byte 数组方式下载大文件时，基本上不会出现 OOM。除非是“10MB × 下载次数”超过了 Java 内存。

7.4 Spring MVC 中多种主题的使用

在前面的组件讲解中，提到了 ThemeResolver 主题样式的解析器，那么如何使用该解析器呢？本节通过 Session 解析主题名称来切换主题加载不同的 CSS 样式，进行文字颜色的切换（还可以进行其他样式的配置）。**提示**：鉴于前后端分离的大趋势，大无线端已有很多主题切换的成熟方案，建议读者了解这部分内容即可。

Spring MVC 中的主题是由 org.springframework.ui.context.Theme 接口表示的。Theme 是由 ThemeSource 根据主题的名称来解析并返回的。

7.4.1 主题的配置与使用

下面在 Spring 中创建 Bean 名称为 themeSource 的 ResourceBundleThemeSource 类。

接下来在项目中配置主题，在实现了 WebMvcConfigurer 接口的 MvcConfig 类中添加方法，创建 ResourceBundleThemeSource 类，如代码清单 7-31 所示。

代码清单 7-31　创建 ResourceBundleThemeSource 类

```
/** 创建主题资源、配置主题资源路径 */
@Bean
public ResourceBundleThemeSource themeSource() {
        ResourceBundleThemeSource resourceBundleThemeSource = new
ResourceBundleThemeSource();
      /** 配置主题名的前缀路径，注意最后的“.”不能少，因为会当做目录分隔符拼接主题名称的。由于配置
了静态目录在resources/static/，所以只需要配置theme.即可*/
    resourceBundleThemeSource.setBasenamePrefix("theme.");
    return resourceBundleThemeSource;
}
```

setBasenamePrefix("theme.") 配置会从 WEB-INF/classes/theme/ 目录下查找主题资源。ResourceBundleThemeSource 解析 Theme 时会通过解析器解析请求，返回主题的名称。ThemeResolver 解析器有三个实现。这里使用 SessionThemeResolver 解析器，从 Session 中解析主题的名称。

下面配置 Bean 名称为 themeResolver 的 SessionThemeResolver，如代码清单 7-32 所示。

代码清单 7-32　配置 SessionThemeResolver

```
/** 配置主题解析器 */
@Bean
public SessionThemeResolver sessionThemeResolver() {
    SessionThemeResolver sessionThemeResolver = new SessionThemeResolver();
      /** 配置默认的theme名称。在不进行指定默认的主题时，默认的名称为theme。这里会查找WEB-INF/
classes/theme/default.properties主题文件*/
    sessionThemeResolver.setDefaultThemeName("default");
    return sessionThemeResolver;
}
```

定义 themeSource 和 themeResolver 之后，可以开始使用 Theme 了。接下来配置两个主题文件，分别加载不同的 CSS 文件，以及配置 CSS 文件。

代码清单 7-33　配置 CSS 文件

```
文件red.css中内容
.font-color{
    color: red;
}
文件blue.css中内容
.font-color{
    color: blue;
}
```

两个 CSS 文件存放的路径如图 7-11 所示。

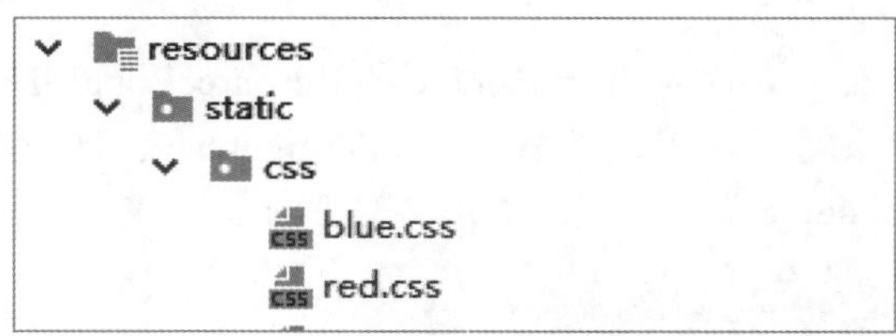

图 7-11　CSS 文件存放目录

下面来创建主题资源文件，如代码清单 7-34 所示。分别配置这两个 CSS 文件的路径，在 resources/theme 目录下创建 default.properties 和 red.properties 文件。

代码清单 7-34　主题资源文件

```
default.properties文件内容：
font.color=/static/css/blue.css
red.properties文件内容：
font.color=/static/css/red.css
```

接下来在 JSP 页面中使用 CSS 文件，需要先引入 Spring 的 taglib，然后在 JSP 代码中通过 Spring 标签使用即可，如代码清单 7-35 所示。

代码清单 7-35　主题展示页面

```
<%@ page contentType="text/html;charset=UTF-8" language="java" %>
<%@ taglib prefix="c" uri="http://java.sun.com/jsp/jstl/core" %>
<%--引入Sprign的taglib--%>
<%@ taglib prefix="spring" uri="http://www.springframework.org/tags"%>
<html>
<head>
<title>主题演示页面</title>
<link rel="stylesheet" href="<spring:theme code='font.color'/>">
</head>
<body>
<h1>主题演示页面</h1>
<h2 class="font-color">字体颜色的变化演示</h2>
```

```
        使用Session设置主题:<br/>
<div>
<a href="<c:url value='/themeCut/red’ />">使用Session切换为红色字体主题</a>
</div>
<div>
<a href="<c:url value='/themeCut/default’ />">使用Session切换为蓝色字体主题</a>
</div>
        下面使用Spring MVC中内置的拦截器切换主题:<br/>
<div>
<a href="<c:url value='/theme?themeName=default'/>">使用拦截器切换为蓝色字体主题</a>
</div>
<div>
<a href="<c:url value='/theme?themeName=red'/>">使用拦截器切换为红色字体主题</a>
</div></body></html>
```

注意：在 addViewControllers 方法中配置 theme 页面的映射，也就是在 addViewControllers 方法的最后添加如代码清单 7-36 所示的代码。

代码清单 7-36　添加页面映射

```
// 映射到主题切换演示页面
registry.addViewController("/theme").setViewName("/theme");
```

这个时候运行项目，访问 http://localhost:8080/theme，则该页面将使用默认的主题，也就是使用 default.properties 中配置的样式。<link rel="stylesheet" href="<spring:theme code='font.color'/>"> 解析成页面以后其实对应的代码就是 <link rel="stylesheet" href="/static/css/blue.css">。简单地理解就是从 properties 文件中获取 name 的 value 并渲染到页面中。

7.4.2　通过 SessionThemeResolver 切换主题

这里使用 SessionThemeResolver 主题解析器切换主题，所以可以切换每个 Session 的主题，如代码清单 7-37 所示。

代码清单 7-37　主题切换

```
/** 使用SessionThemeResolver进行主题的切换 */
@Controller
public class ThemeCutController {
     private static Logger logger = LoggerFactory.getLogger(ThemeCutController.
class);
    @Autowired
    private SessionThemeResolver sessionThemeResolver;
    /** 通过主题名称进行主题的切换 */
    @RequestMapping("/themeCut/{name}")
    public ModelAndView themeCut(HttpServletRequest request,
                                 HttpServletResponse response,
                                 @PathVariable("name") String name) {
        logger.info("通过主题名称，进行主题的切换,themeName={}",name);
        sessionThemeResolver.setThemeName(request,response,name);
```

```
            return new ModelAndView("theme");
        }
    }
```

通过单击 http://localhost:8080/theme 页面上的两个链接即可实现主题的切换了。或者访问 http://localhost:8080/themeCut/red 和 http://localhost:8080/themeCut/default 页面，就会应用对应的主题。如果使用了未存在的主题名资源，则会报 ServletException 异常。

7.4.3 通过 Spring MVC 内置拦截器切换主题

使用 Spring MVC 内置的拦截器进行主题的切换非常方便，只需配置一个拦截器即可实现。在 MvcConfig 类中重写 addInterceptors 方法，如代码清单 7-38 所示。

代码清单 7-38 addInterceptors 方法

```
/** 配置拦截器，这里只是配置主题切换拦截器，可以使用addInterceptor方法配置多个拦截器 */
@Override
public void addInterceptors(InterceptorRegistry registry) {
    ThemeChangeInterceptor themeChangeInterceptor = new ThemeChangeInterceptor();
     /**设置主题名的参数名称为themeName，默认的名称为theme。这样可以在URL上传递一个名称为
themeName的参数来指定使用的主题名称*/
    themeChangeInterceptor.setParamName("themeName");
     /** addPathPatterns()方法指定拦截路径，未指定时拦截所有路径;excludePathPatterns()方法
指定不拦截路径 */
    registry.addInterceptor(themeChangeInterceptor).addPathPatterns("/theme");
}
```

该方法可以配置多个拦截器，这里只演示主题切换的拦截器的配置。另外，可以使用 addPathPatterns 方法指定要拦截哪些请求路径，使用 excludePathPatterns 方法会排除那些不需要拦截的路径。excludePathPatterns 方法会排除优先级别高于拦截路径的路径。

另外，使用 SessionThemeResolver 设置主题，可能没有效果。这是由于拦截器优先级的问题，后面执行的会覆盖前面所设置的主题。

接下来访问 http://localhost:8080/theme，通过使用拦截器来实现主题切换。或者访问 http://localhost:8080/theme?themeName=default 和 http://localhost:8080/theme?themeName=red 切换链接。

如果是多个主题，则其配置类似配置多个主题文件，一个主题名称对应一个主题文件，主题资源解析器只能有一个。如果没有指定主题资源解析器，则会使用默认的主题解析器 FixedThemeResolver。

第 8 章

Spring MVC 核心应用

本章要点

1. 通过 HandlerExceptionResolver 处理全局异常
2. 通过 SimpleMappingExceptionResolver 处理全局异常
3. 通过 ExceptionHandler 处理全局异常
4. 拦截器与过滤器的区别
5. 在 Spring MVC 中实现拦截器
6. 通过 DelegatingFilterProxy 在过滤器中注入 Bean
7. 通过 HandlerInterceptor 实现拦截器
8. Spring MVC 中的 JSON 数据交互形式
9. Spring MVC 国际化配置

Sping MVC 核心应用包括全局异常处理、拦截器与过滤器、JSON 数据交互和 Sping MVC 国际化配置。

8.1 全局异常处理

在项目中难免会遇到各种异常，一般来说，项目的 Dao、Service、Controller 层出现的异常都通过 throws Exception 抛出，最后交给 Spring MVC 的异常处理器来统一处理。6.3 节已讲过使用 HandlerExceptionResolver 处理全局异常，本章讲解 3 种处理全局异常的方式（本质上就两种，一种使用注解，另一种使用继承）。

8.1.1 通过 HandlerExceptionResolver 处理全局异常

6.3 节分析了 HandlerExceptionResolver 异常解析器，接下来实现 HandlerExceptionResolver 接口处理行异常。

5.5 节中，在 ExceptionAdvice 类中使用了 ExceptionHandler 注解对异常进行拦截。首先注释 ExceptionAdvice 类上的 Controller 注解，防止该类生效。

接下来定义异常的返回数据实体类，以及枚举类（通过枚举类来定义状态码），如代码清单 8-1 和代码清单 8-2 所示。

代码清单 8-1　返回实体类

```
/** 返回结果 */
public class ResultModel {
    /** 状态码 */
    private Integer code;
    /** 消息 */
    private String message;
    public ResultModel(Integer code, String message) {
        this.code = code;
        this.message = message;
    }
    /** 返回失败，使用默认状态码 */
    public static ResultModel error(String message) {
        return ResultModel.error(ResultCodeEnum.ERROR.getCode(), message);
    }
    /** 返回失败 */
    public static ResultModel error(Integer code, String message) {
        return new ResultModel(code, message);
    }
    /** 返回成功,使用默认状态码 */
    public static ResultModel success(String message) {
        return ResultModel.success(ResultCodeEnum.SUCCESS.getCode(), message);
    }
```

```java
    /** 返回成功 */
    public static ResultModel success(Integer code, String message) {
        return new ResultModel(code, message);
    }
//...getter/setter和toString方法不要忘记了。
}
```

在其中定义了一些静态方法，方便调用。

代码清单 8-2　返回状态码枚举

```java
/** 返回状态码枚举 */
public enum ResultCodeEnum {
    /** 返回成功状态码 */
    SUCCESS(200),
    /** 失败状态码 */
    ERROR(-1);
    /** 状态码 */
    private Integer code;
// 不要忘记getter()方法和带code参数的构造函数
}
```

下面创建全局异常处理的 ExceptionResolver 类实现 HandlerExceptionResolver 接口，如代码清单 8-3 所示。

代码清单 8-3　全局异常处理的 ExceptionResolver 类

```java
/** 创建全局异常处理类实现HandlerExceptionResolver接口 */
public class ExceptionResolver implements HandlerExceptionResolver {
    private static Logger logger = LoggerFactory.getLogger(ExceptionResolver.
class);
    /** 拦截所有的异常 */
    @Override
    public ModelAndView resolveException(HttpServletRequest request,
HttpServletResponse response, Object handler, Exception ex) {
        logger.error("异常信息:{}", ex.getMessage());
        Map<String, Object> model = new HashMap<>(4);
        model.put("data", ResultModel.error(ex.getMessage()));
        // 根据不同异常转向不同页面(建议自定义异常)
        // 选择的这两个异常均继承RuntimeException异常类
        if (ex instanceof HttpMessageNotWritableException) {
                return new ModelAndView("error/httpMessageNotWritableException",
model);
        } else if (ex instanceof HttpMessageNotReadableException) {
                return new ModelAndView("error/httpMessageNotReadableException",
model);
        }
        return new ModelAndView("error/otherError", model);
    }
}
```

在 WebMvcConfigurer 接口中通过 configureHandlerExceptionResolvers 方法配置异常处理器链，所以在 MvcConfig 类中，重写该方法来拦截异常，如代码清单 8-4 所示。

代码清单 8-4　重写 configureHandlerExceptionResolvers 方法

```
/** 配置异常处理器链 */
@Override
public void configureHandlerExceptionResolvers(List<HandlerExceptionResolver>
resolvers) {
    //将ExceptionResolver添加到异常处理器集合中
    resolvers.add(new ExceptionResolver());
}
```

上面对所有的异常进行了拦截，下面创建错误提示的页面，在 resources/views/error 目录下分别创建三个 JSP 页面，如代码清单 8-5 所示。

代码清单 8-5　错误展示页面

```
<%@ page contentType="text/html;charset=UTF-8" language="java" %>
<html>
<head>    <title>错误页面</title></head>
<body>
<pre>定制的HttpMessageNotReadableException异常页面</pre>
<br/>
        ${data.code},${data.message}
</body></html>
```

${data.code} 和 ${data.message} 用来读取 Model 中传到 JSP 页面的值。最后创建一个 Controller 访问 JSP 页面，并创建一个 ExceptionResolver 类抛出异常，如代码清单 8-6 所示。

代码清单 8-6　创建 ExceptionController 类抛出异常

```
/** 演示异常的抛出Controller */
@Controller
@RequestMapping("exception")
public class ExceptionController {
    private Logger logger = LoggerFactory.getLogger(ExceptionController.class);
    /** 根据传进来的code，抛出不同的运行时异常 */
    @RequestMapping("/code/{code}")
    @ResponseBody
    public String exception(@PathVariable("code") Integer code) {
        logger.info("code={}", code);
        if (code.equals(1)) {
           throw new HttpMessageNotWritableException("code等于1,抛出BindException异常");
        } else if (code.equals(2)) {
            throw new HttpMessageNotReadableException("code等于2，抛出HttpMessageNot
ReadableException异常");
        } else if(code.equals(3)){
            throw new RuntimeException("code等于3，其他的运行时异常");
        }
        return "success";
```

```
    }
}
```

接下来访问下面的链接，查看是否返回不同的错误提醒页面。

（1）http://localhost:8080/exception/code/1。

（2）http://localhost:8080/exception/code/2。

（3）http://localhost:8080/exception/code/3。

其中一个展示页面如图 8-1 所示。

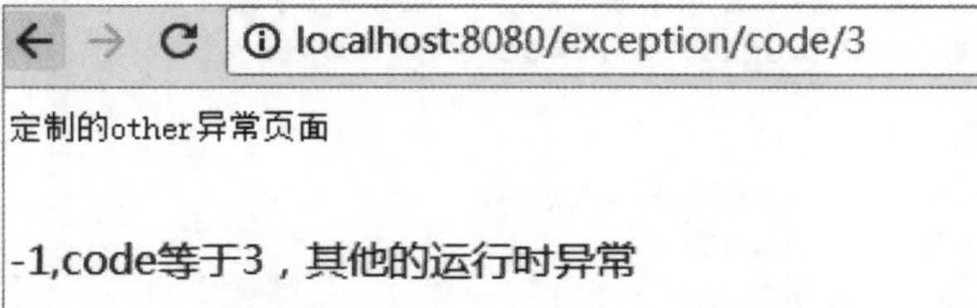

图 8-1　异常返回的展示页面

使用这种方式会拦截程序运行中抛出的所有异常。

8.1.2　通过 SimpleMappingExceptionResolver 处理全局异常

SimpleMappingExceptionResolver 其实就是 Spring MVC 中 HandlerExceptionResolver 的实现类。SimpleMappingExceptionResolver 类的部分源码如代码清单 8-7 所示。

代码清单 8-7　SimpleMappingExceptionResolver 类的部分源码

```
public class SimpleMappingExceptionResolver extends AbstractHandlerExceptionResolver {
    // 存储异常和viewName的映射关系
    @Nullable
    private Properties exceptionMappings;
    // 设置需要排除的异常
    @Nullable
    private Class<?>[] excludedExceptions;
    // 如果没有找到对应的错误视图，则返回默认的错误视图
    @Nullable
    private String defaultErrorView;
    // 该方法就是由AbstractHandlerExceptionResolver类中resolveException方法调用而来的
    @Override
    @Nullable
    protected ModelAndView doResolveException(
            HttpServletRequest request, HttpServletResponse response, @Nullable Object
handler, Exception ex) {
        // 通过设置的异常映射，获取异常的页面名称
        String viewName = determineViewName(ex, request);
        if (viewName != null) {
            // 返回错误视图应用HTTP状态代码。仅在处理最上层的请求时才应用它
            Integer statusCode = determineStatusCode(request, viewName);
            if (statusCode != null) {
```

```
                // 设置HTTP状态码到响应中
                applyStatusCodeIfPossible(request, response, statusCode);
            }
            // 通过视图名称、异常和请求返回ModelAndView
            return getModelAndView(viewName, ex, request);
        } else {
            return null;
        }
    }
    /** 确定给定异常的视图名称，首先检查异常是否需要排除，然后搜索exceptionMappings中的异常映射，
如果没有找到，那么使用defaultErrorView作为最后的viewName返回
     */
    @Nullable
    protected String determineViewName(Exception ex, HttpServletRequest request) {
        String viewName = null;
        // 排除不需要拦截的异常
        if (this.excludedExceptions != null) {
            for (Class<?> excludedEx : this.excludedExceptions) {
                if (excludedEx.equals(ex.getClass())) {
                    return null;
                }
            }
        }
        // 检查拦截的异常映射是否为空
        if (this.exceptionMappings != null) {
        // 通过异常和映射去查找视图名称
            viewName = findMatchingViewName(this.exceptionMappings, ex);
        }
        //没找到则返回默认的视图
        if (viewName == null && this.defaultErrorView != null) {
            //...
        }
        viewName = this.defaultErrorView;
    }
      return viewName;
}
/** 在给定的异常映射中查找匹配的视图名称 */
    @Nullable
    protected String findMatchingViewName(Properties exceptionMappings, Exception ex) {
        String viewName = null;
        String dominantMapping = null;
        int deepest = Integer.MAX_VALUE;
        // 遍历匹配
           for (Enumeration<?> names = exceptionMappings.propertyNames(); names.
hasMore Elements(); ) {
            String exceptionMapping = (String) names.nextElement();
            // 获取超类匹配的深度
            int depth = getDepth(exceptionMapping, ex);
            // 判断，匹配最接近的异常，也就是说，只要写了拦截异常，最上层的Exception异常类肯定会被拦截
            //
```

```
            if (depth >= 0 && (depth < deepest || (depth == deepest &&
                    dominantMapping != null && exceptionMapping.length() > dominant
Mapping.length()))) {
                deepest = depth;
                dominantMapping = exceptionMapping;
                // 通过异常获取视图名称
                viewName = exceptionMappings.getProperty(exceptionMapping);
            }
        }
        //...
    }
    return viewName;
}
/**
 * 获取超类的匹配深度，一直没找到则返回-1
 */
protected int getDepth(String exceptionMapping,Exception ex){
        return getDepth(exceptionMapping,ex.getClass(),0);
        }
private int getDepth(String exceptionMapping,Class<?> exceptionClass,int depth){
        if(exceptionClass.getName().contains(exceptionMapping)){
    // 若已匹配，直接返回
        return depth;
        }
        // Throwable类是Java语言中所有错误和异常的超类
        if(exceptionClass==Throwable.class){
        return-1;
        }
        return getDepth(exceptionMapping,exceptionClass.getSuperclass(), depth+1);
  }
/**
  * 确定要应用于给定错误视图的HTTP状态代码。默认实现返回给定视图名称的状态代码，没找到则返回默认的
状态码
 */
@Nullable
protected Integer determineStatusCode(HttpServletRequest request,String viewName){
        if(this.statusCodes.containsKey(viewName)){
        return this.statusCodes.get(viewName);
        }
        return this.defaultStatusCode;
   }
/** 将指定的HTTP状态代码应用于给定的响应。如果不是在该请求中执行的，则为默认状态码 */
protected void applyStatusCodeIfPossible(HttpServletRequest request,
HttpServletResponse response,int statusCode){
        if(!WebUtils.isIncludeRequest(request)){
        ...
        response.setStatus(statusCode);
        request.setAttribute(WebUtils.ERROR_STATUS_CODE_ATTRIBUTE,statusCode);
    }
  }
```

```
/** 通过给定的视图名称、异常和请求返回ModelAndView */
protected ModelAndView getModelAndView(String viewName,Exception ex,HttpServletRequest
request){
        return getModelAndView(viewName,ex);
        }
/**
  * 返回给定视图名称和异常的ModelAndView。默认实现添加指定的异常属性(名称为exception，可以在前
端页面获取异常信息)。可以在子类中重写该方法进行自定义
 */
protected ModelAndView getModelAndView(String viewName,Exception ex){
        ModelAndView mv=new ModelAndView(viewName);
        if(this.exceptionAttribute!=null){
        ...
        }
        // 添加异常属性
        mv.addObject(this.exceptionAttribute,ex);
     }
        return mv;
    }
}
```

通过源码可以发现，使用 SimpleMappingExceptionResolver 类拦截异常可以省略很多步骤，只需要定义一个 SimpleMappingExceptionResolver 类型的 Bean，配置异常与视图的映射即可拦截异常，非常方便快捷。当然，限制比开发者自己实现 HandlerExceptionResolver 接口要多一些。如果需要扩展，则可以继承 SimpleMappingExceptionResolver 来定制需要的功能，如记录日志、配置异常信息属性的名称等。

接下来简单地演示拦截 HttpMessageNotWritableException 和 HttpMessageNotReadableException 异常的过程，默认的异常页面使用前面的 otherError 页面。

在 MvcConfig 类中添加 simpleMappingExceptionResolver 方法，如代码清单 8-8 所示的代码。

代码清单 8-8　在 MvcConfig 类中添加 simpleMappingExceptionResolver 方法

```
/** 配置SimpleMappingExceptionResolver异常拦截解析器 */
public SimpleMappingExceptionResolver simpleMappingExceptionResolver()
{
    SimpleMappingExceptionResolver simpleMappingExceptionResolver = new SimpleMapp
ingExceptionResolver();
    Properties mappings = new Properties();
    // 配置拦截映射,异常与视图对应
    mappings.put("HttpMessageNotWritableException", "error/httpMessageNotWritableEx
ception");
    mappings.put("HttpMessageNotReadableException", "error/httpMessageNotReadableEx
ception");
    // 设置默认的视图，也就是未配置异常映射的视图，其他异常会跳转到该页面
    simpleMappingExceptionResolver.setDefaultErrorView("error/otherError");
    simpleMappingExceptionResolver.setExceptionMappings(mappings);
    return simpleMappingExceptionResolver;
}
```

然后注释 configureHandlerExceptionResolvers 方法中的代码，如果没有注释，则会按照添加进 resolvers 的顺序优先拦截并返回。注册新建的 SimpleMappingExceptionResolver 类，如代码清单 8-9 所示。

代码清单 8-9　configureHandlerExceptionResolvers 类

```
/** 配置异常处理器链 */
@Override
public void configureHandlerExceptionResolvers(List<HandlerExceptionResolver>
resolvers) {
// 配置SimpleMappingExceptionResolver异常拦截器
    resolvers.add(simpleMappingExceptionResolver());
}
```

由于使用的是 SimpleMappingExceptionResolver 异常解析器，ModelAndView 中的属性名称为 exception，还可以重写 getModelAndView 方法，自定义返回到页面的属性名称，不过在这里并没有重写，使用的就是默认的属性名称。在每个错误的 JSP 页面中添加 ${exception}，即可看到错误的输出信息。在项目中，如果使用 SimpleMappingExceptionResolver 类拦截异常，而且需要记录日志或者有其他个性化的需求，那么建议继承 SimpleMappingExceptionResolver 类，重写相应的方法，进行功能增强。

另外，Spring MVC 中实现 HandlerExceptionResolver 接口的实现类还有几个，如 default-Handler ExceptionResolver 类、ResponseStatusExceptionResolver 类等。每个类都有不同的实现功能，读者可以查看源码进行分析。

8.1.3　通过 ExceptionHandler 处理全局异常

使用 ExceptionHandler 注解可以拦截全局异常，另外，ExceptionHandler 注解是可以在 Controller 内部方法上定义的，Controller 内部方法会拦截并处理该 Controller（或者任意子类）中由 ExceptionHandler 注解的方法抛出的异常。

如果 ExceptionHandler 定义在 ControllerAdvice 注解的类中，那么会处理所有控制器抛出的异常。接下来针对这两种方式进行讲解。

首先注释配置异常处理器链的 configureHandlerExceptionResolvers 方法。

1. 类级别的异常拦截

在 ExceptionController 类中添加 httpMessageNotReadableExceptionHandler 方法，如代码清单 8-10 所示。

代码清单 8-10　用 httpMessageNotReadableExceptionHandler 方法拦截异常

```
/** 配置拦截本类中的HttpMessageNotReadableException异常 */
@ExceptionHandler(HttpMessageNotReadableException.class)
public ModelAndView httpMessageNotReadableExceptionHandler(HttpServletRequest
request, HttpMessageNotReadableException ex){
    logger.error("异常信息:{}", ex.getMessage());
```

```
        Map<String, Object> model = new HashMap<>(4);
        model.put("data", ResultModel.error("[ExceptionHandler]"+ex.getMessage()));
        return new ModelAndView("error/httpMessageNotReadableException", model);
    }
```

通过 ExceptionHandler(HttpMessageNotReadableException.class) 注解该方法，可以拦截 ExceptionController 类中抛出的 HttpMessageNotReadableException 异常。

代码清单 8-11 所示为 ExceptionHandler 注解的源码。

代码清单 8-11　ExceptionHandler 的源码

```
@Target(ElementType.METHOD)
@Retention(RetentionPolicy.RUNTIME)
@Documented
public @interface ExceptionHandler {
    Class<? extends Throwable>[] value() default {};
}
```

ExceptionHandler 注解是 Spring 3.0 以后才有的，只有一个属性 value，是一个 class 数组，用来配置拦截的异常，如果为空，则默认拦截异常列表中的所有异常。**提示**：如果同时使用 HandlerExceptionResolver 接口，那么使用 configureHandlerExceptionResolvers 方法注册的异常处理器链会优先拦截异常并返回。

使用 ExceptionHandler 注解在 Controller 中添加方法，局限性太明显，所以需要配合 ControllerAdvice 注解一起使用。

2. 使用 ExceptionHandler 注解拦截全局异常

该种方法在 5.5 节中已经讲解过，可以查看代码清单 5-30。

注意：如果同时使用 HandlerExceptionResolver 接口和 ExceptionHandler 注解拦截全局异常，那么优先匹配 HandlerExceptionResolver 接口的 configureHandlerExceptionResolvers 方法。

8.2　拦截器与过滤器

拦截器和过滤器都是 AOP 理念的体现，虽然两者在实现的功能上差不多，但是实际的技术原理还是有较大差别的。

8.2.1　拦截器与过滤器的区别

过滤器是依赖于 Servlet 存在的，所以不能脱离 Servlet 容器，基于函数的回调，属于 Servlet 规范的一部分（只能应用于 Web 应用）。过滤器几乎可以过滤所有的请求，过滤器的实例在 Servlet 容器初始化时会调用一次。使用过滤器的目的是做一些过滤操作，可以拦截并修改 request 和 response，在请求传入 Servlet 或者 Spring MVC 的前端控制器前修改全局的字符编码、请求过滤和 CORS（Cross-origin Resource Sharing，跨域资源共享）等。

而拦截器是依赖框架的，如果在 Spring MVC 中使用拦截器，那么就要依赖 Spring MVC 框架，其实现原理基于 Java 的反射机制（不局限于在 Web 项目中使用）。例如，拦截器拦截某个 URL 映射的方法（**提示**：在 Spring MVC 中，拦截器能拦截 URL 是因为 Spring MVC 内部有方法映射，实际上还是拦截方法），那么可以使用拦截器在 Controller 方法运行前拦截，或者在方法后运行，输出日志。Spring MVC 中的拦截器只能对 Spring MVC 前端控制器管理的映射请求起作用。

执行顺序：过滤器处理前→拦截器处理前→ Controller 方法→拦截器处理后→过滤器处理后。

基本上使用过滤器能做的事情，使用拦截器都能实现。从使用惯例上看，拦截某些请求方法等，建议使用拦截器。

8.2.2 在 Spring MVC 中实现拦截器

由于都是使用编程式编程，所以不再介绍使用 XML 方法配置拦截器。通过前面的介绍，现在已知 web.xml 中的配置基本是在 WebApplicationInitializer 接口中使用实现类来实现的，所以拦截器也是在该接口中实现注册。

例如，需要实现一个字符的拦截器，将字符编码统一为 UTF-8，为了防止乱码，可以在 onStartup 方法中添加注册拦截器代码，代码清单如 8-12 所示。

代码清单 8-12 添加编码拦截器

```
// 注册Filter，字符编码的过滤器
FilterRegistration.Dynamic dynamic = servletContext.addFilter("characterEncodingFilter",
        new CharacterEncodingFilter("UTF-8", true, true));
// 映射Filter
dynamic.addMappingForUrlPatterns(EnumSet.of(DispatcherType.REQUEST), true, "/*");
```

CharacterEncodingFilter 为 Servlet 过滤器，允许为请求指定字符编码。因为编码一般在 HTML 页面或表单中指定，浏览器通常不会设置字符编码，如果请求尚未指定编码，则此过滤器可以应用其编码，如果将 forceEncoding 设置为 true，则会在任何情况下强制执行此过滤器的编码。

另外 Spring MVC 中还有很多默认的过滤器实现类，如 ServletContextRequestLoggingFilter 日志拦截器等。下面自定义一个过滤器，过滤一些请求中的敏感词。

Spring 中提供了很多过滤器的实现类，如 Filter 接口的实现类，如图 8-2 所示。

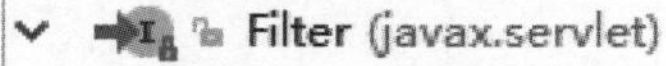

图 8-2 Spring 中实现 Filter 接口的实现类

可以选择继承其中一种进行实现，或者直接实现 Filter 接口。这里选择继承 OncePerRequest-Filter 类，如代码清单 8-13 所示。

代码清单 8-13　SensitiveWordFilter 过滤器

```
/** 敏感词过滤器 */
public class SensitiveWordFilter extends OncePerRequestFilter {
    /** 分隔符，建议可以复杂一些*/
private final static String SEPARATOR = "#-#=#&";
    /** 保证每个请求只会被调用一次 */
    @Override
    protected void doFilterInternal(HttpServletRequest request, HttpServletResponse
response, FilterChain filterChain) throws ServletException, IOException {
        /**在Spring MVC的DispatcherServlet之前执行
        *在过滤器中用装饰模式把原request功能增强
        *拦截调用HttpServletRequest 的getParamter()方法和getParameterValues()方法*/
        SensitiveWordRequest req = new SensitiveWordRequest(request);
        // 放行
        filterChain.doFilter(req, response);
    }
    /** 对HttpServletRequest对象进行增强 */
    class SensitiveWordRequest extends HttpServletRequestWrapper {
        public SensitiveWordRequest(HttpServletRequest request) {
            super(request);
        }
        /** 过滤敏感词，这里获取了模拟敏感词，建议从数据库或者文件中获取敏感词 */
        @Override
        public String getParameter(String name) {
            String str = super.getParameter(name);
            List<String> list = Arrays.asList("骂人", "敏感词");
            for (String word : list) {
                str = str.replaceAll(word, "*");
            }
            return str;
        }
        /**
         * 该方法在Spring MVC中被调用,主要用在Controller层的方法参数绑定上。通过该方 法可获取所
有参数再绑定到方法参数上
         * 在该方法中将敏感词过滤后，即使是通过Spring MVC绑定到方法参数上的值也会被过滤
         */
        @Override
        public String[] getParameterValues(String name) {
            String[] strs = super.getParameterValues(name);
            List<String> list = Arrays.asList("骂人", "敏感词");
            StringBuffer allStrs = new StringBuffer(strs[0]);
             /**提高过滤的效率。否则，假设参数值有20个，待过滤的词有10 000个，那么如果通过两层
循环过滤则需要10×10 000次。另外，建议使用DFA算法过滤敏感词。本项目使用的敏感词过滤性能待检测，而
且缺点非常明显，就是需要分隔符*/
            for (int i = 1; i < strs.length; i++) {
                allStrs.append(SEPARATOR + strs[i]);
            }
        for (String word : list) {
```

```java
                allStrs = new StringBuffer(allStrs.toString().replaceAll(word, "*"));
            }
            return allStrs.toString().split(SEPARATOR);
        }
    }
}
```

接下来将过滤器注册到 ServletContext 中，在 WebInitializer 类的 onStartup 方法最后添加如代码清单 8-14 所示的代码。

代码清单 8-14　注册 Filter

```java
// 注册Filter，敏感字过滤器
FilterRegistration.Dynamic sensitiveWordDynamic = servletContext.addFilter("sensit
iveWordFilter", new SensitiveWordFilter());
// 映射Filter
sensitiveWordDynamic.addMappingForUrlPatterns(EnumSet.of(DispatcherType.REQUEST),
true, "/*");
```

关于为什么要重写 HttpServletRequest 类中的 getParameterValues 方法，可以从 Spring MVC 中的参数绑定解析器中得知。RequestParamMethodArgumesntResolver 类负责解析 RequestParam 注解的方法参数，类型为 MultipartFile 的参数和 Spring 的 MultipartResolver 抽象类、javax.servlet.http.Part 类型的参数及 Servlet 3.0 中部分请求的连接参数。

该解析器即使未使用 RequestParam 注解，一些 Controller 层方法的简单类型参数也会使用 RequestParamMethodArgumentResolver 解析器来对请求参数进行绑定。由 resolveName 方法将给定的参数类型和值名称解析为方法的参数值。RequestParamMethodArgumentResolver 类中的 resolveName 方法实现如代码清单 8-15 所示。

代码清单 8-15　RequestParamMethodArgumentResolver 类中的 resolveName 方法实现

```java
@Nullable
protected Object resolveName(String name, MethodParameter parameter,
NativeWebRequest request) throws Exception {
// 由于过滤器先运行，所以这里传进来的参数已经是增强的HttpServletRequest，也就是
SensitiveWordRequest类型的HttpServletRequest
    HttpServletRequest servletRequest = request.getNativeRequest(HttpServletRequest.class);
    if (servletRequest != null) {
         Object mpArg = MultipartResolutionDelegate.resolveMultipartArgument(name,
parameter, servletRequest);
        if (mpArg != MultipartResolutionDelegate.UNRESOLVABLE) {
            return mpArg;
        }
    }
    Object arg = null;
// 返回请求对象
    MultipartHttpServletRequest multipartRequest = request.getNativeRequest(Multipa
rtHttpServletRequest.class);
    if (multipartRequest != null) {
```

```
            List<MultipartFile> files = multipartRequest.getFiles(name);
            if (!files.isEmpty()) {
                arg = (files.size() == 1 ? files.get(0) : files);
            }
        }
        if (arg == null) {
    // 通过getParameterValues方法获取参数值
            String[] paramValues = request.getParameterValues(name);
            if (paramValues != null) {
                arg = (paramValues.length == 1 ? paramValues[0] : paramValues);
            }
        }
        return arg;
    }
```

通过代码清单 8-15 可知，需要重写 getParameterValues 来过滤敏感词。

下面访问链接进行验证，查看敏感词是否已经被屏蔽。访问以上项目的链接：http://localhost:8080/forward/login?username=%E9%AA%82%E4%BA%BA123test%E6%95%8F%E6%84%9F%E8%AF%8D%E6%B5%8B%E8%AF%95，结果如图 8-3 所示。

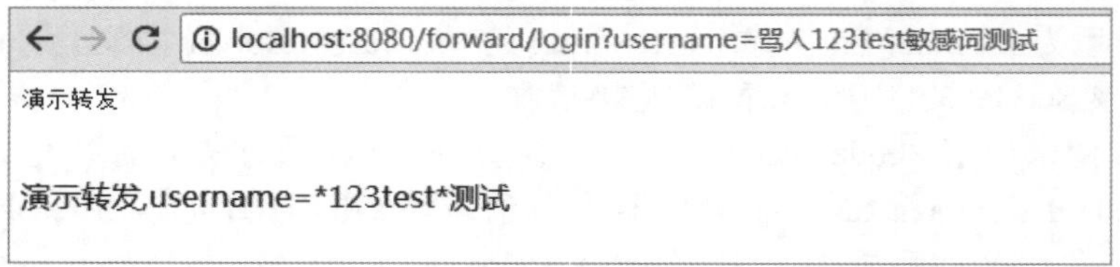

图 8-3　敏感词过滤演示

如果在项目中有多个过滤器，则拦截顺序与添加过滤器到 ServletContext 时的顺序有关。

8.2.3　通过 DelegatingFilterProxy 在过滤器中注入 Bean

一般情况下，如果需要从数据中获取敏感词，就需要调用 Service 层，既然需要从 Service 层获取敏感词，那么肯定涉及在过滤器中注入 Spring 中的 Bean。虽然 Spring 的 Bean 初始化先于过滤器初始化，可以通过在过滤器的 init 中获取 Spring 容器的 Bean 得到验证，但在过滤器中无法直接通过 org.springframework.beans.factory.annotation.Autowired 等注解注入 Bean。因为过滤器是 Servlet 规范级别的，Spring 容器也是基于 Servlet 的，所以两者平级，Spring 容器的上下文无法直接管理过滤器，所以无法注入。在 Spring 3.1 以前，可以通过实现 Filter 接口，然后在 init 方法中通过上下文如 WebApplicationContext.getBean 方法来获取 Spring 容器中的 Bean。但是在 Spring 3.1 以后，Spring 提供了一个过滤器代理类，也就是 DelegatingFilterProxy 类，DelegatingFilterProxy 的父类实现了 Filter 接口。

DelegatingFilterProxy 类是标准 Servlet 过滤器代理类，可以代理 Spring 容器中实现了 Filter 接口的类，可以通过 DelegatingFilterProxy 获取 Spring 的依赖注入机制并管理过滤器 Bean 的生命周期。

在使用 DelegatingFilterProxy 时需要注入 DelegatingFilterProxy 类，并添加名称为 targetBeanName 的 Filter（如果没有指定 targetBeanName，则使用 DelegatingFilterProxy 注册时的 Filter 名称）。

还需要设置 targetFilterLifecycle 过滤器的值为 true，强制目标 Bean 的生命周期由 Spring 管理。

接下来编写 Service 层的敏感词接口和敏感词实现类代码，模拟从数据库中读取敏感词，如代码清单 8-16 和代码清单 8-17 所示。

代码清单 8-16　Service 层的敏感词接口

```
/** 获取敏感词列表接口 */
public interface SensitiveWordService {
    /** 获取所有的敏感词*/
    List<String> selectAllSensitiveWord();
}
```

代码清单 8-17　Service 层的敏感词实现类

```
/** 获取敏感词 */
@Service
public class SensitiveWordServiceImpl implements SensitiveWordService {
    @Override
    public List<String> selectAllSensitiveWord() {
        // 模拟从数据库中获取敏感词
        return Arrays.asList("骂人", "敏感词");
    }
}
```

接下来编写敏感词过滤器，新建一个 SensitiveWordServiceFilter 过滤器类，如代码清单 8-18 所示。

代码清单 8-18　SensitiveWordServiceFilter 过滤器

```
/**
 * 注入Spring容器中的Bean来实现敏感词过滤
 * 此处要加入Configuration(Component)注解，在ServletContext中无须再注册，可交由Spring容
器管理该Filter的生命周期
 */
@Configuration
public class SensitiveWordServiceFilter extends OncePerRequestFilter {
    /** 注入Spring中的Bean */
    @Autowired
    private SensitiveWordService sensitiveWordService;
    /** 分隔符，建议可以复杂一些 */
    private final static String SEPARATOR = "#-#=#&";
    /** 保证每个请求只会被调用一次 */
    @Override
    protected void doFilterInternal(HttpServletRequest request, HttpServletResponse
response, FilterChain filterChain) throws ServletException, IOException {
        // 在Spring MVC的DispatcherServlet之前执行
        // 在过滤器中用装饰模式将原request的功能增强
        // 拦截调用HttpServletRequest的getParamter方法和getParameterValues方法
        SensitiveWordServiceRequest req = new SensitiveWordServiceRequest(request);
        filterChain.doFilter(req, response);
```

```
    }
    /** 对HttpServletRequest对象进行增强 */
    class SensitiveWordServiceRequest extends HttpServletRequestWrapper {
        ...
        @Override
        public String getParameter(String name) {
            String str = super.getParameter(name);
            List<String> list = sensitiveWordService.selectAllSensitiveWord();
            ...
        }
        @Override
        public String[] getParameterValues(String name) {
            String[] strs = super.getParameterValues(name);
            List<String> list = sensitiveWordService.selectAllSensitiveWord();
            ...
            return allStrs.toString().split(SEPARATOR);
        }
    }
}
```

虽然 SensitiveWordServiceFilter 类无须注册，但需要将 DelegatingFilterProxy 类在 Servlet-Context 中注册。接下来在 WebInitializer 类 onStartup 方法中将代码清单 8-14 的代码注释。添加注册 DelegatingFilterProxy 类，如代码清单 8-19 所示。

代码清单 8-19　注册 DelegatingFilterProxy 类

```
// 注册DelegatingFilterProxy,并且设置targetBeanName的值为sensitiveWordServiceFilter
DelegatingFilterProxy delegatingFilterProxy = new DelegatingFilterProxy("sensitiveW
ordServiceFilter");
// 配置此参数为true，将此filter的生命周期交由server容器管理，包括执行其init、destroy方法
delegatingFilterProxy.setTargetFilterLifecycle(true);
FilterRegistration.Dynamic filterRegistration = servletContext.addFilter("delegatin
gFilterProxy", delegatingFilterProxy);
// 映射Filter
filterRegistration.addMappingForUrlPatterns(EnumSet.of(DispatcherType.REQUEST),
true, "/*");
```

接下来访问 8.2.2 小节中的链接，结果与图 8-3 一致。在 initFilterBean 方法中完成 Filter 的初始化操作，通过 WebApplicationContext 获取名称为 targetBeanName 的 Bean（注意该 Bean 必须实现了 Filter 接口），通过给定请求和响应在 invokeDelegate 方法中调用委托的 Filter（也就是以 targetBeanName 为名称的 Bean）。

另外，一个 DelegatingFilterProxy 只能对应一个自定义的过滤器，如果有多个过滤器，则需要注册多个 DelegatingFilterProxy。可以选择 org.springframework.web.filter.CompositeFilter 复合过滤器配置多个自定义的过滤器，然后使用 DelegatingFilterProxy 代理该复合过滤器。

8.2.4　通过 HandlerInterceptor 实现拦截器

下面讲解拦截器使用的实例，拦截的应用场景非常多，如登录拦截、权限认证等，主要就是

拦截用户的请求并进行相应的处理。而且相对于过滤器，拦截器可以直接注入 Spring 容器中的 Bean。

拦截器是基于 Java 的动态代理实现的，关于动态代理，可参阅 2.3 节。

HandlerInterceptor 接口的源码如代码清单 8-20 所示。

代码清单 8-20　HandlerInterceptor 接口的源码

```
public interface HandlerInterceptor {
      default boolean preHandle(HttpServletRequest request, HttpServletResponse
response, Object handler)throws Exception {
         return true;
    }
      default void postHandle(HttpServletRequest request, HttpServletResponse
response, Object handler,@Nullable ModelAndView modelAndView) throws Exception {
    }
      default void afterCompletion(HttpServletRequest request, HttpServletResponse
response, Object handler,@Nullable Exception ex) throws Exception {
    }
}
```

在 HandlerInterceptor 接口中定义了 3 个方法，分别为 preHandle、postHandle 和 afterCompletion。如果需要实现拦截器，那么就通过实现这 3 个方法来进行拦截处理。

（1）preHandle 方法：带参数时写法为 preHandle(HttpServletRequest request, HttpServletResponse esponse, Object handler)，在请求处理之前调用，在该方法中一般对请求做一些预处理或者初始化的操作，该方法的返回值为 Boolean 类型。当返回值为 false 时，表示不会继续往下执行，请求结束，后续的拦截器和处理方法都不会再执行。在 Spring MVC 中，拦截器是链式调用的。也就是说，可以同时存在多个拦截器，每个拦截器都会根据设置的 Order 顺序调用（一般不设置 Order，则按照声明顺序依次调用）。当返回值为 true 时，会依次调用下一个拦截器的 preHandle 方法，最后才执行被拦截的处理方法。

（2）postHandle 方法：带参数时写法为 postHandle(HttpServletRequest request, HttpServlet Responseresponse, Object handler,@Nullable ModelAndView modelAndView)，当 preHandle 方法返回值为 true 时才会调用该方法。postHandle 方法在请求的处理方法执行完成后才执行，即在调用的方法调用完成之后，DispatcherServlet 呈现视图之前执行。preHandle 方法可以通过给定的 ModelAndView 将其他模型对象暴露给视图。也就是说，该方法是在 Controller 处理之后，对 ModelAndView 进行操作的。例如，可以在该方法中增加 Model 值等操作。该方法以相反的顺序在每个拦截器上调用，也就是第 1 个拦截器将是最后一个被加载的拦截器。

（3）afterCompletion 方法：带参数时写法为 afterCompletion(HttpServletRequest request, Http ServletResponse response, Object handler,@Nullable Exception ex)，该方法也是在 preHandle 方法返回 true 时才会被调用，是整个请求处理结束之后的回调方法，也就是渲染视图之后的回调方法。无论处理程序返回什么结果，最后都会被调用，所以可以在该方法中对一些资源进行适当的回收清理。与 postHandle 方法相同，调用顺序与链上的拦截器顺序相反。

在 Spring 中，还提供了该接口的功能扩展类，也就是 AsyncHandlerInterceptor 接口和

HandlerInterceptorAdapter 抽象类。

AsyncHandlerInterceptor 接口的源码如代码清单 8-21 所示。

代码清单 8-21　AsyncHandlerInterceptor 接口的源码

```
public interface AsyncHandlerInterceptor extends HandlerInterceptor {
    default void afterConcurrentHandlingStarted(HttpServletRequest request,
HttpServletResponse response,Object handler) throws Exception {
    }
}
```

该接口中添加了一个新方法，即 afterConcurrentHandlingStarted。该方法是用来处理异步的并发请求的，当 Controller 中有异步请求方法时会触发 afterConcurrentHandlingStarted 方法，异步请求先调用 preHandle 方法，然后执行 afterConcurrentHandlingStarted 方法，异步的线程执行完成后会依次再执行 preHandle、postHandle 和 afterCompletion 方法。afterConcurrentHandlingStarted 方法的典型用法是清理线程的局部变量。一般来说，该接口使用不多。

HandlerInterceptorAdapter 抽象类实现了 AsyncHandlerInterceptor 接口，如代码清单 8-22 所示。

代码清单 8-22　HandlerInterceptorAdapter 抽象类

```
public abstract class HandlerInterceptorAdapter implements AsyncHandlerInterceptor
{
    @Override
     public boolean preHandle(HttpServletRequest request, HttpServletResponse
response, Object handler)throws Exception {
        return true;
    }
    @Override
     public void postHandle(HttpServletRequest request, HttpServletResponse
response, Object handler,@Nullable ModelAndView modelAndView) throws Exception {
    }
    @Override
     public void afterCompletion(HttpServletRequest request, HttpServletResponse
response, Object handler,@Nullable Exception ex) throws Exception {
    }
    @Override
     public void afterConcurrentHandlingStarted(HttpServletRequest request,
HttpServletResponse response,Object handler) throws Exception {
    }
}
```

从以上源码中可以看出，除了 preHandle 方法总是返回 true，其他的方法都是空方法，所以在平时的开发过程中，可以选择实现 HandlerInterceptor 接口或者继承 HandlerInterceptorAdapter 类来实现拦截器。

接下来通过 HandlerInterceptor 接口实现一个登录拦截器。新建一个 LoginHanderInterceptor 类，如代码清单 8-23 所示。

代码清单 8-23　LoginHanderInterceptor 类

```
/** 登录拦截器 */
public class LoginHanderInterceptor implements HandlerInterceptor {
    /** 一般情况下，只会重写该方法 */
    @Override
     public boolean preHandle(HttpServletRequest request, HttpServletResponse
response, Object handler) throws Exception {
        // 判断用户是否登录
        User user = (User) request.getSession().getAttribute("user");
        if(user==null){ // 用户尚未登录则为return false; }
        return true;
    }
}
```

接下来需要在 Spring MVC 中注册拦截器，先定位到 WebMvcConfigurer 接口的 MvcConfig 类，在其 addInterceptors 方法中使用 InterceptorRegistry 注册拦截器，如代码清单 8-24 所示。

代码清单 8-24　注册拦截器

```
/** 配置拦截器，这里只配置主题切换拦截器，可以使用addInterceptor方法配置多个拦截器 */
@Override
public void addInterceptors(InterceptorRegistry registry) {
    ...
    // 注册登录拦截器
    LoginHanderInterceptor loginHanderInterceptor = new LoginHanderInterceptor();
    // addPathPatterns-- 添加需要拦截的路径;excludePathPatterns -- 不进行拦截的路径
     registry.addInterceptor(loginHanderInterceptor).addPathPatterns("/*").
excludePathPatterns("/login");
}
```

这里对所有的路径进行拦截，并排除“/login”路径。这里仅作为演示使用，在日常中，可根据项目实际情况进行配置。

8.3　JSON 数据交互

由于 JSON 类型的数据格式简单、解析方便，所以 JSON 类型的数据在接口调用、前后端传递参数过程中非常常见。在 Spring MVC 中进行 JSON 交互，主要用到了 Spring MVC 中的两个注解（RequestBody 注解和 ResponseBody 注解）。

8.3.1　Spring MVC 中 JSON 交互形式

Spring MVC 与前端的数据交互方式主要有两种，第 1 种是浏览器请求通过表单形式的 K-V 结构传递参数到后端，第 2 种是浏览器通过 JSON 字符串（JSON 格式的字符串，本质上也是 K-V 结构）传递参数到后端。这里主要讲解第 2 种方式，Spring MVC 与前端 JSON 数据的交互如图 8-4 所示。

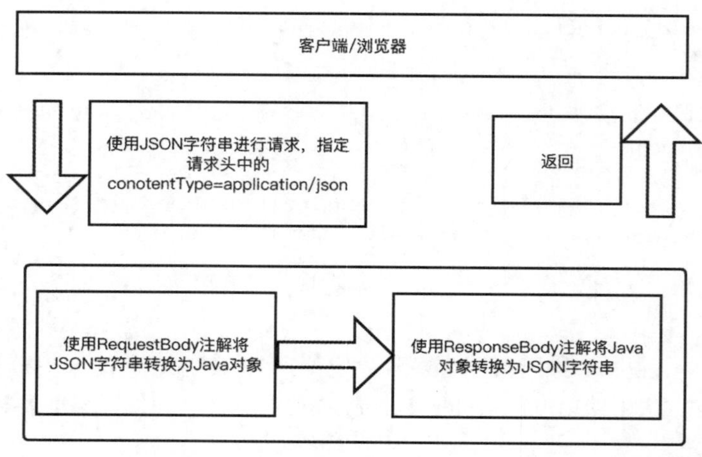

图 8-4　Spring MVC 与前端 JSON 数据的交互

8.3.2　应用实例

5.2 节已经讲解了 RequestBody 注解和 ResponseBody 注解，这里仅讲解如何使用。将 User 对象的 JSON 字符串传入 Controller 方法，并且返回 JSON 字符串到页面中展示。**提示**：MappingJackson2HttpMessageConverter 转换器需要引入两个 .jar 包，也就是在 5.2.3 小节中添加的 Jackson 包，然后新建 JsonController 类。JSON 字符串交互实例如代码清单 8-25 所示。

代码清单 8-25　JSON 字符串交互实例

```
/** JSON数据交互实例 */
@Controller
@RequestMapping("/json")
public class JsonController {
    /** 将JSON字符串转换为对象，并且将对象转换成JSON格式数据返回到前端 */
    @PostMapping(value = "/user")
    @ResponseBody
    public User user(@RequestBody User user) {
        System.out.println("-----user=" + user);
        return user;
    }
}
```

这里选择 Postman 作为演示工具。访问 http://127.0.0.1:8080/json/user，并且设置访问方式为 POST，在 Body 选项卡中选择 raw 单选项，填入 JSON 字符串，在 Content-Type 中选择 JSON (application/json) 选项，如图 8-5 所示。

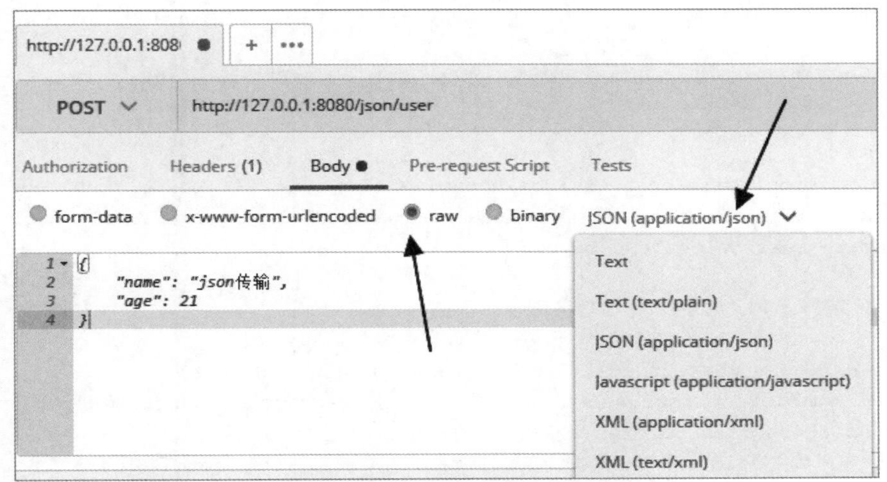

图 8-5　使用 POST 窗口接口

对象返回的结果如图 8-6 所示。

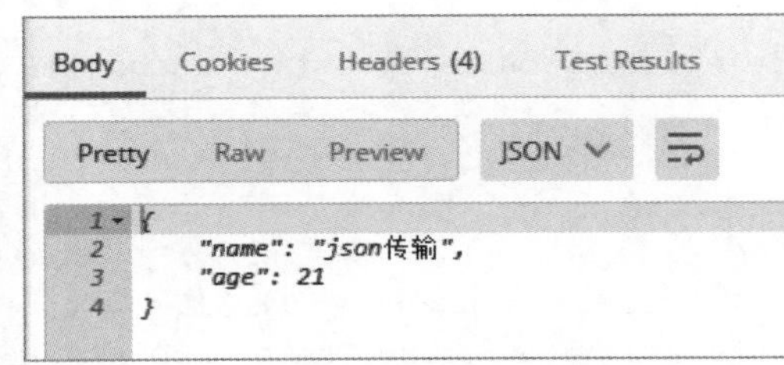

图 8-6　对象返回的结果

8.4　Spring MVC 国际化配置

6.6 节介绍了 LocaleResolver 国际化解析器，但是如何在项目中使用国际化配置还没有进行实例讲解，本节介绍如何使用国际化配置，主要有以下 2 个步骤：

（1）加载国际化的资源。

（2）输出国际化的资源。

输出国际化的资源，可以选择在页面中使用 Spring MVC 的标签库，另外还可以在 Controller 处理方法中输出国际化的数据，此时需要使用 RequestContext 中的 getMessage 方法。

实现国际化的方式有多种，在接下来的实例中，将基于 Session 实现国际化，并且在页面中展示使用 Spring MVC 的标签库实现输出国际化的资源的案例。

从 6.6 节可知，首先需要在 Spring 容器中注入名称为 localeResolver 和 messageSource 的两个 Bean 实例。这里的 localeResolver 的 Bean 实例选择 SessionLocaleResolver 对象，messageSource 的 Bean 实例选择 ReloadableResourceBundleMessageSource 对象。

在 MvcConfig 类中配置 Bean 实例，如代码清单 8-26 所示。

代码清单 8-26　配置名称为 localeResolver 和 messageSource 的两个 Bean 实例

```
/** 通过SessionLocaleResolver解析器实现国际化 */
@Bean
public SessionLocaleResolver localeResolver() {
    SessionLocaleResolver sessionLocaleResolver = new SessionLocaleResolver();
    return sessionLocaleResolver;
}
/** 加载国际化资源，配置国际化资源路径 */
@Bean
public ReloadableResourceBundleMessageSource messageSource() {
     ReloadableResourceBundleMessageSource reloadResourceBundleMessageSource = new
ReloadableResourceBundleMessageSource();
    // 前缀名称为i18n，资源包基名(globalization/i18n)，如果当前语言环境下的资源包没有，则默认
为i18n.properties文件
    // 如果为中文环境，则为i18n_zh_CN.properties文件
    reloadResourceBundleMessageSource.setBasename("classpath:globalization/i18n");
    // 编码
    reloadResourceBundleMessageSource.setDefaultEncoding("utf-8");
    // 缓存时间，单位为s，就是隔多久检查一次文件是否修改，如果修改，则动态加载；默认为-1，不动态更新
    reloadResourceBundleMessageSource.setCacheSeconds(10);
     /*设置useCodeAsDefaultMessage，默认为false，如果Spring在ResourceBundle中找不到
messageKey，则抛出NoSuchMessageException，把它设置为True；如果在ResourceBundle中找不到
messageKey，则不会抛出异常，而是使用messageKey作为返回值 */
    reloadResourceBundleMessageSource.setUseCodeAsDefaultMessage(true);
    return reloadResourceBundleMessageSource;
}
```

接下来添加国际化的资源。在 resources 目录下创建 globalization 目录，然后创建 properties 文件，国际化资源文件如图 8-7 所示。

图 8-7　国际化资源文件

每个文件对应的资源如代码清单 8-27 所示。

代码清单 8-27　每个文件对应的资源

```
i18n.properties文件内容如下
title=默认的国际化配置
welcome=你好，欢迎访问
name=Spring MVC学习

i18n_en_US.properties文件内容如下
title=Default international configuration
welcome=Hello,Welcome to visit
name=Spring MVC learning

i18n_zh_CN.properties文件内容如下
title=中文的国际化配置
welcome=你好，欢迎访问
name=Spring MVC学习
```

虽然可以通过 Session 来切换语言，但那样比较麻烦，最好的方式是使用国际化的拦截器，通过 URL 传递参数来实现语言切换，也就是添加 LocaleChangeInterceptor 拦截器。在 MvcConfig 类的 addInterceptors 方法中注册拦截器，如代码清单 8-28 所示。

代码清单 8-28　添加 LocaleChangeInterceptor 拦截器

```
// 注册国际化拦截器
LocaleChangeInterceptor localeChangeInterceptor = new LocaleChangeInterceptor();
// 配置国际化语言参数的名称为locale。如不指定，默认为locale
localeChangeInterceptor.setParamName("localeName");
registry.addInterceptor(localeChangeInterceptor);
```

接下来创建前台页面，前台页面在 views/i18n/index.jsp 中，如代码清单 8-29 所示。

代码清单 8-29　国际化页面

```
<!-- 国际化界面演示 -->
<%@ page contentType="text/html;charset=UTF-8" language="java" %>
<%@ taglib uri="http://www.springframework.org/tags" prefix="spring" %>
<%@ taglib prefix="c" uri="http://java.sun.com/jsp/jstl/core" %>
<html>
<head>
<title><spring:message code="title" scope="session"></spring:message></title>
</head>
<body>
<h1><spring:message code="welcome" scope="session"></spring:message></h1>
<spring:message code="name" scope="session"></spring:message><br/>
<a href="<c:url value='/i18n/index?localeName=zh_CN' />">中文</a>
<a href="<c:url value='/i18n/index?localeName=en_US' />">English</a>
</body>
</html>
```

由于这里只是做页面的跳转，所以可以直接在 MvcConfig 类的 addViewControllers 方法中添加映射，如代码清单 8-30 所示。

代码清单 8-30　添加页面映射

```
// 国际化页面展示
registry.addViewController("/i18n/index").setViewName("/i18n/index");
```

最后在浏览器中访问地址 http://localhost:8080/i18n/index，可以通过单击“中文”链接和“English”链接切换语言，如图 8-8 所示。

图 8-8　国际化页面演示

8.5　总结

至此，Spring MVC 的讲解告一段落。Spring MVC 中最关键的部分就是第 6 章介绍的九大组件，虽然列出了大量的源码，但笔者觉得还是远远不够，毕竟只是列出了关键组件类的源码，还有很多辅助的组件及子类没有覆盖，建议读者在阅读本书的时候，配合 Spring 的源码进行实践。

第Ⅲ篇　MyBatis 框架

本篇讲解 MyBatis 框架中的组件源码，XML 配置和代码注解配置，MyBatis 中的缓存机制源码和缓存的应用，动态 SQL，以及 MyBatis Generator 逆向代码生成工具的实战应用。

通过本篇的学习，希望读者能够熟练掌握 MyBatis 的相关知识，能娴熟地进行数据库的增、删、改、查操作。希望读者能记住：在企业项目开发中，数据库的操作方式是固定的，但业务是变化的，所以尽量在工作中积累一套自己的数据库操作代码自动生成工具。在这里为读者推荐 GitHub 中的框架，该框架可以全自动生成从 Controller 层到 Dao 层的所有数据操作代码。

第9章 MyBatis 四大核心组件

本章要点

1. MyBatis 组件的源码分析
2. SqlSessionFactoryBuilder 源码分析
3. XMLConfigBuilder 源码分析
4. SqlSessionFactory 源码分析
5. SqlSession 源码分析
6. MapperAnnotationBuilder 源码分析

9.1 MyBatis 四大核心组件简介

在前面的 1.2.1 小节中介绍了 MyBatis。MyBatis 是一个持久层的框架，可将它理解为与数据库做交互的框架。与 MyBatis 功能类似的框架还有 Hibernate 等。MyBatis 对 JDBC 操作数据库的过程进行了封装，使开发者更加专注于 SQL 本身，而不需要处理如注册驱动、创建 Connection（与数据库的连接）、创建 Statement（负责执行 SQL）及结果集检索等烦琐的过程代码。可以通过 XML 或者注解的方式来配置各种 Statement，并且通过 Java 对象和 Statement 中的 SQL 进行映射，生成最终需要执行的 SQL 语句，最后通过 MyBatis 执行 SQL 并且将 ResultSet（结果集）映射成 Java 对象进行返回。

MyBatis 的核心组件主要有以下 4 个：

（1）SqlSessionFactoryBuilder（工厂构造器）：主要根据代码或者配置来生成对应的 SqlSessionFactory 对象，采用的是构造者模式（构造者模式是一种对象的创建模式。能够将复杂对象的内部构成特征和对象的构建过程完全分开）。

（2）SqlSessionFactory（工厂接口）：负责生成 SqlSession 实例的工厂类，一般作为单例使用，在 MyBatis 中默认的实现为 DefaultSqlSessionFactory。

（3）SqlSession（会话）：执行 SQL 语句并返回结果集，通过会话可以获取 Mapper 接口。现在一般不会在业务逻辑代码中写会话，而是使用 MyBatis 提供的 SQL Mapper 接口编程技术。

（4）SQL Mapper（映射器）：由一个 Java 接口和一个 XML 文件（或 Mapper 方法上的注解）构成，通过对应的 SQL 和映射规则，可以执行 SQL 语句并返回结果。

接下来针对这 4 个组件进行源码分析。

9.2 SqlSessionFactoryBuilder 源码分析

SqlSessionFactoryBuilder（工厂构造器）根据开发者的配置信息或者代码生成 SQL Session-Factory。MyBatis 可以通过 XML 或者代码这两种方式配置 SqlSessionFactory。

MyBatis 的初始化阶段就是 SqlSessionFactory 的创建阶段，也就是将数据库账户密码等信息的键值对设置到 Configuration 的相关变量中的阶段。利用初始化好的 Configuration 对象创建默认的 SqlSessionFactory（DefaultSqlSessionFactory）。

SqlSessionFactoryBuilder 的源码如代码清单 9-1 所示。

代码清单 9-1　SqlSessionFactoryBuilder 源码

```
public class SqlSessionFactoryBuilder {
    public SqlSessionFactory build(Reader reader) {
        return build(reader, null, null);
    }
    public SqlSessionFactory build(Reader reader, String environment) {
```

```
        return build(reader, environment, null);
    }
    public SqlSessionFactory build(Reader reader, Properties properties) {
        return build(reader, null, properties);
    }
     public SqlSessionFactory build(Reader reader, String environment, Properties
properties) {
        try {
              XMLConfigBuilder parser = new XMLConfigBuilder(reader, environment,
properties);
        // 传入配置，创建SqlSessionFactory对象
            return build(parser.parse());
        } catch (Exception e) {
            throw ExceptionFactory.wrapException("Error building SqlSession.", e);
        } finally {
            ErrorContext.instance().reset();
            try {
                reader.close();
            } catch (IOException e) {
            }
        }
    }
    public SqlSessionFactory build(InputStream inputStream) {
        return build(inputStream, null, null);
    }
    public SqlSessionFactory build(InputStream inputStream, String environment) {
        return build(inputStream, environment, null);
    }
    public SqlSessionFactory build(InputStream inputStream, Properties properties)
{
        return build(inputStream, null, properties);
    }
    /**
     创建SqlSessionFactory单例
     inputStream:MyBatis的全局XML配置文件输入流
     environment:指定该SqlSessionFactory 的数据库环境，默认是default
     properties:设置动态常量，这些动态常量将会和XML的properties文件中的常量合并在一起
     */
     public SqlSessionFactory build(InputStream inputStream, String environment,
Properties properties) {
        try {
           // 创建XMLConfigBuilder，XML文件解析器
           XMLConfigBuilder parser = new XMLConfigBuilder(inputStream, environment,
properties);
            // 获取Configuration对象，也就是MyBatis配置文件中的信息
            // 传入Configuration对象，创建SqlSessionFactory对象
            return build(parser.parse());
        } catch (Exception e) {
            throw ExceptionFactory.wrapException("Error building SqlSession.", e);
        } finally {
            ErrorContext.instance().reset();
```

```
            try {
                inputStream.close();
            } catch (IOException e) {
            }
        }
    }
    // 从这里可知，MyBatis中默认的SqlSessionFactory实现为DefaultSqlSessionFactory
    public SqlSessionFactory build(Configuration config) {
        return new DefaultSqlSessionFactory(config);
    }
}
```

从源码中可以看到，SqlSessionFactoryBuilder 的构造函数有很多，但是最后都通过 XMLConfigBuilder 解析 XML 流，然后进行 new DefaultSqlSessionFactory(config) 操作。接下来针对 XMLConfigBuilder，进行一系列的源码分析。

9.3 XMLConfigBuilder 源码分析

XMLConfigBuilder 类的作用如下：

（1）解析 XML 文件流，生成 XML 文件中元素节点的描述信息（XNode 对象，即 XML 文件的 Java 描述对象）。

（2）通过 root 根节点 XNode 对象，解析每个子节点的 XNode 对象，获取属性或子节点属性，生成键值对。在 XMLConfigBuilder 中创建并初始化 Configuration，并将节点的键值对设置到 Configuration 实例中。

另外，再简单地理解一下 Configuration，其相当于数据类包含了几乎所有的 MyBatis 配置信息。可以查看 Configuration 类的源码，Configuration 类中有很多 Map、配置信息等。

9.3.1 XMLConfigBuilder 构造函数

使用 XMLConfigBuilder 读取 XML 文件并解析，是 MyBatis 初始化的第 1 步。接下来查看 XML ConfigBuilder 中的部分实例化源码，如代码清单 9-2 所示。

代码清单 9-2　XMLConfigBuilder 构造函数

```
// 创建XMLConfigBuilder的过程
public XMLConfigBuilder(Reader reader, String environment, Properties props) {
// 通过XML的文件流，创建XPathParser解析对象
      this(new XPathParser(reader, true, props, new XMLMapperEntityResolver()),
environment, props);
}
private XMLConfigBuilder(XPathParser parser, String environment, Properties props)
{
    // 通过XPathParser 对象，创建Configuration，注册一些别名供后面解析XML使用，如JDBC等
    super(new Configuration());
```

```
    ErrorContext.instance().resource("SQL Mapper Configuration");
// 将动态的Properties键值对，设置到Configuration对象variables中
    this.configuration.setVariables(props);
    this.parsed = false;
// 指定数据库环境
    this.environment = environment;
    this.parser = parser;
}
```

至此，Configuration 实例就创建好了，在代码清单 9-1 中，传入的是 parser.parse 返回的对象。

9.3.2 parse 与 parseConfiguration 方法

XMLConfigBuilder 类中 parse 方法的源码如代码清单 9-3 所示。

代码清单 9-3　XMLConfigBuilder 类中 parse 方法

```
public Configuration parse() {
    if (parsed) {
        throw new BuilderException("Each XMLConfigBuilder can only be used once.");
    }
    parsed = true;
// 利用XPathParser的evalNode方法从XML文档中抽取“/configuration”对应的内容，创建XNode对象
// parseConfiguration方法负责将解析处理的键值对设置到对应的Configuration实例中
    parseConfiguration(parser.evalNode("/configuration"));
    return configuration;
}
```

parseConfiguration 方法的源码如代码清单 9-4 所示。

代码清单 9-4　parseConfiguration 方法

```
//解析XNode对象，生成Configuration对象，将解析处理的键值对设置到对应的Configuration实例中
private void parseConfiguration(XNode root) {
    try {
            /* propertiesElement 方法是解析“properties”节点，存放到 configuration 对象的
variables 对象中，如配置数据库的 username，password 等信息 */
        propertiesElement(root.evalNode("properties"));
            /* 解析 XML 中的“settings”节点并进行验证。这些值配置了 MyBatis 的运行方式，如开启缓
存，懒加载等。没有设置的值将会使用默认值。这里还没有设置，在后面的 settingsElement 方法才会设置
settings 元素的配置信息到 Configuration 中，改写 Configuration 中相关的默认值 */
        Properties settings = settingsAsProperties(root.evalNode("settings"));
        loadCustomVfs(settings);
        /* 解析 XML 中的“typeAliases”节点，起别名（指定包名的类别名）并设置到 Configuration */
        typeAliasesElement(root.evalNode("typeAliases"));
          /* 解析 XML 中的“plugins”节点并设置到 Configuration，该节点定义一些插件，用来拦截某
些类，进行一些特定的操作，如 SqlSession 中的四个核心插件：ParameterHandler、ResultSetHandler、
StatementHandler、Executor*/
        pluginElement(root.evalNode("plugins"));
         /* 解析 XML 中的“objectFactory”节点并设置到 Configuration，定义对象工厂。对象工厂用
来创建 MyBatis 的返回结果对象 */
```

```
        objectFactoryElement(root.evalNode("objectFactory"));
        /* 解析 XML 中的“objectWrapperFactory”节点并设置到 Configuration，对象的包装工厂类，包装 Object 实例 */
        objectWrapperFactoryElement(root.evalNode("objectWrapperFactory"));
        /* 解析 XML 中的“reflectorFactory”节点并设置到 Configuration，创建 Reflector 类的反射工厂 */
        reflectorFactoryElement(root.evalNode("reflectorFactory"));
        /* 将获取的 Properties，也就是前面解析到的“settings”，将键值对设置到 Configuration 的对应变量中，这个是比较关键的一个方法 */
        settingsElement(settings);
        /* 解析 XML 中的“environments”节点，定义数据库的环境，并设置到 Configuration 对象中。可配置多个 Environment。每个配置节点都会有对应的 transactionManager 和 dataSource*/
        environmentsElement(root.evalNode("environments"));
        /* 解析 XML 中的“databaseIdProvider”节点，设置数据库的 Id，其实就是设置数据库的厂商唯一标识，并设置到 Configuration 对象中。也就是说，MyBatis 可以根据不同的数据库厂商标识来执行其支持的 SQL 语句 */
        databaseIdProviderElement(root.evalNode("databaseIdProvider"));
        /* 解析 XML 中的“typeHandlers”节点，定义类型处理器，也就是将数据库中的类型的值，对应到 Java 中对象的类型，并进行转换 */
        typeHandlerElement(root.evalNode("typeHandlers"));
        /* 解析 XML 中的“mappers”节点，定义映射器，也就是接口方法对应的 SQL 映射语句。只需要定义好 XML 映射文件的路径即可 */
        mapperElement(root.evalNode("mappers"));
    } catch (Exception e) {
        throw new BuilderException("Error parsing SQL Mapper Configuration. Cause: " + e, e);
    }
}
```

可以看出来，parseConfiguration 方法分别读取 XML 配置文件中的节点信息，并且将节点中的信息设置到对应的 Configuration 对象变量中。

对于 Properties 对象，这里不展开介绍，将 Properties 简单地理解为键值对 Map 即可。

下面分析其中的 loadCustomVfs 方法和 settingsElement 方法（建议在这些关键部分添加断点，当运行项目时，可以通过 DEBUG 查看运行情况，以及对象的值）。

9.3.3 loadCustomVfs 方法

loadCustomVfs 方法源码如代码清单 9-5 所示。

代码清单 9-5 loadCustomVfs 方法

```
/**
 * 加载虚拟文件系统的配置，读取服务器的资源
 * 获取用户自定义的VFS实现，配置在“settings”元素中
 * setting中放name为vfsImpl的，值为VFS实现类的全限定名称，多个则用“,”号隔开
 * @throws ClassNotFoundException 类未找到异常
 */
private void loadCustomVfs(Properties props) throws ClassNotFoundException {
    //获取名称为vfsImpl的值
    String value = props.getProperty("vfsImpl");
```

```
        if (value != null) {
            //实现类可以有多个，多个用“,”号隔开
            String[] clazzes = value.split(",");
            for (String clazz : clazzes) {
                if (!clazz.isEmpty()) {
                    @SuppressWarnings("unchecked")
                    //反射加载VFS实现类
                    Class<? extends VFS> vfsImpl = (Class<? extends VFS>) Resources.class
    ForName(clazz);
                    //设置到configuration实例中
                    configuration.setVfsImpl(vfsImpl);
                }
            }
        }
    }
```

该方法运行完后，还没有将 settings 的配置信息设置到 configuration 对象中，这是由于其他节点的元素解析结果会在设置 settings 时用到，如 typeAilases 的别名、plugins 插件的拦截器等。

9.3.4 settingsElement 方法

settingsElement 方法的作用是将 settings 配置信息设置到 configuration 对象中，其源码如代码清单 9-6 所示。

代码清单 9-6 settingsElement 方法

```
//将XML配置文件中解析settings节点出来的键值对设置到configuration实例中
private void settingsElement(Properties props) throws Exception {
    /**
     * 该属性是用来指定MyBatis应该如何自动映射数据库的列到Java POJO对象的属性
     * 默认值为PARTIAL。从AutoMappingBehavior枚举类中可以知道设置的值
     * 一共有三个值:NONE-表示禁用自动映射
     * PARTIAL-表示默认的值，仅自动映射内部没有嵌套结果的映射，映射简单的结果集
     * FULL-表示自动映射任何复杂度的结果映射(包含嵌套或其他)
     */
    configuration.setAutoMappingBehavior(AutoMappingBehavior.valueOf(props.getProp
erty("autoMappingBehavior", "PARTIAL")));
    /**
     * 设置MyBatis检测到自动映射时遇到未知列(或未知属性类型)时该怎么处理
     * 默认为NONE。参考AutoMappingUnknownColumnBehavior枚举类中的值
     * “NONE” - 表示不做任何的处理
     * “WARNING” - 表示输出警告日志，org.apache.ibatis.session.AutoMappingUnknownColumnBehavior
类的日志级别必须设置为WARN
     * “FAILING” - 表示映射失败，抛出SqlSessionException异常
     */
    configuration.setAutoMappingUnknownColumnBehavior(AutoMappingUnknownColumnBehavi
or.valueOf(props.getProperty("autoMappingUnknownColumnBehavior", "NONE")));
    /**
     * 设置是否开启二级缓存(Mapper级别的缓存)
```

```
    * MyBatis的一级缓存(即SqlSession内的HashMap)默认是开启的(二级缓存与一级缓存区别在于二级缓存的范围更大，不同SqlSession可以共享一个Mapper中的二级缓存区域)
     * 二级缓存默认也是开启的(但是还需要在Mapper的XML文件中进行使用cache开启二级缓存才会生效)，注意缓存的对象实现Serializable接口
    */
     configuration.setCacheEnabled(booleanValueOf(props.getProperty("cacheEnabled"), true));
     /* 指定MyBatis创建懒加载(即延迟加载)对象(核心技术就是使用代理模式)时，使用的动态代理框架。CGLIB或者JAVASSIST(这两个都是能够操作Java字节码的框架)*/
     configuration.setProxyFactory((ProxyFactory) createInstance(props.getProperty("proxyFactory")));

    /* 是否开启懒加载。如果开启，所有的关联对象都会进行懒加载。默认是关闭的，也就是所有的对象都是即时加载*/
    configuration.setLazyLoadingEnabled(booleanValueOf(props.getProperty("lazyLoadingEnabled"), false));

    /* 访问有延迟加载属性对象的任意方法都会加载对象内的所有属性。默认为false，只有在使用时才会进行加载，也就是按需加载*/
    configuration.setAggressiveLazyLoading(booleanValueOf(props.getProperty("aggressiveLazyLoading"), false));

    //允许多种结果集从一个单独的语句中返回(需要合适的驱动支持)，默认true代表允许
    configuration.setMultipleResultSetsEnabled(booleanValueOf(props.getProperty("multipleResultSetsEnabled"), true));

    //使用列标签代替列名，默认true代表替代
     configuration.setUseColumnLabel(booleanValueOf(props.getProperty("useColumnLabel"), true));

    /*是否允许JDBC自动生成主键。默认为false不允许。需要合适的驱动支持，如果设置为true，则有些驱动可能会不兼容*/
    configuration.setUseGeneratedKeys(booleanValueOf(props.getProperty("useGeneratedKeys"), false));

    /**
     * 设置MyBatis默认的执行器。取值参考ExecutorType枚举。默认为SIMPLE
       * SIMPLE 普通的执行器。默认执行简单SQL语句，也就是一次数据库的连接只进行发送一条SQL语句，在做批量执行的时候需要用<foreach> 标签拼接成一条长的SQL语句，然后一次执行完毕。相当于JDBC的Statement的execute(sql)
       * REUSE 重用预处理语句(prepared statements)。在MyBatis执行Dao方法时，相当于JDBC重用一条SQL语句，再通过Statement传入多项参数值，然后执行Statement的executeUpdate()或executeBatch()
       * BATCH 重用语句并执行批量更新。在MyBatis执行Dao方法时，相当于JDBC语句的 Statement的addBatch(sql)，即仅仅是将执行SQL加入到批量计划。只有在commit之前才会执行Statement的execteBatch()。所以，当使用批量更新时，如果不进行commit，则SQL是不会执行的
     * 建议defaultExecutorType使用默认的即可。否则和平时开发的不同，可能会出现一些与你预期不同的结果
     */
    configuration.setDefaultExecutorType(ExecutorType.valueOf(props.getProperty("defaultExecutorType", "SIMPLE")));
```

```
    //设置超时时间，也就是设置数据库驱动等待数据库响应的时间，单位：秒
    configuration.setDefaultStatementTimeout(integerValueOf(props.getProperty("def
aultStatementTimeout"), null));

    //为驱动的结果集获取数量设置一个提示值，防止出现OOM。也就是说，执行SQL的最大结果集行数
    configuration.setDefaultFetchSize(integerValueOf(props.getProperty("defaultFetc
hSize"), null));

    //开启自动驼峰命名(camel case)规则映射，即将数据库列名user_column映射为Java属性名
userColumn
    configuration.setMapUnderscoreToCamelCase(booleanValueOf(props.getProperty("ma
pUnderscoreToCamelCase"), false));

    //允许在嵌套语句中使用分页(RowBounds)。如果允许使用则设置为false
    configuration.setSafeRowBoundsEnabled(booleanValueOf(props.getProperty("safeRo
wBoundsEnabled"), false));

    /**
     * 设置本地缓存机制。本地缓存用来加速嵌套查询和防止循环引用
     * 如果值为SESSION，则缓存本SqlSession中的所有查询语句，也就是一个会话中的所有查询都会缓存
     * 如果值为STATEMENT，则相同SqlSession的同一个调用语句才进行缓存，也就是本地会话仅仅作用在
同一个调用语句上，对于相同的SqlSession，如果是不同的调用语句也不会共享缓存
     */
     configuration.setLocalCacheScope(LocalCacheScope.valueOf(props.getProperty
("localCacheScope", "SESSION")));

    /*当没有为参数提供特定的JDBC类型时，为Java中的空值(NULL)指定JDBC类型。某些驱动需要指定列的
JDBC类型，多数情况直接用一般类型即可，如 NULL、VARCHAR 或 OTHER */
     configuration.setJdbcTypeForNull(JdbcType.valueOf(props.getProperty
("jdbcTypeForNull", "OTHER")));

    //指定哪些方法触发延迟加载关联的对象。多个方法名之间用逗号隔开，如"equals,clone,
hashCode,toString"
    configuration.setLazyLoadTriggerMethods(stringSetValueOf(props.getProperty("la
zyLoadTriggerMethods"), "equals,clone,hashCode,toString"));

    //允许在嵌套语句中使用分页(ResultHandler)。如果允许使用则设置为false
    configuration.setSafeResultHandlerEnabled(booleanValueOf(props.getProperty("sa
feResultHandlerEnabled"), true));

    /*指定用来生成动态SQL语句的默认语言。值为一个类型别名或完全限定类名。默认为org.apache.
ibatis.scripting.xmltags.XMLLanguageDriver*/
    configuration.setDefaultScriptingLanguage(resolveClass(props.getProperty("defa
ultScriptingLanguage")));

    /*从MyBatis 3.4.5开始，指定Enum使用默认的TypeHandler。TypeHandler是一个Java的全限定
名:org.apache.ibatis.type.EnumTypeHandler*/
    @SuppressWarnings("unchecked")
     Class<? extends TypeHandler> typeHandler = (Class<? extends TypeHandler>)
resolveClass(props.getProperty("defaultEnumTypeHandler"));
```

```
    configuration.setDefaultEnumTypeHandler(typeHandler);

    /*设置callSettersOnNulls，指定结果集中的值为null值的时候是否调用Java映射对象的setter（如果是Map对象，则为put）方法，该属性对于有Map.keySet()方法依赖或者说null值初始化的时候是很有用的。但是注意，基本类型（int，boolean等）是不能被设置成null的*/
    configuration.setCallSettersOnNulls(booleanValueOf(props.getProperty("callSettersOnNulls"), false));

    /*允许使用方法签名中的形参名称作为SQL语句的参数名称。为了使用该特性，你的工程必须采用Java 8+编译，并且加上-parameters选项。（从MyBatis 3.4.1开始）*/
    configuration.setUseActualParamName(booleanValueOf(props.getProperty("useActualParamName"), true));
    /*当返回行的所有列都是空时，MyBatis默认返回null。 当开启这个设置时，MyBatis会返回一个空实例（也就是返回一个new的空对象）。也适用于嵌套的结果集。（从MyBatis 3.4.2开始）*/
    configuration.setReturnInstanceForEmptyRow(booleanValueOf(props.getProperty("returnInstanceForEmptyRow"), false));

    //指定 MyBatis 增加到日志名称的前缀。设置后每一条日志都会添加这个前缀
    configuration.setLogPrefix(props.getProperty("logPrefix"));

    /*指定 MyBatis 所用日志的具体实现，未指定时将自动查找。值为SLF4J | LOG4J | LOG4J2 | JDK_LOGGING | COMMONS_LOGGING | STDOUT_LOGGING | NO_LOGGING*/
    @SuppressWarnings("unchecked")
    Class<? extends Log> logImpl = (Class<? extends Log>) resolveClass(props.getProperty("logImpl"));
    configuration.setLogImpl(logImpl);

    /* 指定一个提供Configuration实例的类（生成Configuration对象的工厂类的类型别名或者全类名）。这个被返回的Configuration实例用来加载被反序列化对象的懒加载属性值。 这个类必须包含一个签名方法static Configuration getConfiguration(). （从MyBatis的3.2.3 版本开始）*/
    configuration.setConfigurationFactory(resolveClass(props.getProperty("configurationFactory")));
}
```

settingsElement 方法就是将一系列的配置设置到 Configuration 实例中。**提示**：配置的默认值可能会随着 MyBatis 版本的不同而改变。本书所有实例中 MyBatis 的版本为 3.4.6。

将 settingsElement 方法分析完后，相信读者应对 MyBatis 的配置已经熟练掌握了。关于 XML 解析的代码，设置别名的代码，以及读取并解析 plugin 元素部分不进行讲解。接下来分析 environmentsElement 方法，也就是定义数据库的环境。

9.3.5 environmentsElement 方法

environmentsElement 方法读取并解析 environments 元素（environments 是存储有关数据库连接数据的配置信息，可以配置多个数据库的连接环境，提高 SQL 语句的适用性。因为不同的数据库，实际的 SQL 语句一般都不同），并设置到 Configuration 实例中。接下来查看源码，如代码清单 9-7 所示。

代码清单 9-7 environmentsElement 方法源码

```
private void environmentsElement(XNode context) throws Exception {
```

```
    if (context != null) {
         /*environment的值为前面SqlSessionFactoryBuilder().build(InputStream inputStream,
String environment, Properties properties)时传入的environment值，该值进行指定了
SqlSessionFactory所用的数据库环境*/
        if (environment == null) {
            //如果不声明，则使用XML中default元素对应的数据库环境
            environment = context.getStringAttribute("default");
        }
        for (XNode child : context.getChildren()) {
            String id = child.getStringAttribute("id");
              /*isSpecifiedEnvironment方法是判断environment是否是我们指定的environment。
每个environment对应一个SqlSessionFactory实例，运行不同environment环境中的SQL，对应不同的
SqlSessionFactory进行执行*/
            if (isSpecifiedEnvironment(id)) {
                 /*获取transactionManager元素，创建TransactionFactory事务管理器实例，并进
行初始化*/
                TransactionFactory txFactory = transactionManagerElement(child.eval
Node("transactionManager"));
                // 获取dataSource元素，创建DataSourceFactory实例，并进行初始化
                DataSourceFactory dsFactory = dataSourceElement(child.
evalNode("dataSource"));
                //从DataSourceFactory工厂类中获取DataSource数据源
                DataSource dataSource = dsFactory.getDataSource();
                /*Builder设计模式创建Environment对象,将id,transactionFactory,dataSourc
e 成员变量设置到Environment对象中*/
                        Environment.Builder environmentBuilder = new Environment.
Builder(id)
                          .transactionFactory(txFactory)
                          .dataSource(dataSource);
                //将创建好的Environment对象设置到configuration实例中
                configuration.setEnvironment(environmentBuilder.build());
            }
        }
    }
}
```

还有 databaseIdProviderElement 方法，根据不同的数据库执行不同的 SQL 语句。其实 environments 一般用的场景不多，因为在一个项目中同时使用 MySQL 和 Oracle 数据库的情况很少。如果要使用多数据源，就需要为每个数据源都配置一个 SqlSessionFactory（获取不同的 SqlSessionFactory 需要指定 environment，不指定则会获取默认的）。关于多数据驱动的部分不再进行讲解。

9.3.6 typeHandlerElement 方法

typeHandlerElement 方法用来读取并解析 XML 中的 typeHandler 元素（typeHandler 的作用就是将参数从 javaType 转化为 jdbcType，或者从数据库取出结果时把 jdbcType 转化为 javaType），配置类型处理器，并且注册到 typeHandlerRegistry 中，如代码清单 9-8 所示。

代码清单 9-8 typeHandlerElement 方法

```
private void typeHandlerElement(XNode parent) throws Exception {
    if (parent != null) {
        for (XNode child : parent.getChildren()) {
            if ("package".equals(child.getName())) {
                String typeHandlerPackage = child.getStringAttribute("name");
                //调用TypeHandlerRegistry.register，遍历包下所有添加了相关注解的类
                typeHandlerRegistry.register(typeHandlerPackage);
            } else {
                /**
                 * 这里是配置的是typeHandler子节点，如XML配置
                 *  <typeHandlers>
                 *    <typeHandler handler="com.uifuture.handler.DemoTypeHandler" .../>
                 *  </typeHandlers>
                 * 读取javaType属性，对应Java类型
                 */
                String javaTypeName = child.getStringAttribute("javaType");
                //读取jdbcType属性，对应数据库中的类型
                String jdbcTypeName = child.getStringAttribute("jdbcType");
                 /*读取handler属性，为类处理器。例如，要将int转换为Java中的Integer，就需要一
个IntegerTypeHandler*/
                String handlerTypeName = child.getStringAttribute("handler");
                Class<?> javaTypeClass = resolveClass(javaTypeName);
                JdbcType jdbcType = resolveJdbcType(jdbcTypeName);
                Class<?> typeHandlerClass = resolveClass(handlerTypeName);
                if (javaTypeClass != null) {
                    if (jdbcType == null) {
                        typeHandlerRegistry.register(javaTypeClass, typeHandlerClass);
                    } else {
                         typeHandlerRegistry.register(javaTypeClass, jdbcType, type
HandlerClass);
                    }
                } else {
                    typeHandlerRegistry.register(typeHandlerClass);
                }
            }
        }
    }
}
```

在 typeHandlerElement 方法中，主要读取解析了 typeHandlers 元素。typeHandlers 元素的子元素可以为 package 或 typeHandler。根据 javaType、jdbcType 和 handler 属性的值（别名或者全限定类名），创建并初始化对应的实例，最后把实例注册到 typeHandlerRegistry 中。

9.3.7 mapperElement 方法

mappers 节点元素定义了配置包含 SQL 语句的 XML 文件的路径。XML 文件的路径有多种配置方法，可以使用相对路径，也可以使用绝对 URL 路径，还可以通过 Java 类名和使用包名设置等。

在代码中会分别罗列 4 种情况，如代码清单 9-9 所示。

代码清单 9-9　mapperElement 方法

```
private void mapperElement(XNode parent) throws Exception {
    if (parent != null) {
        for (XNode child : parent.getChildren()) {
            if ("package".equals(child.getName())) {
                String mapperPackage = child.getStringAttribute("name");
                // ① 使用包名自动扫描设置
                //自动扫描该包下所有的映射器
                configuration.addMappers(mapperPackage);
            } else {
                //获取mappers节点下mapper节点的resource属性值
                String resource = child.getStringAttribute("resource");
                //获取mappers节点下mapper节点的url属性值
                String url = child.getStringAttribute("url");
                //获取mappers节点下mapper节点的class属性值
                String mapperClass = child.getStringAttribute("class");
                //从下面的代码中可以看出class,url,resource只能设置一个
                if (resource != null && url == null && mapperClass == null) {
                    /**
                     * ② 使用类路径(相对路径)设置
                     * 如果是使用resource ，也就是使用了类路径设置。则在XML表示为
                     *  <mappers>
                     *    <mapper resource="com/uifuture/mapper/UserMapper.xml"/>
                     *        ...可以配置多个<mapper>
                     *  </mappers>
                     */
                    ErrorContext.instance().resource(resource);
                    //读取resource属性指定的Mapper.XML配置文件成InputStream
                    InputStream inputStream = Resources.get Resource As Stream (resource);
                    /*调用XMLMapperBuilder，需要注意的是这里是for循环，所以每个Mapper文件都
会新建一个XMLMapperBuilder对象进行解析*/
                    XMLMapperBuilder mapperParser = new XMLMapperBuilder(inputStream,
configuration, resource, configuration.getSqlFragments());
                    mapperParser.parse();
                } else if (resource == null && url != null && mapperClass == null) {
                    /**
                     * ③ 使用绝对路径设置
                     * 读取url属性指定的路径设置。在XML中对应的配置为
                     *  <mappers>
                     *    <mapper url="file:///D:/var/mappers/AuthorMapper.xml"/>
                     *  </mappers>
                     */
                    ErrorContext.instance().resource(url);
                    //加载文件
                    InputStream inputStream = Resources.getUrlAsStream(url);
                    //使用XMLMapperBuilder 解析文件
```

```
                    XMLMapperBuilder mapperParser = new
XMLMapperBuilder(inputStream, configuration, url, configuration.getSqlFragments());
                    mapperParser.parse();
                } else if (resource == null && url == null && mapperClass != null) {
                    /**
                     * ④ 使用Java类名设置
                     * 读取class属性的Java类名，例如XML中配置
                     *  <mappers>
                     *     <mapper class="com.uifuture.entity.UserMapper"/>
                     *  </mappers>
                     */
                    Class<?> mapperInterface = Resources.classForName(mapperClass);
                    configuration.addMapper(mapperInterface);
                } else {
                    throw new BuilderException("A mapper element may only specify a
url, resource or class, but not more than one.");
                }
            }
        }
    }
}
```

在该方法中，进行了 Mapper 文件的解析，通过 package 和 class 的方式都可以直接添加到 configuration 实例中。使用 url 和 resource 方式的，都需要创建 XMLMapperBuilder 实例，并通过 XMLMapperBuilder 实例中的 parse() 方法，解析并添加到 configuration 实例中。

到此，关于 parseConfiguration 方法基本都解析完成了。里面进行的都是一些 MyBatis 初始化工作，并且在运行中也会用到的配置。SqlSessionFactory 的创建也基本完成。

9.4 SqlSessionFactory 源码分析

SqlSessionFactory（工厂接口）类的最核心的作用是生成 SqlSession 对象。SqlSessionFactory 的默认实现是 DefaultSqlSessionFactory。MyBatis 提供了构造器 SqlSessionFactoryBuilder 来生成 SqlSessionFactory。通过 9.2 节可知，只需提供 XML 或者通过代码创建 SqlSessionFactory，SqlSessionFactoryBuilder 即可根据具体的 XML 文件或者代码来生成 SqlSessionFactory（建议使用 XML，有利于以后的维护）。

9.4.1 SqlSessionFactory 源码

SqlSessionFactory 的源码如代码清单 9-10 所示。

代码清单 9-10 SqlSessionFactory

```
public interface SqlSessionFactory {
    SqlSession openSession();
```

```
    SqlSession openSession(boolean autoCommit);
    SqlSession openSession(Connection connection);
    SqlSession openSession(TransactionIsolationLevel level);
    SqlSession openSession(ExecutorType execType);
    SqlSession openSession(ExecutorType execType, boolean autoCommit);
    SqlSession openSession(ExecutorType execType, TransactionIsolationLevel level);
    SqlSession openSession(ExecutorType execType, Connection connection);
    //获取Configuration配置信息
    Configuration getConfiguration();
}
```

从源码中可以看到，SqlSessionFactory 工厂方法中提供了好几种生成 SqlSession 实例的方法。在 MyBatis 中，SqlSessionFactory 接口有两个实现类：SqlSessionMapper 和 DefaultSqlSessions-Factory。默认使用的是DefaultSqlSessionFactory实现类，SqlSessionMapper是用于多线程的环境中的。

9.4.2 DefaultSqlSessionFactory 分析

DefaultSqlSessionFactory 是 SqlSessionFactory 的实现类：其中只有一个最核心的方法（open-SessionFromDataSource 方法），如代码清单 9-11 所示。

代码清单 9-11 openSessionFromDataSource 方法

```
private SqlSession openSessionFromDataSource(ExecutorType execType,
TransactionIsolationLevel level, boolean autoCommit) {
    Transaction tx = null;
    try {
        /*通过configuration实例获取Environment对象，Environment对象中包含了数据源和事务的
一些配置*/
        final Environment environment = configuration.getEnvironment();
        /*通过environment实例获取事务工厂，如果开发者没有进行配置，getTransactionFactoryFro
mEnvironment方法默认返回的是ManagedTransactionFactory对象*/
        final TransactionFactory transactionFactory = getTransactionFactoryFromEnvir
onment(environment);
        //通过事务工厂获取Transaction对象
        tx = transactionFactory.newTransaction(environment.getDataSource(), level,
autoCommit);
        /*通过事务和ExecutorType（执行器类型）获取执行器。Executor是对JDBC中Statement
（MappedStatement）的封装*/
        final Executor executor = configuration.newExecutor(tx, execType);
        return new DefaultSqlSession(configuration, executor, autoCommit);
    } catch (Exception e) {
        //可能已经获取了一个连接，在这里进行关闭事务操作，并抛出异常
        closeTransaction(tx);
        throw ExceptionFactory.wrapException("Error opening session.  Cause: " + e,
e);
    } finally {
        //将执行信息进行清空重设，ErrorContext为线程单例类
        ErrorContext.instance().reset();
    }
```

```
}
//获取TransactionFactory事务工厂
private TransactionFactory getTransactionFactoryFromEnvironment(Environment
environment) {
    if (environment == null || environment.getTransactionFactory() == null) {
        //默认返回ManagedTransactionFactory
        return new ManagedTransactionFactory();
    }
    //返回environment中配置的TransactionFactory
    return environment.getTransactionFactory();
}
```

其实还有一个 openSessionFromConnection(ExecutorType execType, Connection connection) 方法，与 openSessionFromDataSource 方法类似，不再进行介绍。

从代码清单 9-11 可以看出，创建 SqlSession 分以下 5 个步骤：

（1）从 configuration 配置中获取 Environment 实例。

（2）通过 Environment 实例获取事务工厂。

（3）通过事务工厂、DataSource 来获取事务对象。

（4）通过事务和执行器类型获取 Executor（SqlSession 的所有操作，如增、删、改、查操作都是通过 Executor 对象完成的）。

（5）创建一个 SqlSession 对象。

SqlSessionFactory 接口主要的作用是生成 SqlSession，接下来继续分析 SqlSession。

9.5 SqlSession 源码分析

SqlSession（会话）对应着一次和数据库之间的会话。从这里可以看出，SqlSession 的生命周期和项目 MyBatis 的生命周期是不一样的（一次会话以 SqlSession 对象的创建开始，以 SqlSession 对象的关闭结束）。每次会话访问数据库时都需要创建 SqlSession。关于 SqlSession 的创建在前面的 9.3 节已经介绍了。

接下来分析 SqlSession 接口的源码。SqlSession 接口中的方法有一定的规律，这里分成 4 部分进行讲解。由于 SqlSession 接口在 MyBatis 中使用的默认实现是 DefaultSqlSession 类，下面将以 DefaultSqlSession 类中的源码进行讲解。

9.5.1 多参数 select 方法

在 DefaultSqlSession 类中，有几个查询的方法，可以返回对象行数的限制，提供自定义的结果控制，如代码清单 9-12 所示。

代码清单 9-12　DefaultSqlSession 类中多参数的 select 方法

```
public class DefaultSqlSession implements SqlSession {
```

```
    /*selectMap根据生成的对象中的一个属性将结果列表转换为Map，属性的值作为Map中的key，对象作为Map中的value。例如，执行selectMap("selectAuthors","id")，则会返回Map[Integer,Author]。statement参数为映射SQL语句的标识字符串。例如，com.uifuture.dao.UserMapper.findOne，是映射到XML中具体的SQL语句。这里的statement参数代表具体去执行哪个select节点，statement参数的值就是XML文件中该select节点的id值。通过rowBounds可以限制返回行数的范围 */
    @Override
    public <K, V> Map<K, V> selectMap(String statement, Object parameter, String mapKey, RowBounds r
owBounds) {
        final List<? extends V> list = selectList(statement, parameter, rowBounds);
          final DefaultMapResultHandler<K, V> mapResultHandler = new DefaultMapResultHandler<K, V>(mapKey, configuration.getObjectFactory(), configuration.getObjectWrapperFactory(), configuration.getReflectorFactory());
        final DefaultResultContext<V> context = new DefaultResultContext<V>();
        for (V o : list) {
            //将List中的对象依次设置到DefaultResultContext中的结果对象中
            context.nextResultObject(o);
            mapResultHandler.handleResult(context);
        }
        //返回这个Map结果集
        return mapResultHandler.getMappedResults();
    }

    @Override
      public <T> Cursor<T> selectCursor(String statement, Object parameter, RowBounds rowBounds) {
        try {
            //获取对应statement的映射器声明
            MappedStatement ms = configuration.getMappedStatement(statement);
            /*通过执行器执行查询。在前面9.2.4小节中已经介绍了执行器的类型，在MyBatis中对应着实例化不同的Executor接口的实现类*/
             Cursor<T> cursor = executor.queryCursor(ms, wrapCollection(parameter), rowBounds);
            //将结果添加到DefaultSqlSession的cursorList中
            registerCursor(cursor);
            return cursor;
        } catch (Exception e) {
             throw ExceptionFactory.wrapException("Error querying database.  Cause: " + e, e);
        } finally {
            ErrorContext.instance().reset();
        }
    }

    @Override
     public <E> List<E> selectList(String statement, Object parameter, RowBounds rowBounds) {
        try {
```

```
//获取对应statement的映射器声明
            MappedStatement ms = configuration.getMappedStatement(statement);
//通过执行器执行查询
                return executor.query(ms, wrapCollection(parameter), rowBounds,
Executor.NO_RESULT_HANDLER);
        } catch (Exception e) {
            throw ExceptionFactory.wrapException("Error querying database.  Cause:
" + e, e);
        } finally {
            ErrorContext.instance().reset();
        }
    }

    //通过该select方法中的ResultHandler参数，可以自定义返回结果的控制逻辑
    @Override
    public void select(String statement, Object parameter, ResultHandler handler) {
        select(statement, parameter, RowBounds.DEFAULT, handler);
    }

    @Override
     public void select(String statement, Object parameter, RowBounds rowBounds,
ResultHandler handler) {
        try {
            //获取对应statement的映射器声明
            MappedStatement ms = configuration.getMappedStatement(statement);
            //调用执行器的查询方法
            executor.query(ms, wrapCollection(parameter), rowBounds, handler);
        } catch (Exception e) {
            throw ExceptionFactory.wrapException("Error querying database.  Cause:
" + e, e);
        } finally {
            ErrorContext.instance().reset();
        }
    }
    // ... 省略部分代码
}
```

在代码清单 9-12 的方法参数中都会有 RowBounds 对象，RowBounds 对象中提供了指定数量的记录值，另外还有返回结果数量的参数。通过该对象可以控制返回行数的范围，如代码清单 9-13 所示。

代码清单 9-13　RowBounds 对象部分代码

```
public class RowBounds {
    public static final int NO_ROW_OFFSET = 0;
    public static final int NO_ROW_LIMIT = Integer.MAX_VALUE;
    public static final RowBounds DEFAULT = new RowBounds();
    //偏移数量值
    private final int offset;
    //查询数量值
    private final int limit;
```

```
    public RowBounds() {
        this.offset = NO_ROW_OFFSET;
        this.limit = NO_ROW_LIMIT;
    }
    //构造函数中指定offset 和limit的值
    public RowBounds(int offset, int limit) {
        this.offset = offset;
        this.limit = limit;
    }
    ...
}
```

代码清单 9-12 中的 selectMap 方法有调用 DefaultMapResultHandler 的 handleResult 方法，代码清单 9-14 列出了该方法的源码。

代码清单 9-14　DefaultMapResultHandler 的 handleResult 方法

```
public void handleResult(ResultContext<? extends V> context) {
    final V value = context.getResultObject();
    //通过结果、对象工厂、对象包装工厂以及反射工厂，获取MetaObject
    final MetaObject mo = MetaObject.forObject(value, objectFactory,
objectWrapperFactory, reflectorFactory);
    //注意，这里的key可能为null
    final K key = (K) mo.getValue(mapKey);
    mappedResults.put(key, value);
}
```

9.5.2　带参数的增、删、改、查方法

除了 9.4.3 小节列出的方法，SqlSession 接口还提供了 RowBounds 参数，可使用默认值的查询方法，也就是一些不包含 RowBounds 参数的方法。另外还有一些带参数的增、删、改方法，如代码清单 9-15 所示。

代码清单 9-15　DefaultSqlSession 类中带参数的方法

```
public class DefaultSqlSession implements SqlSession {
    //执行查询，返回唯一的一个对象，内部实现就是调用了selectList方法
    @Override
    public <T> T selectOne(String statement, Object parameter) {
        //返回集合的第一个结果或null。如果有多个结果，则抛出异常
        List<T> list = this.<T>selectList(statement, parameter);
        if (list.size() == 1) {
            return list.get(0);
        } else if (list.size() > 1) {
            throw new TooManyResultsException("Expected one result (or null) to be
returned by selectOne(), but found: " + list.size());
        } else {
            return null;
        }
```

```
    }
    //使用默认的RowBounds值进行查询
    @Override
     public <K, V> Map<K, V> selectMap(String statement, Object parameter, String
mapKey) {
        return this.selectMap(statement, parameter, mapKey, RowBounds.DEFAULT);
    }
    //游标查询
    @Override
    public <T> Cursor<T> selectCursor(String statement) {
        return selectCursor(statement, null);
    }
    @Override
    public <T> Cursor<T> selectCursor(String statement, Object parameter) {
        return selectCursor(statement, parameter, RowBounds.DEFAULT);
    }
    /*查询出对象的集合。selectOne和selectList的唯一不同就是selectOne仅仅返回一个对象。如果调
用selectOne返回多个对象，那么就会抛出TooManyResultsException异常*/
    @Override
    public <E> List<E> selectList(String statement) {
        return this.selectList(statement, null);
    }
    @Override
    public <E> List<E> selectList(String statement, Object parameter) {
        return this.selectList(statement, parameter, RowBounds.DEFAULT);
    }
    //查询出一个结果绑定到结果处理器,可以自定义一些结果处理逻辑
    @Override
    public void select(String statement, Object parameter, ResultHandler handler) {
        select(statement, parameter, RowBounds.DEFAULT, handler);
    }
    @Override
    public void select(String statement, ResultHandler handler) {
        select(statement, null, RowBounds.DEFAULT, handler);
    }
    //插入数据
    @Override
    public int insert(String statement) {
        return insert(statement, null);
    }
    @Override
    public int insert(String statement, Object parameter) {
        return update(statement, parameter);
    }
    //修改数据
    @Override
    public int update(String statement) {
        return update(statement, null);
    }
```

```
    @Override
    public int update(String statement, Object parameter) {
        try {
            dirty = true;
            //在configuration配置中获取MappedStatement
            MappedStatement ms = configuration.getMappedStatement(statement);
            //通过执行器进行修改操作
            return executor.update(ms, wrapCollection(parameter));
        } catch (Exception e) {
            throw ExceptionFactory.wrapException("Error updating database.  Cause: " + e, e);
        } finally {
            ErrorContext.instance().reset();
        }
    }
    //删除操作
    @Override
    public int delete(String statement) {
        return update(statement, null);
    }
    @Override
    public int delete(String statement, Object parameter) {
        return update(statement, parameter);
    }
}
```

另外，除了增、删、改、查方法以外，还有另外几个与配置相关的方法，如代码清单 9-16 所示。

代码清单 9-16　与配置相关的方法

```
//事务提交的方法
void commit();
void commit(boolean force);
//事务回滚的方法
void rollback();
void rollback(boolean force);
/*刷新批处理语句。这在MyBatis 3.0.6版本后增加的。刷新(执行)存储在JDBC驱动类中的批量更新语句。当
将ExecutorType.BATCH作为ExecutorType使用时可以采用此方法 */
List<BatchResult> flushStatements();
//关闭该session
@Override
void close();
//清除本地会话缓存
void clearCache();
//获取当前的配置
Configuration getConfiguration();
//根据Mapper接口Class获取Mapper
<T> T getMapper(Class<T> type);
//返回数据库内部连接
Connection getConnection();
```

SqlSession 接口中的方法很多，但是基本上可以理解为对数据库进行增、删、改、查的操作，这些方法用来执行 SQL 映射的 XML 文件中的增、删、改、查语句。

SqlSession 对象执行是线程不安全的，所以在一次与数据库会话的连接中只会有一个 SqlSession，在会话结束前，SqlSession 会调用 close 方法进行关闭。当关闭 SqlSession 时，会通过 isCommitOrRollbackRequired 方法判断当前 SqlSession 是否已经被提交，如果没有提交，那么会进行该 SqlSession 的回滚再关闭；如果已经提交，那么 SqlSession 直接关闭。如果需要再一次和数据库进行会话，那么需要重新创建一个 SqlSession。

另外，在前面的查询方法中，都会有 ResultHandler 参数的方法。ResultHandler 接口允许以自定义方式处理每一行结果，可以将结果添加到 List、Set 或者 Map 中，或者抛出每一个结果，而不仅仅保留总数。ResultHandler 接口的源码很简单，如代码清单 9-17 所示。

代码清单 9-17　ResultHandler 接口

```
public interface ResultHandler<T> {
    void handleResult(ResultContext<? extends T> resultContext);
}
```

ResultHandler 接口只有一个方法，通过 ResultContext 参数可以访问结果对象本身的方法。如果不需要加载大量的结果对象，那么 ResultContext 对象可以使用 stop 方法让 MyBatis 停止加载。

从源码可以看出来，SqlSession 是 MyBatis 工作的主要接口，通过这些接口可以执行增、删、改、查命令，来获取 Mapper 和事务管理。SqlSession 实现类中最后都是调用 executor 执行器中的方法进行增、删、改、查操作。

9.6　MapperAnnotationBuilder 源码分析

MapperAnnotationBuilder（Mapper 注解构建器）全类名为 org.apache.ibatis.builder.annotation.MapperAnnotationBuilder，从名字可知，其用来解析指定的 Mapper 接口对应方法上的一些注解。

虽然 MapperAnnotationBuilder 会解析 Class 对象中包含的注解，但是 MapperAnnotationBuilder 会优先解析 XML 配置文件，并且该 XML 文件的命名空间需要与 Class 对象所在的路径一致，且文件名需要和类名一致，只有在解析完 XML 配置文件之后，才会解析 Class 类方法上的注解（如 Select、Insert 等注解）。

由于 MapperAnnotationBuilder 类代码比较多，下面进行分块讲解。首先分析该类中的属性和构造函数，如代码清单 9-18 所示。

代码清单 9-18　MapperAnnotationBuilder 类中属性与构造函数

```
/*从构造函数中可以看出，sqlAnnotationTypes中只添加了4个Class对象。也就是Select.class,
Insert.class, Update.class, Delete.class */
private final Set<Class<? extends Annotation>> sqlAnnotationTypes = new
HashSet<Class<? extends Annotation>>();
```

```
/*在构造函数中添加了4个元素:SelectProvider，InsertProvider，UpdateProvider，DeleteProvider。这4个元素中有Class和方法名，即SQL语句保存在注解中的指定类的指定方法中 */
private final Set<Class<? extends Annotation>> sqlProviderAnnotationTypes = new HashSet<Class<? extends Annotation>>();
//贯穿MyBatis核心的配置对象
private final Configuration configuration;
/* Mapper构建助手。用于组装解析出来的配置，生成Cache，ResultMap，MappedStatement等对象。并且将生成的对象添加到configuration配置对象中 */
private final MapperBuilderAssistant assistant;
//需要解析目标Mapper接口的Class对象
private final Class<?> type;
//构造函数
public MapperAnnotationBuilder(Configuration configuration, Class<?> type) {
    String resource = type.getName().replace('.', '/') + ".java (best guess)";
    this.assistant = new MapperBuilderAssistant(configuration, resource);
    this.configuration = configuration;
    this.type = type;
//初始化sqlAnnotationTypes
    sqlAnnotationTypes.add(Select.class);
    sqlAnnotationTypes.add(Insert.class);
    sqlAnnotationTypes.add(Update.class);
    sqlAnnotationTypes.add(Delete.class);
//初始化sqlProviderAnnotationTypes
    sqlProviderAnnotationTypes.add(SelectProvider.class);
    sqlProviderAnnotationTypes.add(InsertProvider.class);
    sqlProviderAnnotationTypes.add(UpdateProvider.class);
    sqlProviderAnnotationTypes.add(DeleteProvider.class);
}
```

可以看到该类负责解析接口方法上的增、删、改、查注解，并映射为执行具体的 SQL 语句，返回结果集。

9.6.1 parse 方法

接下来查看 MapperBuilderAssistant 类中的解析逻辑，如代码清单 9-19 所示。

代码清单 9–19　MapperBuilderAssistant 中的解析方法

```
public void parse() {
      /*获取到Class对象的唯一标识，类似全类名。例如，UserMapper接口对应的全类名字符串为：interface com.uifuture.mapper.UserMapper*/
    String resource = type.toString();
    //判断该对象是否已经解析过
    if (!configuration.isResourceLoaded(resource)) {
        //加载指定的XML配置文件，后面会分析该方法
        loadXmlResource();
        //将Class对应的标识添加到已经加载的资源列表中去
        configuration.addLoadedResource(resource);
```

```
        //设置当前的命名空间为接口的全类名
        assistant.setCurrentNamespace(type.getName());
        //解析缓存对象
        parseCache();
        //解析缓存的引用，如果有，则会覆盖上面方法解析的缓存对象
        parseCacheRef();
        //获取类中所有的方法，后面就是解析方法上的注解了。生成MappedStatement和ResultMap对象
        Method[] methods = type.getMethods();
        //遍历获取的方法
        for (Method method : methods) {
            try {
                //当且仅当此方法是Java语言规范定义的bridge方法时才返回true。
                if (!method.isBridge()) {
                    //解析方法，生成对应的MappedStatement对象
                    parseStatement(method);
                }
            } catch (IncompleteElementException e) {
                configuration.addIncompleteMethod(new MethodResolver(this, method));
            }
        }
    }

    //解析待定的方法
    parsePendingMethods();
}
```

9.6.2 loadXmlResource 方法

接下来分析 loadXmlResource 方法，如代码清单 9-20 所示。

代码清单 9-20 loadXmlResource 方法

```
private void loadXmlResource() {
    /*configuration中是否加载了XML资源要通过一个标志来判断，这个标志是在XMLMapperBuilder类
的bindMapperForNamespace方法中设置的。标志规则为:namespace:类的全限定名*/
    if (!configuration.isResourceLoaded("namespace:" + type.getName())) {
        //根据Class对象的类名称获取XML配置文件的路径
        String xmlResource = type.getName().replace('.', '/') + ".xml";
        InputStream inputStream = null;
        try {
            //获取配置文件的字节流
            inputStream = Resources.getResourceAsStream(type.getClassLoader(), xmlResource);
        } catch (IOException e) {
            // 忽略，不需要资源
        }
        //判断XML文件流是否存在
        if (inputStream != null) {
```

```
                    XMLMapperBuilder xmlParser = new XMLMapperBuilder(inputStream, assistant.getConfiguration(), xmlResource, configuration.getSqlFragments(), type.getName());
                xmlParser.parse();
            }
        }
    }
```

这里的XML路径名称只是一种MyBatis中默认的约定，也就是路径与Class全类名一致。当然，也可以不一致，如果不一致，需要开发者配置Mapper的XML文件路径。为了避免重复地加载XML文件，XMLMapperBuilder对象在解析完XML文件以后，会调用bindMapperForNamespace方法，尝试加载配置文件中的根元素上的命名空间属性来加载Class对象，并且会调用configuration.addLoadedResource("namespace:" + builderAssistant.getCurrentNamespace())将标识加载到已加载的资源列表中，以此来通知MapperAnnotationBuilder类中loadXmlResource方法此处是否为重复加载。

9.6.3 parseCache 方法与 parseCacheRef 方法

接下来分析解析缓存对象的方法，也就是parseCache方法，如代码清单9-21所示。

代码清单 9-21 parseCache 方法

```
private void parseCache() {
    //获取Mapper接口类的CacheNamespace注解
    CacheNamespace cacheDomain = type.getAnnotation(CacheNamespace.class);
     /*只有CacheNamespace注解存在，才会进行缓存操作，也就是MyBatis的二级缓存。在XML中也可以通过cache标记开启*/
    if (cacheDomain != null) {
        Integer size = cacheDomain.size() == 0 ? null : cacheDomain.size();
        //获取缓存的刷新间隔
        Long flushInterval = cacheDomain.flushInterval() == 0 ? null : cacheDomain.flushInterval();
        Properties props = convertToProperties(cacheDomain.properties());
        //调用Mapper构建助手创建缓存对象
        assistant.useNewCache(cacheDomain.implementation(), cacheDomain.eviction(), flushInterval, size, cacheDomain.readWrite(), cacheDomain.blocking(), props);
    }
}
```

这里缓存对应着MyBatis中的二级缓存，CacheNamespace对应着Mapper的XML配置文件中的<cache>标记。在MyBatis 3.4.2以后，支持properties自定义属性的配置。

接下来分析前面的解析缓存引用方法，如代码清单9-22所示。

代码清单 9-22 parseCacheRef 方法

```
private void parseCacheRef() {
    //获取对应接口Class上的CacheNamespaceRef注解
```

```
    CacheNamespaceRef cacheDomainRef = type.getAnnotation(CacheNamespaceRef.class);
    if (cacheDomainRef != null) {
        /*CacheNamespaceRef注解用于引用缓存的命名空间类型(命名空间名称为指定类型的FQCN）*/
        Class<?> refType = cacheDomainRef.value();
        //引用缓存的命名空间名称
        String refName = cacheDomainRef.name();
        if (refType == void.class && refName.isEmpty()) {
            throw new BuilderException("Should be specified either value() or name()
attribute in the @CacheNamespaceRef");
        }
        if (refType != void.class && !refName.isEmpty()) {
            throw new BuilderException("Cannot use both value() and name() attribute
in the @CacheNamespaceRef");
        }
        //获取namespace
        String namespace = (refType != void.class) ? refType.getName() : refName;
        //调用Mapper构建助手添加引用关系。以Class对象的全类名为目标的namespace
        assistant.useCacheRef(namespace);
    }
}
```

CacheNamespaceRef 注解在 Mapper 的 XML 文件中对应的是 <cache-ref namespace=""/> 标记。

9.6.4 parseStatement 方法

接下来介绍 parseStatement 方法的实现。MapperAnnotationBuilder 类的 parse 方法会遍历 Mapper 接口对象中的所有方法，一个 Method 对象对应着一个 MappedStatement 对象，ResultMap 的定义与 XML 配置文件方式不同，配置文件由单独的 <resultMap> 标记进行定义，而注解方式直接定义在方法上。每个方法都可以创建新的 ResultMap 对象，也可以引用已经存在的 ResultMap 对象的唯一 id，如代码清单 9-23 所示。

代码清单 9-23　parseStatement 方法

```
void parseStatement(Method method) {
    //根据method获取方法中参数类型
    Class<?> parameterTypeClass = getParameterType(method);
    //通过方法上的org.apache.ibatis.annotations.Lang注解获取SQL语言的驱动
    LanguageDriver languageDriver = getLanguageDriver(method);
    //通过方法上的注解(增、删、改、查等注解)以及驱动获取SqlSource
     SqlSource sqlSource = getSqlSourceFromAnnotations(method, parameterTypeClass,
languageDriver);
    //如果获取到了SqlSource
    if (sqlSource != null) {
        //获取方法上的Options注解对象
        Options options = method.getAnnotation(Options.class);
        //获取到映射语句的id。从这里可以知道规则为"类的全名.方法名称"
        final String mappedStatementId = type.getName() + "." + method.getName();
```

```
        Integer fetchSize = null;
        Integer timeout = null;
        StatementType statementType = StatementType.PREPARED;
        ResultSetType resultSetType = ResultSetType.FORWARD_ONLY;
        //通过方法上的增、删、改、查注解获取SQL命令的类型
        SqlCommandType sqlCommandType = getSqlCommandType(method);
        boolean isSelect = sqlCommandType == SqlCommandType.SELECT;
        boolean flushCache = !isSelect;
        boolean useCache = isSelect;
        KeyGenerator keyGenerator;
        String keyProperty = "id";
        String keyColumn = null;
        if (SqlCommandType.INSERT.equals(sqlCommandType) || SqlCommandType.UPDATE.
equals(sqlCommandType)) {
            //获取方法上的SelectKey注解，SelectKey注解相当于XML配置中的<selectKey>标记
            SelectKey selectKey = method.getAnnotation(SelectKey.class);
            //...省略部分代码
        }
        //...省略部分代码
        String resultMapId = null;
          /* 获取ResultMap注解(ResultMap注解给@Select或者@SelectProvider提供在XML映射中的
<resultMap>的id。这使得select注解可以复用那些定义在XML中的ResultMap。从代码的顺序可以知道，如果
同一select注解中还存在@Results或者@ConstructorArgs，那么这两个注解将被ResultMap注解覆盖）*/
        ResultMap resultMapAnnotation = method.getAnnotation(ResultMap.class);
        //如果方法上配置了ResultMap注解，则直接生成引用id即可
        if (resultMapAnnotation != null) {
            //获取指定的resultMap的id，可以为多个
            String[] resultMaps = resultMapAnnotation.value();
            StringBuilder sb = new StringBuilder();
            //遍历id，进行拼接，使用逗号分隔
            for (String resultMap : resultMaps) {
                if (sb.length() > 0) {
                    sb.append(",");
                }
                sb.append(resultMap);
            }
            resultMapId = sb.toString();
            //如果不存在ResultMap注解，并且注解为Select类型
        } else if (isSelect) {
            //则通过解析Arg、Result、TypeDiscriminator注解来生成新的ResultMap。同时返回resultMapId
            resultMapId = parseResultMap(method);
        }
        //构建MappedStatement，并且添加到MapperBuilderAssistant对象中
        assistant.addMappedStatement(
                //...
                options != null ? nullOrEmpty(options.resultSets()) : null);
    }
}
```

从以上代码中可以看出，如果在 Mapper 接口方法上不使用 @Select、@Insert、@Update、@Delete 或者对应的 @xxxProvider 中的任意一个，那么只会有 Lang 注解能够生效，其他注解（如 ResultMap 注解）是无效的。

代码清单 9-23 中 resultMapId = parseResultMap(method) 的 parseResultMap 方法，将生成新的 ResultMap 对象，如代码清单 9-24 所示。

代码清单 9-24　parseResultMap 方法

```
private String parseResultMap(Method method) {
    //获取返回值Bean的类型
    Class<?> returnType = getReturnType(method);
     /*获取方法上的ConstructorArgs注解，相当于XML配置中的<constructor>标记。该注解包含了Arg注解数组。Arg注解相当于一个单独的构造方法参数，里面包含了id、column、javaType、typeHandler属性*/
    ConstructorArgs args = method.getAnnotation(ConstructorArgs.class);
    Results results = method.getAnnotation(Results.class);
    //获取方法上的TypeDiscriminator注解，该注解对于XML配置中的<discriminator>标记
    TypeDiscriminator typeDiscriminator = method.getAnnotation(TypeDiscriminator.class);
    /*根据Method生成resultMap唯一的标识。格式是类全名.方法名—参数类型的简单名称，在generate-ResultMapName方法中有实现细节*/
    String resultMapId = generateResultMapName(method);
     /*resultMapId和接口方法的返回值类型解析完成。这里是解析构造方法映射，属性映射等，最后将添加结果映射到配置的对象中*/
     applyResultMap(resultMapId, returnType, argsIf(args), resultsIf(results), typeDiscriminator);
    return resultMapId;
}

//构建ResultMap的唯一id
private String generateResultMapName(Method method) {
    Results results = method.getAnnotation(Results.class);
    if (results != null && !results.id().isEmpty()) {
        return type.getName() + "." + results.id();
    }
    StringBuilder suffix = new StringBuilder();
    for (Class<?> c : method.getParameterTypes()) {
        suffix.append("-");
        suffix.append(c.getSimpleName());
    }
    if (suffix.length() < 1) {
        //如果没有参数，则添加-void
        suffix.append("-void");
    }
    return type.getName() + "." + method.getName() + suffix;
}
```

注意：注解的方式只支持在属性映射时使用另外的 select 语句的情况，但是不支持嵌套的属

性映射配置。

到这里，MyBatis 的组件源码基本上已经讲解了一遍，虽然看源码非常枯燥，但是认真看完的读者，在后面的章节中使用 MyBatis 时会觉得非常简单。

框架就是如此，如果读者能将源码看懂三分之一以上，那么使用起来就没有任何难度。第 10 章开始进行 MyBatis 的 XML 配置文件的讲解。

第10章 MyBatis的XML配置文件

本章要点

1. properties 元素
2. settings 元素
3. typeAliases 元素
4. typeHandlers 元素
5. objectFactory 元素
6. transactionManager 元素
7. dataSource 元素
8. databaseIdProvider 元素
9. mappers 元素

10.1 MyBatis 依赖

项目需要对 MyBatis 的一些行为和属性进行配置，方法是在 resources 目录下添加 MyBatis 的 XML 配置文件，在该配置文件中进行配置即可。在配置之前，先回顾一下前面的 9.2.4 小节，在 settingsElement 方法中已经将每个 XML 标记和属性讲解了一遍。

接下来，将使用一个 MyBatis 的项目例子来对 MyBatis 依赖进行讲解。如果是使用 Maven 来构建项目，则将代码清单 10-1 放到 pom.xml 的 <dependencies></dependencies> 标记中即可。

代码清单 10-1　为项目添加 MyBatis 依赖

```
<!-- 添加MyBatis的核心包 -->
<dependency>
    <groupId>org.mybatis</groupId>
    <artifactId>mybatis</artifactId>
    <version>3.4.6</version>
</dependency>
```

10.2 properties 元素

properties（属性）元素主要用来配置一些动态的 key-value（键值对）。例如,数据库的用户名、密码等配置，可以通过在 MyBatis 的配置文件中导入 properties 文件（Java 属性文件）实现，也可以直接在 MyBatis 的配置文件中定义 <properties></properties> 标记设置 key-value 键值对，如代码清单 10-2 所示。

代码清单 10-2　引入 properties 文件与设置 key-value 键值对

```
<!-- 引入的属性文件 classpath就在resource下-->
<properties resource="classpath:jdbc.properties">
    <property name="username" value="root"/>
    <property name="password" value="1234"/>
</properties>
```

jdbc.properties 文件内容如代码清单 10-3 所示。

代码清单 10-3　jdbc.properties 文件内容

```
driver=com.mysql.jdbc.Driver
url=jdbc:mysql://localhost:3306/chapter-10-ssm?useUnicode=true&characterEncoding=U
TF-8&zeroDateTimeBehavior=convertToNull
```

其中配置的属性可以替换整个 MyBatis 配置文件中需要动态配置的值。如果要在 MyBatis 中用到这里配置的属性值，则使用“${}”表达式引用即可，如代码清单 10-4 所示。

代码清单 10-4　数据库配置

```
<environments default="development">
    <environment id="development">
        <transactionManager type="JDBC"/>
        <!--数据库连接实例的数据源(DataSource)配置-->
        <dataSource type="POOLED">
            <property name="driver" value="${driver}"/>
            <property name="url" value="${url}"/>
            <property name="username" value="${username}"/>
            <property name="password" value="${password}"/>
        </dataSource>
    </environment>
</environments>
```

其中的 ${driver}、${url}、${username}、${password} 就是引用的配置文件中设置的值。

如果在多个地方配置了同一个 key 值的属性值，那么 MyBatis 将会按照下面的顺序进行加载，后面加载的值会覆盖前面同名的值。

（1）<properties></properties> 标记中的值最先被读取加载。

（2）根据 <properties></properties> 标记的 resource 属性读取对应路径下的属性文件，或者是根据 url 属性指定的路径读取属性文件，覆盖在（1）中加载的同名属性值。

（3）当读取的属性作为方法参数时，其优先级最高。例如，通过 SqlSessionFactoryBuilder.build 方法直接赋值，此时读取作为方法参数传递的属性，会覆盖之前读取过的同名属性。

在 MyBatis 3.4.2 以后的版本中，可以在引用的配置中指定一个默认值，如代码清单 10-5 所示。

代码清单 10-5　指定配置的默认值

```
<!-- 引入属性文件，默认就在resource下-->
<properties resource="jdbc.properties">
    <-- ... -->
    <!-- 开启默认值特性 -->
    <property name="org.apache.ibatis.parsing.PropertyParser.enable-default-value"
value="true"/>
    <!-- 改变默认的分隔字符 -->
     <property name="org.apache.ibatis.parsing.PropertyParser.default-value-
separator" value="?:"/>
</properties>
<environments default="development">
    <!-- ... -->
    <property name="username" value="${username?:root}"/>
    <!-- ... -->
</environments>
```

默认是通过冒号（:）来指定默认值的，这时需要在 properties 中配置 org.apache.ibatis.parsing.PropertyParser.enable-default-value，value 为 true。

但是，在配置 properties 时是可以使用 ":" 作为属性的 key 的（如 jdbc:username）。这种情况下，使用默认的 ":" 显然不行。MyBatis 针对这种情况，允许开发者自定义默认值的字符，也就是配置 org.apache.ibatis.parsing.PropertyParser.default-value-separator，在 value 处指定默认值分隔的字符。

10.3 settings 元素

settings（设置）元素都是在 <settings></settings> 标记中进行配置的。前面的 9.3.4 小节，在 settingsElement 方法中已经将每个 settings 的属性讲了一遍。

settings 的设置会改变 MyBatis 运行时的行为，其主要是配置缓存、映射、日志、执行器等，完整的 settings 配置如代码清单 10-6 所示。在编写代码时注意标记的顺序，<settings> 标记需要放在 <properties></properties> 标记后面。

代码清单 10-6　完整的 settings 配置

```
<settings>
    <!--全局开启或关闭配置文件中的所有映射器已经配置的任何缓存。默认值为true-->
    <setting name="cacheEnabled" value="true"/>
     <!--延迟加载的全局开关。当开启时，所有关联对象都会延迟加载。特定关联关系中可以通过设置
fetchType属性来覆盖该项的开关状态。默认值为false-->
    <setting name="lazyLoadingEnabled" value="true"/>
     <!--当开启时，任何方法的调用都会加载该对象的所有属性，否则每个属性会按需加载(参考
lazyLoadTriggerMethods)。默认值在MyBatis=3.4.1版本以前为true，3.4.1版本以后默认值为false-->
    <setting name="aggressiveLazyLoading" value="false" />
    <!--是否允许单一语句返回多结果集(需要兼容驱动，如果驱动不支持单一语句返回多结果，则该配置无用)。
默认值为true-->
    <setting name="multipleResultSetsEnabled" value="true"/>
    <!--使用列标签代替列名。不同的驱动在这方面会有不同的表现，具体可参考相关驱动文档或通过测试这两
种不同的模式来观察所用驱动的结果。默认值为true-->
    <setting name="useColumnLabel" value="true"/>
    <!--允许JDBC 支持自动生成主键，需要驱动兼容。如果设置为true则这个设置强制使用自动生成主键，尽管
一些驱动不能兼容，但仍能正常工作(如Derby)。默认为false-->
    <setting name="useGeneratedKeys" value="false"/>
    <!--指定MyBatis应如何自动映射列到字段或属性。
     NONE ：取消自动映射；
     PARTIAL ：只会自动映射没有定义嵌套结果集映射的结果集；
     FULL ：会自动映射任意复杂的结果集(无论是否嵌套)；
     默认值为PARTIAL-->
    <setting name="autoMappingBehavior" value="PARTIAL"/>
    <!--指定发现自动映射目标未知列(或者未知属性类型)的行为。
    NONE：不作任何反应
     WARNING：输出提醒日志 (org.apache.ibatis.session.AutoMappingUnknownColumnBehavior
的日志等级必须设置为WARNING)；
    FAILING：映射失败 (抛出SqlSessionException)；
    默认值为NONE-->
    <setting name="autoMappingUnknownColumnBehavior" value="WARNING"/>
    <!--配置默认的执行器。
    SIMPLE ：普通的执行器；
    REUSE ：执行器，会重用预处理语句(prepared statements)；
    BATCH ：执行器，会重用语句并执行批量更新；
    默认值为SIMPLE-->
```

```xml
    <setting name="defaultExecutorType" value="SIMPLE"/>
    <!--设置超时时间，它决定驱动等待数据库响应的秒数。没有默认值-->
    <setting name="defaultStatementTimeout" value="25"/>
    <!--为驱动的结果集获取数量(fetchSize)设置一个提示值。此参数只可以在查询设置中被覆盖，没有默认值-->
    <setting name="defaultFetchSize" value="100"/>
    <!--允许在嵌套语句中使用分页(RowBounds)。如果允许使用则设置为false。默认值为false-->
    <setting name="safeRowBoundsEnabled" value="false"/>
    <!--允许在嵌套语句中使用分页(ResultHandler)。如果允许使用则设置为false。默认值为true-->
    <setting name="safeResultHandlerEnabled" value="false"/>
    <!--是否开启自动驼峰命名规则(camelcase)映射，即从经典数据库列名A_COLUMN 到经典Java 属性名aColumn的类似映射。默认值为false-->
    <setting name="mapUnderscoreToCamelCase" value="false"/>
    <!--MyBatis 利用本地缓存机制(LocalCache)防止循环引用(circular references)和加速重复嵌套查询，默认值为SESSION，这种情况下会缓存一个会话中执行的所有查询。若设置值为STATEMENT，则本地会话仅用在语句执行上，对相同SqlSession的不同调用将不会共享数据-->
    <setting name="localCacheScope" value="SESSION"/>
    <!--当没有为参数提供特定的JDBC 类型时，为空值指定JDBC 类型。某些驱动需要指定列的JDBC 类型，大多数情况直接用一般类型即可，如NULL、VARCHAR 或OTHER。默认值为OTHER-->
    <setting name="jdbcTypeForNull" value="OTHER"/>
    <!--指定对象的哪个方法触发一次延迟加载。可以配置多个，使用逗号隔开-->
    <setting name="lazyLoadTriggerMethods" value="equals,clone,hashCode,toString"/>
    <!--指定动态SQL生成的默认语言，值为一个类型别名或完全限定类名。默认值为org.apache.ibatis.scripting.xmltags.XMLLanguageDriver-->
    <setting name="defaultScriptingLanguage" value="org.apache.ibatis.scripting.xmltags.XMLLanguageDriver"/>
    <!--指定Enum使用的默认TypeHandler (从MyBatis 3.4.5版本开始) 。默认值为org.apache.ibatis.type.EnumTypeHandler -->
    <setting name="defaultEnumTypeHandler" value="org.apache.ibatis.type.EnumTypeHandler"/>
    <!--指定当结果集中值为null时是否调用映射对象的setter (map对象时为put)方法，这对于有Map.keySet() 依赖或null值初始化的时候是有用的。注意基本类型(int、boolean等)是不能设置成null的。默认值为false-->
    <setting name="callSettersOnNulls" value="false"/>
    <!--当返回行的所有列都是空时，MyBatis默认返回null。当开启这个设置时，MyBatis会返回一个空实例。请注意，它也适用于嵌套的结果集 (i.e. collectioin and association)(从MyBatis 3.4.2版本开始)。默认值为false-->
    <setting name="returnInstanceForEmptyRow" value="false"/>
    <!--指定MyBatis所用日志的具体实现方法，未指定时将自动查找。默认值没有指定。SLF4J | LOG4J | LOG4J2 | JDK_LOGGING | COMMONS_LOGGING | STDOUT_LOGGING | NO_LOGGING-->
    <setting name="logPrefix" value="SLF4J"/>
    <!--指定MyBatis创建具有延迟加载能力的对象所用到的代理工具，如CGLIB | JAVASSIST。在MyBatis 3.3以上为JAVASSIST -->
    <setting name="proxyFactory" value="JAVASSIST"/>
    <!--指定VFS(虚拟文件系统，用来读取服务器里的资源)的实现方法，自定义VFS实现的类全限定名，以逗号分隔。在应用服务中提供非常简单的API访问资源文件。不同的应用服务器有不同的文件结构，有些特殊的需要进行适配，但基本不会用到，除非有特定的环境要求，需要不同的资源反射-->
    <!--<setting name="vfsImpl" value=""/>-->
```

```
    <!--允许使用方法签名中的名称作为语句参数名称。为了使用该特性，要求工程必须采用Java 8编译，并
且加上-parameters选项(从MyBatis 3.4.1版本开始) -->
    <setting name="useActualParamName" value="true"/>
    <!--指定一个提供Configuration实例的类(Configuration类主要用来存储对MyBatis的配置文件及
Mapper文件解析后的数据)。这个被返回的Configuration实例用来加载被反序列化对象的懒加载属性值。这
个类必须包含一个签名方法static Configuration getConfiguration() (从MyBatis 3.2.3版本开
始)的类型别名或者全类名-->
    <!--<setting name="configurationFactory" value=""/>-->
</settings>
```

这个完整的配置已经差不多概括了 MyBatis 运行时的全部配置。

10.4 typeAliases 元素

给一些 Java 类起类型别名主要是为了精简变量值。typeAliases（类型别名）元素存在的意义只是使 MyBatis 匹配时减少类的完全限定名冗余，如代码清单 10-7 所示。

代码清单 10-7 为 typeAliases 配置 typeAlias

```
...
<settings>
    ...
</settings>
//必须在settings标记之后
<typeAliases>
    <!--type是指Java Bean的完全限定名; alias是指别名-->
    <typeAlias type="com.uifuture.chapter10.entity.Users" alias="users"/>
</typeAliases>
...
```

这样配置以后，在映射的 Mapper XML 文件的 parameterType 和 resultType 中可以使用 users 来替换 com.uifuture.chapter10.entity.Users。

另外，还可以在 <typeAliases></typeAliases> 标记中设置 <package> 标记，见代码清单 10-8，这样 MyBatis 会自动映射到 package 下面的实体类，搜索到需要的 JavaBean。

代码清单 10-8 为 <typeAliases> 设置 package

```
<typeAliases>
    <package name="com.uifuture.chapter10.entity"/>
</typeAliases>
```

默认情况是使用该 Bean 首字母小写的非限定类名作为别名。例如，com.uifuture.chapter10.entity.User 类的别名会使用 user，也可以在类上单独指定别名，使用 @Alias 即可。

表 10-1 是一些 Java 类中内置的相应类型别名，注意它们都是大小写不敏感的。如果是由于基本类型名称重复导致的问题，则需要特殊处理。

表 10-1　Java 类中内置的相应类型别名

别　　名	映射的类型	别　　名	映射的类型
_byte	byte	double	Double
_long	long	float	Float
_short	short	boolean	Boolean
_int	int	date	Date
_integer	int	decimal	BigDecimal
_double	double	bigdecimal	BigDecimal
_float	float	object	Object
_boolean	boolean	map	Map
string	String	hashmap	HashMap
byte	Byte	list	List
long	Long	arraylist	ArrayList
short	Short	collection	Collection
int	Integer	iterator	Iterator
integer	Integer		

在开发中，需要特别注意别名重复的情况。

10.5　typeHandlers 元素

当 MyBatis 在预处理语句（PreparedStatement）中设置一个参数时或者从结果集中取出一个值时，都会使用类型处理器将获取的值用合适的方式转换成 Java 类型。

typeHandlers（类型处理器）元素的作用就是处理数据库类型和 Java 类型之间的转换。

10.6　objectFactory 元素

MyBatis 在每次创建结果对象实例的时候，都会使用对象工厂实例。默认的对象工厂（DefaultObjectFactory）仅仅只是实例化目标类，可以通过默认空构造方法，或者在有参数映射的时候通过参数构造方法来进行实例化。如果需要覆盖默认的对象工厂，则可以通过创建自己的对象工厂来继承默认的对象工厂。最后在 MyBatis 中配置 <objectFactory> 标记即可。

10.7 environments 元素

MyBatis 的 environments（环境）元素配置实际就是数据源的配置。MyBatis 可以配置多个环境，帮助开发者将 SQL 映射到多个不同的数据库。例如，开发、测试和生产环境需要有不同的配置，或者共享相同 Schema 的多个生产数据库使用相同的 SQL 映射。

但需要注意的是，尽管可以配置多个环境，每个 SqlSessionFactory 实例只能对应一个数据库，并且有几个数据库就需要创建几个 SqlSessionFactory 实例。可以使用以下几种方式创建 SqlSessionFactory 实例，如代码清单 10-9 所示。

代码清单 10-9　创建 SqlSessionFactory 实例

```
SqlSessionFactory factory=new SqlSessionFactoryBuilder().build(reader,
environment);
SqlSessionFactory factory=new SqlSessionFactoryBuilder().build(reader,
environment, properties);
//如果忽略了环境参数，那么会加载默认环境
SqlSessionFactory factory=new SqlSessionFactoryBuilder().build(reader);
SqlSessionFactory factory=new SqlSessionFactoryBuilder().build(reader,
properties);
```

在 MyBatis 的 XML 配置文件中配置环境，如代码清单 10-10 所示。

代码清单 10-10　配置环境

```
<environments default="development">
    <environment id="development">
        <transactionManager type="JDBC">
            <!--在这里可以配置property键值对-->
        </transactionManager>
        <!--数据库连接实例的数据源(DataSource)配置-->
        <dataSource type="POOLED">
            <property name="driver" value="${driver}"/>
            <property name="url" value="${url}"/>
            <property name="username" value="${username?:root}"/>
            <property name="password" value="${password}"/>
        </dataSource>
    </environment>
</environments>
```

默认的环境 id 为 development，如代码清单 10-10 中定义的 default="development"，每个 environments 需要定义环境 id，开发者也可以自定义。另外还有事务管理器的配置（如 type="JDBC"）和数据源的配置（如 type="POOLED"）。

默认的环境和环境 id 是自解释的。可以对环境随意命名，但是必须要保证默认的环境匹配其中的一个环境 id。

10.7.1 transactionManager 元素

<transactionManager>(事务管理器)标记的 type 在 MyBatis 中有两种事务管理器,也就是 type= "[JDBC|MANAGED]"：

- JDBC：直接使用 JDBC 的提交和回滚设置,依赖于从数据源得到的连接来管理事务作用域。
- MANAGED：使用该配置表示不进行提交或者回滚连接，让容器来管理事务的整个生命周期（如 Java EE 应用服务器的上下文）。

本书讲解的是 Spring+MyBatis，在 MyBatis 中是没有配置事务管理器的，因为 Spring 模块会使用自带的管理器来覆盖前面自行配置的事务管理器。当然，可以使用 TransactionFactory 接口的实现类的完全限定名或者类型别名在 type 下进行配置，替换这两种事务管理器类型。

10.7.2 dataSource 元素

dataSource（数据源）元素使用标准的 JDBC 据源接口来配置 JDBC 连接对象的资源。type 有以下 3 种内置的数据源类型供选择（也就是 type=" [UNPOOLED|POOLED|JNDI] "）。一般情况下，都是进行 POOLED 配置，这里主要讲解 POOLED。

（1）UNPOOLED：UNPOOLED 的作用只是每次在被请求时进行打开和关闭连接。对于简单的应用来说，在要不要使用连接池这种不重要的情况下，适合采用这种类型。该类型只可以配置以下 5 种属性：

- driver：JDBC 驱动的 Java 类的完全限定名。
- url：数据库的 JDBC URL 地址。
- Username：登录数据库的用户名。
- Password：登录数据库的密码。
- defaultTransactionIsolationLevel：默认的连接事务隔离级别。

（2）POOLED：POOLED 支持 JDBC 数据源连接池，利用“池”的概念将 JDBC 连接对象组织起来，避免了创建新的连接实例时所必需的初始化和认证时间。除了有 UNPOOLED 包含的属性外，还包括以下的一些属性用来配置 POOLED 数据源：

- poolMaximumActiveConnections：在任意时间可以存在的活动(也就是正在使用)连接数量，默认值为 10（最大活动连接数）。
- poolMaximumIdleConnections：任意时间存在的空闲连接数（最大空闲连接数）。
- poolMaximumCheckoutTime：在被强制返回之前,池中连接被检测出（checked out）的时间，默认值为 20000 ms。
- poolTimeToWait：这是一个底层的设置，如果获取连接耗费了相当长的时间，连接池会打印状态日志并重新尝试获取一个连接（避免在误配置的情况下一直保持安静的失败状态），默认值为 20000 ms（获取连接超时时间）。
- poolMaximumLocalBadConnectionTolerance：这是一个关于坏连接容忍度的底层设置，作用于每一个尝试从缓存池获取连接的线程。如果一个线程获取到的是一个坏的连接，那么数据源允许这个线程尝试重新获取一个新的连接，但是这个重新尝试的次数不应该超过 poolMaximumIdleConnections 与 poolMaximumLocalBadConnectionTolerance 之和。默认值为 3（新增于 MyBatis 3.4.5）。

- poolPingQuery：发送到数据库的侦测查询，用来检验连接是否正常工作并准备接收请求。默认为“NO PING QUERY SET”，这样会导致多数数据库驱动失败时带有一个恰当的错误提示消息（对数据库进行 ping 时所使用的 SQL）。
- poolPingEnabled: 是否启用侦测查询。若开启，需要设置 poolPingQuery 属性为一个可执行的 SQL 语句（最好是一个速度非常快的 SQL 语句），默认值为 false（当默认值为 true 时，将开启 ping 机制）。
- poolPingConnectionsNotUsedFor：配置 poolPingQuery 的频率。其可以被设置为与数据库连接超时时间一样以避免不必要的侦测，默认值为 0，单位为 ms（即所有连接每一时刻都被侦测，仅当 poolPingEnabled 为 true 时适用）。它不能被设置为默认值，不能在每次使用连接池之前都使用 ping 机制，否则，会使每一条 SQL 的执行都要额外执行一次 ping 语句，所以使用此属性来避免这种不合理做法时。只针对闲置时间超过某个时间的连接进行 ping，例如，当值设置为 3600000 时，如果从连接池中拿出的连接闲置超过 1h，则会对它进行 ping。

（3）JNDI：基本不会用到，这里不进行介绍。

10.8 databaseIdProvider 元素

MyBatis 可以根据不同的数据库厂商执行不同的语句，这种多厂商的支持基于映射语句中的 databaseId 属性。MyBatis 会加载不带 databaseId 属性和带有匹配当前数据库 databaseId 属性的所有语句。如果同时找到带有 databaseId 和不带 databaseId 的相同语句，则后者会被舍弃。可以简单地理解为，通过配置 databaseId，可以针对同一个 Mapper 方法写两个 SQL 语句。通过配置不同的 databaseId，可以匹配到不同的数据库。例如，在 MySQL 中获取系统时间的函数为 NOW() 函数，而在 Oracle 中获取系统时间的函数为 to_char(susdate, 'yyyy-mm-dd hh24:mi:ss')。如果在 SQL 中要用到系统时间的函数，那么针对不同的数据库，写法肯定是不同的。此时 databaseIdProvider（数据库厂商标识）元素就派上用场了。

如果要支持多厂商特性，只要像代码清单 10-11 那样在 mybatis-config.xml 文件中按照合适的顺序加入 <databaseIdProvider> 标记即可。

代码清单 10-11　添加 <databaseIdProvider> 标记

```
<databaseIdProvider type="DB_VENDOR" />
```

这里配置的 DB_VENDOR 会通过 DatabaseMetaData 类中的 getDatabaseProductName 方法返回的字符串进行设置。

由于通常情况下这个字符串都非常长且相同产品的不同版本会返回不同的值，因此最好通过设置属性别名来使其变短，如代码清单 10-12 所示。

代码清单 10-12　配置 DB_VENDOR 属性别名

```
<databaseIdProvider type="DB_VENDOR">
    <!--设置属性别名-->
    <property name="SQL Server" value="sqlserver"/>
    <property name="DB2" value="db2"/>
    <property name="Oracle" value="oracle" />
</databaseIdProvider>
```

在提供了属性别名后，DB_VENDOR 的 databaseIdProvider 会被设置为第 1 个能匹配数据库产品名称的属性键对应的值，如果没有匹配的属性，则会设置为 null。在代码清单 10-12 中，getDatabaseProductName 方法如果返回 Oracle，则 databaseId 会被设置为 oracle。所以如果有多个数据源，不能只设置一个别名，需要全部设置别名。

当然，开发者也可以通过实现 org.apache.ibatis.mapping.databaseIdProvider 接口并在 mybatis-config.xml 中进行配置来构建个人的 databaseIdProvider。databaseIdProvider 接口的源码如代码清单 10-13 所示。

代码清单 10-13　databaseIdProvider 接口源码

```
public interface databaseIdProvider {
    //设置属性
    void setProperties(Properties p);
    //获取databaseId
    String getDatabaseId(DataSource dataSource) throws SQLException;
}
```

10.9　mappers 元素

前面都是在配置 MyBatis 的运行环境和行为。现在开始定义 SQL 语句，并需要与 Java 的接口对应起来，那么开发者需要告诉 MyBatis 到哪里能找到这些 SQL 语句进行执行。也许开发者认为可以让 Java 自动遍历所有的 XML 文件进行查找，但是 Java 并没有提供很好的方法，而且这种方式也比较消耗性能。最好的方式是开发者主动告诉 MyBatis 去哪里能够找到映射文件。配置 mapper（映射器）元素可以指定相对于类路径的资源引用，也可以使用完全限定资源的定位符（如 file:/// 的 URL），或者是使用类的完全限定名和包名等。使用相对于类路径的资源引用配置 mapper 如代码清单 10-14 所示。

代码清单 10-14　使用相对于类路径的资源引用配置 mapper

```
<!-- 使用相对于类路径的资源引用，类似相对路径 -->
<mappers>
    <mapper resource="com/uifuture/chapter10/mapper/UsersMapper.xml"></mapper>
</mappers>
```

使用完全限定资源的定位符，也就是文件中的绝对路径配置 mapper，如代码清单 10-15 所示。

代码清单 10-15　使用绝对路径配置 mapper

```
<!-- 使用完全限定资源的定位符(URL)配置mapper-->
<mappers>
    <mapper url="file:///D:/github/uifuture-ssm/chapter-10-mybatis-xml-config/src/
main/java/com/uifuture/chapter10/mapper/UsersMapper.xml"/>
</mappers>
```

使用映射器接口实现类的完全限定名配置 mapper，如代码清单 10-16 所示。

代码清单 10-16　使用映射器接口实现类的完全限定名配置 mapper

```
<!-- 使用映射器接口实现类的完全限定类名 -->
<mappers>
    <mapper class="com.uifuture.chapter10.mapper.UsersMapper"/>
</mappers>
```

使用包名配置 mapper，将包内的映射器接口实现全部注册为映射器，如代码清单 10-17 所示。

代码清单 10-17　使用包名配置 mapper

```
<!--将包内的映射器接口实现全部注册为映射器 -->
<mappers>
    <package name="com.uifuture.chapter10"/>
</mappers>
```

需要注意的是，上面几种形式在不使用注解注入的情况下，都需要接口和 XML 配置文件在一个包下，且确保名称相同。所以开发者可以通过代码清单 10-18 所示的方式配置接口和 XML 配置文件。

代码清单 10-18　配置接口和 XML 映射

```
<!--在不同路径下不使用注解，可以这样配置 -->
<mappers>
    <!-- 扫描路径下的mapper映射文件 -->
    <mapper resource="com/uifuture/chapter10/mapper/UsersMapper.xml" />
    <!-- 扫描包下的接口文件 -->
    <package name="com.uifuture.chapter10.dao.UsersMapper" />
</mappers>
```

Users 实体类代码如代码清单 10-19 所示。

代码清单 10-19　Users 实体类代码

```
import com.uifuture.chapter10.base.BaseEntity;
/** table name: users */
public class Users extends BaseEntity {
    /** fields name: users.username */
    private String username;
    /** fields name: users.password */
```

```
    private String password;
    /** fields name: users.age */
    private Integer age;
    ...// 省略settter/getter方法，请自行添加
    @Override
    public String toString() {
        final StringBuilder sb = new StringBuilder("Users{");
        sb.append(super.toString());
        sb.append(",");
        sb.append("username='").append(username).append('\'');
        sb.append(", password='").append(password).append('\'');
        sb.append(", age=").append(age);
        sb.append('}');
        return sb.toString();
    }
}
```

BaseEntity 类是所有数据库实体类的基类，如代码清单 10-20 所示。

代码清单 10-20　BaseEntity 类代码

```
/** 所有实体类的超类 */
public class BaseEntity implements Serializable {
    private static final long serialVersionUID = -4282387477058447315L;
    private Integer id;
    private Date createTime;
    private Date updateTime;
    //...省略setter/getter方法
    @Override
    public String toString(){
        final StringBuilder sb = new StringBuilder("BaseEntity{");
        sb.append("id=").append(id);
        sb.append(",createTime=").append(createTime);
        sb.append(",updateTime=").append(updateTime);
        sb.append('}');
        return sb.toString();
    }
}
```

下面是 Dao 层的接口，UsersMapper 接口类如代码清单 10-21 所示。

代码清单 10-21　UsersMapper 接口类代码

```
package com.uifuture.chapter10.dao;
import com.uifuture.chapter10.entity.Users;
public interface UsersMapper{
    /** 根据id删除数据 */
    int deleteByPrimaryKey(Integer id);
    /** 插入所有数据 */
    int insert(Users record);
```

```
    /** 插入非空数据 */
    int insertSelective(Users record);
    /** 通过id查询对象 */
    Users selectByPrimaryKey(Integer id);
    /** 通过id修改非空数据*/
    int updateByPrimaryKeySelective(Users record);
    /** 通过id修改所有的数据 */
    int updateByPrimaryKey(Users record);
}
```

接下来讲解 SQL 映射文件，前面这几个类是为后面测试做准备的。本书中所有实例代码都可以在书中列出的 GitHub 项目地址中找到。

第11章

MyBatis的XML映射文件

本章要点

1. MyBatis 的 XML 映射文件分析
2. 查询元素
3. 增、改、删元素
4. 安全传参与字符串替换
5. MyBatis 结果集、高级结果映射与嵌套结果
6. MyBatis 中的鉴别器
7. MyBatis 的缓存属性
8. MyBatis 的缓存机制
9. MyBatis 的缓存使用方式以及实例

11.1 顶级元素简介

本章所讲的都是 MyBatis 的真正强大之处——映射语句，它们是实际场景中使用最多的部分。在映射文件中，有以下 9 个顶级元素：

（1）select：映射 SQL 中的查询语句。

（2）sql：可以被重用的 SQL 语句块。

（3）insert：映射 SQL 中的插入语句。

（4）update：映射 SQL 中的更新语句。

（5）delete：映射 SQL 中的删除语句。

（6）parameteMap：以前的参数映射，新版本中已废弃，这里不多介绍。

（7）resultMap：描述如何从数据库结果集加载对象，其就是 Java 中用来接收结果集的对象。

（8）cache：用来配置给定空间的二级缓存。

（9）cache-ref：用来引用其他命名空间的缓存配置。

下面将针对每个顶级元素来进行讲解。

11.2 查询

11.2.1 通过 select 元素实现简单查询

查询（select）可以说是在 MyBatis 中使用最多的元素了。如果没有查询数据的功能，那么只把数据插入数据库是没有多大价值的。多数的应用都是查询操作比修改操作执行得要频繁。通常情况下，每个插入、更新或者删除操作都对应着多个查询操作，这也是 MyBatis 将焦点放在查询和结果映射的原因。如果只是简单的查询，那么代码非常简单，如代码清单 11-1 所示。

代码清单 11-1　简单使用查询语句

```
<select id="selectUsersByPrimaryKey" parameterType="java.lang.Integer" resultType="hashmap">
    select *
    from users
    where id=#{id,jdbcType=INTEGER}
</select>
```

这个 SQL 语句的映射 id 是 selectUsersByPrimaryKey。如果有配置映射 Java 类和这个 XML 文件对应，那么对应的接口方法名也是 selectUsersByPrimaryKey。另外就是接收一个 Integer 类型的参数，并且返回一个 HashMap 类型的对象，其中 Map 中的 Key 是列名，值就是查询出来的结果。

注意这里 id 传参使用的是 #{id,jdbcType=INTEGER}。其用于告诉 MyBatis 创建一个预处理语句参数，通过 JDBC，这个参数在 SQL 中会使用一个“?”（俗称占位符）来标识，并且会被传递到一个新的预处理语句中，如代码清单 11-2 所示。

代码清单 11-2 JDBC 预处理语句

```
// SimilarJDBCcode, NOT MyBatis…
String selectPerson="SELECT * FROM users WHEREID=?";
PreparedStatement ps=conn.prepareStatement(selectPerson);
ps.setInt(1,id);
```

如果想通过 JDBC 实现该 selectUsersByPrimaryKey 所做的事，还得编写非常多的 JDBC 代码，使用 MyBatis 可以为开发者节省时间。

<select></select> 标记中还有很多的属性，它们可以决定每一条 SQL 语句的细节。<select></select> 标记中的属性如代码清单 11-3 所示。

代码清单 11-3 <select></select> 标记中的属性

```
<select
        id="selectPerson"
        parameterType="int"
        parameterMap="deprecated"
        resultType="hashmap"
        resultMap="personResultMap"
        flushCache="false"
        useCache="true"
        timeout="10000"
        fetchSize="256"
        statementType="PREPARED"
        resultSetType="FORWARD_ONLY">
</select>
```

下面对每个属性进行讲解，具体细节参考表 11-1（databaseId、resultOrdered、resultSets 是扩展属性，一般很少用到）。

表 11-1 <select></select> 标记的属性

属　性	描　述
id	命名空间中唯一的标识符，可以被用来引用对应的语句和对应接口的方法名
parameterType	指定参数类型。这个属性是可选的，因为 MyBatis 可以通过 TypeHandler 推断出具体传入语句的参数，默认值为 unset（未设置，依赖驱动）
parameterMap	引用外部参数 map，已经被废弃。通过内联参数映射和 parameterType 属性替换来实现
resultType	指定返回值的类型。**提示：**如果是集合类型，那应该是集合内元素的类型，而不是集合类本身，返回值类型使用 resultType 或 resultMap，但不能同时使用
resultMap	指定返回值是外部 Map 类型。结果集的映射是 MyBatis 最强大的特性，许多复杂映射的情形都能迎刃而解
flushCache	将其设置为 true，任何时候只要语句被调用，本地缓存和二级缓存都会被清空。默认值为 false
useCache	如果设置为 true，开启二级缓存。默认值为 true
timeout	数据库驱动程序等待数据库返回请求结果的秒数。默认值为 unset
fetchSize	这是驱动程序每次批量返回的结果行数。默认值为 unset
statementType	值是 STATEMENT、PREPARED 或 CALLABLE 中的一个，这样会让 MyBatis 分别使用 Statement、PreparedStatement 或 CallableStatement。默认值为 PREPARED

续表

属　性	描　述
resultSetType	值是 FORWARD_ONLY、SCROLL_SENSITIVE 或 SCROLL_INSENSITIVE 中的一个，默认值为 unset
databaseId	如果配置了 databaseIdProvider，MyBatis 会加载所有的不带 databaseId 或匹配当前 databaseId 的语句
resultOrdered	这个设置仅针对嵌套结果 select 语句适用。如果其值为 true，假设包含嵌套结果集，当返回一个主结果行时，就不会有对前面结果集引用的情况，这样就不会使其在获取嵌套的结果集时导致内存不够用。默认值为 false
resultSets	这个设置仅适用多结果集的情况，它将列出语句执行后返回的结果集中每个结果集的名称，名称是逗号分隔的

11.2.2　sql 元素在查询时的重要作用

开发者在编写 XML 映射查询时，如果每编写一个查询都需要将查询字段复制一份是不太合理的。因为如果字段有修改，则需要修改所有地方，这维护成本无疑是巨大的。有没有更方便的方法，使其只需要修改一处就可以呢？答案是肯定的，那就是使用 <sql> 标记。

sql 元素可以定义可重用的 SQL 代码片段，可以被包含在其他的语句中，可以被静态地（在加载参数时）参数化。不同的属性值还可以通过包含的实例变化注入值。sql 元素定义如代码清单 11-4 所示。

代码清单 11-4　sql 元素定义

```
<sql id="userColumns"> ${alias}.id,${alias}.username,${alias}.password </sql>
```

在这里定义了动态的 alias 属性。此外，该属性值可以在重用时动态注入不同的值，如代码清单 11-5 所示。

代码清单 11-5　动态注入 sql 元素中“${}”的值

```
<select id="selectUsers" resultType="map">
    select
    <include refid="userColumns"><property name="alias" value="t1"/></include>,
    <include refid="userColumns"><property name="alias" value="t2"/></include>
    from some_table t1
    cross join some_table t2
</select>
```

上面就是使用 <include> 标记重用了 userColumns 的 SQL 语句，并且使用 <property> 设置了 name 属性的值。

另外，<include> 标记中的 refid 属性也是可以使用“${}”进行动态注入的，如代码清单 11-6 所示。

代码清单 11-6　<include> 标记嵌套使用

```
<!--建立一个默认的root的logger-->
<root level="DEBUG">
    <appender-ref ref="console"/>
</root>
```

```
<sql id="sometable">
    ${prefix}Table
</sql>
<sql id="someinclude">
    from
    <include refid="${include_target}"/>
</sql>
<select id="select" resultType="map">
    select
    field1, field2, field3
    <include refid="someinclude">
        <property name="prefix" value="Some"/>
        <property name="include_target" value="sometable"/>
    </include>
</select>
```

在这里，<select> 标记中最终的 SQL 语句如代码清单 11-7 所示。

代码清单 11-7　<select> 标记嵌套 <include> 标记最终的结果

```
<select id="select" resultType="map">
    select
    field1, field2, field3
    from
    SomeTable
</select>
```

11.3　增、改、删

增、改、删（insert、update、delete）语句的标记实现都非常相似，在这里进行统一介绍。这 3 个元素的标记内部分别是编写增、改、删的 SQL 语句。下面先看这 3 个标记的属性，如代码清单 11-8 所示。

代码清单 11-8　增、改、删标记的属性

```
<insert
        id="insertAuthor"
        parameterType="domain.blog.Author"
        flushCache="true"
        statementType="PREPARED"
        keyProperty=""
        keyColumn=""
        useGeneratedKeys=""
        timeout="20">
</insert>
<update
        id="updateAuthor"
        parameterType="domain.blog.Author"
```

```
        flushCache="true"
        statementType="PREPARED"
        timeout="20">
</update>
<delete
        id="deleteAuthor"
        parameterType="domain.blog.Author"
        flushCache="true"
        statementType="PREPARED"
        timeout="20">
</delete>
```

上面给出了 insert、update、delete 的一些属性配置，表 11-2 列出了其属性描述。

表 11-2　增、改、删标记的属性描述

属　性	描　述
id	命名空间中的唯一标识符
parameterType	指定参数类型。这个属性是可选的，因为 MyBatis 可以通过 TypeHandler 推断出具体传入语句的参数。默认值为 unset
flushCache	将其设置为 true，任何时候只要语句被调用，本地缓存和二级缓存都会被清空。默认值为 true
statementType	值是 STATEMENT、PREPARED 或 CALLABLE 中的一个。这会让 MyBatis 分别使用 Statement、PreparedStatement 或 CallableStatement。默认值为 PREPARED
keyProperty	（仅对 insert 和 update 有用）唯一标记一个属性，MyBatis 会通过 getGeneratedKeys 的返回值或者通过 insert 语句的 selectKey 子元素设置它的键值，默认值为 unset。如果希望得到多个生成的列，则使用以逗号分隔的属性名称列表
keyColumn	（仅对 insert 和 update 有用）通过生成的键值设置表中的列名，这个设置仅在某些数据库（如 PostgreSQL）是必需的，当主键列不是表中的第 1 列时需要设置。如果希望得到多个生成的列，也可以使用以逗号分隔的属性名称列表
useGeneratedKeys	（仅对 insert 和 update 有用）MyBatis 使用 JDBC 的 getGeneratedKeys 方法来取出由数据库内部生成的主键（如 MySQL 和 SQL Server 这样的关系型数据库管理系统的自动递增字段）。默认值为 false
timeout	数据库驱动程序等待数据库返回请求结果的秒数。默认值为 unset

下面是 MyBatis 映射文件中 insert、update 和 delete 语句的示例，如代码清单 11-9 所示。

代码清单 11-9　insert、update 和 delete 语句的示例

```
<insert id="insert" parameterType="com.uifuture.chapter10.entity.Users">
    insert into users (username, password, age,create_time, update_time)
      values (#{username,jdbcType=VARCHAR}, #{password,jdbcType=VARCHAR},
#{age,jdbcType=INTEGER}, #{createTime,jdbcType=TIMESTAMP}, #{updateTime,jdbcType=TIMESTAMP})
</insert>
<update id="updateByPrimaryKey" parameterType="com.uifuture.chapter10.entity.
Users">
    update users
    set username=#{username,jdbcType=VARCHAR},
    password=#{password,jdbcType=VARCHAR},
    age=#{age,jdbcType=INTEGER},
    create_time=#{createTime,jdbcType=TIMESTAMP},
    update_time=#{updateTime,jdbcType=TIMESTAMP}
```

```
    where id=#{id,jdbcType=INTEGER}
</update>
<delete id="deleteByPrimaryKey">
    delete from users where id=#{id}
</delete>
```

通过代码清单 11-9 的 XML，编写对应的接口，就可以完成增、改、删操作。

11.3.1 insert 与 update 处理主键自动生成

在表 11-2 中，有几个仅对 insert 和 update 有用的属性，它们用来处理主键的生成，并且有多种生成方式可供配置。

举个例子，如果想让数据库支持自动生成主键的字段（如 MySQL 和 SQL Server），就可以设置 userGeneratedKeys="true"，然后将自动生成主键的 JavaBean 名称设置到 keyProperty（**提示**：这里的值是 JavaBean 的属性名称，而不是数据库的）属性上。代码清单 11-9 的 insert 语句可以修改为代码清单 11-10 所示的语句，在插入之后，MyBatis 会将 keyProperty 属性对应的主键设置到对象中。

通过调用该接口执行的对象，在插入数据库之后，对象中的 id 会被设置值。

代码清单 11-10　插入对象后自动生成的主键被对象带回

```
<insert id="insert" parameterType="com.uifuture.chapter10.entity.Users"
        useGeneratedKeys="true" keyProperty="id">
    insert into users (username, password, age, create_time, update_time)
    values (#{username,jdbcType=VARCHAR}, #{password,jdbcType=VARCHAR}, #{age,jdbc
Type=INTEGER},#{createTime,jdbcType=TIMESTAMP}, #{updateTime,jdbcType=TIMESTAMP})
</insert>
```

11.3.2 通过 insert 获取所有对象的主键

在项目中，可能经常会遇到插入多行数据的情况，MySQL 是支持多行插入的。在这里的实例中可以传入 Users 集合或者数据，而插入对象集合之后，对象的主键（一般为 id，后面所说的 id 即为表的主键）将会在后面使用到。开发者可以通过下面的方式来获取所有对象的 id，如代码清单 11-11 所示。

代码清单 11-11　获取所有插入对象的 id

```
<insert id="insert" parameterType="com.uifuture.chapter10.entity.Users"
        useGeneratedKeys="true" keyProperty="id">
    insert into users (username, password, age,create_time, update_time)
    values
    <foreach item="item" collection="list" separator=",">
        (#{item.username,jdbcType=VARCHAR}, #{item.password,jdbcType=VARCHAR},
#{item.age,jdbcType=INTEGER},#{item.createTime,jdbcType=TIMESTAMP}, #{item.
updateTime,jdbcType=TIMESTAMP})
    </foreach>
</insert>
```

与前面类似的写法是，在 <insert> 标记上加上 useGeneratedKeys 和 keyProperty 属性，MyBatis 会在插入数据后，自动将数据库自增长的值复制给对象 Bean 的 keyProperty 属性值。因此，在插入数据以后，可以调用对象的 getter 方法获取到自增长主键的值。

11.3.3 通过 insert 获取自增长主键

还有另外一种写法可以获取自增长主键，但只能获取到单条插入语句的主键，即使用 selectKey 元素获取自增长主键，如代码清单 11-12 所示。

代码清单 11-12 使用 selectKey 元素获取自增长主键

```
<insert id="insert" parameterType="com.uifuture.chapter10.entity.Users">
    <selectKey keyProperty="id" order="AFTER" resultType="java.lang.Integer">
        SELECT LAST_INSERT_ID()
    </selectKey>
    insert into users (username, password, age, create_time, update_time)
    values (#{username,jdbcType=VARCHAR}, #{password,jdbcType=VARCHAR}, #{age,jdbc
Type=INTEGER},#{createTime,jdbcType=TIMESTAMP}, #{updateTime,jdbcType=TIMESTAMP})
</insert>
```

需要注意的是，插入语句的接口方法返回的值并不是 id，而是影响的行数，那么这个 id 到哪里去了，它是被设置到插入的实体中去了？例如，调用 mapper.insert(users)，在执行了该条语句之后，可以通过 users.getId 获取主键 id 的值。

在上面的实例中，insert 语句会先执行，然后运行 selectKey 元素的语句，查询出 id，并且设置到对象中。

selectKey 元素的描述如代码清单 11-13 所示。

代码清单 11-13 selectKey 元素的描述

```
<selectKey
        keyProperty="id"
        resultType="java.lang.Integer"
        order="AFTER"
        statementType="PREPARED">
```

接下来通过表 11-3 来了解 selectKey 元素中的属性描述。

表 11-3 selectKey 元素中的属性描述

属　性	描　述
keyProperty	selectKey 语句返回的结果值被设置的目标属性
resultType	结果的类型。MyBatis 通常可以推算出来，但是为了更加确定这里直接写明。MyBatis 允许任何将简单类型用作主键的类型（包括字符串）。如果希望作用于多个生成的列，则可以使用一个包含期望属性的 Object 或一个 Map
order	其可以被设置为 BEFORE 或 AFTER。如果设置为 BEFORE，那么它会首先选择主键，设置 keyProperty，然后执行插入语句。如果设置为 AFTER，那么先执行插入语句，然后是 selectKey 元素
statementType	MyBatis 支持 STATEMENT、PREPARED 和 CALLABLE 语句的映射类型，分别代表 PreparedStatement 和 CallableStatement 类型

使用该元素语句的场景比较少，一般也就是在获取返回主键 id 时使用。

11.4 参数

前面的增、删、改、查实例都是 MyBatis 中简单参数的例子。实际上，这些参数的作用是非常强大的，接下来看看参数（Parameters）是如何使用的。

11.4.1 安全传参

代码清单 11-14 简单演示了一个 id 的传参。

代码清单 11-14 简单传参

```
<select id="selectByPrimaryKey" parameterType="java.lang.Integer">
    select *
    from users
    where id=#{id}
</select>
```

上面使用 #{id} 传入了最简单的命名参数映射，id 参数类型被设置为 int。Java 原生的类型或者简单的数据类型（如整型和字符串）因为没有相关的属性，会完全使用参数值来代替。但是，如果使用的是一个复杂的对象，那么会有些不同，如代码清单 11-15 所示。

代码清单 11-15 复杂传参

```
<insert id="insert" parameterType="com.uifuture.chapter10.entity.Users">
    insert into users (username, password, age,create_time, update_time)
    values (#{username}, #{password}, #{age},  #{createTime}, #{updateTime})
</insert>
```

如果参数是 JavaBean，例如，代码清单 11-15 中的 Users 对象传递到语句中，那么传进去的 username、password、age、create-Time、updateTime 属性会被查找出来，然后将它们的值传入预处理语句的参数中，这样使用起来相对于向语句中传参来说是非常方便的。但参数的映射功能远不止于此。

MyBatis 的参数也可以指定一个特殊的数据类型，如代码清单 11-16 所示。

代码清单 11-16 指定参数的类型

```
#{age,javaType=int,jdbcType=INTEGER}
```

其中，javaType 通常可以由参数对象来确定，除非该对象是一个 HashMap。这时所使用的 typeHandler 应该明确指明 javaType。

如果有特殊的定制类型处理，那么可以自定义类型处理器，在 typeHandler 中指定自定义的类型处理器类（或者别名）。另外，还可以指定 mode 属性，resultMap 属性指定映射结果集到参数类型。但是一般情况下不需要开发者进行一些复杂的配置。

大多数情况下只需要简单地指定属性名，其他情况下 MyBatis 会自行去推断，开发者最多需要为可能是空的列指定 jdbcType。

11.4.2 字符串替换

上面讲解了“#{}”格式的使用情况，使用“#{}”格式的语法会使 MyBatis 创建 PreparedStatement 参数并安全地设置参数（就如使用“?”一样）。这样更加安全，一般情况下都是使用这种写法。其实在 MyBatis 中还有另外一种传参的方式，即直接在 SQL 语句中插入一个不进行转义的字符串，如下面的语句：

```
ORDER BY ${columnName}
```

通过参数来控制排序，这里 MyBatis 不会修改或者转义字符串。

但是要注意，通过“${}”这种方式接收用户的输入，并且将其参数用于语句中是非常不安全的，如可能会遭到潜在的 SQL 注入攻击，所示读者要谨慎使用。

11.5 结果集

MyBatis 绝大部分的工作都体现在结果集（ResultMap）上。所以毫无疑问，resultMap 元素是 MyBatis 中最重要、最强大的元素。它可以让开发者从 90% 的 JDBC ResultSets 数据提取代码中解放出来，并且在一些情况下允许开发者进行一些 JDBC 不支持的操作。实际上，当对复杂语句进行联合映射时，它可能代替数千行同等功能的代码。

ResultMap 的设计思想是，简单的语句不需要明确的结果映射，复杂的语句只需要描述它们之间的关系。

代码清单 11-17 演示了简单的映射语句示例，其中没有明确的 resultMap。

代码清单 11-17　简单的映射语句

```
<select id="selectUsers" resultType="map">
    Select  id, username, password, age, create_time, update_time
    from users
    where id=#{id}
</select>
```

上面的代码只是简单地将所有的列映射到 HashMap 的键值对上，这里是由 resultType 属性指定的。虽然绝大多数情况下使用 HashMap 都是够用的，但是 HashMap 不是一个很好的领域模型。另外，开发者在项目中可能会使用 JavaBean 或者 POJO（Plain Old Java Object，普通的 Java 对象）作为领域模型，MyBatis 对这两者都支持。接下来看代码清单 11-18 和代码清单 11-19。

代码清单 11-18　BaseEntity JavaBean

```
public class BaseEntity implements Serializable {
    private static final long serialVersionUID=-4282387477058447315L;
    private Integer id;
    private Date createTime;
    private Date updateTime;
```

```
    //getter、setter、toString方法在这里省略
    }
```

代码清单 11-19　Users JavaBean

```
    public class Users extends BaseEntity {
        /** fields name: users.username */
        private String username;
        /** fields name: users.password */
        private String password;
        /** fields name: users.age */
        private Integer age;
    //getter、setter、toString方法在这里省略
    }
```

基于 JavaBean 的规范，上面有几个属性会对应到 select 语句中的列名。这样的一个 JavaBean 可以被映射到 ResultSet 中，就像映射到 HashMap 中一样，如代码清单 11-20 所示。

代码清单 11-20　映射结果集到 JavaBean

```
    <select id="selectUsersPrimaryKey" resultType="com.uifuture.chapter10.entity.
    Users">
        select
        id, username, password, age, create_time, update_time
        from users
        where id=#{id,jdbcType=INTEGER}
    </select>
```

可以看到，resultType 的值非常长，写起来很麻烦，这个时候就需要使用类型别名了，因为通过别名就可以不用输入类的全名。下面使用别名映射结果集到 Users，如代码清单 11-21 所示。

代码清单 11-21　使用别名映射结果集到 Users

```
    //在mybatis-config.xml配置文件中配置别名
    <typeAliases>
        <!--type是指JavaBean的完整限定名，alias是指别名-->
        <typeAlias type="com.uifuture.chapter10.entity.Users" alias="users"/>
    </typeAliases>

    //在映射文件XML中配置select语句
    <select id="selectUsersPrimaryKey" resultType="users">
        Select id, username, password, age, create_time, update_time
        from users
        where id=#{id,jdbcType=INTEGER}
    </select>
```

这种情况下，MyBatis 会自动创建一个 ResultMap，再基于属性名映射到 JavaBean 的属性上。

提示：MyBatis 会自动处理驼峰命名和下划线命名的转换。但是如果列和属性没有按照规则命名，开发者可以在 select 语句中对不规范的列名使用别名（基本的 SQL 特性），别名对应于 JavaBean 中的属性名即可。

当然，开发者还可以以显式的resultMap来声明数据库列和JavaBean的对应关系，如代码清单11-22所示。虽然这不是必需的，但这是解决列名不匹配的另外一种方式。

代码清单 11-22　定义 resultMap 元素

```
<resultMap id="BaseResultMap" type="com.uifuture.chapter10.entity.Users">
    <id column="id" jdbcType="INTEGER" property="id"/>
    <result column="username" jdbcType="VARCHAR" property="username"/>
    <result column="password" jdbcType="VARCHAR" property="password"/>
    <result column="age" jdbcType="INTEGER" property="age"/>
    <result column="create_time" jdbcType="TIMESTAMP" property="createTime"/>
    <result column="update_time" jdbcType="TIMESTAMP" property="updateTime"/>
</resultMap>
```

在以上这种情况下，可以在select元素中使用resultMap属性（注意去掉resultType属性），如代码清单11-23所示。

代码清单 11-23　使用 resultMap 映射

```
<select id="selectByPrimaryKey" parameterType="java.lang.Integer"
resultMap="BaseResultMap">
    Selectid, username, password, age, create_time, update_time
    from users
    where id=#{id,jdbcType=INTEGER}
</select>
```

11.5.1　高级结果映射

有些时候，项目中需要用一条SQL语句查询一对多或者多对一的对象。例如，查询一篇博客，该博客由某个作者所写，同时还有很多博客，每篇博客都有多条评论和标签等信息。先看需要编写的查询博客信息的SQL语句，如代码清单11-24所示。

代码清单 11-24　查询博客信息

```
<select id="selectBlogDetails" resultMap="detailedBlogResultMap">
    select
    B.id as blog_id,
    B.title as blog_title,
    B.author_id as blog_author_id,
    A.id as author_id,
    A.username as author_username,
    A.password as author_password,
    A.email as author_email,
    A.bio as author_bio,
    A.favourite_section as author_favourite_section,
    P.id as post_id,
    P.blog_id as post_blog_id,
    P.author_id as post_author_id,
    P.created_on as post_created_on,
```

```
    P.section as post_section,
    P.subject as post_subject,
    P.draft as draft,
    P.body as post_body,
    C.id as comment_id,
    C.post_id as comment_post_id,
    C.name as comment_name,
    C.comment as comment_text,
    T.id as tag_id,
    T.name as tag_name
    from Blog B
    left outer join Author A on B.author_id=A.id
    left outer join Post P on B.id=P.blog_id
    left outer join Comment C on P.id=C.post_id
    left outer join Post_Tag PT on PT.post_id=P.id
    left outer join Tag T on PT.tag_id=T.id
    where B.id=#{id}
</select>
```

开发者可以将通过一个对象模型查询出来的值映射到该对象中，这个对象表示一篇博客，它由某位作者所写。另外，还有很多博客，每篇博客都有 0 或多条评论和标签。

下面来看一个比较复杂对象模型的 ResultMap（假设作者、博客、评论和标签都是类型的别名），如代码清单 11-25 所示。

代码清单 11-25　复杂对象模型的 ResultMap

```
<resultMap id="detailedBlogResultMap" type="Blog">
    <constructor>
        <idArg column="blog_id" javaType="int"/>
    </constructor>
    <result property="title" column="blog_title"/>
    <association property="author" javaType="Author">
        <id property="id" column="author_id"/>
        <result property="username" column="author_username"/>
        <result property="password" column="author_password"/>
        <result property="email" column="author_email"/>
        <result property="bio" column="author_bio"/>
        <result property="favouriteSection" column="author_favourite_section"/>
    </association>
    <collection property="posts" ofType="Post">
        <id property="id" column="post_id"/>
        <result property="subject" column="post_subject"/>
        <association property="author" javaType="Author"/>
        <collection property="comments" ofType="Comment">
            <id property="id" column="comment_id"/>
        </collection>
        <collection property="tags" ofType="Tag" >
            <id property="id" column="tag_id"/>
        </collection>
```

```
        <discriminator javaType="int" column="draft">
            <case value="1" resultType="DraftPost"/>
        </discriminator>
    </collection>
</resultMap>
```

resultMap 元素有很多子元素，下面是 resultMap 元素的子元素说明：

- constructor：用于在实例化类时，注入结果到构造方法中。其中，idArg 为 ID 参数，标记出作为 ID 的结果可以帮助提高整体性能，arg 为被注入到构造方法的一个普通结果。
- result：注入到字段或 JavaBean 属性的普通结果。
- association：用于一对一关联；也可以用于复杂对象中多个字段与类型映射的包装。其可以指定为一个 resultMap 元素，或者引用一个 resultMap 元素。
- collection：一个复杂类型的集合。其可以指定为一个 resultMap 元素，或者引用一个 resultMap 元素。
- discriminator：使用结果值来决定使用哪个 resultMap。其中，case 为基于某些值的结果映射。一个 case 也是一个映射它本身的结果，因此可以包含很多相同的元素，或者它可以参照一个外部的 resultMap。

resultMap 元素中的属性说明见表 11-4。

表 11-4　resultMap 元素中的属性说明

属　性	描　述
id	当前命名空间中的唯一标识，用于标识一个 resultMap
type	类的全名，或者一个类型别名
autoMapping	如果设置这个属性，MyBatis 会为这个 ResultMap 开启或者关闭自动映射。这个属性会覆盖全局的属性 autoMappingBehavior，默认值为 unset

不推荐使用复杂的结果映射进行一次关联查询，强烈推荐全部使用单表操作，原因如下：

- 绝大多数项目在发展的后期，性能的瓶颈都是在数据库上，将关联查询拆分为多条单表查询有利于后期在应用上直接进行优化。例如，集群、分布式、缓存等，查询拆分后可以轻易地将计算压力分解到 N 台服务器上。
- 减少了耦合（数据库与代码之间的耦合）。若使用关联查询，会使后期的扩展和维护变得更加困难。
- 关联查询会给后期的分库、分表带来非常大的成本和风险。

如果一定需要使用复杂的结果映射，为了确保你实现的行为和想要的一致，最好进行单元测试。

11.5.2　id 与 result 元素

id 与 result 元素如代码清单 11-26 所示。

代码清单 11-26　id 与 result 元素

```
<id property="id" column="post_id"/>
<result property="subject" column="post_subject"/>
```

id 和 result 都将一个列的值映射到一个简单数据类型（字符串型、整型、双精度浮点数、日期型等）的属性或字段。

id 和 result 元素中的属性说明见表 11-5。

表 11-5　id 和 result 元素中的属性说明

属　性	描　述
property	映射到列结果的字段或属性
column	数据库中的列名，或者是列的别名。一般情况下，其与递给 resultSet.getString(columnName) 方法的参数一样
javaType	一个 Java 类的全名或一个类型别名（参考上面内置类型别名的列表）。如果映射到一个 JavaBean，MyBatis 通常可以断定类型。然而，如果映射到的是 HashMap，那么应该明确地指定 javaType 来保证获得期望的行为
jdbcType	设置 JDBC 类型。只需要在可能执行插入、更新和删除的允许空值的列上指定 JDBC 类型，这是 JDBC 的要求而非 MyBatis 的要求。如果直接面向 JDBC 编程，开发者则需要对可能为 null 的值指定该类型
typeHandler	前面讨论过默认类型处理器，使用这个属性可以覆盖默认的类型处理器。该属性的值是一个类型处理器实现类的全名，或者是类型别名

11.5.3　关联元素

关联元素包含 association、collection，其处理有一个对应的关系。在关联方面，SQL 中有嵌套关联和联合查询（join 查询）。MyBatis 中也是一样，有以下两种不同的方式。

- 嵌套查询：通过执行另外一个 SQL 映射语句来返回预期的复杂类型。
- 嵌套结果：使用嵌套结果映射来处理复杂、重复的联合结果子集。它与普通的只由 select 和 resultMap 属性的结果映射不同。

下面来了解关联元素中的部分属性。javaType、jdbcType、typeHandler 属性的功能可参考表 11-5。property 为映射到列结果的字段或属性。如果用来匹配的 JavaBean 存在给定名称的属性，那么它将会被使用，否则 MyBatis 将会寻找与给定名称相同的字段。

11.5.4　关联的嵌套结果

在这里对嵌套查询不进行讲解，读者简单理解即可。例如，有两个查询语句，一个用来加载博客，另一个用来加载作者，每篇博客都有对应的作者，使用嵌套查询会产生 N+1 查询问题。

N+1 查询问题，也就是指执行一条单独的 SQL 语句来获取结果列表（这里为博客列表），为 1。然后对返回的每条记录（也就是博客）再执行一条查询语句来加载细节（也就是查询作者，在这里就是 N）。如果使用嵌套查询会导致执行速度非常慢。所以在实际场景中，可通过多次应用 SQL 语句来拆解一条嵌套 SQL 语句，从而提升 SQL 语句执行效率。

在使用嵌套结果查询时，可能会用到关联元素中的属性，见表 11-6。

表 11-6 嵌套结果查询的部分相关元素

属 性	描 述
resultMap	结果映射的 ID，它可以映射关联的嵌套结果到一个合适的对象
columnPrefix	当连接多表时，开发者将不得不使用列别名来避免 ResultSet 中的重复列名。指定 columnPrefix 允许映射列名到一个外部的结果集中（也就是增加一个前缀来与另外的列名进行区分）
notNullColumn	默认情况下，子对象仅在至少一个列映射到其属性非空时才创建。通过对这个属性指定非空的列将改变默认行为，这样做之后，MyBatis 将仅在这些列非空时才创建一个子对象。指定多个列名时，可以使用逗号分隔。默认值为 unset
autoMapping	当映射结果到当前属性时，MybBatis 将启用或者禁用自动映射。该属性覆盖全局的自动映射行为。注意它对外部结果集无影响，所以它在 select 或 resultMap 属性中是毫无意义的。默认值为 unset

在代码清单 11-25 中编写了一个非常复杂的嵌套关联代码。接下来，通过一个非常简单的示例来讲解它是如何工作的。代码清单 11-27 将博客表和作者表连接在一起查询。

代码清单 11-27 简单的关联查询

```
<select id="selectBlog" resultMap="blogResult">
    select
    B.id            as blog_id,
    B.title         as blog_title,
    B.author_id     as blog_author_id,
    A.id            as author_id,
    A.username      as author_username,
    A.password      as author_password,
    A.email         as author_email,
    A.bio           as author_bio
    from Blog B left outer join Author A on B.author_id = A.id
    where B.id = #{id}
</select>
```

上面采取别名的方式确保了所有的结果被唯一且清晰的名称命名。接下来使用 MyBatis 来映射这个结果，方法非常简单，如代码清单 11-28 所示。

代码清单 11-28 XML 映射关系

```
<resultMap id="blogResult" type="Blog">
    <id property="id" column="blog_id" />
    <result property="title" column="blog_title"/>
     <association property="author" column="blog_author_id" javaType="Author"
resultMap="authorResult"/>
</resultMap>

<resultMap id="authorResult" type="Author">
    <id property="id" column="author_id"/>
    <result property="username" column="author_username"/>
    <result property="password" column="author_password"/>
    <result property="email" column="author_email"/>
```

```
    <result property="bio" column="author_bio"/>
</resultMap>
```

id 元素在嵌套结果映射中非常重要，开发者至少应该指定一个可唯一标识结果的属性。如果不进行指定，将会出现严重的性能问题。在可唯一标识结果的情况下，尽可能选择少的属性。

在上面，使用 resultMap id="authorResult" 来创建一个 resultMap，这使得 authorResult 的结果映射可重用。当然，如果不需要重用，可以将所有的结果映射到一个单独描述的结果映射中，如代码清单 11-29 所示。

代码清单 11-29　单独的结果映射

```
<resultMap id="blogResult" type="Blog">
    <id property="id" column="blog_id" />
    <result property="title" column="blog_title"/>
    <association property="author" javaType="Author">
        <id property="id" column="author_id"/>
        <result property="username" column="author_username"/>
        <result property="password" column="author_password"/>
        <result property="email" column="author_email"/>
        <result property="bio" column="author_bio"/>
    </association>
</resultMap>
```

下面来讲解 columnPrefix 属性的用法。

如果在博客中有并列作者怎么办？答案是增加 co-author 字段。并列作者查询语句如代码清单 11-30 所示。

代码清单 11-30　并列作者查询语句

```
<select id="selectBlog" resultMap="blogResult">
    select
    B.id            as blog_id,
    B.title         as blog_title,
    A.id            as author_id,
    A.username      as author_username,
    A.password      as author_password,
    A.email         as author_email,
    A.bio           as author_bio,
    CA.id           as co_author_id,
    CA.username     as co_author_username,
    CA.password     as co_author_password,
    CA.email        as co_author_email,
    CA.bio          as co_author_bio
    from Blog B
    left outer join Author A on B.author_id = A.id
    left outer join Author CA on B.co_author_id = CA.id
    where B.id = #{id}
</select>
```

Author 对象的 resultMap 映射如代码清单 11-31 所示。

代码清单 11-31　Author 对象的 resultMap 映射

```
<resultMap id="authorResult" type="Author">
    <id property="id" column="author_id"/>
    <result property="username" column="author_username"/>
    <result property="password" column="author_password"/>
    <result property="email" column="author_email"/>
    <result property="bio" column="author_bio"/>
</resultMap>
```

由于结果中的列名与 resultMap 中的列名不同，因此需要 columnPrefix 指定前缀来重用映射 co-author 结果的 resultMap，如代码清单 11-32 所示。

代码清单 11-32　重用 authorResult

```
<resultMap id="blogResult" type="Blog">
    <id property="id" column="blog_id" />
    <result property="title" column="blog_title"/>
    <association property="author"
                resultMap="authorResult" />
    <association property="coAuthor"
                resultMap="authorResult"
                columnPrefix="co_" />
</resultMap>
```

关于 MyBatis 中的一对一关系查询，开发者熟悉上面的知识已经完全够用了。一对多的关系在下面进行讲解。

11.5.5　集合（一对多查询）

集合元素和前面关联元素的作用基本是相同的。集合查询如代码清单 11-33 所示。

代码清单 11-33　集合查询

```
<collection property="posts" ofType="domain.blog.Post">
    <id property="id" column="post_id"/>
    <result property="subject" column="post_subject"/>
    <result property="body" column="post_body"/>
</collection>
```

这里继续引用前面的博客作者例子。一个博客只有一个作者，但是博客中可以有很多文章，所以博客与文章的关系为一对多的关系，那么可以用代码清单 11-34 的写法在博客中定义文章。

代码清单 11-34　文章集合

```
private List<Post> posts;
```

如果在 MyBatis 中需要映射嵌套结果集合到博客中，使用集合元素 collection 即可。

11.5.6　集合嵌套查询与嵌套结果

先来创建两个表，再进行下面的讲解。博客与文章表如代码清单 11-35 所示。

代码清单 11-35　创建博客与文章表

```
DROP TABLE IF EXISTS `blog`;
CREATE TABLE `blog` (
`id` int(11) NOT NULL AUTO_INCREMENT COMMENT '主键',
`title` varchar(255) NOT NULL,
`author_id` int(11) DEFAULT NULL COMMENT '作者id',
PRIMARY KEY (`id`)
) ENGINE=InnoDB AUTO_INCREMENT=2 DEFAULT CHARSET=utf8mb4;

-- ----------------------------
-- Records of blog
-- ----------------------------
INSERT INTO `blog` VALUES ('1','测试', '1');

-- ----------------------------
-- Table structure for post
-- ----------------------------
DROP TABLE IF EXISTS `post`;
CREATE TABLE `post` (
`id` int(11) NOT NULL AUTO_INCREMENT,
`subject` varchar(255) NOT NULL,
`body` varchar(255) NOT NULL,
`blog_id` int(11) NOT NULL,
PRIMARY KEY (`id`)
) ENGINE=InnoDB AUTO_INCREMENT=4 DEFAULT CHARSET=utf8mb4 COMMENT='文章';

-- ----------------------------
-- Records of post
-- ----------------------------
INSERT INTO `post` VALUES ('1', '1', '1', '1');
INSERT INTO `post` VALUES ('2', '2', '2', '1');
INSERT INTO `post` VALUES ('3', '3s', '3', '1');
```

接下来需要写好Dao接口、实体类、配置等。下面进行mapper接口开发，如代码清单11-36所示。

代码清单 11-36　使用嵌套查询为博客加载文章

```
<?xml version="1.0" encoding="UTF-8"?>
<!DOCTYPE mapper PUBLIC "-//mybatis.org//DTD Mapper 3.0//EN" "http://mybatis.org/
dtd/mybatis-3-mapper.dtd">
<mapper namespace="com.uifuture.chapter11.dao.BlogMapper">
    <resultMap id="blogResult" type="com.uifuture.chapter11.entity.BlogExt">
        <collection property="posts" javaType="ArrayList" column="id"
                    ofType="com.uifuture.chapter11.entity.Post" select="selectPostsForBlog"/>
    </resultMap>
    <select id="selectBlogExtById" resultMap="blogResult">
        SELECT * FROM BLOG WHERE ID = #{id}
    </select>
    <select id="selectPostsForBlog" resultType="com.uifuture.chapter11.entity.Post">
        SELECT * FROM POST WHERE BLOG_ID = #{id}
    </select>
</mapper>
```

<collection> 标记中 property 属性的值对应的是实体类中 List<Post> 的名称，ofType 属性用来区分 JavaBean（或者字段）属性类型和集合包含的类型。

<collection property="posts" javaType="ArrayList" column="id" ofType="com.uifuture.chapter11.entity.Post" select="selectPostsForBlog"/> 表示名称为 posts、Post 类型的 ArrayList 集合。

BlogExt 实体类如代码清单 11-37 所示。

代码清单 11-37　BlogExt 实体类

```
public class BlogExt extends Blog {
    private List<Post> posts;
//省略getter/setter/toString
}
```

接下来看嵌套结果的写法。

首先看在 MyBatis 映射文件中 SQL 语句的写法，如代码清单 11-38 所示。

代码清单 11-38　selectBlogExt 映射接口的 SQL 语句

```
<resultMap id="blogExtResult" type="BlogExt">
    <id property="id" column="blog_id" />
    <result property="title" column="blog_title"/>
    <collection property="posts" ofType="Post">
        <id property="id" column="post_id"/>
        <result property="subject" column="post_subject"/>
        <result property="body" column="post_body"/>
    </collection>
</resultMap>
<select id="selectBlogExt" resultMap="blogExtResult">
    select
    B.id as blog_id,
    B.title as blog_title,
    B.author_id as blog_author_id,
    P.id as post_id,
    P.subject as post_subject,
    P.body as post_body
    from Blog B
    left join Post P on B.id = P.blog_id
    where B.id = 1
</select>
```

关于结果集的一对一、一对多的查询就介绍到这里了。如果是大型项目，或者说以后有缓存、分布式扩展的，不推荐使用级联的写法。

11.5.7　鉴别器

MyBatis 中的鉴别器，可以简单理解为 Java 中的 switch 语句。有时候，一个单独的数据库查询可能返回很多不同数据类型的结果集（有些关联），鉴别器元素在 MyBatis 中就是被设计来处理这种情况的，还包括类的继承层次结构。鉴别器在实际项目中使用得不多，下面对其进行简单介绍。

例如，有一个交通工具类 Vehicle，另外还有子类 Car（汽车）和 Boat（船），Car 和 Boat 分别继承了 Vehicle 类。在 Vehicle 中有一个 type 字段，用来区分交通工具类型。根据 type 字段，在 MyBatis 中实现将查询的 Vehicle 数据自动封装成不同的类型对象（Car 或者 Boat）。XML 映射文件代码如代码清单 11-39 所示。

代码清单 11-39　XML 映射文件代码

```
<resultMap type="Vehicle" id="vehicleResult">
    <id property="id" column="id"/>
    <result property="color" column="color"/>
    <discriminator javaType="int" column="type">
        <case value="1" resultType="Car">
            <result property="doorCount" column="door_count"></result>
        </case>
        <case value="2" resultType="Boat">
            <result property="quant" column="quant"></result>
        </case>
    </discriminator>
</resultMap>
<select id="getVehicleById" resultMap="vehicleResult">
    select * from vehicle where id= #{id}
</select>
```

利用上面的 MyBatis 元素代码就可以通过 getVehicleById 方法来查询出不同类型的 Vehicle 对象。由于 type 可能会有很多种，实际项目中的数据库又比较复杂，因此，在映射查询中通过鉴别器来查询不同类型的方式实际上并不常用，在这里也就不详细介绍了。

11.6　自动映射

在前面的几个章节中，在一些不是很复杂的情况下，MyBatis 会自动映射（Auto-mapping）查询结果，但是如果遇到复杂的场景，则需要构建一个 resultMap。在本节中，将混合使用这两种方式，以便了解自动映射是怎么工作的。

当使用自动映射查询结果时，MyBatis 能够自动获取 SQL 返回的列名，并且在 Java 类中查询相同名称的属性（忽略大小写，默认驼峰命名和下划线命名可互相转换）。

一般情况下，数据库列名使用大写单词命名，单词之间使用下划线分割。Java 属性名遵循驼峰命名方式，MyBatis 默认是能够在这两种命名方式之间自动映射的，也就是 mapUnderscoreToCamelCase 的值默认为 true。

在某些 resultMap 中，如果有没有被写到代码中映射的列，将会自动映射，自动映射会优先处理，处理完后才会是显式地映射。在代码清单 11-40 所示的例子中 id 和 username 等列会自动映射，而 create_time 列由于配置了映射，则会根据配置映射。

代码清单 11-40　自动映射与配置映射

```
<resultMap id="userResultMap" type="com.uifuture.chapter11.entity.Users">
```

```
        <result column="create_time" jdbcType="TIMESTAMP" property="createTime"/>
    </resultMap>
    <select id="selectuserResultMap" parameterType="java.lang.Integer"
    resultMap="userResultMap">
        select
        id, username, password, age, create_time, update_time
        from users
        where id = #{id,jdbcType=INTEGER}
    </select>
```

关于自动映射，有以下 3 种等级：

（1）NONE：禁用自动映射，也就是仅仅支持手动映射属性。

（2）PARTIAL：将自动映射除了那些有内部定义嵌套结果映射的属性。

（3）FULL：自动映射所有属性。

MyBatis 的默认自动映射等级为 PARTIAL。使用 PARTIAL 是有原因的，例如，当使用 FULL 时，自动映射在处理 join 结果时执行，并且 join 将获取若干相同行的不同实体数据，有可能导致非预期的映射结果。简单理解就是，如果 join 的两个表中都有 id 属性，如果是自动映射，那么后面的实体 id 属性会覆盖前面实体的 id 设置，导致实际结果和预期的结果不同。所以在这里，建议使用默认的自动映射等级即可。

使用 autoMapping 属性可以忽略自动映射等级的配置，可以在 resultMap 上启动或者禁用自动映射指定的 resultMap，禁用后需要手动配置映射。

11.7 缓存

在 MyBatis 中有非常强大的查询缓存，而且其配置和定制非常方便。默认情况下，一级缓存是默认开启的，二级缓存没有开启。如果需要开启二级缓存，则直接在 SQL 映射文件中添加 <cache/> 即可。针对一级缓存和二级缓存的区别，在本节中会有更多的介绍。

11.7.1 缓存的属性

下面简单说明 <cache> 标记的作用。

- 可以将映射文件中的所有 select 语句进行缓存。
- 映射文件中的所有 insert、update 和 delete 语句都会刷新缓存。
- 缓存默认使用 LRU（Least Recently Used，最近最少使用）算法进行数据回收。
- 根据时间表（如 No Flush Interval，没有时间间隔），缓存不会以任何时间顺序来刷新。
- 缓存会存储集合或者对象的引用。
- 缓存是被视为 read/write（可读写）的缓存，意味着对象检索不是共享的，可以被调用者安全地修改，而不干扰其他调用者的修改操作。

有一点需要注意，使用该 <cache> 标记只是缓存所在映射文件中声明的语句，如果是结合

Java API 和 XML 映射文件使用，那么在 Java 接口中声明的 SQL 接口语句默认情况下是不会被缓存的，但是可以使用 @CacheNamespaceRef 引用缓存区域。

上面所说的几点作用，都可以通过 <cache> 标记属性来进行修改，如代码清单 11-41 所示。

代码清单 11-41　<cache> 属性

```
<cache
        eviction="FIFO"
        flushInterval="60000"
        size="512"
        readOnly="true"/>
```

上面 <cache> 标记的配置创建了一个 FIFO 可回收策略，每隔 60s 刷新一次，将存储结果对象或者集合的 512 个引用，并且返回的对象仅仅认为是可读对象，因此在不同线程中的调用者之间进行修改可能会导致冲突。

可用的回收策略有以下 4 种：

（1）LRU：最近最少使用，移除最长时间不被使用的对象。默认的回收策略。

（2）FIFO：先进先出策略，按照对象进入缓存的顺序来移除它们。

（3）SOFT：软引用，移除基于垃圾回收器状态中软引用规则的对象。

（4）WEAK：弱引用，移除基于垃圾回收器状态中弱引用规则的对象。

flushInterval 为刷新间隔，默认时间单位为 ms。它可以被设置为任意的正整数，默认情况下不设置，也就是没有刷新时间。缓存仅仅在调用语句时进行刷新。

size 为引用数量，它可以被设置为任意的正整数，注意缓存的对象数量要与实际的运行环境可用内存结合配置，默认值为 1024。

readOnly 为可读属性，它可以被设置为 true 或者 false。只读的缓存将会给所有的调用者返回相同的缓存实例对象，因此不要去修改这些对象。可读写的缓存会返回缓存对象的副本（通过序列化），这样虽然速度会稍微慢一点，但是线程是安全的，所以 MyBatis 的默认值是 false。

11.7.2　自定义缓存

实现自定义的二级缓存比较常用。例如，使用 Redis 作为 MyBatis 的二级缓存，这是目前在很多企业项目中常用的方式。

在 SQL 的 XML 配置文件中添加以下语句：

```
<cache type="com.***.RedisCache"/>
```

type 指向自定义缓存类，自定义缓存类必须实现 org.mybatis.cache.Cache 接口，也就是在 RedisCache 中实现 Cache 接口，Cache 接口的源码如代码清单 11-42 所示。

代码清单 11-42　Cache 接口的源码

```
public interface Cache {
    String getId();
    int getSize();
    void putObject(Object key, Object value);
```

```
    Object getObject(Object key);
    boolean hasKey(Object key);
    Object removeObject(Object key);
    void clear();
}
```

该接口很简单，包括设置值、获取大小、判断是否有值、移除对象等方法。如果要自定义缓存，重写这几个方法即可。在后面将会介绍如何自定义 Redis 缓存，在这里不进行介绍了。

缓存配置与缓存实例是绑定在 SQL 映射命名空间上的，因此，所有在相同命名空间中的语句共用缓存。单个语句可以修改与缓存进行交互的方式，或不进行缓存或不刷新缓存。默认情况下的配置如代码清单 11-43 所示。

代码清单 11-43　单个语句默认的缓存配置

```
<select… flushCache="false" useCache="true"/>
<insert…flushCache="true"/>
<update…flushCache="true"/>
<delete…flushCache="true"/>
```

上面的配置是默认的增、删、改、查缓存配置。当 flushCache 为 true 时表示执行该语句之后刷新缓存；当 useCache 为 true 时表示优先使用缓存。在这里，可以修改默认的配置，可以排除从缓存中查询特定语句的结果，或者需要用查询语句来刷新缓存或需要执行更新语句但不需要刷新缓存，都可以通过修改单条语句的缓存配置来进行定制。

在项目中，可能需要在另外的 SQL 映射文件中与某个命名空间内缓存的配置和缓存的实例共享，在这种情况下可以使用 cache-ref 元素来引用另外一个缓存，如代码清单 11-44 所示。

代码清单 11-44　使用 cache-ref 元素来引用另外一个缓存

```
<cache-ref namespace="com.uifuture.dao.SomeMapper"/>
```

11.8　MyBatis 缓存机制

从前面内容可以知道，MyBatis 中自带的缓存有一级缓存和二级缓存。下面针对这两种缓存再进行深入介绍。

11.8.1　一级缓存

MyBatis 的一级缓存是指 Session 的缓存。一级缓存的作用域默认是一个 SqlSession。一级缓存是默认开启的。也就是说，在同一个 SqlSession 中，执行相同的 SQL 查询，第 1 次会去数据库查询，并写到缓存中，第 2 次以后直接去缓存中获取。当执行 SQL 查询中间发生了增、删、改的操作时，MyBatis 会把 SqlSession 的缓存清空。

一级缓存的范围有 SESSION 和 STATEMENT 两种，默认是 SESSION。在有些特定的场景下，不需要使用一级缓存，可以把一级缓存的范围指定为 STATEMENT，这样每次执行完一个 Mapper

中的语句后都会将一级缓存清除。如果需要更改一级缓存的范围，则可以在 MyBatis 的配置文件中通过 localCacheScope 指定，如代码清单 11-45 所示。

代码清单 11-45　修改一级缓存的范围

```
<setting name="localCacheScope" value="STATEMENT"/>
```

这样即可将一级缓存的作用范围修改。当然，一般情况下不需要修改。MyBatis 和 Spring 整合以后，关于 SqlSession 需要注意的是，通过 Spring 集成 MyBatis 后，Spring 每个事务的每个 Mapper 中的每次查询,都对应着一个全新的 SqlSession 实例。在这种情况下,就不会命中一级缓存,所以需要开启二级缓存。

11.8.2　二级缓存

MyBatis 的二级缓存是指 mapper 映射文件。二级缓存的作用域是同一个 namespace 下的 mapper 映射文件内容，多个 SqlSession 共享。MyBatis 需要手动设置启动二级缓存。

二级缓存是默认启用的（要生效需要对每个 Mapper 进行配置），如果想取消，要通过 MyBatis 配置元素下的子元素来将 cacheEnabled 指定为 false，如代码清单 11-46 所示。

代码清单 11-46　禁用二级缓存

```
<settings>
    <setting name="cacheEnabled" value="false" />
</settings>
```

cacheEnabled 默认是启用的，只有当值为 true 时，底层使用的 Executor 才是支持二级缓存的 CachingExecutor。具体可参考 MyBatis 的核心配置类 org.apache.ibatis.session.Configuration 的 newExecutor 方法实现。newExecutor 方法的源码如代码清单 11-47 所示。

代码清单 11-47　newExecutor 方法的源码

```
//...
public Executor newExecutor(Transaction transaction) {
    return this.newExecutor(transaction, this.defaultExecutorType);
}
public Executor newExecutor(Transaction transaction, ExecutorType executorType) {
    executorType=executorType==null ? this.defaultExecutorType : executorType;
    executorType=executorType==null ? ExecutorType.SIMPLE : executorType;
    Object executor;
    if (ExecutorType.BATCH ==executorType) {
        executor = new BatchExecutor(this, transaction);
    } else if (ExecutorType.REUSE == executorType) {
        executor = new ReuseExecutor(this, transaction);
    } else {
        executor = new SimpleExecutor(this, transaction);
    }
    if (this.cacheEnabled) {// 设置为true才执行
        executor = new CachingExecutor((Executor)executor);
    }
```

```
        Executor executor = (Executor)this.interceptorChain.pluginAll(executor);
        return executor;
    }
    ...
```

除了上面一个配置外，还需要在每个 Dao 对应的 Mapper.xml 文件中定义需要使用的 cache，如代码清单 11-48 所示。

代码清单 11-48　配置 <cache> 标记

```
<mapper namespace="…UserMapper">
    <cache/><!-- 加上该句即可，使用默认配置，还有另外一种方式在后面写出 -->
</mapper>
```

具体可以看一下 org.apache.ibatis.executor.CachingExecutor 类的实现，其中使用的 cache 就是在对应的 Mapper.xml 中定义的 cache。CachingExecutor 源码如代码清单 11-49 所示。

代码清单 11-49　CachingExecutor 源码

```
public <E> List<E> query(MappedStatement ms, Object parameterObject, RowBounds
rowBounds, ResultHandler resultHandler) throws SQLException {
    BoundSql boundSql=ms.getBoundSql(parameterObject);
    CacheKey key=this.createCacheKey(ms, parameterObject, rowBounds, boundSql);
    return this.query(ms, parameterObject, rowBounds, resultHandler, key, boundSql);
}
public <E> List<E> query(MappedStatement ms, Object parameterObject, RowBounds
rowBounds, ResultHandler resultHandler, CacheKey key, BoundSql boundSql) throws
SQLException {
    Cache cache=ms.getCache();
    if (cache != null) {//第1个条件，定义需要使用的cache
        this.flushCacheIfRequired(ms);
        if (ms.isUseCache() && resultHandler==null) {/*第2个条件,需要当前的查询语句是配置
了使用cache的，即useCache()是返回true的，useCache配置项默认是true*/
            this.ensureNoOutParams(ms, parameterObject, boundSql);
            List<E> list= (List)this.tcm.getObject(cache, key);
            if (list == null) {
                list = this.delegate.query(ms, parameterObject, rowBounds, result
Handler, key, boundSql);
                this.tcm.putObject(cache, key, list);
            }
            return list;
        }
    }
    return this.delegate.query(ms, parameterObject, rowBounds, resultHandler, key,
boundSql);
}
```

还有一个条件就是需要当前的查询语句是配置了使用 cache 的，即上面源码的 useCache 是返回 true 的。默认情况下所有 select 语句的 useCache 都是 true。如果在启用了二级缓存后，有某个查询语句是不想缓存的，则可以通过指定其 useCache 为 false 来实现对应的效果，如代码

清单 11-50 所示。

代码清单 11-50 禁用单条语句的缓存

```
<select id="selectByPrimaryKey" resultMap="BaseResultMap" parameterType="java.
lang.String" useCache="false">
    select
    <include refid="Base_Column_List"/>
    from tuser
    where id = #{id,jdbcType=VARCHAR}
</select>
```

11.9 定义要使用的 cache 的两种方式

要使用的 cache 有两种方式定义：一种是通过 cache 元素定义；另一种是通过 cache-ref 元素定义。需要注意的是，对于同一个 Mapper 来讲，只能使用一个 cache。当同时使用时，定义的优先级更高（后面的代码会给出原因）。Mapper 使用的 cache 是与 Mapper 对应的 namespace 绑定的，一个 namespace 最多只会有一个 cache 与其绑定。

11.9.1 通过 cache 元素定义

通过 cache 元素来定义要使用的 Cache 时，最简单的做法是直接在对应的 Mapper.xml 文件中指定一个空的元素（看前面的代码）。这时 MyBatis 会按照默认配置创建一个 Cache 对象，准确地说是 PerpetualCache 对象，更准确地说是 LruCache 对象（底层用了装饰器模式）。

具体可看 org.apache.ibatis.builder.xml.XMLMapperBuilder 中的 cacheElement 方法解析 cache 元素的逻辑，如代码清单 11-51 所示。

代码清单 11-51 cacheElement 方法解析 cache 元素的逻辑

```
...
private void configurationElement(XNode context) {
    try {
        String namespace=context.getStringAttribute("namespace");
        if (namespace.equals("")) {
            throw new BuilderException("Mapper's namespace cannot be empty");
        } else {
            this.builderAssistant.setCurrentNamespace(namespace);
            this.cacheRefElement(context.evalNode("cache-ref"));
            this.cacheElement(context.evalNode("cache"));// 执行在后面
            this.parameterMapElement(context.evalNodes("/mapper/parameterMap"));
            this.resultMapElements(context.evalNodes("/mapper/resultMap"));
            this.sqlElement(context.evalNodes("/mapper/sql"));
            this.buildStatementFromContext(context.evalNodes("select|insert|update|d
elete"));
```

```
        }
    } catch (Exception var3) {
        throw new BuilderException("Error parsing Mapper XML. Cause: " + var3, var3);
    }
}
...
private void cacheRefElement(XNode context) {
    if (context != null) {
            this.configuration.addCacheRef(this.builderAssistant.get
CurrentNamespace(), context.getStringAttribute("namespace"));
           CacheRefResolver cacheRefResolver=new CacheRefResolver(this.builder
Assistant, context.getStringAttribute("namespace"));
        try {
            cacheRefResolver.resolveCacheRef();
        } catch (IncompleteElementException var4) {
            this.configuration.addIncompleteCacheRef(cacheRefResolver);
        }
    }

}
private void cacheElement(XNode context) throws Exception {
    if (context != null) {
        String type=context.getStringAttribute("type", "PERPETUAL");
        Class<? extends Cache> typeClass=this.typeAliasRegistry.resolveAlias(type);
        String eviction=context.getStringAttribute("eviction", "LRU");
        Class<? extends Cache> evictionClass=this.typeAliasRegistry.resolveAlias(eviction);
        Long flushInterval=context.getLongAttribute("flushInterval");
        Integer size=context.getIntAttribute("size");
        boolean readWrite=!context.getBooleanAttribute("readOnly", false).booleanValue();
        Properties props=context.getChildrenAsProperties();
        this.builderAssistant.useNewCache(typeClass, evictionClass, flushInterval,
size, readWrite, props);
//如果同时存在cache和cache-ref，则这里的设置会覆盖前面cache-ref的缓存
    }
}
```

<cache>标记定义会生成一个使用最近最少使用算法且最多只能存储1024个元素的缓存，而且是可读写的缓存。该缓存是全局共享的，任何一个线程在拿到缓存结果后对数据的修改都将影响其他线程获取的缓存结果。

cache元素可指定以下属性，每种属性的指定都是针对底层Cache的一种装饰，采用装饰器的模式。

- blocking：默认为false，当指定为true时将采用BlockingCache进行封装，使用BlockingCache会在查询缓存时锁住对应的Key，如果缓存命中则会释放对应的锁，否则会在查询数据库以后再释放锁，这样可以阻止并发情况下多个线程同时查询数据，详情可参考BlockingCache的源码。

- eviction：元素驱逐算法，默认是 LRU，对应的就是 LruCache。其默认只保存 1024 个 Key，超出时按照最近最少使用算法进行驱逐，详情参考 LruCache 的源码。如果想使用自定义的算法，则可以将该值指定为自定义的驱逐算法实现类，只需要自定义的类实现 Mybatis 的 Cache 接口即可。除了 LRU 以外，系统还提供了 FIFO（先进先出，对应 FifoCache）、SOFT（采用软引用存储 Value，便于垃圾回收，对应 SoftCache）和 WEAK（采用弱引用存储 Value，便于垃圾回收，对应 WeakCache）三种策略。
- flushInterval：清空缓存的时间间隔，单位是 ms，默认是不会清空的。当指定了该值时会再用 ScheduleCache 包装一次，其会在每次对缓存进行操作时判断距离最近一次清空缓存的时间是否超过了 flushInterval 指定的时间，如果超出了，则清空当前的缓存，详情可参考 ScheduleCache 的实现。
- readOnly：是否只读，默认为 false。当指定为 false 时，底层会用 SerializedCache 包装一次，其会在写缓存的时候将缓存对象进行序列化，然后在读缓存的时候进行反序列化，这样每次读到的都将是一个新的对象，即使更改了读取到的结果，也不会影响原来缓存的对象，即非只读，每次获取这个缓存结果都可以进行修改，而不会影响原来的缓存结果；当指定为 true 时，那就是每次获取的都是同一个引用，对其修改会影响后续的缓存数据获取，这种情况下不建议对获取到的缓存结果进行更改，意为只读(不建议设置为 true)。这是 MyBatis 二级缓存读写和只读的定义，可能与通常情况下的只读和读写意义有点不同。每次都进行序列化和反序列化无疑会影响性能，但是这样的缓存结果更安全，不会被随意更改，具体可根据实际情况进行选择。详情可参考 SerializedCache 的源码。
- size：用来指定缓存中最多可保存的 Key 数量。其是针对 LruCache 而言的，LruCache 默认最多只存储 1024 个 Key，可通过该属性来改变默认值。当然，如果通过 eviction 指定了自定义的驱逐算法，同时自己的实现里面也有 setSize 方法，那么也可以通过 cache 的 size 属性给自定义的驱逐算法中的 size 赋值。
- type：用来指定当前底层缓存实现类，默认为 PerpetualCache。如果想使用自定义的 Cache，则可以通过该属性来指定，对应的值是自定义的 Cache 的全路径名称。

11.9.2 通过 cache-ref 元素定义

cache-ref 元素可以用来指定其他 Mapper.xml 中定义的 cache，有的时候可能多个不同的 Mapper 需要共享同一个缓存，希望在 MapperA 中缓存的内容在 MapperB 中可以直接命中，这个时候就可以考虑使用 cache-ref。这种场景只需要保证缓存的 Key 是一致的即可命中，二级缓存的 Key 是通过 Executor 接口的 createCacheKey 方法生成的，其实现基本都是 BaseExecutor，createCacheKey 方法源码如代码清单 11-52 所示。

代码清单 11-52　createCacheKey 方法源码

```
public CacheKey createCacheKey(MappedStatement ms, Object parameterObject,
RowBounds rowBounds, BoundSql boundSql) {
    if (this.closed) {
        throw new ExecutorException("Executor was closed.");
```

```
    } else {
        CacheKey cacheKey=new CacheKey();
        cacheKey.update(ms.getId());
        cacheKey.update(rowBounds.getOffset());
        cacheKey.update(rowBounds.getLimit());
        cacheKey.update(boundSql.getSql());
        List<ParameterMapping> parameterMappings=boundSql.getParameterMappings();
          TypeHandlerRegistry typeHandlerRegistry=ms.getConfiguration().getType
HandlerRegistry();
        for(int i=0; i<parameterMappings.size(); ++i) {
            ParameterMapping parameterMapping= (ParameterMapping)parameterMappings.
get(i);
            if (parameterMapping.getMode()!= ParameterMode.OUT) {
                String propertyName=parameterMapping.getProperty();
                Object value;
                if (boundSql.hasAdditionalParameter(propertyName)) {
                    value=boundSql.getAdditionalParameter(propertyName);
                } else if (parameterObject==null) {
                    value=null;
                } else if (typeHandlerRegistry.hasTypeHandler(parameterObject.getClass())) {
                    value=parameterObject;
                } else {
                            MetaObject metaObject=this.configuration.newMetaObject
(parameterObject);
                    value=metaObject.getValue(propertyName);
                }
                cacheKey.update(value);
            }
        }
        return cacheKey;
    }
}
```

打个比方，若要在 MenuMapper.xml 中的查询都使用在 UserMapper.xml 中定义的 cache，则可以通过 cache-ref 元素的 namespace 属性指定需要引用的 Cache 所在的 namespace，即 UserMapper.xml 中定义的 namespace。假设在 UserMapper.xml 中定义的 namespace 是 com.uifuture.dao.UserMapper，则在 MenuMapper.xml 的 cache-ref 应该定义如下：

```
<cache-ref namespace="com.uifuture.dao.UserMapper"/>
```

这样两个 Mapper 就共享同一个缓存了。

11.10　二级缓存实例

缓存是 MyBatis 中非常重要的特性。前面已经介绍了缓存的一部分基础知识，在这里会讲解一些重要的点，以及编写一个实例并进行测试。

11.10.1 二级缓存的测试

首先，在 MyBatis 配置文件中将 cacheEnabled 设置为 true（默认即为 true）。接下来，在需要缓存的 SQL 映射文件中添加 <cache> 标记，如图 11-1 所示。

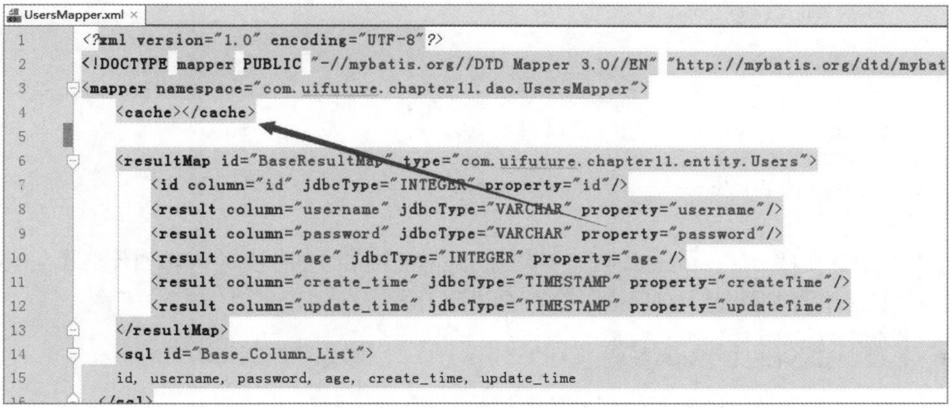

```
UsersMapper.xml ×
<?xml version="1.0" encoding="UTF-8"?>
<!DOCTYPE mapper PUBLIC "-//mybatis.org//DTD Mapper 3.0//EN" "http://mybatis.org/dtd/mybat
<mapper namespace="com.uifuture.chapter11.dao.UsersMapper">
    <cache></cache>

    <resultMap id="BaseResultMap" type="com.uifuture.chapter11.entity.Users">
        <id column="id" jdbcType="INTEGER" property="id"/>
        <result column="username" jdbcType="VARCHAR" property="username"/>
        <result column="password" jdbcType="VARCHAR" property="password"/>
        <result column="age" jdbcType="INTEGER" property="age"/>
        <result column="create_time" jdbcType="TIMESTAMP" property="createTime"/>
        <result column="update_time" jdbcType="TIMESTAMP" property="updateTime"/>
    </resultMap>
    <sql id="Base_Column_List">
    id, username, password, age, create_time, update_time
  </sql>
```

图 11-1　添加 cache 标签

接着可以添加二级缓存测试方法，如代码清单 11-53 所示。

代码清单 11-53　二级缓存测试方法

```
private static final Logger LOGGER = LoggerFactory.getLogger(UsersMapperTest.
class);
@Test
public void selectByPrimaryKey()throws IOException {
    String resource="mybatis-config.xml";
    InputStream inputStream=Resources.getResourceAsStream(resource);
      SqlSessionFactory sqlSessionFactory=new SqlSessionFactoryBuilder().
build(inputStream);
    try (SqlSession session=sqlSessionFactory.openSession();) {
        UsersMapper mapper=session.getMapper(UsersMapper.class);
        LOGGER.info("============{}",mapper.selectByPrimaryKey(1));
        // 同一个sqlSession中运行的是一级缓存
        LOGGER.info("============{}",mapper.selectByPrimaryKey(1));
    }
    //只有前面的SqlSession关闭，并新获取一个SqlSession，在开启二级缓存的情况下，才会启动二级缓存
    try (SqlSession session=sqlSessionFactory.openSession();) {
        UsersMapper mapper=session.getMapper(UsersMapper.class);
        LOGGER.info("============{}",mapper.selectByPrimaryKey(1));
        LOGGER.info("============{}",mapper.selectByPrimaryKey(1));
    }
}
```

如果想要看到缓存效果，需要添加日志配置输出。这里集成了 logback，并添加配置文件。首先在 pom.xml 中添加日志依赖，如代码清单 11-54 所示。

代码清单 11-54　添加日志依赖

```
<dependency>
    <groupId>ch.qos.logback</groupId>
    <artifactId>logback-classic</artifactId>
    <version>1.2.3</version>
    <scope>compile</scope>
</dependency>
<dependency>
    <groupId>org.slf4j</groupId>
    <artifactId>jul-to-slf4j</artifactId>
    <version>1.7.25</version>
    <scope>compile</scope>
</dependency>
```

然后添加配置文件，在 resources 目录下新建 logback.xml 文件，文件内容如代清单 11-55 所示。关于 logback 日志的详细配置，在后面的章节会讲到。

代码清单 11-55　logback 日志简单配置

```
<?xml version="1.0" encoding="utf-8" ?>
<!-- 当scan属性设置为true时，配置文件如果发生改变，会被重新加载，默认值为true
 scan="true" scanPeriod="60 seconds" -->
<configuration debug="true">
    <property name="FILE_LOG_PATTERN"
              value="${FILE_LOG_PATTERN:-%d{yyyy-MM-dd HH:mm:ss.SSS} ${LOG_LEVEL_
PATTERN:-%5p} ${PID:- } --- [%t] %-40.40logger{39} : %m%n}"/>
     <!-- 彩色日志格式:"%black", "%red", "%green","%yellow","%blue", "%magenta",
"%cyan", "%white", "%gray", "%boldRed","%boldGreen", "%boldYellow", "%boldBlue",
"%boldMagenta""%boldCyan", "%boldWhite" and "%highlight"
    -->
    <property name="CONSOLE_LOG_PATTERN"
               value="%highlight([%-5level]) %white(%d{yyyy-MM-dd HH:mm:ss SSS})
%green([%thread]) %magenta(%logger) - %blue(%msg%n)"/>
    <!--设置应用名-->
    <contextName>uifuture-ssm</contextName>
    <!--日志输出到控制台-->
    <appender name="console" class="ch.qos.logback.core.ConsoleAppender">
        <encoder>
            <pattern>${CONSOLE_LOG_PATTERN}</pattern>
            <charset>utf8</charset>
        </encoder>
    </appender>
    <!--建立一个默认root的logger-->
    <root level="DEBUG">
        <appender-ref ref="console"/>
    </root>
</configuration>
```

接下来运行测试类，可得到图 11-2 所示的输出。

可以看到，在第 1 次查询时会执行 SQL 语句；在第 2 次查询时，由于是同一个 SqlSession，运行的是 MyBatis 的一级缓存，因此命中率也为 0.0；在第 3 次查询时，开始运行二级缓存，这时候的缓存命中率为 0.33；第 4 次也是运行的二级缓存。

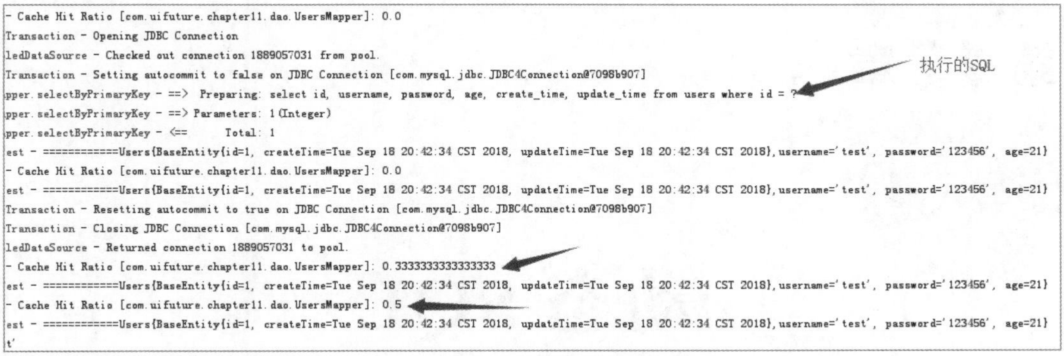

```
- Cache Hit Ratio [com.uifuture.chapter11.dao.UsersMapper]: 0.0
Transaction - Opening JDBC Connection
ledDataSource - Checked out connection 1889057031 from pool.
Transaction - Setting autocommit to false on JDBC Connection [com.mysql.jdbc.JDBC4Connection@7098b907]
pper.selectByPrimaryKey - ==>  Preparing: select id, username, password, age, create_time, update_time from users where id = ?
pper.selectByPrimaryKey - ==> Parameters: 1(Integer)
pper.selectByPrimaryKey - <==      Total: 1
est - ============Users{BaseEntity{id=1, createTime=Tue Sep 18 20:42:34 CST 2018, updateTime=Tue Sep 18 20:42:34 CST 2018},username='test', password='123456', age=21}
- Cache Hit Ratio [com.uifuture.chapter11.dao.UsersMapper]: 0.0
est - ============Users{BaseEntity{id=1, createTime=Tue Sep 18 20:42:34 CST 2018, updateTime=Tue Sep 18 20:42:34 CST 2018},username='test', password='123456', age=21}
Transaction - Resetting autocommit to true on JDBC Connection [com.mysql.jdbc.JDBC4Connection@7098b907]
Transaction - Closing JDBC Connection [com.mysql.jdbc.JDBC4Connection@7098b907]
ledDataSource - Returned connection 1889057031 to pool.
- Cache Hit Ratio [com.uifuture.chapter11.dao.UsersMapper]: 0.3333333333333333
est - ============Users{BaseEntity{id=1, createTime=Tue Sep 18 20:42:34 CST 2018, updateTime=Tue Sep 18 20:42:34 CST 2018},username='test', password='123456', age=21}
- Cache Hit Ratio [com.uifuture.chapter11.dao.UsersMapper]: 0.5
est - ============Users{BaseEntity{id=1, createTime=Tue Sep 18 20:42:34 CST 2018, updateTime=Tue Sep 18 20:42:34 CST 2018},username='test', password='123456', age=21}
t'
```

执行的SQL

图 11-2　运行测试类的输出结果

CacheHitRatio 表示二级缓存命中率。开启二级缓存后，每执行一次查询，系统都会计算一次二级缓存的命中率。

第 1 次查询也是先从缓存中查询，只不过缓存中一定是没有数据的，所以会再从 DB 中查询。第 2 次由于运行的是一级缓存，二级缓存中不存在该数据，因此命中率为 0.0。但第 3 次查询是从二级缓存中读取的，所以这一次的命中率为 1/3 ≈ 0.3333。第 4 次查询，命中率为 2/4=0.5。

注意：增、删、改操作无论是否提交 sqlSession.commit，均会清空一级缓存和二级缓存，使查询再次从 DB 中执行 select 语句。二级缓存的清空，实质上是对所查找的 key 对应的 value 设置为 null，而非将 entry 对象删除。从 DB 中进行 select 查询的条件是：缓存中根本不存在这个 key 或者缓存中存在该 key 所对应的 entry 对象，但 value 为 null。

设置增、删、改操作不刷新二级缓存。若要使某个增、删、改操作不清空二级缓存，则需要在其中添加属性 flushCache="false"，默认为 true。

11.10.2　二级缓存使用原则

（1）只能在一个命名空间下使用二级缓存。由于二级缓存中的数据是基于 namespace 的，即不同 namespace 中的数据互不干扰，在多个 namespace 中若均存在对同一个表的操作，那么这多个 namespace 中的数据可能就会出现不一致现象。

（2）在单表上使用二级缓存。如果一个表与其他表有关联关系，那么就有可能存在多个 namespace 对同一数据的操作。而不同 namespace 中的数据互不干扰，所以就有可能出现多个 namespace 中的数据不一致现象。

（3）查询多于修改时使用二级缓存。在查询操作远远多于增、删、改操作的情况下可以使用二级缓存。因为任何增、删、改操作都将刷新二级缓存，对二级缓存的频繁刷新将降低系统性能。

动态 SQL

本章要点

1. MyBatis 中动态 SQL 的编写
2. 多数据库厂商支持
3. 动态 SQL 中的可插拔脚本语言

12.1 动态 SQL 简介

MyBatis 中的另外一个强大特性就是它的动态 SQL（英文全称为 Dynamic SQL）。如果之前没用过类似的框架，可能已经体会到了根据不同条件拼接 SQL 的痛苦，如拼接时添加必要的空格，还要注意去掉列表最后一个列名的逗号。在 MyBatis 中，使用动态 SQL 可以彻底摆脱这种痛苦。

MyBatis 提供了强大的动态 SQL 语言，得以在任意的 SQL 映射语句中生成动态 SQL，其实就是 XML 解析，了解 JSP 中 JSTL 的读者会很容易理解。同理，MyBatis 也自定义了标记库，通过标记库可建立对象关系映射（ORM）。动态 SQL 使用起来比较简单，下面先从几个相关元素的实例讲解入手，后面会针对动态 SQL 讲解原理。

12.2 if 元素

动态 SQL 通常要做的就是根据条件来执行不同的 SQL 语句，如 if 元素动态拼接，如代码清单 12-1 所示。

代码清单 12-1　if 元素动态拼接

```
<select id="findUserResultMap" resultMap="userResultMap">
    SELECT * FROM users WHERE
    username = #{username}
    <if test="age != null">
        AND age = #{age}
    </if>
</select>
```

这种语句在 MyBatis 中十分常见，它提供了动态查询数据的功能。通常情况下，可能不仅根据 username 来查询用户，还需要其他列来组合，所以根据上面的代码，可以使用 <if> 标记。举例说明，如果没有传入 age，就不会将 age 作为条件进行查询。如果有更多或者更少的列，则增加或者减少 <if> 标记即可。

12.3 choose 元素、when 元素、otherwise 元素

某些场景下，可能不会用到所有的条件语句。MyBatis 提供了 choose 元素，当只想选择某一项执行时，可以使用 choose 元素。choose 元素有点像 Java 中的 switch 语句。

下面来模拟一种情形：当填写了 username 时，通过 username 查询用户对象；若没有填写 username，而只填写了 age，则通过 age 查询用户对象；如果都没有填写，则通过创建时间来查询

用户对象。可使用 choose 元素实现，如代码清单 12-2 所示。

代码清单 12-2　使用 choose 元素

```
<select id="findUsers"
        resultMap="userResultMap">
    SELECT * FROM users WHERE
    <choose>
        <when test="username != null">
            AND username = #{username}
        </when>
        <when test="age != null">
            AND age = #{age}
        </when>
        <otherwise>
            AND create_time  <![CDATA[ > ]]>  #{createTime}
        </otherwise>
    </choose>
</select>
```

上面的 <![CDATA[>]]> 语句只是为了原样输出“>”符号。

12.4　trim 元素、where 元素、set 元素

前面的两个例子在一般情况下就够用了。但是仔细想一想，前面的 if 示例如果用在 select 中，会变成什么样呢？ if 元素的 select 语句如代码清单 12-3 所示。

代码清单 12-3　if 元素的 select 语句

```
<select id="findUserResultMap"
        resultMap="userResultMap">
    SELECT * FROM users WHERE
    <if test="username != null">
        username=#{username}
    </if>
    <if test="age != null">
        AND age=#{age}
    </if>
</select>
```

在这种情况下，如果 username 和 age 都为 null，则没有一个条件能够匹配上，最后生成的 SQL 语句如代码清单 12-4 所示。

代码清单 12-4　有问题的 SQL 语句

```
SELECT * FROM users
WHERE
```

最后生成的 SQL 语句肯定是无法执行的，因为多出了 WHERE 关键字，肯定会导致失败。如

果第 1 个 username 为 null，age 不为 null，则第 2 个条件匹配的 SQL 语句如代码清单 12-5 所示。

代码清单 12-5　第 2 个条件匹配的 SQL 语句

```
SELECT * FROM users
WHERE
AND age=#{age}
```

这个查询也会失败，所以在这种情况下，不能简单地使用条件语句来解决问题。

在 MyBatis 中，有一种简单的处理方式，至少保证在绝大多数情况下都能应对，即使用 where 元素，如代码清单 12-6 所示。

代码清单 12-6　使用 where 元素

```
<select id="findUserResultMap"
        resultMap="userResultMap">
    SELECT * FROM users
    <where>
        <if test="username != null">
            username=#{username}
        </if>
        <if test="age != null">
            AND age=#{age}
        </if>
    </where>
</select>
```

where 元素只会在至少有一个子元素条件返回 SQL 子句的情况下才会插入 where 子句；而且，如果语句的开头为 AND 或者 OR，则 where 元素会将它们移除。如果不想使用 where 元素，也可以使用 trim 元素达到相同的目的，如代码清单 12-7 所示。

代码清单 12-7　与 where 元素等效的 trim 元素用法

```
<select id="findUserResultMap"
        resultMap="userResultMap">
    SELECT * FROM users
    <trim prefix="WHERE" prefixOverrides="AND |OR ">
        <if test="username != null">
            username=#{username}
        </if>
        <if test="age != null">
            AND age=#{age}
        </if>
    </trim>
</select>
```

prefixOverrides 属性会忽略通过“|”分隔的文本序列（注意空格是必要的）。它的作用是移除初始字符中在 prefixOverrides 属性中的内容。

类似的用于动态更新语句的元素为 set，set 元素可以用于动态包含需要更新的列，而舍去其他的列，如代码清单 12-8 所示。

代码清单 12-8　set 元素演示

```
<update id="updateByPrimaryKeySelective" parameterType="com.uifuture.chapter12.
entity.Users">
    update users
    <set>
        <if test="username != null">
            username = #{username,jdbcType=VARCHAR},
        </if>
        <if test="password != null">
            password = #{password,jdbcType=VARCHAR},
        </if>
        <if test="age != null">
            age = #{age,jdbcType=INTEGER},
        </if>
    </set>
    where id = #{id,jdbcType=INTEGER}
</update>
```

在这里，set 元素会动态前置 set 关键字，同时也会删除无关的逗号，因为使用条件语句后很可能会在生成的 SQL 语句最后留下逗号。如果使用 trim 元素来替换 set 元素，则如代码清单 12-9 所示。

代码清单 12-9　与 set 元素等效的 trim 元素

```
<update id="updateByPrimaryKeySelectiveTrim" parameterType="com.uifuture.
chapter12.entity.Users">
    update users
    <trim prefix="SET" suffixOverrides=",">
        <if test="username != null">
            username = #{username,jdbcType=VARCHAR},
        </if>
        <if test="password != null">
            password = #{password,jdbcType=VARCHAR},
        </if>
        <if test="age != null">
            age = #{age,jdbcType=INTEGER},
        </if>
    </trim>
    where id = #{id,jdbcType=INTEGER}
</update>
```

注意： 在这里使用的 suffixOverrides，其含义就是需要删除的后缀值为“,”，同时前缀值为“SET”。

12.5　foreach 元素

动态 SQL 的另一个经常需要用到的操作就是对一个集合进行遍历，通常是在构建 in 条件语句的时候。例如，foreach 元素构建 select 语句如代码清单 12-10 所示。

代码清单 12-10　foreach 元素构建 select 语句

```
<select id="selectUserIn" resultType="users">
    SELECT *
    FROM users
    WHERE ID in
    <foreach item="item" index="index" collection="list"
             open="(" separator="," close=")">
        #{item}
    </foreach>
</select>
```

foreach 元素的功能非常强大，它允许指定一个集合，并可以在 foreach 元素体内使用集合项（item）和索引（index）变量，也允许指定开头与结尾的字符串以及在迭代结果之间放置分隔符。这个元素是非常智能的，因此它不会附加多余的分隔符。例如，这里的 open 与 close 即开头和结尾的字符串，separator 即迭代结果之间的分隔符。

使用 foreach 元素可以迭代集合（list、set 等）、Map 对象或者数组对象。当使用可迭代对象或者数组时，index 是当前迭代的次数，item 的值是本次迭代获取到的元素。当使用 Map 对象时，index 是键，item 是值。

12.6　bind 元素

bind 元素可以从 OGNL 表达式中创建一个变量并将其绑定到上下文，如代码清单 12-11 所示。

代码清单 12-11　将一个对象中的变量绑定到上下文

```
Java接口：
List<Blog> selectBlogsLike(Blog blog);

XML:
<select id="selectBlogsLike" resultMap="BaseResultMap">
<bind name="pattern" value="'%'+_parameter.getTitle()+'%'" />
        SELECT * FROM blog
        WHERE title LIKE #{pattern}
</select>
```

上面是官方文档的写法，还可以写成如代码清单 12-12 所示的形式。

代码清单 12-12　将一个变量绑定到上下文

```
Java接口：
List<Blog> selectBlogsTitleLike(@Param("title") String title);

XML:
<select id="selectBlogsTitleLike" resultMap="BaseResultMap">
```

```
<bind name="pattern" value="'%'+title+'%'" />
        SELECT * FROM blog
        WHERE title LIKE #{pattern}
</select>
```

注意：这里是加了 Param 注解的。当只有一个 Stirng 参数时，如果不指定 @Param，则 MyBatis 会在 <if>/<bind> 标记属性值中将 parameterType 参数默认成接口的参数类型，然后用 XML 中的 title 参数去调用该类型下参数 title 的 get/set 方法。使用了 @param ，MyBatis 就会一一对应赋值，不会导致错误出现。另外，当只有一个参数时，可在代码中使用 _parameter（MyBatis 内置参数）字符串。

另外还有一个注意的点，就是 #{pattern} 不要写成 ${pattern}，因为“$”是直接注入值。在当前场景下，值需要使用“''”单引号括下来。除非是表名或者字段，否则不要使用“${}”。

12.7 多数据库支持与可拔插 SQL 脚本语言

12.7.1 多数据库厂商支持

开发者可以方便地在 MyBatis 中配置对多数据库厂商的支持，如代码清单 12-13 所示。通过预先定义每个数据库的 id，就可以根据不同的数据库厂商来构建特定的语句。

代码清单 12-13 多数据厂商支持

```
全局配置：
<databaseIdProvider type="DB_VENDOR">
    <!--为不同的数据库厂商起别名-->
    <property name="SQL Server" value="sqlserver"/>
    <property name="DB2" value="db2"/>
    <property name="Oracle" value="oracle"/>
    <property name="MySQL" value="mysql"/>
</databaseIdProvider>
映射文件：
<insert id="insert" parameterType="com.uifuture.chapter12.entity.Blog">
    <selectKey keyProperty="id" order="AFTER" resultType="java.lang.Integer">
        <if test="_databaseId == 'oracle'">
            select seq_users.nextval from dual
        </if>
        <if test="_databaseId == 'mysql'">
            SELECT LAST_INSERT_ID()
        </if>
    </selectKey>
    insert into blog (title, author_id)
    values (#{title,jdbcType=VARCHAR}, #{authorId,jdbcType=INTEGER})
</insert>
```

```
Java接口:
int insert(Blog record);
```

这样在执行 insert 接口时，不同数据库都可以正确地返回最近更新的自动增长 id。**提示**：整个语句返回的值也是数据库中受影响的行数，自动增长 id 则绑定在插入对象中，这是 MyBatis 处理的。在执行插入或者更新语句成功后，直接从对象中获取 id 即可。

12.7.2 动态 SQL 中的可插拔脚本语言

从 MyBatis 3.2 版本开始，MyBatis 提供了可插拔脚本的功能。也就是说，可以用自定义语句拼接规则，并编写动态 SQL，而不是只能使用默认的脚本驱动。在 MyBatis 中有两个内置的语言，见表 12-1。

表 12-1 MyBatis 中的两个内置语言及驱动

别 名	驱 动
xml	XmlLanguageDriver
raw	RawLanguageDriver

官方的默认脚本驱动写在了 org.apache.ibatis.scripting.xmltags.XmlLanguageDriver 中，默认的别名是 xml。XML 语言是默认的，能够运用前面章节中所有的动态标记。实际上，使用 RawLanguageDriver（就是使用原始语言）时，在 MyBatis 执行参数替换语句传递到数据库的驱动程序过程中，原始语言的速度远远超过可扩展标志语言（XML）。可以在 MyBatis-config.xml 配置文件中修改默认脚本语言，如代码清单 12-14 所示。

代码清单 12-14 修改默认脚本语言

```
<setting name="defaultScriptingLanguage" value="org.apache.ibatis.scripting.
defaults.RawLanguageDriver"/>
```

但需要注意的是，使用 raw 之后，就无法使用动态 SQL 了，所以绝大多数情况下还是要使用 XML 脚本语言。

如果需要实现自定义的语言驱动程序，可以实现以下的接口，如代码清单 12-15 所示。

代码清单 12-15 LanguageDriver 接口

```
public interface LanguageDriver {
ParameterHandler createParameterHandler(MappedStatement mappedStatement, Object
  parameterObject, BoundSql boundSql);
SqlSource createSqlSource(Configuration configuration, XNode script, Class<?>
parameterType);
    SqlSource createSqlSource(Configuration configuration, String script, Class<?>
parameterType);
}
```

关于自定义脚本语言驱动，本书不详细讲解。如果有兴趣，读者可以查看 RawLanguageDriver 及 XmlLanguageDriver 的源码，也可以看看 ApacheVelocity，参考 MyBatis-Velocity 项目（在 MyBatis 的开源库中）。

第 13 章

代码生成器

本章要点

1. MyBatis Generator 逆向代码生成工具
2. MyBatis Generator 概述
3. MyBatis Generator 快速入门
4. 使用 Maven 运行 MyBatis Generator
5. 使用 Java 程序运行 MyBatis Generator

13.1　MBG 概述

看完前面的章节，如果开发者想要连接数据库并对表进行增、删、改、查操作，那么需要编写很多映射配置，操作比较烦琐。那么有没有简便的方法？答案是肯定的，那就是利用代码生成器（MyBatis Generator，MBG）来自动构建映射文件、接口及数据库实体。

MBG 检查一个（或多个）数据库表，并生成可用于访问表的文件，这样就减少了设置对象和配置文件与数据库表交互的最初麻烦。MBG 减少了对数据库简单的 CRUD（插入、查询、更新、删除）操作，极大节省了开发者的时间，但仍然需要开发者手工编写 SQL 和对象以用于连接查询或存储过程。

13.1.1　MBG 会生成的代码

（1）表结构的 Java POJO，可能包括以下几点：

- 一个和表主键匹配的类 [如果存在主键（注：只有联合主键会有）]。
- 一个包含了非主键字段的类 [BLOB 字段除外 (注：单字段做主键时该类会包含)]。
- 一个包含了 BLOB 类型字段的类 (如果表包含了 BLOB 字段)。
- 一个允许动态查询、更新和删除的类（注：指的是 Example 查询）。

这些类之间会有适当的继承关系。**提示**：可以配置生成器来生成不同类型 POJO 的层次结构。例如，可以选择针对每个表生成一个单独的实体对象。

（2）MyBatis 兼容的 SQL 映射 XML 文件。MBG 在配置中为每个表进行简单的 CRUD 操作生成 SQL，生成的 SQL 语句可能包括以下几点：

- insert：插入。
- update by primary key：根据主键更新记录。
- update by example：根据条件更新记录。
- delete by primary key：根据主键删除记录。
- delete by example：根据条件删除记录。
- select by primary key：根据主键查询记录。
- select by example：根据条件查询记录。
- count by example：根据条件查询记录总数。

根据表的结构生成的这些语句会有不同的变化（举例来说，如果表中没有主键，那么 MBG 将不会生成 update by primary key 方法），并且开发者可以手动配置是否生成根据非主键条件 CRUD 操作的 SQL。

（3）Java 类会适当地使用上面的对象，生成的 Java 类是可选的。MBG 会为 MyBatis 3.x 生成 Java 类：一个可以与 MyBatis 3.x 一起使用的 mapper 接口类。

MyBatis Generator 可以在迭代开发环境中良好地运行，在持续的构建环境中作为一个 Ant 任务或 Maven 插件。

运行 MBG 时要记住以下 2 个重要事项：

（1）MBG 会自动合并已经存在且与新生成文件重名的 XML。MBG 默认不会覆盖已经生成的 XML 所做的修改，开发者可以反复地运行而不必担心失去自定义的更改。

（2）MBG 不会合并 Java 文件，但可以覆盖已经存在的文件或者保存新生成的文件为唯一的名称。开发者可以选择手动合并这些更改。

13.1.2 MBG 依赖项

MBG 依赖于 JRE，并且需要 JRE 6.0 及以上的版本才能运行，还需要数据库的驱动，继承了 java.sql.DatabaseMetaData 接口的 JDBC 驱动。该接口中的 getColumns 和 getPrimaryKeys 两个方法是必须在驱动中实现的。也就是说，MBG 必须依赖 JRE 与数据库驱动才能正常运行。

13.2 MBG 快速入门

在这里举一个最简单的例子。启动并快速运行 MBG 可以按照以下 4 个步骤进行：

①准备好 JRE 环境和数据库驱动包。

②创建并填写适当的配置文件数据。必填的配置项有以下 4 个：

- <jdbcConnection> 标记：定义如何连接目标数据库。
- <javaModelGenerator> 标记：指定生成 Java 模型对象所属的包。
- <sqlMapGenerator> 标记：指定生成 SQL 映射文件所属的包和目标项目。
- <javaClientGenerator> 标记（可选的）：指定目标包和目标项目生成的 Java 接口及类（如果不想生成 Java 接口代码，可以省略 <javaClientGenerator> 标记；正常情况下不省略）。

③在命令行中输入以下命令运行 MBG：

java -jar mybatis-generator-core-1.3.2.jar -configfile generator.xml -overwrite

其中，java -jar 为数据库驱动 Jar 包路径，-configfile generator 为文件路径，-overwrite 为覆盖原文件。

以上命令会告知 MBG 让配置文件运行。MBG 还会覆盖已经存在的同名 Java 文件，如果想保留已经存在的 Java 文件，可以不使用 -overwrite 参数。如果存在冲突 MBG 会用唯一的名称保存新生成的文件。

④ MBG 运行后，需要创建或修改标准 MyBatis 配置文件来使用新生成的代码或接口。

下面讲解 XML 配置文件的实例。在常见的案例中，MBG 是由一个 XML 配置文件驱动的。配置文件告知 MBG 以下信息：

- 如何连接数据库。
- 生成什么对象，以及如何生成它们。
- 哪些表需要生成对象。

代码清单 13-1 所示的是一个非常简单的 MBG 配置文件的例子。

代码清单 13-1 MBG 简单配置

```
<?xml version="1.0" encoding="UTF-8"?>
<!DOCTYPE generatorConfiguration PUBLIC "-//mybatis.org//DTD MyBatis Generator
Configuration 1.0//EN" "http://mybatis.org/dtd/mybatis-generator-config_1_0.dtd">
<generatorConfiguration>
    <!-- 数据库驱动包位置 -->
    <classPathEntry location="mysql-connector-java-5.1.39-bin.jar" />
    <context id="DB2Tables" targetRuntime="MyBatis3">
        <commentGenerator>
            <property name="suppressAllComments" value="true" />
        </commentGenerator>
        <!-- 数据库链接URL、用户名、密码 -->
         <jdbcConnection driverClass="com.mysql.jdbc.Driver" connectionURL="jdbc:mysql://
localhost:3306/数据库名" userId="用户名" password="密码">
        </jdbcConnection>
        <javaTypeResolver>
            <property name="forceBigDecimals" value="false" />
        </javaTypeResolver>
        <!-- 生成模型的包名和位置 -->
          <javaModelGenerator targetPackage="com.uifuture.entity" targetProject="\
generator">
            <property name="enableSubPackages" value="true" />
           <property name="trimStrings" value="true" />
        </javaModelGenerator>
        <!-- 生成的映射文件包名和位置 -->
           <sqlMapGenerator targetPackage="com.uifuture.mapping" targetProject="\
generator">
            <property name="enableSubPackages" value="true" />
        </sqlMapGenerator>
        <!-- 生成Dao的包名和位置 -->
           <javaClientGenerator type="XMLMAPPER" targetPackage="com.uifuture.dao"
targetProject="\generator">
            <property name="enableSubPackages" value="true" />
        </javaClientGenerator>
         <!-- 要生成的表,可以编写很多的table元素[更改tableName(表名)和domainObjectName(生
成的类名)就可以]-->
         <table tableName="user" domainObjectName="User" enableCountByExample="false"
enableUpdateByExample="false" enableDeleteByExample="false"
enableSelectByExample="false" selectByExampleQueryId="false" />
    </context>
</generatorConfiguration>
```

有关此文件的说明可以在代码清单 13-1 中了解，此文件是最简单的生成代码。

13 代码生成器

13.3 使用 Maven 运行 MBG

MBG 可以通过以下 4 种方式运行：

（1）通过命令提示符使用 XML 配置文件。

（2）将作为 Ant 任务使用 XML 配置文件。

（3）将作为 Maven 插件进行使用。

（4）在 Java 程序中使用 XML 配置文件。

由于第 1 种命令提示和 Ant 任务在 Java 中目前已经不常用了，所以在这里仅讲解将 MBG 作为 Maven 插件，以及直接使用 Java 程序运行的方式。

13.3.1 将 MBG 作为 Maven 插件使用

MBG 包含了一个可以集成到 Maven 的 Maven 插件。按照 Maven 的配置惯例，将 MBG 集成到 Maven 很容易。最简配置如下：

首先，直接在 pom.xml 中添加类似代码清单 13-2 所示的代码。

代码清单 13-2 用基础 Maven 插件配置 MBG

```
<build>
    <plugins>
        <plugin>
            <groupId>org.mybatis.generator</groupId>
            <artifactId>mybatis-generator-maven-plugin</artifactId>
            <version>1.3.0</version>
        </plugin>
    </plugins>
</build>
```

然后，引入依赖，如代码清单 13-3 所示。

代码清单 13-3 引入 MyBatis 依赖和 MySQL 依赖

```
<dependencies>
    <!-- MyBatis依赖 -->
    <dependency>
        <groupId>org.mybatis</groupId>
        <artifactId>mybatis</artifactId>
        <version>3.4.6</version>
    </dependency>
    <!-- MySQL依赖 -->
    <dependency>
        <groupId>mysql</groupId>
        <artifactId>mysql-connector-java</artifactId>
        <version>5.1.46</version>
    </dependency>
</dependencies>
```

另外，需要在 resources 下配置数据库账号密码，以及需要 MBG 的配置文件 generatorConfig.xml。

注意：配置文件的名称必须为 generatorConfig.xml（可以进行配置），否则会报 [ERROR] Failed to execute goal org.mybatis.generator:mybatis-generator-maven-plugin:1.3.7:generate (default-cli) on project ssm-mybatis-generator: configfile D:\github\uifuture-ssm\ssm-mybatis-generator\src\main\resources\generatorConfig.xml does not exist -> [Help 1] 错误。

数据库配置文件如代码清单 13-4 所示。

代码清单 13-4　数据库配置文件

```
jdbc.driverLocation=C:\\Users\\Administrator\\.m2\\repository\\mysql\\mysql-
connector-java\\5.1.46\\mysql-connector-java-5.1.46.jar
jdbc.driverClass=com.mysql.jdbc.Driver
# useSSL=false&serverTimezone=GMT在MySql高版本下需要
# remarks=true
jdbc.connectionURL=jdbc:mysql://localhost:3306/chapter-10-ssm?useUnicode=true&chara
cterEncoding=UTF-8&useSSL=false&serverTimezone=GMT&remarks=true
jdbc.userId=root
jdbc.password=1234
model.package=com.uifuture.entity
mapper.xml.package=com.uifuture.mapper
mapper.impl.package=com.uifuture.dao

#main方法运行插件路径不需要返回上层路径
#model.project=ssm-mybatis-generator/src/main/java
#mapper.xml.project=ssm-mybatis-generator/src/main/java
#mapper.impl.project=ssm-mybatis-generator/src/main/java

#使用maven插件运行需要有/
model.project=../ssm-mybatis-generator/src/main/java
mapper.xml.project=../ssm-mybatis-generator/src/main/java
mapper.impl.project=../ssm-mybatis-generator/src/main/java
```

在这里需要注意的是 jdbc.driverLocation。其在每个人计算机中的位置可能会不同，需要改成开发者的数据库驱动包路径，绝对路径和相对路径都行。

13.3.2　完整的 MBG 配置文件

MBG 配置文件如代码清单 13-5 所示。

代码清单 13-5　MBG 配置文件

```
<?xml version="1.0" encoding="UTF-8"?>
<!DOCTYPE generatorConfiguration
        PUBLIC "-//mybatis.org//DTD MyBatis Generator Configuration 1.0//EN"
        "http://mybatis.org/dtd/mybatis-generator-config_1_0.dtd">
<generatorConfiguration>
    <!--导入属性配置 -->
    <properties resource="generator.properties"/>
```

```
    <!--指定特定数据库的jdbc驱动jar包的位置 -->
    <classPathEntry location="${jdbc.driverLocation}"/>

     <!--<context id="MySQLTables" targetRuntime="MyBatis3Simple"> MyBatis3可以生成
selestive-->
    <context id="prod" targetRuntime="MyBatis3">
          <!-- 自动识别数据库关键字，默认值为false，如果设置为true，则为SqlReservedWords中定
义的关键字列表；一般保留默认值，遇到数据库关键字（Java关键字），使用columnOverride覆盖 -->
        <property name="autoDelimitKeywords" value="false"/>
        <!-- 生成的Java文件的编码 -->
        <property name="javaFileEncoding" value="UTF-8"/>
        <!-- 格式化java代码 -->
          <property name="javaFormatter" value="org.mybatis.generator.api.dom.De
faultJavaFormatter"/>
        <!-- 格式化XML代码 -->
          <property name="xmlFormatter" value="org.mybatis.generator.api.dom.
DefaultXmlFormatter"/>
          <!--为生成的Java model添加序列化接口，并生成serialVersionUID字段：<plugin
type="org.mybatis.generator.plugins.SerializablePlugin" />;
           生成tostring方法：<plugin type="org.mybatis.generator.plugins.ToStringPlugin"/>;
           配置 jdbc的数据库连接 -->
         <jdbcConnection driverClass="${jdbc.driverClass}" connectionURL="${jdbc.
connectionURL}" userId="${jdbc.userId}" password="${jdbc.password}">
        </jdbcConnection>
          <!--非必需，类型处理器，在数据库类型和java类型之间的转换控制。 java类型处理器：
用于处理DB中的类型到Java中的类型，默认使用 org.mybatis.generator.internal.types.
JavaTypeResolverDefaultImpl。注意一点，默认会先使用Integer、Long、Short等来对应DECIMAL和
NUMERIC数据类型 -->
           <javaTypeResolver type="org.mybatis.generator.internal.types.
JavaTypeResolverDefaultImpl">
           <!--Value的取值有以下几种情况：
               true：使用BigDecimal，对应DECIMAL和NUMERIC数据类型
               false：默认值，
                   scale>0;length>18：使用BigDecimal;
                   scale=0;length[10,18]：使用Long;
                   scale=0;length[5,9]：使用Integer;
                   scale=0;length<5：使用Short
           -->
           <property name="forceBigDecimals" value="false"/>
        </javaTypeResolver>

        <!--Model模型生成器，用来生成含有主键key的类，记录类及查询Example类
            targetPackage：指定生成的model生成所在的包名；
            targetProject：指定在该项目下所在的路径
        -->
            <javaModelGenerator targetPackage="${model.package}"
targetProject="${model.project}">
            <!-- 是否对model添加构造函数。自动为每一个生成的类创建一个构造方法，构造方法包含了
所有的field，而不是使用setter -->
```

```
            <property name="constructorBased" value="false"/>
            <!-- 是否允许子包，即targetPackage.schemaName.tableName。在targetPackage的基础上，根据数据库的schema再生成一层package，最终生成的类放在这个package下，默认为false -->
            <property name="enableSubPackages" value="false"/>
            <!-- 建立的Model对象是否不可改变，如果为true，则生成的Model对象不会有 setter方法，只有构造方法 -->
            <property name="immutable" value="false"/>
            <!-- 设置是否在setter方法中，对String类型字段调用trim()方法 -->
            <property name="trimStrings" value="true"/>
        </javaModelGenerator>

        <!--Mapper映射文件生成所在的目录，为每一个数据库的表生成对应的SqlMap文件 -->
            <sqlMapGenerator targetPackage="${mapper.xml.package}" targetProject="${mapper.xml.project}">
             <!-- 在targetPackage的基础上，根据数据库的schema再生成一层package，最终生成的类放在这个package下，默认为false -->
            <property name="enableSubPackages" value="false"/>
        </sqlMapGenerator>

        <!-- 客户端代码，生成易于使用的针对Model对象和XML配置文件的代码
      (1) type="ANNOTATEDMAPPER",生成Java Model 和基于注解(SQL生成在annotation中)的Mapper对象；
      (2) type="MIXEDMAPPER"，使用混合配置，会生成Mapper接口，并适当添加合适的Annotation，但是XML会生成在XML中；
      (3) type="XMLMAPPER",生成SQLMap XML文件和独立的Mapper接口。
            注意，如果context是MyBatis3Simple，只支持ANNOTATEDMAPPER和XMLMAPPER
        -->
        <javaClientGenerator targetPackage="${mapper.impl.package}"
                             targetProject="${mapper.impl.project}" type="XMLMAPPER">
             <!-- 在targetPackage的基础上，根据数据库的schema再生成一层package，最终生成的类放在这个package下，默认为false -->
            <property name="enableSubPackages" value="false"/>
            <!-- 定义Maper.java 源代码中的ByExample方法的可视性，可选的值有：
                 public、private、protected、default。
                 如果 targetRuntime="MyBatis3"，则此参数被忽略
                 <property name="exampleMethodVisibility" value=""/>，方法名计数器
                 提示：如果目标运行时是MyBatis3，则忽略此属性
                 <property name="methodNameCalculator" value=""/>
            -->
        </javaClientGenerator>

        <!--选择一个table来生成相关文件，可以有一个或多个table，必须要有table元素
            选择的table会生成以下文件：
           (1) SQL map文件；
           (2) 生成一个主键类；
           (3) 除了BLOB和主键的其他字段的类；
           (4) 包含BLOB的类；
           (5) 一个用户生成动态查询的条件类(selectByExample, deleteByExample)，可选；
```

(6) Mapper接口(可选);

(7) tableName(必要):要生成对象的表名。

注意:大小写是敏感的。正常情况下,MBG会自动的去识别数据库标识符的大小写敏感度,在一般情况下,MBG会根据设置的schema、catalog或tablename去查询数据表,按照下面的流程进行:

① 如果schema、catalog或tablename中有空格,那么设置的是什么格式,就精确地使用指定的大小写格式去查询;

② 否则,如果数据库的标识符使用大写的,那么MBG自动把表名变成大写再查询;

③ 否则,如果数据库的标识符使用小写的,那么MBG自动把表名变成小写再查询;

④ 否则,使用指定的大小写格式查询。

另外,当创建表时,如果使用的""规定了数据库对象的大小写,则就算数据库标识符使用了大写,在这种情况下也会使用规定的大小写来创建表名。不过,设置delimitIdentifiers="true"即可保留大小写格式。可选:

(1) schema:数据库的schema;

(2) catalog:数据库的catalog;

(3) alias:为数据表设置的别名,如果设置了alias,那么生成的所有的SELECT SQL语句中,列名会变成alias_actualColumnName;

(4) domainObjectName:生成的domain类的名字,如果不设置,直接使用表名作为domain类的名字;可以设置为somepck.domainName,那么会自动把domainName类再放到somepck包里面;

(5) enableInsert(默认true):指定是否生成insert语句;

(6) enableSelectByPrimaryKey(默认true):指定是否生成按照主键查询对象的语句(就是getById或get);

(7) enableSelectByExample(默认true):MyBatis3Simple为false,指定是否生成动态查询语句;

(8) enableUpdateByPrimaryKey(默认true):指定是否生成按照主键修改对象的语句(即update);

(9) enableDeleteByPrimaryKey(默认true):指定是否生成按照主键删除对象的语句(即delete);

(10) enableDeleteByExample(默认true):MyBatis3Simple为false,指定是否生成动态删除语句;

(11) enableCountByExample(默认true):MyBatis3Simple为false,指定是否生成动态查询总条数语句(用于分页的总条数查询);

(12) enableUpdateByExample(默认true):MyBatis3Simple为false,指定是否生成动态修改语句(只修改对象中不为空的属性);

(13) modelType:参考context元素的defaultModelType,相当于覆盖;

(14) delimitIdentifiers:参考tableName的解释。注意,默认的delimitIdentifiers是双引号,如果类似MYSQL这样的数据库,使用的是"`"(反引号),那么还需要设置context的beginningDelimiter和endingDelimiter属性;

(15) delimitAllColumns:设置是否所有生成的SQL中的列名都使用标识符引起来。默认为false,delimitIdentifiers参考context的属性。

注意,table里面很多参数都是对javaModelGenerator、context等元素的默认属性的一个复写

-->

<tabletableName="users" enableCountByExample="false" enableDeleteByExample="false" enableSelectByExample="false" enableUpdateByExample="false">

<!--用于指定自动生成主键的属性。如果指定该元素,MBG会在生成insert的SQL映射文件中插入一个合适的<selectKey>元素。简单来说,就是可以用来做自增主键的设置。必选属性:

(1) column : 生成列的列名;

(2) sqlStatement : 返回新值的SQL语句。一些特殊值如 Mysql,转化为`SELECT LAST_INSERT_ID()。JDBC在MyBatis 3.0中,生成正确的代码,且脱离数据库的限制。

可选属性:

(1) identity : 默认为false。如果为true,被标记为identity列,且<selectKey>元素后被插入在insert后面。如果为false,则会被插入在insert之前;

(2) type : 为selectKey元素指定类型,pre或post。指定类型后,selectKey永远在insert语句之前

```
        -->
        <generatedKey column="id" sqlStatement="Mysql" identity="true"></generatedKey>
    </table>
    <table tableName="blog" enableCountByExample="false" enableDeleteByExample="false"
enableSelectByExample="false" enableUpdateByExample="false">
        <generatedKey column="id" sqlStatement="Mysql" identity="true"></generatedKey>
    </table>
  </context>
</generatorConfiguration>
```

在 IDEA 中查看右边的插件，可以直接单击 mybatis-generator:generate 选项运行，如图 13-1 所示。

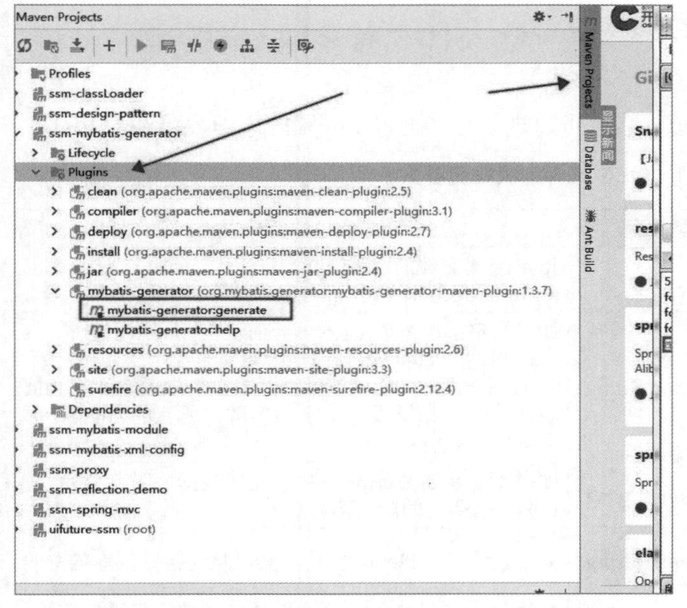

图 13-1　MBG 插件位置

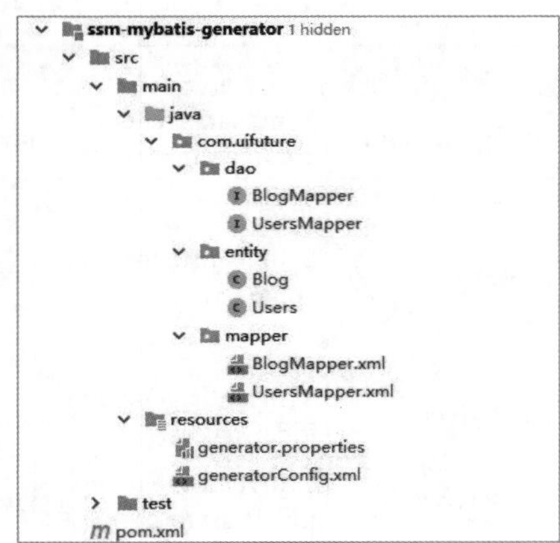

图 13-2　生成代码后的文件结构

当完成前面的配置并运行 MBG 插件后，可以看到图 13-2 所示的文件结构。

在其中，可以看到 dao、entity、mapper 包下均生成了对应的接口、类和 XML 映射文件。

默认情况下，MBG 重新运行是不会覆盖文件的，而是生成新的文件（并以在文件后面增加数字的形式进行命名）。例如，已知 Blog.java 文件，重新运行 MBG 是不会覆盖 Blog.java 文件的，而是会另起名称为 Blog.java.1。前面如果使用命令行形式，需要增加 overwrite=true 参数。如果使用插件的形式，只需要在 POM 文件中增加配置项即可。详解见 13.3.3 小节。

13.3.3　MBG 其他配置

前面遇到的几个问题，如配置文件名称自定义、重复生成文件如何选择生成规则，以及其他的配置等，修改方式很简单，只需要修改 POM 插件配置即可解决以上遇到的问题。在这里，将配置文件名称进行自定义，以及将文件生成规则修改为覆盖，如代码清单 13-6 所示。

代码清单 13-6　增加自定义配置

```
<plugins>
    <plugin>
        <groupId>org.mybatis.generator</groupId>
        <artifactId>mybatis-generator-maven-plugin</artifactId>
        <version>1.3.7</version>
        <configuration>
            <configurationFile>${basedir}/src/main/resources/generatorConfig.xml</
configurationFile>
            <overwrite>true</overwrite>
        </configuration>
    </plugin>
</plugins>
```

这里增加了 configuration 元素，在该元素中可以进行一些配置，配置的详细参数见表 13-1。

表 13-1　MBG 插件形式配置参数

参　数	表达式	类　型	注　释
configurationFile	${mybatis.generator.configurationFile}	java.io.File	指定配置文件的名称。默认值：${basedir}/src/main/resources/generatorConfig.xml
contexts	${mybatis.generator.contexts}	java.lang.String	如果指定了该参数，逗号隔开的这些 context 会被执行。这些指定的 context 必须与配置文件中 <context> 标记的 id 属性一致。只有指定的这些 contextid 才会被激活执行。如果没有指定该参数，所有的 context 都会被激活执行
jdbcDriver	${mybatis.generator.jdbcDriver}	java.lang.String	如果指定了 sqlScript 参数，连接数据库时其值是 JDBC 驱动类的限定名称
jdbcPassword	${mybatis.generator.jdbcPassword}	java.lang.String	如果指定了 sqlScript 参数，其值是连接数据库的密码
jdbcURL	${mybatis.generator.jdbcURL}	java.lang.String	如果指定了 sqlScript 参数，其值连接的是数据库的 JDBCURL
jdbcUserId	${mybatis.generator.jdbcUserId}	java.lang.String	如果指定了 sqlScript 参数，其值连接的是数据库的用户 id
outputDirectory	${mybatis.generator.outputDirectory}	java.io.File	放置 MBG 所生成文件的目录。当 targetProject 在配置文件中设置特殊值“MAVEN”时使用（大小写敏感）。默认值为 ${project.build.directory}/generated-sources/mybatis-generator
overwrite	${mybatis.generator.overwrite}	boolean	如果指定了该参数，当存在已经同名的文件时，新生成的文件则会覆盖原有的文件。如果没有指定该参数，当存在同名的文件时，MBG 则会给新生成的文件生成唯一的名称（如 MyClass.java.1、MyClass.java.2 等）。**提示：** 生成器一定会自动合并或覆盖已经生成的 XML 文件。默认值为 false

续表

参　数	表达式	类　型	注　释
sqlScript	${mybatis.generator.sqlScript}	java.lang.String	要在生成代码之前运行的 SQL 脚本文件的位置。如果为空，则不会执行任何脚本。如果不为空，则必须提供 jdbcDriver 参数、jdbcURL 参数。另外，如果连接数据库需要认证，也需要提供 jdbcUserId 和 jdbcPassword 参数。值可以是一个文件系统的绝对路径或者是一个以“classpath:”开头放在构建的类路径下的路径
tableNames	${mybatis.generator.tableNames}	java.lang.String	如果指定了该参数，则逗号隔开的这个表会被运行，这些表名必须与 <table> 配置中的表名完全一致。如果没有指定该参数，则所有的表都会被执行。按以下方式指定表名：tableschema.tablecatalog…table 等

表 13-1 中的参数值为 POM 插件配置中的元素名称，表中的表达式是在 MBG 配置文件中使用的，该表达式表示引用 ${} 中该配置的值。表 13-1 中的所有参数都是可选的，在没有特殊情况下，使用默认值即可。

13.4　使用 Java 程序运行 MBG

MBG 也可以直接使用 Java 进行调用。对于 MBG 的配置，可以使用 XML 配置文件或者完全使用 Java 进行配置，也可以混合使用。

首先将 generator.properties 修改为使用 main 方法运行的方式，这是由于插件和 Java 主程序运行的路径不同，所以包的生成路径配置不同（使用绝对路径可以避免该问题）。

将 generator.properties 修改为如代码清单 13-7 所示的代码。

代码清单 13–7　修改后 generator.properties

```
... //省略部分代码
# main方法运行插件路径不需要返回上层路径
model.project=ssm-mybatis-generator/src/main/java
mapper.xml.project=ssm-mybatis-generator/src/main/java
mapper.impl.project=ssm-mybatis-generator/src/main/java
```

只需要修改上面的 3 个值路径即可。

代码清单 13-8 所示的代码展示了如何通过 XML 配置文件从 Java 运行 MBG。

代码清单 13–8　Java 运行 MBG

```
/**
 * 运行方法一
 */
public class GeneratorMain {
    public static void main(String[] args) throws Exception {
```

13
代码生成器

```
            File configFile=new File("ssm-mybatis-generator\\src\\main\\resources\\
generatorConfig.xml");
        List<String> warnings=new ArrayList<>();
        ConfigurationParser cp=new ConfigurationParser(warnings);
        Configuration config=cp.parseConfiguration(configFile);
        DefaultShellCallback callback=new DefaultShellCallback(true);
          MyBatisGenerator myBatisGenerator=new MyBatisGenerator(config, callback,
warnings);
        myBatisGenerator.generate(null);
    }
}
```

配置文件属性可以通过ConfigurationParser构造函数的参数传递给解析器。如果没有显式传递，配置文件的属性将会从 JVM 的系统属性中搜索。例如，属性 generated.source.dir 可以在配置文件中通过 ${generated.source.dir} 被访问。如果没有指定配置文件中的一个属性，这个属性将会原样输出。

此外，还可以使用 ShellRunner 来运行 MBG，如代码清单 13-9 所示。

代码清单 13-9　使用 ShellRunner 运行 MBG

```
/** generator使用ShellRunner方式运行
 */
public class GeneratorMain2 {
    public static void main(String[] args) {
            args=new String[] { "-configfile", "ssm-mybatis-generator\\src\\main\\
resources\\generatorConfig.xml", "-overwrite" };
        ShellRunner.main(args);
    }
}
```

MyBatis 框架的介绍到这里就结束了。可以回想一下 MyBatis 中的组件知识:XML 配置,缓存，XML 映射增、删、改、查，MBG 的内容。目前所有的业务系统的操作无非就是针对数据库的增、删、改、查，所以 MyBatis 框架是很重要的。第Ⅳ篇开始讲解 Spring 框架。

第Ⅳ篇 Spring 框架

本篇通过生活中的案例的类比来介绍和讲解 IoC、DI、AOP。介绍了 Spring 的核心机制，包括初始化过程和 Spring 中的 Bean，也介绍了创建 Bean 的三种方式及如何通过多种方式加载属性文件，对事件机制、定时器任务等知识点均有讲解并配有实战案例，对 Spring 中的几个核心注解进行了源码解析。在本篇最后，讲解了数据库事务管理以及 Spring 对事务管理的支持，通过小案例对 Spring 事务管理器进行应用，也讲解了 Transactional 注解的用法和实现原理。

希望读者通过本篇的学习，对 IoC 和 AOP 有一些深刻的认识，对 Spring 中的容器、Bean，甚至是 Spring 都有一些自己独到的见解。在企业项目开发中，事务是十分常见的，因此在本篇中，也单独用一章讲解了事务。对于事务，建议读者着重理解一个词：原子性。

第14章 IoC 与 DI 详解

本章要点

1. IoC 概述
2. 用代码理解 IoC
3. Spring 中 IoC 容器的实现
4. 传统 OOP 与 IoC 的对比
5. DI 与 IoC 的关系
6. Spring 中依赖注入方式
7. Spring IoC 入门案例

14.1 IoC 概述

本书第 1 章，简单地介绍了 Spring 中的 IoC(控制反转) 和 DI(依赖注入) 这两个概念，对于初学 Spring 的人来说，会比较难以理解。

控制反转模式，其基本概念是：不创建对象，但是描述创建它们的方式。在代码中不直接与对象和服务连接，但在配置文件中描述组件需要哪一项服务。在 Spring 框架中 IoC 容器负责将这些联系在一起。在 IoC 出现以前，组件之间的协调关系是由程序内部代码来控制的，或者说，以前使用 new 关键字来实现两组间之间的依赖关系，这种方式就造成了组件之间的互相耦合。

IoC 就是解决这个问题的，它将实现组件间的关系从程序内部提到了程序外部，由容器来管理。也就是说，由容器在运行期将组件间的某种依赖关系动态地注入到组件中。

在典型的 IoC 场景中，容器创建了所有对象，并设置必要的属性将它们连接在一起，决定什么时候调用什么方法。

IoC 有 3 种实现模式：

（1）服务需要设计专门的接口，通过该接口，由对象提供这些服务，可以从对象查询依赖性。

（2）通过 JavaBean 的属性（如 setter 方法）分配依赖性。

（3）依赖性以构造函数的形式提供，不以 JavaBean 属性的形式公开。

在 Spring 框架的 IoC 容器中，采用（2）和（3）方式进行实现。

IoC 本身并不能算为一种技术，应该算是一种思想，它使开发者从烦琐的对象交互中解脱出来，专注于对象本身，更进一步突出面向对象。

14.2 深入理解 IoC

IoC 的思想最核心的地方在于，资源不由使用资源的双方管理，而由第三方管理，这可以带来很多好处。第一，资源集中管理，让资源可配置和易管理；第二，降低了使用资源双方的依赖程度，也就是耦合度。

可简单地理解为甲方要达成某种目的不需要直接依赖乙方，只需要将诉求告知第三方机构。

例如，甲方需要租房子，而乙方有房子需要出租，乙方并不需要找租户，直接找第三方，并告诉第三方有房子需要出租即可。如此甲乙双方互相不依赖，只有在进行交易时才产生联系。反之亦然。

这样做的好处就是，甲乙双方可以在对方不真实存在的情况下独立存在，而且保证不交易时无联系，若要交易时可以很容易联系。因为交易由第三方来负责联系，而且甲乙都认为第三方可靠，交易就很可靠很灵活了。

这就是 IoC 的核心思想，生活中这种例子比比皆是。例如，支付宝在整个淘宝体系里就是庞大的 IoC 容器，是交易双方之外的第三方，提供可靠、可依赖、可灵活变更交易方的资源管理中心。

此外,人事代理也是雇佣机构和个人之外的第三方。关于 IoC,还可以通过 DI 来理解,后面会讲到。

在以上的描述中，诞生了两个专业词汇：IoC 和 DI。

所谓的 DI 就是甲方开放接口，在需要的时候，能够将乙方的房子传递进来（注入）。

所谓的 IoC 就是甲乙双方不相互依赖，交易活动的进行不依赖于任何一方，整个活动的进行由第三方负责管理。依赖第三方是不可避免的，只是将甲方和乙方的进行解耦，完全解耦是没有意义的。

14.3　通过代码理解 IoC

下面用一个简单的实例来理解 IoC 思想。假设一个场景：人 (Person) 都需要一个对象（ObjectTarget），如代码清单 14-1 所示。

代码清单 14-1　不使用 IoC 获取对象

```
public class Person {
    /**获取一个对象*/
    public ObjectTarget getObjectTarget() {
        ObjectTarget objectTarget = new ObjectTarget();
        //...
        System.out.println("I find a objectTarget:"+objectTarget.toString());
        return objectTarget;
    }
}
```

以上代码说明在获取对象之前必须先“new”一个对象，否则会找不到该对象。如果使用 IoC 会怎么样呢？如代码清单 14-2 所示。

代码清单 14-2　使用 IoC 获取对象

```
public class Person {
    private ObjectTarget objectTarget;
    /**通过IoC思想获取一个对象*/
    public ObjectTarget getObjectTarget() {
        System.out.println("I find a objectTarget:"+objectTarget.toString());
        return objectTarget;
    }
}
```

用小明想吃蔬菜来举例,在没有使用 IoC 的情况下,小明想吃蔬菜,需要自己去买菜、洗菜切菜,经过烹饪后小明才能吃到炒熟的蔬菜。如果使用了 IoC 思想（找厨师），那么小明只需要告诉厨师小李,想吃蔬菜了。小李自然会将买菜、洗菜、切菜等工作做好,最后将炒熟的蔬菜端到小明的面前,小明只需要拿起筷子吃就可以了。未使用 IoC 和使用 IoC 的对比如图 14-1 所示。

图 14-1 表达了 IoC 的核心思想，也是它要解决的问题：让开发者免去对依赖对象的维护，需要的时候可以去取，不用关心依赖对象的任何过程。

接下来的问题是如何将依赖的对象准备好（依赖注入），常用的方式有两种：构造方法注入和 setter 注入。

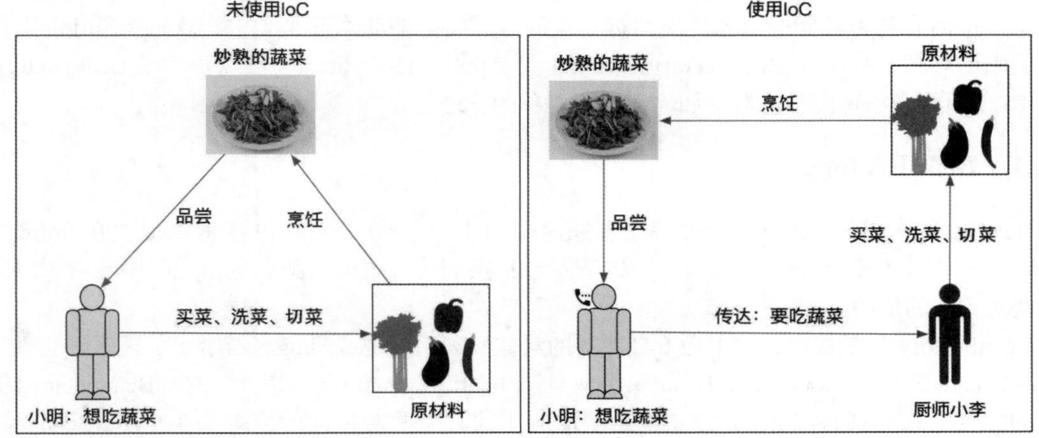

图 14-1 未使用 IoC 和使用 IoC 的对比

使用构造方法注入，当 Person 对象生成时，对象就准备好了，如代码清单 14-3 所示。

代码清单 14-3 使用构造方法注入

```
public Person(ObjectTarget objectTarget) {
    this.objectTarget = objectTarget;
}
```

setter 注入有所不同，对象只在进行设置时，才会匹配，如代码清单 14-4 所示。

代码清单 14-4 使用 setter 方法进行注入

```
public void setObjectTarget(ObjectTarget objectTarget) {
    this.objectTarget = objectTarget;
}
```

但无论哪一种注入方法，都是由第三方来执行。

14.4 Spring 中 IoC 容器的实现

IoC 中的第三方看起来非常美好，但实现起来并不容易，需要借助一系列技术。它需要知道服务的对象是谁，以及需要为对象提供什么样的服务。在 14.3 节的例子中提供的服务是先要完成对象的构建，并在将其送到服务对象前完成对象的绑定。

要应用 IoC 需要实现以下 2 步：

（1）对象的构建。

（2）对象的绑定。

对于这两方面，实现技术有很多方式，如硬编码（IoC 框架都支持）、配置文件（在没有注解之前最常用的方式）和注解（最简洁的方式）。但无论哪种方式都是在 IoC 容器（可以理解为一个大池子，里面有各种各样的对象，能通过一定的方式将它们联系起来）中实现的。Spring 提供了两种类型的容器，一个是 BeanFactory，另一个是 ApplicationContext（可以认为是 BeanFactory 的扩展）。下面将介绍这两种容器是如何实现对对象的管理的。

14.4.1 BeanFactory

BeanFactory，以 Factory 结尾，表示它是一个工厂类，负责生产和管理 Bean。在 Spring 中，BeanFactory 是 IoC 容器的核心接口，它的职责包括实例化、定位、配置应用程序中的对象以及建立这些对象间的依赖。

BeanFactory 只是接口，并不是 IoC 容器的具体实现，但是 Spring 容器给出了很多种实现，如 DefaultListableBeanFactory、XmlBeanFactory、ApplicationContext 等。其中，XmlBeanFactory 就是常用的一个，该实现将以 XML 方式描述组成应用的对象及对象间的依赖关系。XmlBeanFactory 类将持有 XML 配置元数据，并用它来构建一个完全可配置的系统或应用。BeanFactory 默认采用懒加载策略（Lazy-Load），只有当程序对象需要访问容器中的某个对象时，才对该对象进行初始化以及依赖注入操作。因此，相对来说，容器启动初期速度较快，所需要的资源有限。对于资源有限并且功能要求不是很严格的场景，BeanFactory 是比较合适的 IoC 容器选择。

图 14-2 所示为 BeanFactory 类的简单关系图，只列出了几个最核心的接口和类。

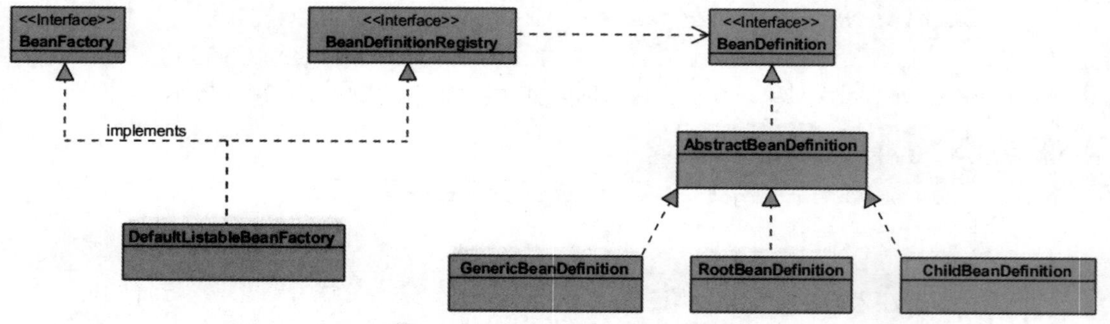

图 14-2　BeanFactory 类关系图

从图 14-2 中可以看到 3 个非常重要的部分：

（1）BeanDefinition 实现 Bean 的定义（即对象的定义），且完成对依赖的定义。

（2）BeanDefinitionRegistry 将定义好的 Bean 注册到容器中（此时会生成一个注册码）。

（3）BeanFactory 是一个 Bean 工厂类，从中可以获取任意定义过的 Bean。

最重要的部分就是 BeanDefinition, 它完成了 Bean 的生成过程。一般情况下通过配置文件（XML、Properties）的方式对 Bean 进行配置，每种文件都需要实现 BeanDefinitionReader 接口，因此是 BeanDefinitionReader 接口的实现类实现了配置文件到 Bean 对象的转换过程。当然，开发者自己也可以实现任意格式的配置文件，只需要实现 BeanDefinitionReader 接口即可。

Bean 的生成大致可以分为两个阶段：容器启动阶段和 Bean 实例化阶段，如图 14-3 所示。

容器启动阶段说明如下：

（1）加载配置文件（一般来说是 XML 文件）。

（2）通过 Reader 生成 BeanDefinition。

（3）将 BeanDefinition 注册到 BeanDefinitionRegistry。

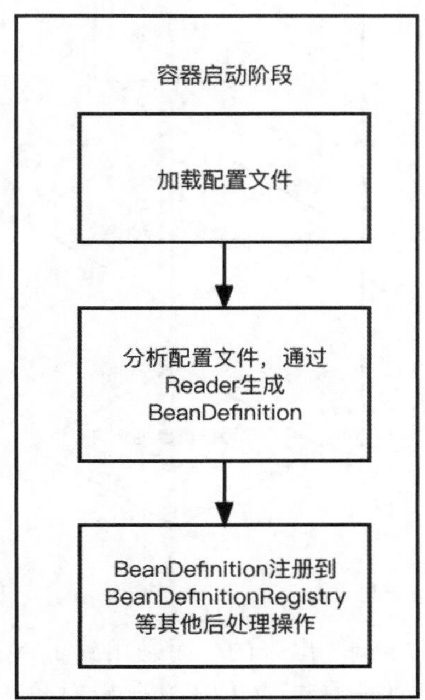

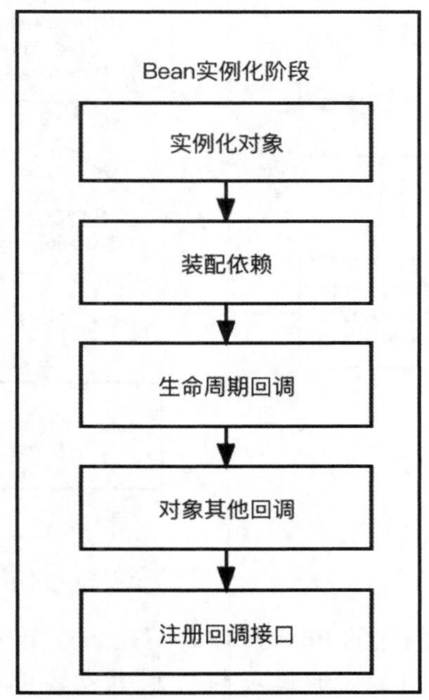

图 14-3 Bean 的生成阶段

Bean 实例化阶段说明如下：

在 Bean 被 getBean 方法调用前，Bean 需要完成初始化，以及其依赖对象的初始化。如果 Bean 本身有回调，则需要调用其相应的回调函数。从上面分析可知，BeanDefinition（容器启动阶段）只完成 Bean 的定义，并未完成 Bean 的初始化。Spring IoC 在初始化完成之后，给开发者提供了一些改变 Bean 定义的方法：

（1）org.springframework.beans.factory.config.PropertyPlaceholderConfigurer：通过配置文件的形式，配置一些参数。

（2）PropertyOverrideConfigurer：通过该对象，可以覆盖原本的 Bean 参数。

（3）CustomEditorConfigurer：负责提供类型转换（配置文件都是 String 型，它需要知道转换成何种类型）。

Bean 的初始化过程如图 14-4 所示。

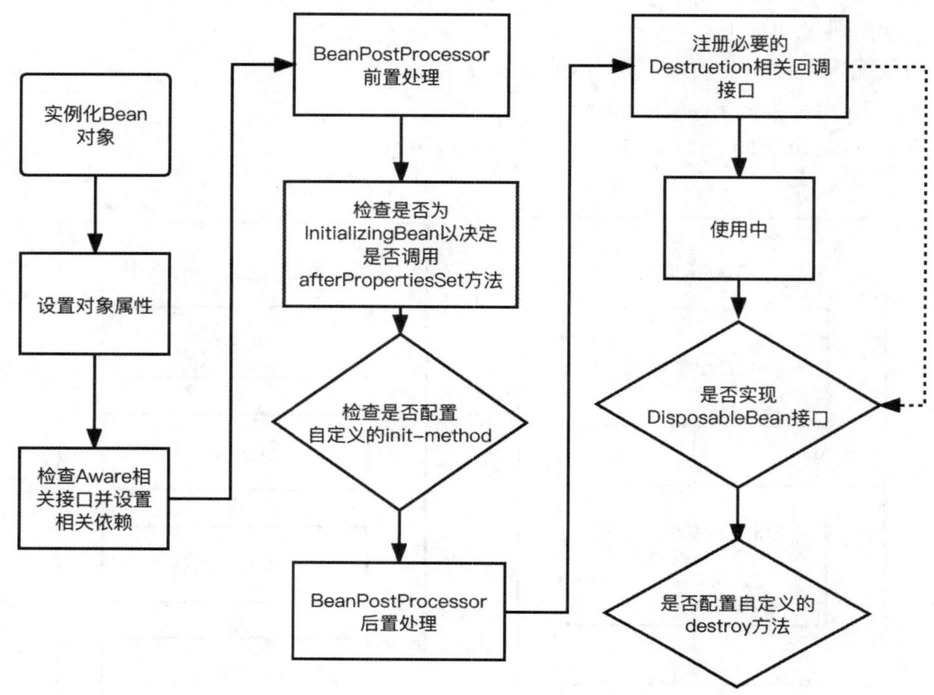

图 14-4　Bean 的初始化过程

这里实例化的 Bean 对象并不是通过定义的类“new”出来的，而是用到了 AOP 机制，生成了其代理对象（通过反射机制生成接口对象，或是通过 CGLIB 生成子对象）。Bean 的具体装载过程是由 beanWrapper 实现的，它继承了 PropertyAccessor（可以对属性进行访问）、PropertyEditorRegistry 和 TypeConverter 接口（实现类型转换）。完成对象属性的设置之后，则会检查是否实现了 Aware 类型的接口，如果实现了，则主动加载。

BeanPostprocessor 是 Spring 的前置后置处理器，可以在初始化 Bean 之前或之后，帮开发者完成一些必要工作。例如，在连接数据库之前将密码存放在一个加密文件，当连接数据库时需要将密码进行解密，则只需要实现相应的接口。

BeanPostProcessor 接口的源码如代码清单 14-5 所示。

代码清单 14-5　BeanPostProcessor 接口源码

```
public interface BeanPostProcessor {
    @Nullable
     default Object postProcessBeforeInitialization(Object bean, String beanName)
throws BeansException {
        return bean;
    }
    @Nullable
```

```
    default Object postProcessAfterInitialization(Object bean, String beanName)
throws BeansException {
        return bean;
    }
}
```

BeanPostProcessor 只有两个方法：postProcessBeforeInitialization 和 postProcessAfterInitialization。postProcessBeforeInitialization 方法会在实例化之前执行处理，postProcessAfterInitialization 方法会在实例化之后进行处理。如果需要在 Spring 容器完成 Bean 的实例化、配置以及在其他初始化方法的前后进行一些业务的逻辑处理，就可以定义一个或者多个 BeanPostProcessor 接口的实现，然后注册到容器中去。

在完成 BeanPostProcessor 之后，就会看到对象是否定义了 InitializingBean 接口，如果是，则会调用其 afterPropertiesSet 方法进一步调整对象实例的状态。

Spring 还提供了另外一种指定初始化的方式，即在 Bean 定义中指定 init-method 。

在这一切完成后，还可以指定对象销毁的一些回调，如数据库的连接池的配置，销毁前需要关闭连接等。相应的方式可以实现 DisposableBean 接口或指定 destroy-method。

14.4.2 ApplicationContext 接口

ApplicationContext 接口由 BeanFactory 接口派生而来，建立在 BeanFactory 上，ApplicationContext 包含 BeanFactory 的所有功能，通常情况下建议比 BeanFactory 优先使用。ApplicationContext 面向的是框架开发者，几乎所有的应用都可以使用 ApplicationContext 而不是 BeanFactory。

ApplicationContext 加载 Bean 的方式如下：

BeanFactory 提供 BeanReader 来从配置文件中读取 Bean 配置，相应的 ApplicationContext 也提供了几个读取配置文件的方式。ApplicationContext 主要的实现容器是 FileSystemXml-Application Context、ClassPathXmlApplicationContext、XmlWebApplicationContext：

（1）FileSystemXmlApplicationContext：该容器从 XML 文件中加载已被定义的 Bean。在这里，你需要提供 XML 文件的完整路径，默认从系统文件中加载配置文件。

（2）ClassPathXmlApplicationContext：该容器从 XML 文件中加载已被定义的 Bean。在这里，你不需要提供 XML 文件的完整路径，只需正确配置 CLASSPATH 环境变量即可。因为，容器会从 CLASSPATH 中搜索 Bean 配置文件，默认从类路径下加载配置文件。

（3）XmlWebApplicationContext：该容器会在一个 Web 应用程序的范围内加载在 XML 文件中已被定义的 Bean，默认从 Web 根目录下加载配置文件。

ApplicationContext 还额外增加了 ApplicationEventPublisher、ResourceLoader、MessageResource 3 个功能：

（1）ApplicationEventPublisher：Spring 的事件发布者接口，定义了发布事件的接口方法 publishEvent。

（2）ResourceLoader：Spring Ioc 借助 ResourceLoader 来实现资源加载，也提供了各种各样的资源加载方式。

（3）MessageResource：提供国际化支持。

14.5 传统 OOP 和 IoC 的对比

IoC 与传统面向对象编程（OOP）是不同的。通常开发者在编程实现某种功能时都需要几个对象相互作用才能完成。从编程的角度来说，就是一个主对象要保存其他类型对象的引用，并通过调用这些引用的方法来实现功能。获得其他类型的对象引用有两种方法。

一种方法是主对象内部主动获得所需引用（也就是 new 一个对象）；另一种方法是在主对象中设置 setter 方法，通过调用 setter 方法或构造方法传入所需引用。后一种方法就叫 IoC，也是常说的 DI。

下面用一个简单的例子来说明传统 OOP 与 IoC 编程的差别。在 Person 类中构造方法，根据时间的不同来输出不同的问候语，如“上午好”“下午好”。接口方法如代码清单 14-6 所示。

代码清单 14-6　服务接口

```
package com.uifuture.ssm.spring.ioc.demo14_5;
public interface Say {
    String sayHello();
}
```

传统 OOP 实现方式如代码清单 14-7 所示。

代码清单 14-7　传统 OOP

```
package com.uifuture.ssm.spring.ioc.demo14_5.oop;
import com.uifuture.ssm.spring.ioc.demo14_5.Say;
import java.util.Calendar;
/** 传统实现(非IoC方式) */
public class Person implements Say {
    /** 需要的引用*/
    private Calendar cal;
    public Person() {
// 主动获取
        cal = Calendar.getInstance();
    }
    @Override
    public String sayHello(){
        if(cal.get(Calendar.AM_PM) == Calendar.AM){
            return "上午好";
        }else{
            return "下午好";
        }
    }
    public static void main(String args[]){
        Person person = new Person();
        System.out.println(person.sayHello());
    }
}
```

使用 IoC 方式来实现，如代码清单 14-8 所示。

代码清单 14-8　IoC 方式

```
package com.uifuture.ssm.spring.ioc.demo14_5.ioc;
import com.uifuture.ssm.spring.ioc.demo14_5.Say;
import java.util.Calendar;
/**IOC方式实现*/
public class Person implements Say {
    /**需要的引用*/
    private Calendar cal;
    /**依赖注入*/
    public void setCal(Calendar cal) {
        this.cal = cal;
    }
    @Override
    public String sayHello(){
        if(cal.get(Calendar.AM_PM) == Calendar.AM){
            return "上午好";
        }else{
            return "下午好";
        }
    }
    public static void main(String args[]){
        Person person = new Person();
        person.setCal(Calendar.getInstance());
        System.out.println(person.sayHello());
    }
}
```

传统 OOP 和 IoC 的实现在一些简单业务的应用上看不出太大差别，并且 IoC 还需要先通过 set 方法创建外部的 Calendar 对象，然后再传到 Person 对象中。但是，假如事先已经在类中用 new Calendar()，但是后来由于需求变更，不再根据该对象来判断，而是使用新的接口来进行，如使用 new NowCalendar()，这样就需要修改所有使用 Calendar 的类，非常麻烦。如果使用 IoC 方法编程并且使用了 Spring 框架，则只需要修改 XML 中的配置文件，不需要修改源代码。

IoC 并不是一种框架，而是一种软件设计模式，它告诉你如何来解除相互依赖模块的耦合。IoC 为相互依赖的组件提供抽象能力，将组件依赖的对象获得交给第三方来控制，即被依赖对象无须在依赖对象模块的类中通过 new 来获取。

14.6　DI 与 IoC 的关系

大家从前文总是看到 DI 与 IoC，那么这两者之间到底是什么关系？

IoC 是面向对象编程中的一种设计原则，可以用来降低代码之间的耦合度。其中最常见的方式为 DI，还有一种方式叫"依赖查找"（Dependency Lookup，DL）。通过控制反转，当创建对象时，由一个调控系统内所有对象的第三方实体，将其所依赖的对象的引用传递给它。

在软件工程中，DI 是一种实现控制反转从而解决依赖性的设计模式。DI 是指在程序运行过程中，当需要调用另一个对象协助时，无须在代码中创建被调用者，而是依赖于第三方的注入。

实现 IoC 主要有两种方式：DI 和 DL。两者的区别在于，前者是被动接收对象，在类 A 的实例创建过程中即创建了依赖的 B 对象，通过类型或名称来判断，将不同的对象注入到不同的属性中，后者是主动索取相应类型的对象。

DI 有以下实现方式：

（1）基于接口。实现特定接口以供外部容器注入所依赖类型的对象。

（2）基于 set 方法。实现特定属性的 public set 方法，来让外部容器调用传入所依赖类型的对象。

（3）基于构造函数。实现特定参数的构造函数，在新建对象时传入所依赖类型的对象。

（4）基于注解。基于 Java 的注解功能，在变量前加 @Autowired 等，不需要显式地定义以上 3 种代码，便可以让外部容器传入对应的对象。该方案相当于定义了 public 的 set 方法，但是因为没有真正的 set 方法，从而不会为了实现 DI 导致暴露了不该暴露的接口（因为 set 方法只会让容器访问来注入，并不会让其他依赖此类的对象访问）。

相比 DI，DL 更加主动，通过调用框架提供的方法来获取对象，获取前需要提供相关的配置文件路径、key 等信息来确定获取对象的状态。

从上面分析可知，DI 是 IoC 的一种实现方式，下面就围绕 DI 基于接口来讲解一个简单的实例。

如果需要实现聚合多种支付平台的支付方式，如通过支付基类实现支付宝和微信的支付，那么，不理想的写法如代码清单 14-9 所示。

代码清单 14-9　不理想的聚合支付实现

```
package com.uifuture.ssm.spring.ioc.demo14_6.terrible;
/**不理想的写法*/
public class PayImpl {
    public void pay(Integer type){
        if(type.equals(1)){
            new WxPay().pay();
        }else if(type.equals(2)){
            new AliPay().pay();
        }else {
            System.out.println("非常抱歉，暂不支持该种支付方式。");
        }
    }
    class WxPay{
        public void pay(){
            System.out.println("微信支付");
        }
    }
    class AliPay{
        public void pay(){
            System.out.println("支付宝支付");
        }
    }
}
```

上面的 PayImpl 是直接依赖于 WxPay 和 AliPay 的，已经高度耦合，对于以后的扩展和维护非常不利，代码清单 14-10 所示是使用了 IoC 思想的写法。

代码清单 14-10　使用接口的形式实现聚合支付

```
package com.uifuture.ssm.spring.ioc.demo14_6.ioc;
/** 接口*/
public interface Pay {
    void pay();
}
package com.uifuture.ssm.spring.ioc.demo14_6.ioc.impl;
import com.uifuture.ssm.spring.ioc.demo14_6.ioc.Pay;
/**支付基类*/
public class PayImpl implements Pay{
    private Pay pay;
    public void setPay(Pay pay) {
        this.pay = pay;
    }
    @Override
    public void pay() {
        pay.pay();
    }
}
package com.uifuture.ssm.spring.ioc.demo14_6.ioc.impl;
import com.uifuture.ssm.spring.ioc.demo14_6.ioc.Pay;
/** 微信支付*/
public class WxPayImpl implements Pay {
    @Override
    public void pay() {
        System.out.println("微信支付");
    }
}
package com.uifuture.ssm.spring.ioc.demo14_6.ioc.impl;
import com.uifuture.ssm.spring.ioc.demo14_6.ioc.Pay;
/** 支付宝支付 */
public class AliPayImpl implements Pay {
    @Override
    public void pay() {
        System.out.println("支付宝支付");
    }
}
```

按照这种方式构建代码，可将依赖项 WxPayImpl 和 AliPayImpl 通过函数 setPay 注入 PayImpl 类中。调用微信支付非常简单，如代码清单 14-11 所示。

代码清单 14-11　调用微信支付

```
public static void main(String[] args) {
    PayImpl pay = new PayImpl();
    pay.setPay(new WxPayImpl());
    pay.pay();
}
```

调用方可以选择使用哪种支付方式，这就完成了对不同支付系统的 DI。如果再引入一个新的支付方式，如京东支付，只需编写京东的支付类实现 Pay 接口。

14.7 Spring 中的 DI 方式

Spring 的核心思想就是 IoC 和 AOP，AOP 在后面再介绍，先说 Spring 中 IoC 的实现方式。将组件的依赖关系用容器实现，那么容器是如何得知一个组件依赖哪些其组件呢？简单地说，容器是如何知道 B 需要依赖 A 的？这就需要组件来提供一系列的回调方法，来告知容器所依赖的对象。根据回调方法的不同，可以将 Spring IoC 分成以下 3 种 DI 方式：

（1）构造方法注入（Constructor Injection）。

（2）setter 方法注入（Setter Injection）。

（3）接口注入（Interface Injection）。

14.7.1 Spring IoC 快速入门案例

接下来看一个简单的入门实例，要使用 Spring 的 IoC 功能必须先引入 Spring 的核心依赖（项目使用 Maven 作为构建工具），如代码清单 14-12 所示。

代码清单 14-12　Spring 核心依赖

```
<dependency>
  <groupId>org.springframework</groupId>
  <artifactId>spring-beans</artifactId>
  <version>5.0.8.RELEASE</version>
  <scope>compile</scope>
</dependency>
<dependency>
  <groupId>org.springframework</groupId>
  <artifactId>spring-context</artifactId>
  <version>5.0.8.RELEASE</version>
  <scope>compile</scope>
</dependency>
<dependency>
  <groupId>org.springframework</groupId>
  <artifactId>spring-core</artifactId>
  <version>5.0.8.RELEASE</version>
  <scope>compile</scope>
</dependency>
```

先创建 Dao 层，如代码清单 14-13 所示。

代码清单 14-13　Dao 层接口

```
package com.uifuture.ssm.spring.ioc.demo14_7.dao;
```

```
/** 演示SpringIoC */
public interface SpringIoCDao {
    void saySpringIoC();
}
```

Dao 层接口实现类的代码如代码清单 14-14 所示。

代码清单 14-14　Dao 层接口实现类

```
package com.uifuture.ssm.spring.ioc.demo14_7.dao.impl;
import com.uifuture.ssm.spring.ioc.demo14_7.dao.SpringIoCDao;
/**实现类*/
public class SpringIoCDaoImpl implements SpringIoCDao {
    @Override
    public void saySpringIoC() {
        System.out.println("SpringIoCDaoImpl->saySpringIoC()");
    }
}
```

下面再看 Service 层的代码，接口和实现类如代码清单 14-15 所示。

代码清单 14-15　Service 层代码

```
package com.uifuture.ssm.spring.ioc.demo14_7.service;
/**service接口*/
public interface SpringIoCService {
    void saySpringIoC();
}
package com.uifuture.ssm.spring.ioc.demo14_7.service.impl;
import com.uifuture.ssm.spring.ioc.demo14_7.dao.SpringIoCDao;
import com.uifuture.ssm.spring.ioc.demo14_7.service.SpringIoCService;
public class SpringIoCServiceImpl implements SpringIoCService {
    /**待注入对象*/
    private SpringIoCDao springIoCDao;
    public void setSpringIoCDao(SpringIoCDao springIoCDao) {
        this.springIoCDao = springIoCDao;
    }
    @Override
    public void saySpringIoC() {
        System.out.println("SpringIoCServiceImpl->saySpringIoC()");
        springIoCDao.saySpringIoC();
    }
}
```

上面代码创建了 Dao 层和 Service 层的接口类及其实现类，其中 Service 层的操作需要依赖 Dao 层的类，接下来就通过 Spring 的 IoC 容器创建并注入这些类。使用的是 XML 配置文件，在 resource 目录下创建 XML 文件，如代码清单 14-16 所示。

代码清单 14-16　Spring IoC 注入配置文件

```
<beans xmlns="http://www.springframework.org/schema/beans"
```

```
        xmlns:xsi="http://www.w3.org/2001/XMLSchema-instance"
        xsi:schemaLocation="
         http://www.springframework.org/schema/beans
         http://www.springframework.org/schema/beans/spring-beans.xsd
       ">
    <!-- 声明springIoCDaoImpl对象，交给Spring创建-->
<bean name="springIoCDaoImpl" class="com.uifuture.ssm.spring.ioc.demo14_7.dao.
impl.
SpringIoCDaoImpl"/>
    <!-- 声明springIoCServiceImpl对象，交给Spring创建-->
<bean name="springIoCServiceImpl" class="com.uifuture.ssm.spring.ioc.demo14_7.
service.
impl.SpringIoCServiceImpl">
        <!-- 注入springIoCDao对象，需要set方法-->
        <property name="springIoCDao" ref="springIoCDaoImpl"/>
    </bean>
</beans>
```

在配置文件中，需要声明一个 <beans> 的顶级标记，同时需要引入核心命名空间，Spring 的功能在使用时都需要声明相对应的命名空间，上述的命名空间是最基本的。

接下来通过 <bean> 子标记声明这些需要 IoC 容器创建的类，其中 name 是指明 IoC 创建后该对象的名称（也可以使用 id 替换 name，后续会讲到），class 则是告知 IoC 这个类的完全限定名称，IoC 则是通过这组信息利用反射技术创建对应的类对象，如代码清单 14-17 所示。

代码清单 14-17　IoC 创建类对象

```
<!-- 声明springIoCDaoImpl对象，交给Spring创建-->
    <bean name="springIoCDaoImpl"
class="com.uifuture.ssm.spring.ioc.demo14_7.dao.impl.SpringIoCDaoImpl"/>
```

在后面的 springIoCServiceImpl 声明中，多出了 <property> 标记，这个标记指向代码清单 14-17 声明的 Dao 对象，它的作用是把 springIoCDaoImpl 对象传递给 springIoCServiceImpl 从而实现类中的 springIoCDao 属性，该属性必须拥有 set 方法才能注入成功，且 property 中的 name 属性的值必须与 SpringIoCServiceImpl 类中 springIoCDao 对象的名称相同。把这种往类 springIoCServiceImpl 对象中注入其他对象（如 springIoCDao）的操作称为 DI。至此，完成了 Spring 中对需要创建对象的声明。

使用这些类需要利用 Spring 提供的核心类 ApplicationContext，通过该类去加载已声明好的配置文件，便可以获取需要的类了，如代码清单 14-18 所示。

代码清单 14-18　演示 IoC

```
package com.uifuture.ssm.spring.ioc.demo14_7.service;
import org.junit.Test;
import org.springframework.context.ApplicationContext;
import org.springframework.context.support.ClassPathXmlApplicationContext;
public class SpringIoCServiceTest {
    @Test
    public void testByXml() throws Exception {
```

```
        //加载配置文件
        ApplicationContext applicationContext=new ClassPathXmlApplicationContext("sp
ring-ioc.xml");
        SpringIoCService springIoCService= applicationContext.getBean("springIoCSer
viceImpl",SpringIoCService.class);
        /*多次获取并不会创建多个springIoCService对象，因为Spring默认创建是单实例的作用域
         SpringIoCService springIoCService= (SpringIoCService) applicationContext.
getBean("springIoCServiceImpl");*/
        springIoCService.saySpringIoC();
    }
}
```

通过这个简单的案例相信大家已能理解 Spring IoC 主要是帮助开发者做什么了，上述的运行结果如图 14-5 所示。

```
» ✔ Tests passed: 1 of 1 test – 517 ms
/Library/Java/JavaVirtualMachines/jdk1.8.0_201.jdk/Contents/Home/bin/java ...
八月 20, 2022 3:02:02 上午 org.springframework.context.support.AbstractApplicationContext prepareRefresh
信息: Refreshing org.springframework.context.support.ClassPathXmlApplicationContext@28864e92: startup date
 hierarchy
八月 20, 2022 3:02:02 上午 org.springframework.beans.factory.xml.XmlBeanDefinitionReader loadBeanDefinitions
信息: Loading XML bean definitions from URL [file:/Users/chenhx/Desktop/github/uifuture-ssm/ssm-spring-ioc-
SpringIoCServiceImpl->saySpringIoC()
SpringIoCDaoImpl->saySpringIoC()
```

图 14-5　Spring IoC 演示结果

14.7.2　Spring 容器通过 XML 和注解方式装配 Bean

继续分析前面的实例，实例采用的是通过 XML 配置文件进行声明和管理 Bean。XML 中的每一个 <bean> 标记都代表着需要被创建的对象，通过 <property> 标记可以为该类注入其他的依赖对象。通过这种方式 Spring 可以知道对象之间的相互依赖，以及需要创建哪些 Bean。

接下来可以通过 ClassPathXmlApplicationContext 加载 Spring 的 XML 配置文件，后面指定 Bean 对象调用其方法即可。

ClassPathXmlApplicationContext 如何加载 classpath 路径下的文件？指明其路径即可。如果存在多个配置文件，则分别传递，ClassPathXmlApplicationContext 是一个可以接收可变参数的构造函数。实际上这里还可以使用 FileSystemXmlApplicationContext，它默认加载的是项目工作路径，即项目名下面的根目录，至于具体使用哪个，区别并不大。下面演示这两个类如何加载文件。

使用 ClassPathXmlApplicationContext 加载文件，如代码清单 14-19 所示。

代码清单 14-19　使用 ClassPathXmlApplicationContext 加载文件

```
//默认查找classpath路径下的文件
ApplicationContext applicationContext=new ClassPathXmlApplicationContext("spring-
ioc.xml");
//多文件，可传递数组
ApplicationContext applicationContext=new ClassPathXmlApplicationContext("spring-
ioc.xml","spring-ioc2.xml",.....);
//classpath
```

```
ApplicationContext applicationContext=new ClassPathXmlApplicationContext("classpath:
spring-ioc.xml");
//多文件，使用classpath*可匹配多个文件
ApplicationContext applicationContext=new ClassPathXmlApplicationContext("classpath*
:spring-ioc*.xml");
```

上面匹配的都是在 classpath 下的路径，也就是编译后的 classes 路径。classpath 与 classpath* 的区别如下：

（1）classpath：只能加载一个配置文件，如果配置了多个，则只加载第一个。

（2）classpath*：可以加载多个配置文件，如果有多个配置文件，则可以使用该字符串。使用 FileSystemXmlApplicationContext 加载文件，如代码清单 14-20 所示。

代码清单 14-20　使用 FileSystemXmlApplicationContext 加载文件

```
//默认为项目工作路径，即项目的根目录
ApplicationContext applicationContext = new FileSystemXmlApplicationContext("/src/
main/resources/spring-ioc.xml");

//也可以读取classpath下的文件
ApplicationContext applicationContext=new FileSystemXmlApplicationContext("classpath
:spring-ioc.xml");

//前缀file 表示是文件的绝对路径
ApplicationContext applicationContext = new FileSystemXmlApplicationContext("
file:/Users/chenhx/Desktop/github/uifuture-ssm/ssm-spring-ioc-demo/src/main/
resources/spring-ioc.xml");
//多文件与ClassPathXmlApplicationContext相同
```

两个类的其他用法大致相同，主要区别是默认的加载路径不同。

接下来将使用注解的方式实现前面 XML 配置的效果，如代码清单 14-21 所示。

代码清单 14-21　使用注解方式注入 Bean

```
package com.uifuture.ssm.spring.ioc.demo14_7.config;
import com.uifuture.ssm.spring.ioc.demo14_7.dao.SpringIoCDao;
import com.uifuture.ssm.spring.ioc.demo14_7.dao.impl.SpringIoCDaoImpl;
import com.uifuture.ssm.spring.ioc.demo14_7.service.SpringIoCService;
import com.uifuture.ssm.spring.ioc.demo14_7.service.impl.SpringIoCServiceImpl;
import org.springframework.context.annotation.Bean;
import org.springframework.context.annotation.Configuration;
/** 使用注解注入Bean*/
@Configuration
public class BeanConfiguration {
    @Bean
    public SpringIoCDao springIoCDao(){
        return new SpringIoCDaoImpl();
    }
    @Bean
    public SpringIoCService springIoCService(){
        SpringIoCServiceImpl bean=new SpringIoCServiceImpl();
```

```
        //注入Dao
        bean.setSpringIoCDao(springIoCDao());
        return bean;
    }
}
```

上述代码中使用了 @Configuration 标明 BeanConfiguration 类，使得 BeanConfiguration 类代替了 XML 文件，也就是说 @Configuration 等价于 <beans> 标记。在该类中，每个使用 @Bean 的公共方法对应着一个 <bean> 标记的定义，即 @Bean 等价于 <bean> 标记。这种基于 Java 的注解配置方式是在 Spring 3.0 中引入的，使用注解的前提是必须确保 spring-context 包已经引入，引入方式如代码清单 14-22 所示。

代码清单 14-22　引入 spring-context 包

```
<!--添加spring-context包-->
<dependency>
    <groupId>org.springframework</groupId>
    <artifactId>spring-context</artifactId>
    <version>5.0.8.RELEASE</version>
</dependency>
```

接下来使用注解方式演示如何加载文件。使用注解方式与使用 XML 配置文件很相似，只不过是使用 AnnotationConfigApplicationContext 来加载 Java 的配置文件，如代码清单 14-23 所示。

代码清单 14-23　使用注解形式注入 Bean

```
/** 通过注解方式演示IoC注入Bean和获取Bean*/
@Test
public void testByAnnotation() {
    AnnotationConfigApplicationContext applicationContext=new AnnotationConfigAppli
cationContext(BeanConfiguration.class);
    //名称必须和BeanConfiguration中工程方法名称一致
       SpringIoCService springIoCService = applicationContext.
getBean("springIoCService", SpringIoCService.class);
    springIoCService.saySpringIoC();
}
```

其实在大部分情况下，开发者更应该倾向于使用 XML 配置文件来配置 Bean 的相关信息，这是因为使用配置文件会更方便管理代码。随着 SpringBoot 的兴起，相信更多的开发者会使用注解方式进行配置，所以后面的例子均会涉及 XML 和部分注解的配置。

14.7.3　构造方法注入

构造方法注入（Constructor Injection），就是被注入对象可以通过在其构造方法中声明依赖对象的参数列表，让外部（通常是 IoC 容器）知道它需要哪些依赖对象。

IoC 服务提供者会检查被加载对象的构造方法，取得它所需要的依赖对象列表，进而为其注入相应的对象，同一个对象是不可能被构造两次的。因此，被注入对象的构造以及整个生命周期都是由 IoC 服务提供者来管理的。

构造方法注入方式比较直观，在对象被构造完成后，即进入构造状态，可以马上使用，如代码清单 14-24 所示。

代码清单 14-24　Service 层构造方法注入

```
package com.uifuture.ssm.spring.ioc.demo14_7.constructor.service.impl;
import com.uifuture.ssm.spring.ioc.demo14_7.dao.SpringIoCDao;
import com.uifuture.ssm.spring.ioc.demo14_7.service.SpringIoCService;
public class SpringIoCServiceImpl implements SpringIoCService {
    /**待注入对象*/
    private SpringIoCDao springIoCDao;
    /**构造函数注入对象*/
    public SpringIoCServiceImpl(SpringIoCDao springIoCDao) {
        this.springIoCDao = springIoCDao;
    }
    @Override
    public void saySpringIoC() {
        System.out.println("SpringIoCServiceImpl->saySpringIoC()");
        springIoCDao.saySpringIoC();
    }
}
```

XML 配置构造函数注入如代码清单 14-25 所示。

代码清单 14-25　XML 配置构造函数注入

```
<beans xmlns:xsi="http://www.w3.org/2001/XMLSchema-instance"
       xmlns="http://www.springframework.org/schema/beans"
       xsi:schemaLocation="
        http://www.springframework.org/schema/beans
        http://www.springframework.org/schema/beans/spring-beans.xsd
        ">
    <!-- 声明springIoCDaoImpl对象，交给Spring创建-->
    <bean name="springIoCDaoImpl" class="com.uifuture.ssm.spring.ioc.demo14_7.dao.
impl.SpringIoCDaoImpl"/>
    <!-- 声明springIoCServiceImpl对象，通过构方法造注-->
    <bean name="springIoCServiceImpl" class="com.uifuture.ssm.spring.ioc.demo14_7.
constructor.service.impl.SpringIoCServiceImpl">
        <!-- 构造方法方式注入accountDao对象-->
        <constructor-arg  ref="springIoCDaoImpl"/>
    </bean>

</beans>
```

构造方法注入可传入简单值类型和集合类型，由于比较简单而且在项目中不常用，此处不再赘述。

需要注意的是，当一个 Bean 定义中有多个 <constructor-arg> 标记时，它们的放置顺序并不重要，因为 Spring 容器会通过传入的依赖参数与类中的构造函数的参数进行比较，尝试找到合适的构造方法。

但在某些情况下可能会出现问题，如代码清单 14-26 所示的 Person 类，带有两个相同类型的参数，在这种情况下，<constructor-arg> 标记的顺序就有影响了，如果顺序不同，那么对于注入的值来说就会引发严重的错误。

代码清单 14-26　Person 类

```
package com.uifuture.ssm.spring.ioc.demo14_7.constructor.entity;
/**演示构造函数注入*/
public class Person {
    private String address;
    private String name;
    /**构造函数*/
    public Person(String name, String address) {
        this.address = address;
        this.name = name;
    }
    @Override
    public String toString() {
        return "Person{" +
                "address='" + address + '\'' +
                ", name='" + name + '\'' +
                '}';
    }
}
```

增加的 Person 注入配置如代码清单 14-27 所示。

代码清单 14-27　增加的 Person 注入配置

```
<!--增加的Person配置-->
<bean id="person"  class="com.uifuture.ssm.spring.ioc.demo14_7.constructor.entity.
Person" >
    <constructor-arg type="java.lang.String" value="中国杭州"/>
    <constructor-arg type="java.lang.String" value="Tom"/>
</bean>
```

增加测试类，如代码清单 14-28 所示。

代码清单 14-28　测试类

```
/** 演示Person注入配置*/
@Test
public void person() {
    ApplicationContext applicationContext = new ClassPathXmlApplicationContext("spri
ng-ioc-constructor.xml");
    Person person = applicationContext.getBean("person", Person.class);
    System.out.println(person);
}
```

当程序运行时，Spring 容器会尝试查找适合的 Person 构造函数创建 Person 对象，由于 <constructor-arg> 标记的属性类型相同，会按照顺序注入，从而导致注入的对象与预期不同。

因此需要给 Spring 容器一些提示，以便它能成功按照顺序注入属性从而创建 person 实例。在 <constructor-arg> 标记中存在一个 index 属性，通过 index 属性告知 Spring 容器传递的依赖参数的顺序，代码清单 14-29 所示的配置将会使 Spring 容器成功找到参数对应的顺序并创建 Person 实例。

代码清单 14-29　修正后的 Person 注入配置

```
<!--增加的Person配置-->
<bean id="person" class="com.uifuture.ssm.spring.ioc.demo14_7.constructor.entity.
Person" >
    <constructor-arg index="1" type="java.lang.String" value="中国杭州"/>
    <constructor-arg index="0" type="java.lang.String" value="Tom"/>
</bean>
```

另外，在构造函数注入中还有一个无法解决的循环依赖的问题。例如，有两个 Bean：A 和 B，它们通过构造函数互为依赖，这种情况下 Spring 容器将无法进行实例化，如代码清单 14-30 所示。

代码清单 14-30　循环依赖

```
public class A{
    private B b;
    public A(B b){
        this.b=b;
    }
}
public class B{
    private A a;
    public B(A a){
        this.a=a;
    }
}
<bean id="a" class="com.uifuture.ssm.spring.ioc.demo14_7.pojo.A">
    <constructor-arg ref="b" />
</bean>
<bean id="b" class="com.uifuture.ssm.spring.ioc.demo14_7.pojo.B">
    <constructor-arg ref="a" />
</bean>
```

这是由于 A 被创建时，希望 B 被注入到自身，然而此时 B 还有没有被创建，而且 B 也依赖于 A，这样将导致 Spring 容器无法执行，最后抛出异常。解决这种困境的方式是使用 setter 依赖，但还是会造成一些不必要的困扰。因此，强烈建议不要在配置文件中使用循环依赖。

14.7.4　setter 方法注入

对于 JavaBean 对象来说，通常会通过 setXxx 方法和 getXxx 方法来访问对应属性。这些 setXxx 方法统称为 settter 方法，getXxx 方法统称为 getter 方法。通过 setter 方法，可以更改相应的对象属性；通过 getter 方法，可以获得相应的对象属性。因此，只要为其依赖对象所对应的属性添加 setter 方法，就可以通过 setter 方法将相应的依赖对象注入到被注入对象中。

setter 方法注入（Setter Injection）不像构造方法注入那样，在对象构造完成后即可使用。相对来说该方式更宽松一些，可以在对象构造完成后再注入。

setter 方法注入顾名思义，被注入的属性需要有 set 方法，setter 注入支持简单类型和引用类型。如代码清单 14-31 所示的案例，对象注入使用了 <property> 标记的 ref 属性。其实该例子跟代码清单 14-16 基本相同。

代码清单 14-31　使用 setter 方法注入 Bean

```
<!-- 声明springIoCDaoImpl对象，交给Spring创建 -->
<bean name="springIoCDaoImpl" class="com.uifuture.ssm.spring.ioc.demo14_7.dao.
impl.SpringIoCDaoImpl"/>
<!-- 声明springIoCServiceImpl对象，交给Spring创建-->
<bean name="springIoCServiceImpl" class="com.uifuture.ssm.spring.ioc.demo14_7.
service.impl.SpringIoCServiceImpl">
    <!-- 注入springIoCDao对象，需要set方法-->
    <property name="springIoCDao" ref="springIoCDaoImpl"/>
</bean>
```

setter 注入方法除了能够注入上述对象之外，还可以注入简单值、Map、集合、数组等对象。代码清单 14-32 简单地演示了简单值和集合的 setter 注入。

代码清单 14-32　简单值和集合的 setter 注入

```
package com.uifuture.ssm.spring.ioc.demo14_7.gather;
import java.util.List;
import java.util.Map;
import java.util.Set;
/** 演示注入简单值和集合*/
public class GatherEntity {
    private String name;
    private List<String> citys;
    private Set<String> friends;
    private Map<Integer,String> books;
    public void setName(String name) {
        this.name = name;
    }
    public void setCitys(List<String> citys) {
        this.citys = citys;
    }
    public void setFriends(Set<String> friends) {
        this.friends = friends;
    }
    public void setBooks(Map<Integer, String> books) {
        this.books = books;
    }
//省略toString方法
}
```

Spring 的 XML 注入配置如代码清单 14-33 所示。

代码清单 14-33　Spring 的 XML 注入配置

```
<beans xmlns:xsi="http://www.w3.org/2001/XMLSchema-instance"
       xmlns="http://www.springframework.org/schema/beans"
       xsi:schemaLocation="
        http://www.springframework.org/schema/beans
        http://www.springframework.org/schema/beans/spring-beans.xsd
        ">
```

```
    <!-- setter方式注入简单值和集合-->
     <bean name="gatherEntity" class="com.uifuture.ssm.spring.ioc.demo14_7.gather.
GatherEntity">
        <property name="name" value="SpringIOC演示" />
        <!-- 注入map -->
        <property name="books">
            <map>
                <entry key="1" value="Java">
                </entry>
                <entry key="2" value="Spring">
                </entry>
                <entry key="3" value="SSM">
                </entry>
            </map>
        </property>
        <!-- 注入set -->
        <property name="friends">
            <set>
                <value>张三</value>
                <value>李四</value>
                <value>王五</value>
            </set>
        </property>
        <!-- 注入list -->
        <property name="citys">
            <list>
                <value>杭州</value>
                <value>上海</value>
                <value>深圳</value>
                <value>北京</value>
            </list>
        </property>
    </bean>
</beans>
```

使用 setter 方法注入简单值如代码清单 14-34 所示。

代码清单 14-34　使用 setter 方法注入简单值

```
/** 演示setter注入简单值和集合*/
@Test
public void testGetEntity() throws Exception {
    //加载配置文件
    ApplicationContext applicationContext=new ClassPathXmlApplicationContext("spri
ng-ioc-gather.xml");
    //从Spring容器中获取需要的Bean
      GatherEntity gatherEntity = applicationContext.getBean("gatherEntity",
GatherEntity.class);
    //输入Bean，需要添加toString方法
    System.out.println(gatherEntity);
}
```

输出结果如图 14-6 所示。

```
» ✓ Tests passed: 1 of 1 test – 538 ms
/Library/Java/JavaVirtualMachines/jdk1.8.0_201.jdk/Contents/Home/bin/java ...
八月 20, 2022 3:04:33 上午 org.springframework.context.support.AbstractApplicationContext prepareRefresh
信息: Refreshing org.springframework.context.support.ClassPathXmlApplicationContext@28864e92: startup date [Sat Aug 20 03:04:33 CS
 hierarchy
八月 20, 2022 3:04:33 上午 org.springframework.beans.factory.xml.XmlBeanDefinitionReader loadBeanDefinitions
信息: Loading XML bean definitions from class path resource [spring-ioc-gather.xml]
GatherEntity[name='SpringIOC演示', citys=[杭州, 上海, 深圳, 北京], friends=[张三, 李四, 王五], books={1=Java, 2=Spring, 3=SSM}]
```

图 14-6　输出注入的值

14.7.5　接口注入

相对于前两种注入方式，接口注入（Interface Injection）没有那么简单明了。被注入对象如果需要 IoC 服务提供者为其注入依赖对象，就必须实现某个接口。这个接口提供一个方法，用来为其注入依赖对象，IoC 服务提供者最终通过这些接口得知应该为消费者注入什么依赖对象。

相较于前两种注入方式，接口注入比较死板和烦琐，如果需要注入依赖对象，被注入对象就必须声明和实现另外的接口。

从注入方式的使用上来说，接口注入因为具有侵入性，它要求组件必须与特定的接口相关联，因此并不被看好，实际使用有限，而构造方法注入和 setter 方法注入则无须如此。

这里，不得不讲接口注入的原始雏形，如代码清单 14-35 所示。

代码清单 14-35　接口注入的原始雏形

```
package com.uifuture.ssm.spring.ioc.demo14_7.interface_injection.code_14_35;
interface InterfaceB {
    void doSomething();
}
/** 接口注入的原始雏形*/
public class ClassA {
    private InterfaceB clzB;
       public void doSomething() throws ClassNotFoundException,
IllegalAccessException, InstantiationException {
        Object obj = Class.forName("从配置文件中获取类名").newInstance();
        clzB = (InterfaceB) obj;
        clzB.doSomething();
    }

}
```

ClassA 依赖于 InterfaceB 的实现，传统的方法是在代码中创建 InterfaceB 实现类的实例，并将其赋予 clzB，来获取 InterfaceB 实现类的实例。但是这样一来，ClassA 在编译期即依赖于 InterfaceB 的实现。

为了将调用者与实现者在编译期分离，便有了代码清单 14-35 所示代码。

根据预先在配置文件中设定的实现类的类名 (Config.BImplementation) 动态加载实现类，并通过 InterfaceB 强制转型后被 ClassA 所用，这就是接口注入的一个最原始的雏形。

对于一个接口注入型的容器来说，加载接口实现并创建其实例的工作由容器完成，如代码清单 14-36 所示。

代码清单 14-36　接口注入加载创建实例

```
package com.uifuture.ssm.spring.ioc.demo14_7.interface_injection.code_14_36;
interface InterfaceB {
    void doSomething();
}
/** 容器型实现接口注入 */
public class ClassA {
    private InterfaceB clzB;
    public void doSomething(InterfaceB b) {
        clzB = b;
        clzB.doSomething();
    }
}
```

在运行期间，InterfaceB 实例将由容器提供。

虽然不推荐使用接口注入方法，但接口注入方法发展较早，在实际中得到了普遍应用，即使在 IoC 的概念尚未确立时，这样的方法也已经频繁出现在的代码中。

代码清单 14-37 所示的代码大家应该都非常熟悉，初学 Java Web 时相信大家都会用到。

代码清单 14-37　HttpServlet 实现类

```
package com.uifuture.ssm.spring.ioc.demo14_7.interface_injection.code_14_37;
import javax.servlet.ServletException;
import javax.servlet.http.HttpServlet;
import javax.servlet.http.HttpServletRequest;
import javax.servlet.http.HttpServletResponse;
import java.io.IOException;
/** HttpServletRequest，HttpServletResponse是应用比较早期和广泛的接口型注入 */
public class MyServlet extends HttpServlet {
    @Override
    protected void doGet(HttpServletRequest req, HttpServletResponse resp) throws
ServletException, IOException {
        //做一些事情
    }
    @Override
    protected void doPost(HttpServletRequest req, HttpServletResponse resp) throws
ServletException, IOException {
        //做一些事情
    }
}
```

注意：此处并不是 HttpServlet 被应用到了接口注入，而是 HttpServletRequest 和 HttpServlet Response 实例由 Servlet 容器在运行期动态注入。

Spring 的核心机制

本章要点

1. Spring 容器中 Bean 的作用域
2. Spring 容器中 Bean 的生命周期
3. ApplicationContext 初始化过程
4. Spring 的 Bean 和 JavaBean 比较
5. Spring 中 Bean 的三种装配方式
6 . 创建 Bean 的三种方式
7. 加载属性文件的四种方式
8. Spring 条件化装配 Bean-Conditional 注解源码解析与实例演示
9. Spring 中的事件机制
10. Spring 中的定时器
11. Spring 表达式语言
12. context:annotation-config 配置项

15.1 Spring 容器中的 Bean

在 Spring 中，组成应用程序的主体及由 Spring IoC 容器所管理的对象，统称为 Bean。简单地讲，Bean 是一个被实例化和组装，并通过 Spring IoC 容器所管理的对象。除此之外，Bean 与应用程序中的其他对象没有太大区别。Bean 的定义以及 Bean 相互间的依赖关系将通过配置元数据来描述。

15.1.1 Bean 的作用域

当在 Spring 中定义 Bean 时，必须声明该 Bean 作用域的选项 (默认作用域是 singleton)。如果需要 Spring 在每次获取 Bean 时都产生一个新的 Bean 实例，要声明 Bean 的作用域属性为 prototype。

Spring 框架支持以下 5 个作用域：

（1）singleton：singleton 作用域的 Bean 在整个 IoC 容器中只生成一个实例（默认）。

（2）prototype：每次通过容器的 getBean 方法获取 prototype 作用域的 Bean 时，都将产生一个新的 Bean 实例。

（3）request：在同一次 HTTP 请求中 request 作用域只产生一个 Bean 实例，作用域为同个 request 请求，只在基于 Web 的 Spring ApplicationContext 中可用。

（4）session：在同一次 HTTP 会话中 session 作用域只产生一个 Bean 实例，作用域为整个 Session 会话，只在基于 Web 的 Spring ApplicationContext 中可用。

（5）global-session：每个全局的 HTTP Session 对应一个 Bean 实例，只在基于 Web 的 Spring ApplicationContext 中可用。

Spring 5 已经将 global-session 去掉了，并将它保留给 Portlet 应用程序。Portlet 是能够生成语义代码（如 HTML）片段的小型 Java Web 插件，关于 Portlet 不做过多介绍。

1. singleton

当一个 Bean 的作用域为 singleton 时，Spring IoC 容器中只会存在一个共享的 Bean 实例，并且所有对 Bean 的请求，只要 id 与该 Bean 定义相匹配，则只会返回 Bean 的同一实例。

singleton 是单例类型（对应于单例模式），在创建容器的同时自动创建一个 Bean 的对象，并且每次获取到的对象都是同一个对象。

注意：singleton 作用域是 Spring 中的默认作用域。要在 XML 文件中将 Bean 定义成 singleton，如代码清单 15-1 配置。

代码清单 15-1　配置作用域为 singleton 的实例

```
<beans xmlns:xsi="http://www.w3.org/2001/XMLSchema-instance"
       xmlns="http://www.springframework.org/schema/beans"
       xsi:schemaLocation="
        http://www.springframework.org/schema/beans
        http://www.springframework.org/schema/beans/spring-beans.xsd
        ">
```

```
    <!--创建作用域为singleton的bean-->
     <bean id="chinese" class="com.uifuture.spring.core.bean.entity.Chinese" scope
="singleton"></bean>
</beans>
package com.uifuture.spring.core.bean.entity;
public class Chinese {
    private String message;
    public void setMessage(String message){

        this.message  = message;
    }
    public void getMessage(){
        System.out.println("Message : " + message);
    }
}
```

代码清单 15-2 求证了两次从 Spring 容器中获取的实例是否为同一个对象。

代码清单 15-2　获取实例

```
package com.uifuture.spring.core.bean.entity;
import org.junit.Test;
import org.springframework.context.ApplicationContext;
import org.springframework.context.support.ClassPathXmlApplicationContext;
public class ChineseTest {
    /*演示获取单例*/
    @Test
    public void singleton() {
        ApplicationContext context = new ClassPathXmlApplicationContext("spring-content.xml");
        Chinese objA = (Chinese) context.getBean("chinese");
        objA.setMessage("I'm object A");
        objA.getMessage();
        Chinese objB = (Chinese) context.getBean("chinese");
        objB.getMessage();
   System.out.print(objA == objB);
       }
    }
```

输出结果如下：

```
Message : I'm object A
Message : I'm object A
true
```

当然，也可以以注解的形式来指定作用域。通过 @Scope（可以显式指定 Bean 的作用范围）指定作用域，如代码清单 15-3 所示。

代码清单 15-3　通过 @Scope 指定作用域

```
package com.uifuture.spring.core.bean.entity;
import org.springframework.context.annotation.Configuration;
import org.springframework.context.annotation.Scope;
```

```
@Configuration
@Scope(scopeName = "singleton")
public class Chinese { //...
}
```

2. prototype

当 Bean 的作用域为 prototype 时，表示 Bean 定义对应多个对象实例。每次对 prototype 作用域的 Bean 请求（将其注入到另一个 Bean 中，或者以程序的方式调用容器的 getBean 方法）时都会创建一个新的 Bean 实例。

prototype 是原型类型，它在创建容器时并没有实例化，而是当获取 Bean 时才会创建一个对象，而且每次获取到的对象都不是同一个对象。基于线程安全考虑，建议对有状态的 Bean 使用 prototype 作用域，而对无状态的 Bean 使用 singleton 作用域（典型的 Dao 层是不会有任何的状态的）。

与其他的作用域相比，Spring 不会完全管理原型 Bean 的生命周期。Spring 容器只会初始化配置以及装载这些 Bean，并传递给客户端。但是之后就不再管理原型 Bean 之后的动作了。也就是说，初始化生命周期回调方法在所有作用域的 Bean 都会调用，但是销毁生命周期回调方法在原型 Bean 不会调用。因此，客户端代码必须注意清理原型 Bean 以及释放原型 Bean 所持有的一些资源。可以通过使用自定义的 Bean post-processor 来让 Spring 释放原型 Bean 所持有的资源。

从某些方面来说，Spring 容器的角色就是取代了 Java 的 new 操作符，所有的生命周期的控制需要由客户端处理。

在 XML 文件中将 Bean 定义成 prototype，配置如代码清单 15-4 所示。

代码清单 15-4　在 XML 中的 prototype 配置

```
<!-- 创建作用域为prototype的Bean-->
<bean id="chinesePrototype" class="com.uifuture.spring.core.bean.entity.Chinese"
scope="prototype"></bean>
```

接下来通过方法进行演示，如代码清单 15-5 所示。

代码清单 15-5　演示 prototype

```
@Test
public void prototype() {
     ApplicationContext context = new ClassPathXmlApplicationContext("spring-
content.xml");
    Chinese objA = (Chinese) context.getBean("chinesePrototype");
    objA.setMessage("I'm object A");
    objA.getMessage();
    Chinese objB = (Chinese) context.getBean("chinesePrototype");
    objB.getMessage();
System.out.print(objA == objB);
}
```

输出结果如下：

```
Message : I'm object A
```

```
Message : null
false
```

从这里的输出可以看出，objA 和 objB 并不是同一个对象。通过 @Scope 的方式实现就不做演示了，在类上添加 @Scope(scopeName = "prototype") 即可。

3. request

request 只适用于 Web 程序，每一次 HTTP 请求都会产生一个新的 Bean，同时该 Bean 仅在当前 HTTP request 内有效。当请求结束时，该对象的生命周期也宣告结束。在 XML 文件中将 Bean 定义成 request 即可。

- 原型作用域在 Spring 的上下文中可用，而请求作用域仅适用于 Web 应用程序。
- 原型 Bean 根据需求进行初始化，而请求 Bean 是在每个请求下构建的。

request 作用域 Bean 在其作用域内有且仅有一个实例。而开发者可以拥有一个或多个原型作用域 Bean 实例。

4. session

session 同样只适用于 Web 程序，session 作用域表示针对每一次 HTTP 请求都会产生一个新的 Bean，同时该 Bean 仅在当前 HTTP session 内有效。与 request 作用域一样，可以根据需要，放心地更改所创建实例的内部状态，而其他 HTTP session 中创建的实例将不会看到这些特定于某个 HTTP session 的状态变化。当 HTTP session 最终被废弃时，在该 HTTP session 作用域内的 Bean 也会被废弃。

session 作用域的 Bean 与 request 作用域的 Bean 没有太大区别，它们也与纯 Web 应用程序上下文相关联。注解为 session 作用域的 Bean 对于每个用户的会话仅创建一次，并在会话结束时被销毁。

由 session 作用域限制的 Bean 可以被认为是面向 Web 的单例，因为给定环境（用户会话）仅存在一个实例。但要注意，无法在 Web 应用程序上下文中使用它们（简单地说，就是一个函数内部自定义变量所在的作用域，函数执行完变量就销毁了）。

在 XML 文件中将 Bean 定义成 session 即可。

15.1.2 ApplicationContext 初始化过程

Spring Bean 是 Spring 应用中最重要的部分。理解 Spring Bean 的生命周期很容易。当一个 Bean 被实例化时，可能需要执行一些初始化使它转换成可用状态。同样，当 Bean 不再被需要，并且从容器中移除时，可能需要做一些清除工作。

要讲解 Bean 生命周期，首先要从 ApplicationContext 的初始化过程讲起，对 ApplicationContext 接口，无论是哪个实现类，都会用到加载配置的方法，也就是 ConfigurableApplicationContext 接口的 refresh 方法，该方法的实现类为 AbstractApplicationContext，Spring 的 AbstractApplicationContext 是 ApplicationContext 抽象实现类，该抽象类的 refresh 方法定义了 Spring 容器在加载配置文件后的各项处理过程，这些处理过程清晰刻画了 Spring 容器启动时所执行的各项操作（创建 Spring 容器如 ClassPathXmlApplicationContext），refresh 方法的实现源码如代码清单 15-6 所示。

代码清单 15-6　AbstractApplicationContext 类中的 refresh 方法

```
@Override
public void refresh() throws BeansException, IllegalStateException {
   synchronized (this.startupShutdownMonitor) {
      // 设置启动日期，活动标志以及对属性源执行初始化
      prepareRefresh();
      // 通知子类刷新内部beanFactory初始化beanFactory工厂                        ①
      ConfigurableListableBeanFactory beanFactory = obtainFreshBeanFactory();
      //准备bean工厂以便在此上下文中使用.
      prepareBeanFactory(beanFactory);
      try {
         //允许在上下文子类中对beanFactory进行后处理
         postProcessBeanFactory(beanFactory);
         // 调用上下文中注册为beanFactory处理器                                  ②
         invokeBeanFactoryPostProcessors(beanFactory);
         // 注册拦截bean创建的bean处理器                                         ③
         registerBeanPostProcessors(beanFactory);
         // 初始化此上下文的消息源                                               ④
         initMessageSource();
         // 初始化上下文事件广播器                                               ⑤
         initApplicationEventMulticaster();
         // 初始化特定上下文子类中的其他特殊bean                                 ⑥
         onRefresh();
         // 注册事件监听器bean                                                   ⑦
         registerListeners();
         // 实例化所有剩余(非lazy init)单例                                      ⑧
         finishBeanFactoryInitialization(beanFactory);
         //发布上下文刷新事件                                                    ⑨
         finishRefresh();
      }
      catch (BeansException ex) {
         if (logger.isWarnEnabled()) {
               logger.warn("Exception encountered during context initialization - "
+"cancelling refresh attempt: " + ex);
         }
         destroyBeans();
         cancelRefresh(ex);
         throw ex;
      }
      finally {
         resetCommonCaches();
      }
   }
}
```

下面对代码清单 15-6 中的①～⑨条注释分别进行讲解。

1. 初始化 beanFactory

根据配置文件实例化 beanFactory，getBeanFactory 方法由具体子类实现。在这里，Spring 将配置文件的信息解析成一个个 BeanDefinition 对象并装入容器的 Bean 定义的注册表（BeanDefinitionRegistry）中，但此时 Bean 还未初始化；obtainFreshBeanFactory 会调用自身的 refreshBeanFactory 方法，而 refreshBeanFactory 方法由子类 AbstractRefreshableApplicationContext 实现，该方法返回了一个创建的 DefaultListableBeanFactory 对象，这个对象就是由 ApplicationContext 管理的 BeanFactory 容器对象。

这步操作相当于在开发者自己编写的应用代码中不用 ApplicationContext 而直接用 beanFactory 创建的 BeanFactory 对象的操作。

AbstractRefreshableApplicationContext 核心代码如代码清单 15-7 所示。

代码清单 15-7　AbstractRefreshableApplicationContext 核心代码

```
public abstract class AbstractRefreshableApplicationContext extends
AbstractApplicationContext {

    /** 该ApplicationContext管理的BeanFactory容器对象*/
    @Nullable
    private DefaultListableBeanFactory beanFactory;
    /** 实现实际刷新上下文的底层BeanFactory，关闭前一个BeanFactory（如果有的话），并为上下文生
命周期的下一个阶段初始化一个新的BeanFactory*/
    @Override
    protected final void refreshBeanFactory() throws BeansException {
        if (hasBeanFactory()) {
            destroyBeans();
            closeBeanFactory();
        }
        try {
            //创建容器对象
            DefaultListableBeanFactory beanFactory = createBeanFactory();
            beanFactory.setSerializationId(getId());
            customizeBeanFactory(beanFactory);
            //装载配置文件，并传入相关联的BeanFactory对象作为BeanDefinition的容器
            loadBeanDefinitions(beanFactory);
            synchronized (this.beanFactoryMonitor) {
                this.beanFactory = beanFactory;
            }
        }
        catch (IOException ex) {
            throw new ApplicationContextException("I/O error parsing bean definition
source for " + getDisplayName(), ex);
        }
    }
    /** 为该上下文创建一个内部BeanFactory，调用每一个refresh方法进行尝试*/
```

```
    protected DefaultListableBeanFactory createBeanFactory() {
        return new DefaultListableBeanFactory(getInternalParentBeanFactory());
    }
    /** 该方法为一个钩子方法，子类可以覆盖它对当前上下文管理的BeanFactory提供客户化操作，也可以
忽略 */
    protected void customizeBeanFactory(DefaultListableBeanFactory beanFactory) {
        if (this.allowBeanDefinitionOverriding != null) {
            beanFactory.setAllowBeanDefinitionOverriding(this.allowBeanDefinitionOverriding);
        }
        if (this.allowCircularReferences != null) {
            beanFactory.setAllowCircularReferences(this.allowCircularReferences);
        }
    }
    /** 装载配置文件的方法，需要子类实现*/
     protected abstract void loadBeanDefinitions(DefaultListableBeanFactory
beanFactory)
            throws BeansException, IOException;

}
```

代码清单中 15-7 中装载配置文件的方法就是 loadBeanDefinitions 方法，由其子类扩展实现，如代码清单 15-8 所示。

代码清单 15-8　loadBeanDefinitions 方法的实现

```
public abstract class AbstractXmlApplicationContext extends AbstractRefreshableConf
igApplicationContext {
    // 通过xmlbeanDefinitionReader加载bean定义/
    @Override
     protected void loadBeanDefinitions(DefaultListableBeanFactory beanFactory)
throws BeansException, IOException {
        // 使用XMLBeanDefinitionReader来载入bean定义信息的XML文件，传入关联的BeanFactory
        XmlBeanDefinitionReader beanDefinitionReader = new XmlBeanDefinitionReader(
beanFactory);
        // 这里配置reader的环境，其中ResourceLoader是我们用来定位bean定义信息资源的
        beanDefinitionReader.setEnvironment(this.getEnvironment());
        // 因为上下文本身实现了ResourceLoader接口，所以可以直接把上下文作为ResourceLoader传入
        beanDefinitionReader.setResourceLoader(this);
        beanDefinitionReader.setEntityResolver(new ResourceEntityResolver(this));
        // 允许子类提供读卡器的自定义初始化，然后继续实际加载bean定义
        initBeanDefinitionReader(beanDefinitionReader);
        // 转到定义好的XmlBeanDefinitionReader中对载入的bean定义进行处理
        loadBeanDefinitions(beanDefinitionReader);
    }

    protected void initBeanDefinitionReader(XmlBeanDefinitionReader reader) {
        reader.setValidating(this.validating);
    }

    /**
```

```
        * 使用给定的xmlbeanDefinitionReader加载bean定义
        * bean工厂的生命周期由refreshBeanFactory方法处理。因此，此方法只加载bean定义
        */
        protected void loadBeanDefinitions(XmlBeanDefinitionReader reader) throws
BeansException, IOException {
            Resource[] configResources = getConfigResources();
            if (configResources != null) {
                reader.loadBeanDefinitions(configResources);
            }
            String[] configLocations = getConfigLocations();
            if (configLocations != null) {
                //这个XmlBeanDefinitionReader用于从XML文件中读取bean定义
                reader.loadBeanDefinitions(configLocations);
            }
        }
}
```


2. 调用 Spring 工厂后置处理器

先讲解 Spring 工厂的后置处理器，简单地说就是有时需要在 Spring 应用程序中实现一些动态行为，就是可以定义一个实现 org.springframework.beans.factory.config.BeanFactory Post Processor 接口的 Bean。实现覆盖的主要方法是 postProcessBeanFactory，在 Spring 初始化 Bean 之前初始化一些 Bean 的动态数值。

Spring 根据反射机制从 BeanDefinitionRegistry 中找出所有 BeanFactoryPostProcessor 类型的 Bean，并调用其 postProcessBeanFactory 接口方法。

经过加载配置文件，已经把配置文件中定义的所有 Bean 装载到 BeanDefinitionRegistry 的 BeanFctory 中，对于 ApplicationContext 应用该 BeanDefinitionRegistry 类型的 BeanFactory 就是 Spring 默认的 DefaultListableBeanFactory，如代码清单 15-9 所示。

代码清单 15-9　DefaultListableBeanFactory

```
public class DefaultListableBeanFactory extends AbstractAutowireCapableBeanFactory
implements ConfigurableListableBeanFactory, BeanDefinitionRegistry, Serializable {
    ...
}
```

在这些被装载的 Bean 中，若有类型为 BeanFactoryPostProcessor 的 Bean（在配置文件中配置的），则将为对应的 BeanDefinition 生成 BeanFactoryPostProcessor 对象。

Spring 容器扫描 BeanDefinitionRegistry 中的 BeanDefinition，使用 Java 反射机制自动识别出 Bean 工厂后处理器（实现 BeanFactoryPostProcessor 接口）的 Bean，然后调用这些 Bean 工厂后处理器对 BeanDefinitionRegistry 中的 BeanDefinition 进行加工处理，可以完成以下两项工作（当然也可以有其他的操作，用户可自定义）：

（1）对使用占位符的 <bean> 标记进行解析，得到最终的配置值，即对一些半成品式的 BeanDefinition 对象进行加工处理可获得成品的 BeanDefinition 对象。

（2）对 BeanDefinitionRegistry 中的 BeanDefinition 进行扫描，通过 Java 反射机制找出所有属

性编辑器的 Bean（实现 java.beans.PropertyEditor 接口的 Bean），并自动将它们注册到 Spring 容器的属性编辑器注册表中（PropertyEditorRegistry），该 Spring 提供了 CustomEditorConfigurer 进行实现，可实现 BeanFactoryPostProcessor，可用该 Spring 来在此注册自定义属性编辑器。

BeanFactoryPostProcessor 是由 Bean 实现的接口，可以修改其他 Bean 的定义。**提示**：BeanFactoryPostProcessor 只能修改定义，即构造函数参数。

实现 BeanFactoryPostProcessor 接口的 Bean 在初始化"正常"Bean 之前被调用，这就是为什么它能修改元数据（meta data）的原因。调用是通过 org.springframework.context.support.AbstractApplicationContext 的 protected void invokeBeanFactoryPostProcessors(ConfigurableListableBeanFactory beanFactory) 来实现的。BeanFactoryPostProcessor 接口实现 Bean 的 postProcessBeanFactory 方法就是在该步骤被执行的。

该部分的实现代码过长，有兴趣的读者可以使用 IDE 进入源码进行查看。

org.springframework.context.support.AbstractApplicationContext 类 的 invokeBeanFactoryPostProcessors 方法实际调用了 org.springframework.context.support.PostProcessorRegistrationDelegate 类的 invokeBeanFactoryPostProcessors 方法。

3. 注册 Bean 后处理器

根据反射机制从 BeanDefinitionRegistry 中找出所有 BeanPostProcessor 类型的 Bean，并将它们注册到容器 Bean 后处理器的注册表中。

对应的实现代码在 PostProcessorRegistrationDelegate 类的 registerBeanPostProcessors 方法中，如代码清单 15-10 所示。

代码清单 15-10　registerBeanPostProcessors 方法的实现

```
public static void registerBeanPostProcessors(
        ConfigurableListableBeanFactory beanFactory, AbstractApplicationContext
applicationContext) {
     String[] postProcessorNames = beanFactory.getBeanNamesForType(BeanPostProcess
or.class, true, false);
    /*当bean在beanPostProcessor实例化期间创建时,注册beanPostProcessorChecker.当bean不符
合所有beanPostProcessor处理的条件时, 记录一条消息 */
     int beanProcessorTargetCount = beanFactory.getBeanPostProcessorCount() + 1 +
postProcessorNames.length;
     beanFactory.addBeanPostProcessor(new PostProcessorRegistrationDelegate.BeanPost
ProcessorChecker(beanFactory, beanProcessorTargetCount));
    // 拆分实现priorityordered、ordered和rest的beanPostProcessor
    List<BeanPostProcessor> priorityOrderedPostProcessors = new ArrayList<>();
    List<BeanPostProcessor> internalPostProcessors = new ArrayList<>();
    List<String> orderedPostProcessorNames = new ArrayList<>();
    List<String> nonOrderedPostProcessorNames = new ArrayList<>();
    for (String ppName : postProcessorNames) {
        if (beanFactory.isTypeMatch(ppName, PriorityOrdered.class)) {
            BeanPostProcessor pp = beanFactory.getBean(ppName, BeanPostProcessor.class);
            priorityOrderedPostProcessors.add(pp);
            if (pp instanceof MergedBeanDefinitionPostProcessor) {
```

```
                internalPostProcessors.add(pp);
            }
        }
        else if (beanFactory.isTypeMatch(ppName, Ordered.class)) {
            orderedPostProcessorNames.add(ppName);
        }
        else {
            nonOrderedPostProcessorNames.add(ppName);
        }
    }
    // 首先，注册实现PriorityOrdered的BeanPostProcessor
    sortPostProcessors(priorityOrderedPostProcessors, beanFactory);
    registerBeanPostProcessors(beanFactory, priorityOrderedPostProcessors);
    // 然后，注册实现Ordered的BeanPostProcessors
    List<BeanPostProcessor> orderedPostProcessors = new ArrayList<>();
    for (String ppName : orderedPostProcessorNames) {
        BeanPostProcessor pp = beanFactory.getBean(ppName, BeanPostProcessor.class);
        orderedPostProcessors.add(pp);
        if (pp instanceof MergedBeanDefinitionPostProcessor) {
            internalPostProcessors.add(pp);
        }
    }
    sortPostProcessors(orderedPostProcessors, beanFactory);
    registerBeanPostProcessors(beanFactory, orderedPostProcessors);
    // 注册所有常规beanPostProcessor
    List<BeanPostProcessor> nonOrderedPostProcessors = new ArrayList<>();
    for (String ppName : nonOrderedPostProcessorNames) {
        BeanPostProcessor pp = beanFactory.getBean(ppName, BeanPostProcessor.class);
        nonOrderedPostProcessors.add(pp);
        if (pp instanceof MergedBeanDefinitionPostProcessor) {
            internalPostProcessors.add(pp);
        }
    }
    registerBeanPostProcessors(beanFactory, nonOrderedPostProcessors);
    // 最后，重新注册所有内部BeanPostProcessor
    sortPostProcessors(internalPostProcessors, beanFactory);
    registerBeanPostProcessors(beanFactory, internalPostProcessors);
    // 将检测内部bean的后处理器重新注册为applicationListener，并将其移动到处理器链的末尾(用于
提取代理等)
    beanFactory.addBeanPostProcessor(new ApplicationListenerDetector(applicationCont
ext));
}
```

整段代码类似于前面的调用 Spring 工厂后置处理器，区别之处在于，Sping 工厂后置处理器在获取后立即调用，而 Bean 后处理器在获取后注册到上下文持有的 BeanFactory 中，供以后操作调用（在用户获取 Bean 的过程中，对已经完成属性设置工作的 Bean 进行后续加工，Bean 后处理器加工的是 Bean，而 Sping 工厂后置处理器加工的是 BeanDefinition）。

4. 初始化消息源

在此步骤容器中进行了国际化信息资源的初始化，该步骤的关键实现代码如代码清单 15-11 所示。

代码清单 15-11 初始化消息源核心代码

```
/**
 * 初始化消息源。如果在此上下文中没有定义，则使用父级
 */
protected void initMessageSource() {
    ConfigurableListableBeanFactory beanFactory = getBeanFactory();
    /**
     * 判断beanFactory中是否有名字为messageSource的bean
     * MESSAGE_SOURCE_BEAN_NAME = "messageSource"
     * 如果没有，新建DelegatingMessageSource类作为messageSource的Bean
     * 如果有，从beanFactory中获取
     */
    if (beanFactory.containsLocalBean(MESSAGE_SOURCE_BEAN_NAME)) {
        this.messageSource = beanFactory.getBean(MESSAGE_SOURCE_BEAN_NAME, MessageSource.class);
        // 使messagesource知道父类messagesource。需要实现MessageSource接口
            if (this.parent != null && this.messageSource instanceof Hierarchical
MessageSource) {
                HierarchicalMessageSource hms = (HierarchicalMessageSource) this.
messageSource;
            if (hms.getParentMessageSource() == null) {
                // 只有在没有注册父messagesource的情况下，才将父上下文设置为父messagesource
                hms.setParentMessageSource(getInternalParentMessageSource());
            }
        }
        if (logger.isDebugEnabled()) {
            logger.debug("Using MessageSource [" + this.messageSource + "]");
        }
    }
    else {
        // 使用空messagesource可以接受getmessage调用
        DelegatingMessageSource dms = new DelegatingMessageSource();
        dms.setParentMessageSource(getInternalParentMessageSource());
        this.messageSource = dms;
        beanFactory.registerSingleton(MESSAGE_SOURCE_BEAN_NAME, this.messageSource);
        if (logger.isDebugEnabled()) {
              logger.debug("Unable to locate MessageSource with name '" + MESSAGE_
SOURCE_BEAN_NAME + "': using default [" + this.messageSource + "]");
        }
    }
}
```

简单地说，该方法的作用就是判断 beanFactory 中是否有名字为 messageSource 的 Bean（需要实现 MessageSource 接口），如果有，则从 beanFactory 中获取并且判断获取的是不是 HierarchicalMessageSource 类型的，如果是，则设置其父级消息源；如果没有，则新建 DelegatingMessageSource

类作为 messageSource 的 Bean。

5. 初始化上下文事件广播器

初始化 ApplicationEventMulticaster 事件，AbstractApplciationContext 拥有一个 application Event Multicaster 的 Bean（需要实现 ApplicationEventMulticaster 接口），applicationEventMulticaster 提供了容器监听器的注册表，称为事件广播器。

广播事件的功能就是做一件事情时，将自动触发广播。该事件对应观察者模式中的具体主题对象，持有观察者对象的集合，即监听器注册表。

核心代码如代码清单 15-12 所示。

代码清单 15-12 初始化上下文事件广播器核心源码

```
/** 初始化事件广播器*/
protected void initApplicationEventMulticaster() {
    ConfigurableListableBeanFactory beanFactory = getBeanFactory();
    // 查找是否存在id为applicationEventMulticaster的bean对象
    if (beanFactory.containsLocalBean(APPLICATION_EVENT_MULTICASTER_BEAN_NAME)) {
        this.applicationEventMulticaster =
                      beanFactory.getBean(APPLICATION_EVENT_MULTICASTER_BEAN_NAME,
ApplicationEventMulticaster.class);
        if (logger.isDebugEnabled()) {
                  logger.debug("Using ApplicationEventMulticaster [" + this.
applicationEventMulticaster + "]");
        }
    }
    else {
        // 默认使用SimpleApplicationEventMulticaster，并注册为单例
        this.applicationEventMulticaster = new SimpleApplicationEventMulticaster(bea
nFactory);
           beanFactory.registerSingleton(APPLICATION_EVENT_MULTICASTER_BEAN_NAME,
this.applicationEventMulticaster);
        if (logger.isDebugEnabled()) {
            logger.debug("Unable to locate ApplicationEventMulticaster with name '" +
                  APPLICATION_EVENT_MULTICASTER_BEAN_NAME +
                  "': using default [" + this.applicationEventMulticaster + "]");
        }
    }
}
```

简单地说，就是 Spring 在此步骤初始化事件广播器，用户可以在配置文件中为容器定义一个自定义的事件广播器，只要实现 ApplicationEventMulticaster 接口就可以，Spring 在此会根据 beanFactory 自动获取。如果没有找到外部配置的事件广播器，Spring 将使用 SimpleApplicationEvent Multicaster 作为事件广播器。

6. 初始化其他特殊的 Bean

onRefresh 方法是一个钩子方法，子类可以借助这个钩子方法执行一些特殊的操作：如 Abstract

RefreshableWebApplicationContext 就使用该钩子方法执行初始化 ThemeSource 的操作。

7. 注册事件监听器

Spring 根据上下文持有的 beanfactory 对象，从它的 BeanDefinitionRegistry 中找出所有实现 org.springfamework.context.ApplicationListener 的 Bean，将 BeanDefinition 对象生成 Bean，注册为容器的事件监听器，实际的操作就是将其添加到初始化事件广播器所提供的监听器注册表中。简单地说就是检查监听器 Bean 并在监听器注册表中进行注册。

registerListeners 方法的源码如代码清单 15-13 所示。

代码清单 15-13　registerListeners 方法的源码

```
/** 添加实现ApplicationListener接口的监听器bean。不会影响其他监听器，这些监听器可以在不使用
bean的情况下添加 */
protected void registerListeners() {
    //获取所有的Listener，把事件的bean放到ApplicationEventMulticaster中
    for (ApplicationListener<?> listener : getApplicationListeners()) {
        getApplicationEventMulticaster().addApplicationListener(listener);
    }
    String[] listenerBeanNames = getBeanNamesForType(ApplicationListener.class, true,
false);
    // 把事件的名称放到ApplicationListenerBean中
    for (String listenerBeanName : listenerBeanNames) {
        getApplicationEventMulticaster().addApplicationListenerBean(listenerBeanName);
    }
    // 按照顺序发布事件
    Set<ApplicationEvent> earlyEventsToProcess = this.earlyApplicationEvents;
    this.earlyApplicationEvents = null;
    if (earlyEventsToProcess != null) {
        for (ApplicationEvent earlyEvent : earlyEventsToProcess) {
            getApplicationEventMulticaster().multicastEvent(earlyEvent);
        }
    }
}
```

ApplicationListener 接口的源码如代码清单 15-14 所示。

代码清单 15-14　ApplicationListener 接口源码

```
@FunctionalInterface
public interface ApplicationListener<E extends ApplicationEvent> extends
EventListener {
   /** 处理应用程序事件*/
   void onApplicationEvent(E event);
}
```

8. 初始化作用域为 singleton 的 Bean

实例化所有 singleton 的 Bean，并将它们放入 Spring 容器的缓存中。这就是和直接在应用中

使用BeanFactory的区别之处，在创建ApplicationContext对象时，不仅创建了一个BeanFactory对象，并且还应用它实例化所有单实例的Bean。

finishBeanFactoryInitialization方法的源码如代码清单15-15所示。

代码清单15-15 finishBeanFactoryInitialization方法的源码

```
/** 完成此上下文的bean工厂的初始化，初始化所有剩余的singleton 的Bean */
protected void finishBeanFactoryInitialization(ConfigurableListableBeanFactory
beanFactory) {
    // 初始化上下文的转换服务
    if (beanFactory.containsBean(CONVERSION_SERVICE_BEAN_NAME) &&
                beanFactory.isTypeMatch(CONVERSION_SERVICE_BEAN_NAME,
ConversionService.class)) {
        beanFactory.setConversionService(
                beanFactory.getBean(CONVERSION_SERVICE_BEAN_NAME, Conversion
Service.class));
    }
    if (!beanFactory.hasEmbeddedValueResolver()) {
        beanFactory.addEmbeddedValueResolver(strVal -> getEnvironment().
resolvePlaceholders(strVal));
    }
    //初始化LoadTimeWeaverAware bean以允许尽早注册其转换器。
    String[] weaverAwareNames = beanFactory.getBeanNamesForType(LoadTimeWeaverAwa
re.class, false, false);
    for (String weaverAwareName : weaverAwareNames) {
        getBean(weaverAwareName);
    }
    // 停止使用临时类加载器进行类型匹配
    beanFactory.setTempClassLoader(null);
    // 允许缓存所有bean定义元数据，不需要进一步更改
    beanFactory.freezeConfiguration();
    // 实例化所有剩余(非lazy init)单例
    beanFactory.preInstantiateSingletons();
}
```

从源码和注释上可以得知，finishBeanFactoryInitialization是实例化beanFactory上的所有残留的beanNames，当然其中并不包括non-lazy-init的beanNames。最主要的是可确保实例化beanName的便是工厂的内部方法preInstantiateSingletons，代码清单15-16是该方法的源码分析。

代码清单15-16 DefaultListableBeanFactory类preInstantiateSingletons方法源码

```
@Override
public void preInstantiateSingletons() throws BeansException {
    if (logger.isDebugEnabled()) {
        logger.debug("Pre-instantiating singletons in " + this);
    }
    // 获取解析过的所有beanNames
    List<String> beanNames = new ArrayList<>(this.beanDefinitionNames);
    // 触发所有非惰性单例bean的初始化
```

```
    for (String beanName : beanNames) {
        //获取对应的RootBeanDefinition，其内部含有BeanDefinition
        RootBeanDefinition bd = getMergedLocalBeanDefinition(beanName);
        //非抽象、单例模式、非lay-init，满足以上条件的进入到实例化
        if (!bd.isAbstract() && bd.isSingleton() && !bd.isLazyInit()) {
            if (isFactoryBean(beanName)) {
                //对FactoryBean的类型实例化
                Object bean = getBean(FACTORY_BEAN_PREFIX + beanName);
                if (bean instanceof FactoryBean) {
                    final FactoryBean<?> factory = (FactoryBean<?>) bean;
                    boolean isEagerInit;
                    if (System.getSecurityManager() != null && factory instanceof SmartFactoryBean) {
                        isEagerInit = AccessController.doPrivileged((PrivilegedAction<Boolean>)
                                        ((SmartFactoryBean<?>) factory)::isEagerInit,
                                getAccessControlContext());
                    }
                    else {
                        isEagerInit = (factory instanceof SmartFactoryBean &&
                                ((SmartFactoryBean<?>) factory).isEagerInit());
                    }
                    if (isEagerInit) {
                        //getBean方法对相应的beanName实例化
                        getBean(beanName);
                    }
                }
            }
            else {
                //getBean方法对相应的beanName实例化
                getBean(beanName);
            }
        }
    }
    //为所有适用的bean触发初始化后回调
    for (String beanName : beanNames) {
/*获取上述的对应的实例化对象，对SmartInitializingSingleton的实现类进行
afterSingletonInstantiated 方法调用*/
        Object singletonInstance = getSingleton(beanName);
        if (singletonInstance instanceof SmartInitializingSingleton) {
            final SmartInitializingSingleton smartSingleton = (SmartInitializingSingleton)
singletonInstance;
            if (System.getSecurityManager() != null) {
                AccessController.doPrivileged((PrivilegedAction<Object>) () -> {
                    smartSingleton.afterSingletonsInstantiated();
                    return null;
                }, getAccessControlContext());
            }
            else {
                //会调用InitializingBean的afterPropertiesSet方法
                smartSingleton.afterSingletonsInstantiated();
            }
```

```
            }
        }
    }
```

此过程主要是实例化 RootDefinition 中的 beanClass 类，并完成相应的属性设置和方法调用等。即使当中的属性没有提前被初始化，也会被调用 getBean 方法（在该方法中对 beanFactory 中 beanName 对应的 RootBeanDefinition 中的 beanClass 进行准确的实例化，包括属性以及方法复用等）来进行初始化后再被设置。然后是调用所有实现 InitializingBean 接口的 afterPropertiesSet 方法。

接下来调用 BeanPostProcessors 接口，如 postProcessBeforeInitialization 和 postProcess After Initialization 实例化前后方法。其中涉及 @Resource、@Autowired 等注解的解析并进行相应的实例注入。

有一个需要注意的点，就是 BeanPostProcessors 接口的处理，要优先处理 InstantiationAware BeanPostProcessor.class 接口实现类，一旦其内部方法 postProcessBeforeInstantiation(Class,String) 执行后返回 null，则不执行 postProcessAfterInitialization 方法。随后处理其他 BeanPostProcessor 接口的 postProcessBeforeInitialization 方法和 postProcessAfterInitialization 方法。

9. 发布上下文刷新事件

在此处时容器已经启动完成，发布容器 refresh 事件（ContextRefreshedEvent），也就是创建上下文刷新事件，事件广播器负责将些事件广播到每个注册的事件监听器中。

finishRefresh 方法源码如代码清单 15-17 所示。

代码清单 15-17　finishRefresh 方法源码

```
/** 完成此上下文的刷新，调用LifecycleProcessor的onRefresh 方法并发布*/
protected void finishRefresh() {
    // 清除上下文级资源缓存(如扫描中的asm元数据)
    clearResourceCaches();
    // 初始化处理器，这是上下文的生命周期
    initLifecycleProcessor();
    // 首先将刷新传播到生命周期处理器
    getLifecycleProcessor().onRefresh();
    /*发布最终事件
     注意：监听器在MessageSource之后初始化，以便能够在监听器实现中访问它。因此，MessageSource
 实现不能发布事件*/
    publishEvent(new ContextRefreshedEvent(this));
    // 在LiveBeansView中注册
    LiveBeansView.registerApplicationContext(this);
}
```

至此，ApplicationContext 对象就完成了初始化工作：创建 beanFactory 来装配 BeanDefiniton，调用 Sping 工厂后置处理器，注册 Bean 后置处理器，初始化消息资源，初始化上下文事件广播器，初始化其他特殊的 Bean 注册事件监听器，初始化所有作用域为 singleton 的 Bean，最后发布上下文刷新事件。

在初始化过程中，ApplicationContext 向 BeanFactory 注册了大量组件，如 BeanFactoryPost Processor、BeanPostProcessor 等，增强了 BeanFactory 的功能。这些特殊的组件可通过 Application-

Context 相应的 add 方法或 set 方法提供，也获取了 BeanFactory 中注册的相应类型组件，所以要扩展相应功能，可手工调用 add 方法、set 方法，也可以通过向容器注册 Bean 实例的方式实现。

ApplicationContext 是 Spring 中的高级容器，提供了比 BeanFactory 更强大的功能，更加易于使用，一般在使用 Ioc 容器时都是使用 ApplicationContext 而非 BeanFactory。

15.1.3 Bean 的生命周期

Spring Bean 的生命周期很容易理解。简单地说，就是当一个 Bean 被实例化时，该 Bean 有可能需要执行一些初始化操作使它转换成可用状态。同样，当 Bean 不再被需要，要从容器中移除时，也可能需要执行一些清除工作。从实例化到不被需要这一整个过程便是 Spring 中 Bean 的生命周期。

需要注意的是，Spring 不能对一个 prototype Bean（也就是原型类型）的整个生命周期负责，容器在初始化、配置、装饰或者是装配完一个 prototype 实例后，将它交给客户端，随后就对该 prototype 实例不闻不问了。无论在 Spring 中配置何种作用域，Spring 容器都会调用所有对象初始化生命周期的回调方法，而对 prototype 作用域的 Bean 而言，任何配置的生命周期的回调方法都不会被调用到。

Spring 容器可以管理非原型类型 Bean 的生命周期，在此类型下，Spring 能够精确地得知 Bean 何时被创建，何时初始化完成，以及何时被销毁。而对于 prototype 作用域的 Bean，Spring 只负责创建，在容器创建了 Bean 的实例后，Bean 的实例就交给了客户端的代码管理，Spring 容器将不再跟踪其生命周期，并且不会管理那些被配置成 prototype 作用域的 Bean 的生命周期。

下面根据 ApplicationContext 初始化过程来将 Bean 的生命周期梳理一下。

（1）实例化 BeanFactoryPostProcessor。

（2）执行 BeanDefinitionRefistryPostProcessor 接口的 postProcessBeanDefinitionRegistry 方法进行用户 Bean 对象的注册（涉及 @Configuration、@Component、@MapperScan 等注解的解析）。

（3）执行 BeanFactoryPostProcessor 接口的 postProcessBeanFactory 方法。

（4）处理 BeanPostProcessor 接口集合。

（5）优先执行 InstantiationAwareBeanPostProcess 接口的 postProcessBeforeInstantiation 方法。

（6）执行 InstantiationAwareBeanPostProcessor 类的 PostProcessPropertyValues 方法，如果涉及一些属性值，利用 set 方法和构造函数设置一些属性值，包含 @Resource、@Autowired 等注解的赋值。

（7）设置 beanFactory 及 beanName 属性。

（8）如果 Bean 实现了 BeanClassLoaderAware 接口，则调用 setBeanClassLoader 方法，传入 Class Loader 对象的实例。

（9）如果 Bean 实现了 BeanFactoryAware 接口，则调用 setBeanClassLoader 方法，传入 Class Loader 对象的实例。

（10）与上面的类似，如果实现了其他 Aware 接口，就调用相应的方法。

（11）执行 BeanPostProcessor 接口的 postProcessBeforeInitialization 方法，其中包括 ApplicationContextAware/EnvironmentAware 等环节的赋值。

（12）如果 Bean 实现了 InitializingBean 接口，执行 afterPropertiesSet 方法。

（13）执行 <Bean> 标记，或者 @Bean 配置的 init-method 属性指定的方法。

（14）执行 BeanPostProcessor 接口的 postProcessAfterInitialization 方法。

（15）执行 InstantiationAwareBeanPostProcessor 类的 postProcessAfterInstantiation 方法。

（16）执行 Bean 对象的业务方法。

（17）当要销毁 Bean 时，如果 Bean 实现了 DisposableBean 接口，执行 destroy 方法。

（18）如果 Bean 在配置文件中的 <bean> 标记或者 @Bean 包含 destroy-method 属性，执行指定的方法。

本节只讨论两个重要的生命周期回调方法，它们在 Bean 的初始化和销毁过程中是必需的。有时需要在 Bean 属性值设置好之后和 Bean 销毁之前做一些事情，如检查 Bean 中某个属性值设置要求。Spring 框架提供了多种方法可以在 Spring Bean 的生命周期中执行 initialization 方法和 pre-destroy 方法。

下面将以接口和 XML 配置的形式来演示。

（1）实现 InitializingBean 接口和 DisposableBean 接口。

在 InitializingBean 接口定义一个方法，当实例化 Bean 时立即调用该方法。同样，在 Disposable Bean 接口定义一个方法，只有从容器中移除 Bean 之后，才能调用该方法。

这两个接口都只包含了一个接口，源码如代码清单 15-18 所示。

代码清单 15-18　InitializingBean 接口与 DisposableBean 接口源码

```
package org.springframework.beans.factory;
public interface DisposableBean {
   /** 回调bean销毁时执行的方法*/
   void destroy() throws Exception;
}
package org.springframework.beans.factory;
public interface InitializingBean {
   /** 初始化bean时执行，可以针对某个具体的bean进行配置*/
   void afterPropertiesSet() throws Exception;
}
```

接下来看一个例子，在类中实现这两个接口，如代码清单 15-19 所示。

代码清单 15-19　实现 InitializingBean 接口与 DisposableBean 接口

```
package com.uifuture.spring.core.bean.life;
import org.springframework.beans.factory.DisposableBean;
import org.springframework.beans.factory.InitializingBean;
/** 实现InitializingBean接口与DisposableBean接口*/
public class LifeService implements InitializingBean, DisposableBean {
    @Override
    public void destroy() throws Exception {
        System.out.println("执行LifeService类、DisposableBean接口的destroy方法");
    }
    @Override
    public void afterPropertiesSet() throws Exception {
     System.out.println("执行LifeService类、InitializingBean接口的afterPropertiesSet方法");
    }
}
```

使用接口实现的方式比较简单，但是在实际中并不推荐使用这种方式，因为这种接口实现方式会使该 Bean 同 Spring 框架耦合在一起。

接下来讲解 XML 文件的配置方式。

（2）在 Bean 的配置文件中指定 init-method 方法和 destroy-method 方法。

Spring 也允许开发者创建自己的初始化方法和销毁方法，只需要在 XML 配置文件的 bean 元素上指定 init-method 和 destroy-method 属性的值就可以在 Bean 初始化时和销毁之前执行一些操作，如代码清单 15-20 所示。

代码清单 15-20　自定义初始化和销毁方法

```
package com.uifuture.spring.core.bean.life;
/**XML配置Bean的初始化动作以及销毁动作*/
public class LifeXmlService{
    /**通过<bean>的destroy-method属性指定的销毁方法*/
    public void destroyMethod() throws Exception {
        System.out.println("执行XML配置的destroy-method方法");
    }
    /** 通过<bean>的init-method属性指定的初始化方法 */
    public void initMethod() throws Exception {
        System.out.println("执行XML配置的init-method方法");}}
```

在 XML 文件中配置 Bean 的自定义初始化方法和销毁方法，如代码清单 15-21 所示。

代码清单 15-21　配置 Bean 的自定义初始化方法和销毁方法

```
<!--配置Bean的初始化方法和销毁方法-->
<bean name="lifeXmlService" class="com.uifuture.spring.core.bean.life.
LifeXmlService" init-method="initMethod" destroy-method="destroyMethod"></bean>
```

在这里需要注意的是 init-method 和 destroy-method 的方法不能有参数。在实际应用中，推荐这种书写方式，不需要让 Bean 的实现直接依赖于 Spring 的框架，可以自定义方法。

（3）使用注解的形式来指定初始化方法和销毁方法。

除了前面两种方法外，还有一种使用得比较多的方法，那就是使用 @PostConstruct 和 @PreDestroy 来指定 init 方法和 destroy 方法。

为了使注解生效，需要在配置文件中定义 org.springframework.context.annotation.CommonAnnotationBeanPostProcessor 的 bean 元素标签或使用 <context:annotation-config/> 标记，使用 context 元素需要在 beans 元素中声明 context 命名空间。整个例子如代码清单 15-22 所示。

代码清单 15-22　使用注解来指定初始化方法和销毁方法

```
package com.uifuture.spring.core.bean.life;
import javax.annotation.PostConstruct;
import javax.annotation.PreDestroy;
/** 使用注解来指定初始化方法和销毁方法 */
public class LifeAnnotationService {
    /** 销毁方法 */
    @PreDestroy
```

```
    public void preDestroy() throws Exception {
        System.out.println("执行preDestroy注解标注的方法");
    }
    /** 初始化方法 */
    @PostConstruct
    public void initPostConstruct() throws Exception {
        System.out.println("执行PostConstruct注解标注的方法");
    }
}
```

注解形式的 XML 配置如代码清单 15-23 所示。

代码清单 15-23 注解形式的 XML 配置

```
<beans xmlns:xsi="http://www.w3.org/2001/XMLSchema-instance"
       xmlns="http://www.springframework.org/schema/beans"
  xmlns:context="http://www.springframework.org/schema/context"
          xsi:schemaLocation="http://www.springframework.org/schema/beans
           http://www.springframework.org/schema/beans/spring-beans.xsd
           http://www.springframework.org/schema/context
           http://www.springframework.org/schema/context/spring-context.xsd">
   ... //省略其他无关配置
      <!--第1种方式，使PostConstruct注解和PreDestroy注解生效-->
      <bean class="org.springframework.context.annotation.CommonAnnotationBeanPostP
rocessor"/>

<!--第2种方式，使用context:annotation-config，注意在beans中声明context命名空间-->
    <!--<context:annotation-config></context:annotation-config>-->
</beans>
```

最后，Spring 为了方便使用，提供了一种支持：如果有很多同名初始化方法和销毁方法，可以直接在 beans 元素上定义方法名称，不需要开发者在每一个 bean 上声明初始化方法和销毁方法。在 XML 配置文件的 beans 元素中使用 default-init-method 属性和 default-destroy-method 属性即可灵活地配置，如代码清单 15-24 所示。

代码清单 15-24 定义所有 Bean 默认的初始化方法和销毁方法

```
<beans ...
        default-init-method="initMethod"
        default-destroy-method="destroyMethod" >
    <!-- 省略其他不相关代码 -->
</beans>
```

注意：单个 Bean 上面指定的方法会覆盖默认的全局初始化方法和销毁方法。

15.1.4 Spring 的 Bean 和 JavaBean 比较

简单来说，JavaBean 是遵循了某些特定规范和约定的 Java 类而已。例如，满足以下 4 点的 Java 类：

（1）所有属性为 private。

（2）提供默认（无参构造）的构造函数。

（3）允许通过访问器（getter 方法和 setter 方法）来访问类的成员属性。

（4）实现 java.io.Serializable 的序列化接口。

JavaBean 不是一种技术，而是一种规范，为了方便复用、通用和向后兼容。而 Spring Bean 概括来说就是在 Spring 中组成应用程序的主体以及由 Spring IoC 容器所管理的对象。

这两种 Bean 的不同之处如下：

（1）JavaBean 更多的是作为值传递参数，而 Spring Bean 的用处几乎无处不在，任何组件都可以被称为 Bean（所有被 Spring 容器实例化并管理的 Java 类都可以称为 Spirng Bean）。

（2）JavaBean 作为值对象，要求每个属性必须都提供 getter 方法和 setter 方法，但 Spring 中的 Bean 只需为接受设置值注入的属性提供 setter 方法即可。

（3）JavaBean 作为值对象传递，不接受任何容器管理其生命周期，Spring 中的 Bean 由 Spring 管理其生命周期。

总结一下，Spring 中的 Bean 和 JavaBean 的不同之处共有三点：用处不同、写法规范不同及生命周期不同。

15.2 Spring 中 Bean 的装配

创建应用对象之间协作关系的行为通常称为装配，这也是 DI 的本质。在 Spring 中，如果需要创建一个 Bean，并不需要使用 new 关键字，这是由于 Bean 的创建是由 Spring 容器进行的，在开发者需要时，从 Spring 容器中获取即可。

在 Spring 中装配 Bean 有以下 3 种方法：

（1）使用 XML 配置文件显式装配 Bean。

（2）使用注解装配 Bean。

（3）使用 Java 类装配 Bean。

Bean 的配置信息中定义了 Bean 的实现以及依赖关系，Spring 容器会根据各种形式的 Bean 配置信息在容器中建立起 Bean 定义的注册表。然后根据注册表加载、实例化 Bean，并建立 Bean 与 Bean 之间的依赖关系。最后将这些准备就绪的 Bean 放到 Bean 容器中，方便外层的应用进行调用。

15.2.1 使用 XML 装配 Bean

在 Spring 中，XML 文件的配置使用的是 Schema 格式使得不同类型的配置拥有了自身的命名空间，使配置文件更具扩展性了。

基本的基于 Schema 格式的 XML 文件配置模板如代码清单 15-25 所示。

代码清单 15-25　基于 Schema 格式的 XML 文件配置模板

```
<?xml version="1.0" encoding="UTF-8"?>
<beans ①xmlns:xsi="http://www.w3.org/2001/XMLSchema-instance"
       ②xmlns="http://www.springframework.org/schema/beans"
```

```
      ③xmlns:context="http://www.springframework.org/schema/context"
      ④xmlns:aop="http://www.springframework.org/schema/aop"
      ⑤xsi:schemaLocation="http://www.springframework.org/schema/beans
       http://www.springframework.org/schema/beans/spring-beans.xsd
       http://www.springframework.org/schema/context
       http://www.springframework.org/schema/context/spring-context.xsd
       http://www.springframework.org/schema/aop
       http://www.springframework.org/schema/aop/spring-aop.xsd" >
</beans>
```


- xmlns:xsi="http://www.w3.org/2001/XMLSchema-instance"：自定义命名空间，xsi 是简称，用于为每个文档中命名空间指定相应的 Schema 样式文件，是标准组织定义的标准命名空间。
- xmlns="http://www.springframework.org/schema/beans"：默认命名空间，没有空间名，用于 Spring Bean 的定义。
- xmlns:context="http://www.springframework.org/schema/context"：自定义命名空间，context 为配置 Spring 中的组件标签。Spring 配置文件里面需要使用到 context 的标签，声明前缀为 context 的命名空间，后面的 URL 用于标识命名空间的地址不会被解析器用于查找信息。其唯一的作用是赋予命名空间一个唯一的名称。当命名空间被定义在元素的开始标签中时，所有带有相同前缀的子元素都会与同一个命名空间相关联。
- xmlns:aop="http://www.springframework.org/schema/aop"：自定义命名空间，是 Spring 配置 AOP 的命名空间，context 类似。
- xsi:schemaLocation：命名空间对应的 Schema 文件。

XML 配置文件中，命名空间的定义分为下面这两个步骤：

①先指定命名空间的名称。

②指定命名空间的 Schema 样式文件的位置，用空格或回车键换行进行分隔。

指定命名空间的 Schema 样式文件地址有两个用途：一是 XML 解析器可以获取 Schema 文件并对文档进行格式合法性校验；二是在 IDE 下可以进行代码提示，自动补全。

接下来看看 XML 中 Bean 的基本配置，在 XML 中，一个 Bean 的基本配置如代码清单 15-26 所示。

代码清单 15-26　XML 中 Bean 的基本配置

```
<beans xmlns:xsi="http://www.w3.org/2001/XMLSchema-instance"
       xmlns="http://www.springframework.org/schema/beans"
       xsi:schemaLocation="http://www.springframework.org/schema/beans
        http://www.springframework.org/schema/beans/spring-beans.xsd" >
    <!--XML中Bean 的基本配置-->
      <bean id="chinese" class="com.uifuture.spring.core.bean.entity.Chinese"></
bean>
</beans>
```

一般情况下，Spring IoC 容器中的一个 Bean 即对应配置文件中的一个 <bean> 标记。在 Spring 容器的配置文件中定义一个简单的 Bean 只需指定 id 和 class。而 id 也可以不指定，如果不指定，Spring 将该 Bean 的全限定类名作为 Bean 的唯一标识。这个 id 是唯一的，外部应用可以通过 getBean(id) 获取 Bean 的实例。例如，在本例中通过 getBean("chinese") 即可获取 Chinese 实例。id 在容器中起定位查找作用，class 属性指定了 Bean 对应的实现类。

使用 XML 配置 Bean 涉及了 Spring 中的依赖注入方式，也就是通过构造方法注入、setter 方法注入，以及接口注入。

15.2.2 使用注解装配 Bean

注解能够有效地减少开发配置时间，Spring 提供了丰富的注解功能，如今在项目中使用注解的场景越来越多。

Spring 容器默认是禁用注解配置的。如果需要打开注解扫描，可以使用 <context:component-scan> 组件扫描或者 <context:annotation-config> 注解配置。

建议选择 <context:component-scan>，因为它支持 <context:annotation-config> 标记的全部功能，而且支持指定包下扫描注册 Bean。

如果要成功启动 Spring 容器，必须具备三大要件，分别是 Bean 的定义信息、Bean 的实现类及 Spring 容器本身。如果 Bean 采用了基于 XML 的配置，那么 Bean 的定义信息和 Bean 的实现类本身是分离的，而当采用基于注解的配置方式配置 Bean 时，Bean 的定义信息通过在 Bean 实现类上标注注解进行实现。

下面使用注解定义一个 Service 层的 Bean，如代码清单 15-27 所示。

代码清单 15-27　使用注解定义一个 Bean

```
package com.uifuture.spring.core.bean.service;
import org.springframework.stereotype.Component;
/** 通过注解定义一个Bean */
@Component("userService")
public class UserServiceImpl {
}
```

别忘了在配置文件中声明扫描包下的注解，如代码清单 15-28 所示。

代码清单 15-28　配置文件

```
<beans xmlns:xsi="http://www.w3.org/2001/XMLSchema-instance"
       xmlns="http://www.springframework.org/schema/beans"
       xmlns:context="http://www.springframework.org/schema/context"
       xsi:schemaLocation="http://www.springframework.org/schema/beans
        http://www.springframework.org/schema/beans/spring-beans.xsd
        http://www.springframework.org/schema/context
        http://www.springframework.org/schema/context/spring-context.xsd">
    <!--配置扫描com.uifuture.spring.core.bean.service包下的Bean-->
      <context:component-scan base-package="com.uifuture.spring.core.bean.
service"></context:component-scan>
</beans>
```

在代码清单 15-27 中，使用 @Component 在 UserServiceImpl 类声明处对类进行标识，它可以被 Spring 容器识别，Spring 容器会自动将 POJO 转换为 Spring 容器管理的 Bean。代码清单 15-27 的配置与代码清单 15-29 的 XML 配置是等效的。

代码清单 15-29 使用 XML 配置 UserServiceImpl

```
<bean id="userService" class="com.uifuture.spring.core.bean.service.UserServiceImpl"/>
```

除了 @Component（通用的构造型注解，标识该类为 Spring 组件）以外，为了让注解与类本身的用途清晰化，Spring 还提供了另外 3 个功能基本与该注解等效的注解，分别用于对 Dao、Service 及 Web 层的 Controller 进行标注，所以这些注解也称为 Bean 的衍型注解（类似于 XML 文件中定义 Bean<bean id=" " class=""/>）。3 种注解如下：

（1）@Repository：用于对 Dao 层实现类进行标注（为数据服务）。

（2）@Service：用于对 Service 层实现类进行标注（为业务服务）。

（3）@Controller：用于对 Controller 层实现类进行标注（为展示服务）。

建议注解不要写到接口上，而应该写在具体的实现类上。

在 XML 配置文件中可通过 Context 命名空间的 <component-scan> 标记的 base-package 属性指定一个需要扫描的基类包，Spring 容器将会扫描这个基类包下的所有类，并且从类的注解信息中获取 Bean 的定义信息。

如果仅希望扫描特定的类而非基包下的所有类，可以使用 resource-pattern 属性过滤特定的类，如代码清单 15-30 所示。

代码清单 15-30 用 resource-pattern 属性过滤特定的类

```
<context:component-scan base-package="com.uifuture.spring.core.bean" resource-
pattern="service/*.class">
</context:component-scan>
```

在代码清单 15-30 中，将基类包设置为 com.uifuture.spring.core.bean，默认情况下 resource-pattern 属性的值为“**/*.class”，即基类包里的所有类。这里设置为“service/*.class”，则 Spring 容器仅会扫描基包里 service 子包中的类。

15.2.3 使用 Java 类装配 Bean

在普通的 POJO 类中开发者只需要标注 @Configuration，就可以为 Spring 容器提供 Bean 的定义信息，每个标注了 @Bean 的类方法都相当于提供了一个 Bean 的定义信息，相当于在 Spring 容器中注入了该 Bean。

使用普通的 POJO 类定义为 Bean 的配置类，如代码清单 15-31 所示。

代码清单 15-31 POJO 类定义为 Bean 的配置类

```
package com.uifuture.spring.core.bean.config;
import com.uifuture.spring.core.bean.dao.UserDao;
import com.uifuture.spring.core.bean.service.UserServiceImpl;
import org.springframework.context.annotation.Bean;
import org.springframework.context.annotation.Configuration;
/** Bean的配置类、Configuration注解将一个POJO标注为定义Bean的配置类*/
```

```
@Configuration
public class SpringBeanConfig {
    /** 定义了userDao，注入到Spring容器中*/
    @Bean
    public UserDao userDao(){
        return new UserDao();
    }
    /** UserServiceImpl依赖了UserDao，所以可以在这里直接显示调用包含UserDao的构造方法*/
    @Bean
    public UserServiceImpl userService(UserDao userDao){
        return new UserServiceImpl(userDao);
    }
    /** UserServiceImpl依赖了UserDao，可以通过在方法中调用setter的方式注入userDao*/
    @Bean
    public UserServiceImpl userService(){
        UserServiceImpl userService = new UserServiceImpl();
        userService.setUserDao(userDao());
        return userService;
    }
}
```

在代码清单 15-31 中，演示了构造方法注入和 setter 方法注入的形式。在 SpringBeanConfig 类（提示：该类需要被 Spring 容器扫描进去，如在 <context:component-scan> 标记的扫描范围内）的定义处标注了 @Configuration，说明这个类可用于为 Spring 提供 Bean 的定义信息。类的方法处可以标注 @Bean，Bean 的类型由方法返回值类型决定，名称不进行说明则默认与方法名相同，可以通过入参（Input Parameters）显示指定 Bean 名称，如 @Bean(name="userDao")。开发者可以在 @Bean 所标注的方法中提供 Bean 的实例化逻辑。

为了使前面的配置生效，UserServiceImpl 类需要进行一定的改造，UserDao 类有默认的构造方法即可，如代码清单 15-32 所示。

代码清单 15-32　UserServiceImpl 改造

```
public class UserServiceImpl {
    public UserServiceImpl() {
    }
    public UserServiceImpl(UserDao userDao) {
        this.userDao = userDao;
    }
    private UserDao userDao;
    public void setUserDao(UserDao userDao) {
        this.userDao = userDao;
    }
}
```

UserServiceImpl 类中增加了有参数的构造函数，以及增加了 setter 方法。代码清单 15-31 的代码配置和代码清单 15-33 的 XML 配置是等效的。

代码清单 15-33　等效的 XML 配置

```
<bean id="userDao" class="com.uifuture.spring.core.bean.dao.UserDao"/>
<bean id="userService" class="com.uifuture.spring.core.bean.service.
UserServiceImpl">
    <property name="userDao" ref="userDao"></property>
</bean>
```

基于 Java 类的配置方式与基于 XML、注解的配置方式相比，前者通过代码的方式更加灵活地实现了 Bean 的实例化及 Bean 之间的装配，后面两者使用的都是配置声明的方式，在灵活性上要稍逊一些，但是配置上要更加简单一些。

15.3　创建 Bean 实例的三种方式

大多数情况下，Spring 容器直接通过 new 关键字调用构造器来创建 Bean 实例，而 <bean> 标记中的 class 属性指定了 Bean 实例的实现类。因此，<bean> 标记必须指定 Bean 实例的 class 属性，但这并不是实例化 Bean 的唯一方法。Spring 支持使用以下 3 种方式来创建 Bean。

（1）使用构造器、setter 创建 Bean 实例。

（2）使用静态工厂方法创建 Bean 实例。

（3）调用实例工厂方法创建 Bean 实例。

15.3.1　使用构造器创建 Bean 实例

默认情况下 Spring 使用无参数的构造方法来实例化 Bean，Spring 对 Bean 实例的所有属性执行默认初始化，即所有的基本类型值初始化为 0 或 false，所有的引用类型值初始化为 NULL。

使用默认的构造方法实例化 Bean，要求 Bean 必须有无参数的构造函数，如果自定义了有参数的构造函数而没有定义无参数的构造函数，Spring 就会报错，因为 Spring 底层用到了反射机制，而反射机制用到了 Bean 的无参数的构造函数。

最简单的使用构造器创建 Bean 的方式，如代码清单 15-34 所示。

代码清单 15-34　用构造器创建 Bean 实例

```
<bean id="userDao" class="com.uifuture.spring.core.bean.dao.UserDao"/>
```

如果 Bean 中定义了属性，当需要注入参数值时，则可以提供 setter 方法，Spring 容器通过该方法为属性注入参数。

15.3.2　使用静态工厂方法创建 Bean 实例

如果要使用静态工厂方法创建 Bean 实例，则在 Spring 配置文件中，<bean> 标记要指定 2 个属性。

（1）class：静态工厂类的类名（使用静态工厂方法实例化 Bean 时，class 属性并不是指定 Bean 实例的实现类，而是静态工厂类）。

（2）factory-method：静态工厂类的方法名（该方法必须是静态方法）。

其实就是应用了简单工厂模式，通过静态工厂方法，根据所传入参数的不同可以创建不同的对象。就是通过工厂来产生类的对象，然后调用工厂的方法来获取对象（对象实例需要实现同一个接口）。

下面通过一个例子了解静态工厂方法如何创建Bean实例。接口与对象实例如代码清单15-35所示。

代码清单 15-35　接口与对象实例

```
package com.uifuture.spring.core.bean.createbean;
/**接口*/
public interface Animal {
    void say();
}
public class Cat implements Animal {
    private String age;
    @Override
    public void say() {
        System.out.println("i am cat.age="+age);
    }
    public void setAge(String age) {
        this.age = age;
    }
}
public class Dog implements Animal {
    private String age;
    @Override
    public void say() {
        System.out.println("i am dog.age="+age);
    }
    public void setAge(String age) {
        this.age = age;
    }
}
```

这里的对象都需要实现同一个接口，否则无法在工厂方法中统一返回。静态工厂类代码如代码清单15-36所示。

代码清单 15-36　静态工厂类

```
package com.uifuture.spring.core.bean.createbean;
/** 静态工厂类*/
public class AnimalBeanFactory {
    /** name参数决定哪个是Bean的实例*/
    public static Animal getAnimal(String name){
        if("dog".equalsIgnoreCase(name)){
            return new Dog();
        }else if ("cat".equalsIgnoreCase(name)){
            return new Cat();
        }
```

```
            return null;
        }
    }
```

在静态工厂方法中，根据传递的参数值 name 不同，返回不同的实例对象。最后在 Spring 的配置文件中配置静态工厂方法创建 Bean 实例，如代码清单 15-37 所示。

代码清单 15-37　配置静态工厂方法创建 Bean

```
<beans xmlns:xsi="http://www.w3.org/2001/XMLSchema-instance"
       xmlns="http://www.springframework.org/schema/beans"
       xmlns:context="http://www.springframework.org/schema/context"
       xsi:schemaLocation="http://www.springframework.org/schema/beans
        http://www.springframework.org/schema/beans/spring-beans.xsd
        http://www.springframework.org/schema/context
        http://www.springframework.org/schema/context/spring-context.xsd">
    <bean id="cat"
          class="com.uifuture.spring.core.bean.createbean.AnimalBeanFactory"
          factory-method="getAnimal">
        <!--给静态工厂方法的参数传值-->
        <constructor-arg value="cat" />
        <!--注入属性值，调用setAge方法 -->
        <property name="age" value="50"></property>
    </bean>
    <bean id="dog"
          class="com.uifuture.spring.core.bean.createbean.AnimalBeanFactory"
          factory-method="getAnimal">
        <!--给静态工厂方法的参数传值-->
        <constructor-arg value="dog" />
        <property name="age" value="5"></property>
    </bean>
</beans>
```

由上述配置文件可看出，cat 和 dog 两个 Bean 配置的 class 属性和 factory-method 属性完全相同，这是因为这两个实例都是由同一个静态工厂类、同一个静态工厂方法生产得到的。配置这两个 Bean 实例时指定的静态工厂方法的参数值不同（一个是 cat，一个是 dog），并以此为依据在工厂方法中生产不同的对象。

一旦为 <bean> 标记指定了 factory-method 属性，Spring 创建 Bean 实例就不再是调用构造器，而是调用静态工厂方法。如果同时指定了 class 和 factory-method 属性，Sring 会调用静态工厂方法来创建 Bean 实例。

下面编写测试代码，并查看测试结果，如代码清单 15-38 所示。

代码清单 15-38　测试用静态工厂方法创建的 Bean 实例

```
@Test
public void animalTest() {
    // 用加载路径下的配置文件创建ClassPathResource实例
    ApplicationContext context = new ClassPathXmlApplicationContext("spring-content-
create-bean.xml");
```

```
        Dog dog = (Dog) context.getBean("dog");
        dog.say();
        Cat cat = (Cat) context.getBean("cat");
        cat.say();
    }
```

运行结果如图 15-1 所示。

```
/Library/Java/JavaVirtualMachines/jdk1.8.0_201.jdk/Contents/Home/b
Connected to the target VM, address: '127.0.0.1:64272', transport:
三月 13, 2019 10:56:01 下午 org.springframework.context.support.Abs
信息: Refreshing org.springframework.context.support.ClassPathXmlAp
三月 13, 2019 10:56:01 下午 org.springframework.beans.factory.xml.Xr
信息: Loading XML bean definitions from class path resource [spring
i am dog.age=5
i am cat.age=50
Disconnected from the target VM, address: '127.0.0.1:64272', trans

Process finished with exit code 0
```

图 15-1　调用静态工厂方法创建 Bean 测试结果

当使用静态工厂方法创建 Bean 实例时，class 属性确定静态工厂类，而不是该 Bean 的类。Spring 需要知道是哪个静态工厂方法创建的 Bean 实例，然后使用 factory-method 属性来确定方法名。Spring 调用静态工厂方法（可以包含一组参数），并返回一个有效的对象。

如果静态工厂方法需要参数，使用 <constructor-arg> 标记传入。class 元素不再是 Bean 的实现类，而是静态工厂类。必须由 factory-method 属性指定产生实例的静态工厂方法。

通过静态工厂方法创建 Bean 实例后，Spring 依然可以管理该 Bean 实例的依赖关系，包括为其注入所需的依赖 Bean、管理其生命周期等。

15.3.3　调用实例工厂方法创建 Bean 实例

调用实例工厂方法创建 Bean 实例与调用静态工厂方法创建 Bean 实例只有一点不同：实例工厂方法需要工厂实例，而静态工厂方法只需要使用工厂类。

在配置时，静态工厂方法使用 class 属性指定静态工厂类，实例工厂方法使用 factory-bean 属性指定工厂实例。

<bean> 标记要指定以下 2 个属性。

（1）factory-bean：工厂实例的 ID。

（2）factory-method：实例工厂的工厂方法。

与前面的静态工厂方法只有配置文件与工厂类不同，工厂类的修改如代码清单 15-39 所示。

代码清单 15-39　工厂类的修改

```
public class AnimalBeanExampleFactory {
    /** name参数决定哪个是Bean的实例*/
    public Animal getAnimal(String name) {
        if ("dog".equalsIgnoreCase(name)) {
```

```
            return new Dog();
        } else if ("cat".equalsIgnoreCase(name)) {
            return new Cat();
        }
        return null;
    }
}
```

将静态工厂方法修改为实例方法。配置实例工厂创建 Bean 实例与配置静态工厂创建 Bean 实例基本类似，只需将原来的静态工厂类改为现在的工厂实例，配置文件如代码清单 15-40 所示。

代码清单 15-40　调用实例工厂方法配置 Bean 实例的配置文件

```
<beans xmlns:xsi="http://www.w3.org/2001/XMLSchema-instance"
       xmlns="http://www.springframework.org/schema/beans"
       xsi:schemaLocation="http://www.springframework.org/schema/beans
        http://www.springframework.org/schema/beans/spring-beans.xsd">
    <!-- 指定工厂Bean 的Id-->
<bean id="animalBeanFactory" class="com.uifuture.spring.core.bean.createbean.
AnimalBeanExampleFactory" />
    <bean id="cat"
          factory-bean="animalBeanFactory"
          factory-method="getAnimal">
        <!--给实例工厂方法的参数传值-->
        <constructor-arg value="cat"/>
        <!--注入属性值，调用setAge方法 -->
        <property name="age" value="50"></property>
    </bean>
    <bean id="dog"
          factory-bean="animalBeanFactory"
          factory-method="getAnimal">
        <!--给实例工厂方法的参数传值-->
        <constructor-arg value="dog"/>
        <property name="age" value="5"></property>
    </bean>
</beans>
```

最后用测试类修改配置文件路径即可，演示结果与图 15-1 一致。

调用实例工厂方法创建 Bean 实例，与调用静态工厂创建 Bean 实例的用法基本类似，但在细节使用上略有不同。

（1）配置实例工厂方法创建 Bean 实例，必须使用实例工厂 Bean 配置成 Bean 实例；而配置静态工厂方法创建 Bean，则无须配置工厂 Bean。

（2）配置实例工厂方法创建 Bean 实例，必须使用 factory-bean 属性确定工厂 Bean 实例；而配置静态工厂方法创建 Bean 实例，则使用 <class> 标记确定静态工厂类。

相同点如下。

（1）使用 factory-method 属性指定产生 Bean 实例的工厂方法。

（2）工厂方法如果需要参数，则都使用 <constructor-arg> 标记指定参数值。

（3）普通的属性值注入，都使用 <property> 标记确定参数值。

当 Bean 之间的业务逻辑比较独立，或与外界第三方关联不多时，使用默认的构造函数创建即可，Spring 通过 Bean 的默认构造函数（org.springframework.beans.factory.support.DefaultListableBeanFactory）创建。

如果想统一管理各个 Bean 的创建，或者是想在某些 Bean 创建之前做一些相同的初始化处理，则可以使用静态工厂方法创建 Bean，在静态工厂方法中进行统一的处理。

使用实例工厂方法创建 Bean 实例，也就是将工厂方法作为业务 Bean 进行控制，可以用于集成其他框架的 Bean 创建管理方法，也可以使用 Bean 和工厂的角色进行互换。

15.4 加载属性文件

在开发中会遇到需要经常修改参数，而且参数不固定的情况，如数据库配置、第三方配置等，开发者最好能够将这些参数进行动态配置，并放到属性（properties）文件中，这样在源代码读取 properties 配置时，当后期需要改动时更改配置文件，不用更改源代码，这样更加方便（使用携程的 Apollo 可以非常方便地动态修改配置）。Spring 提供了 4 种加载 properties 的方式。

15.4.1 通过 <context:property-placeholder> 标记加载

在这里以加载 MySQL 数据源为例进行讲解。首先在 Spring 的 XML 文件中增加，如代码清单 15-41 所示代码。

代码清单 15-41 在 Spring 配置文件中添加标记

```
<!-- 引入属性文件classpath -->
<context:property-placeholder location="classpath:config.properties" ignore-unresolvable="true"/>
```

将 config.properties 文件内容该配置在 resource 下，如代码清单 15-42 所示。

代码清单 15-42 config.properties 配置文件的内容

```
driverClassName=com.mysql.jdbc.Driver
jdbc_url=jdbc:mysql://localhost:3306/uifuture?useUnicode=true&characterEncoding=UTF-8&zeroDateTimeBehavior=convertToNull
jdbc_username=root
jdbc_password=1234
# 初始化连接大小
jdbc_init=50
# 连接池最小空闲
jdbc_minIdle=20
# 获取连接最大等待时间，单位为毫秒
jdbc_maxActive=60000
```

接下来可以在 Spring 的 XML 配置文件中使用 properties 属性值，如代码清单 15-43 所示。

代码清单 15-43 使用 properties 属性值

```
<!-- 配置数据源-->
<bean id="dataSource" class="com.alibaba.druid.pool.DruidDataSource"
      init-method="init" destroy-method="close"
      p:driverClassName="${driverClassName}"
      p:url="${jdbc_url}"
      p:username="${jdbc_username}"
      p:password="${jdbc_password}"
      p:initialSize="${jdbc_init}"
      p:minIdle="${jdbc_minIdle}"
      p:maxActive="${jdbc_maxActive}"
      p:filters="stat,wall">
</bean>
```

由于配置文件已经由 Spring 进行了加载，所以可以将配置文件中的值注入到 Java 类中，使用 org.springframework.beans.factory.annotation.Value 注解即可。使用注解配置值如代码清单 15-44 所示。

代码清单 15-44 使用注解配置值

```
/***注意：这里变量不能定义为static*/
@Value("${jdbc_url}")
private  String jdbcUrl;
```

15.4.2 通过 <util:properties> 标记加载

在 Spring 的 XML 配置文件中除了使用 <context:property-placeholder> 标记加载 properties 文件外，还可以使用 <util:properties> 标记加载 properties 文件。

在使用 <util:properties> 标记之前，需要在 XML 配置文件 <beans> 标记头中增加如代码清单 15-45 所示代码。

代码清单 15-45 增加 util 声明

```
<beans ...
       xsi:schemaLocation="http://www.springframework.org/schema/beans
...
http://www.springframework.org/schema/util
http://www.springframework.org/schema/util/spring-util.xsd">
```

引入配置文件，如代码清单 15-46 所示。

代码清单 15-46 引入配置文件

```
<util:properties id="jdbcConfig"  local-override="true" location="classpath:
config.properties"/>
```

接下来在 Spring 中使用配置文件的属性，如代码清单 15-47 所示。

代码清单 15-47　使用配置文件的属性

```
<!-- 配置数据源 -->
<bean id="dataSource" class="com.alibaba.druid.pool.DruidDataSource"
      init-method="init" destroy-method="close"
      p:driverClassName="#{jdbcConfig['driverClassName']}"
      p:url="#{jdbcConfig['jdbc_url']}"
      p:username="#{jdbcConfig['jdbc_username']}"
      p:password="#{jdbcConfig['jdbc_password']}"
      p:initialSize="#{jdbcConfig['jdbc_init']}"
      p:minIdle="#{jdbcConfig['jdbc_minIdle']}"
      p:maxActive="#{jdbcConfig['jdbc_maxActive']}"
      p:filters="stat,wall">
</bean>
```

在 Java 中使用的情况也与前面有所不同，在这里使用 jdbc_url 来为例子进行配置，如代码清单 15-48 所示。

代码清单 15-48　在 Java 中配置属性值

```
@Value(value="#{jdbcConfig['jdbc_url']}")
private String jdbcUrl;
```

15.4.3　通过 PropertyPlaceholderConfigurer 类加载

通过 PropertyPlaceholderConfigurer 类加载 properties 文件等价于使用 <context:property-placeholder> 标记加载配置文件。配置方式如代码清单 15-49 所示。

代码清单 15-49　使用 PropertyPlaceholderConfigurer 类加载 properties 文件

```
<bean class="org.springframework.beans.factory.config.PropertyPlaceholderConfigurer">
    <property name="locations">
        <list>
            <value>classpath:config.properties</value>
        </list>
    </property>
</bean>
```

15.4.4　通过 PropertySource 注解加载

除了使用 Spring 的 XML 配置文件加载 properties 文件外，还可以直接使用 Java 来加载 properties 文件，如代码清单 15-50 所示。

代码清单 15-50　使用 PropertySource 注解加载

```
import org.springframework.beans.factory.annotation.Value;
import org.springframework.context.annotation.PropertySource;
import org.springframework.stereotype.Component;
@PropertySource(value={"classpath:config.properties"})
@Component
```

```
public class JdbcProperties {
    @Value("${driverClassName}")
    private String driverClassName;
    @Value("${jdbc_url}")
    private String jdbcUrl;
   //省略其他的值注入
}
```

注意：使用 Component 注解将 JdbcProperties 类注册为 Spring 的 Bean 实例。

15.5 Spring 条件化装配 Bean

在某些场景下，可能开发者需要一个或者多个 Bean，只有在某些特定的情况下才被创建。例如，在某个条件成立的情况下才创建某个 Bean。在 Spring 4 之前，开发者无法通过 Spring 实现这种级别的条件化配置，在 Spring 4 之后，引入了一个注解，就是 @Conditional，可以配合 @Bean 或者 @Component 一起使用。

15.5.1 Conditional 注解源码解析

该注解的作用是按照一定的条件进行判断，满足条件则在 Spring 容器注入 Bean。否则不注入。Conditional 注解源码如代码清单 15-51 所示。

代码清单 15-51 Conditional 注解源码

```
@Target({ElementType.TYPE, ElementType.METHOD})
@Retention(RetentionPolicy.RUNTIME)
@Documented
public @interface Conditional {
   Class<? extends Condition>[] value();
}
```

通过源码可知，需要传入一个 Class 的数组对象 value，并且 Class 需要继承 Condition 接口。Condition 接口源码如代码清单 15-52 所示。

代码清单 15-52 Condition 接口源码

```
@FunctionalInterface
public interface Condition {
   boolean matches(ConditionContext context, AnnotatedTypeMetadata metadata);
}
```

matches 方法的返回值是一个布尔值，当返回 true 时，Spring 就会注册 Bean 实例；当返回 false 时就不会注册。matches 方法有两个参数，参数类型分别是 ConditionContext 和 AnnotatedType-Metadata，这两个参数中包含了大量的信息，ConditionContext 中有 Environment、ClassLoader 等信息，AnnotatedTypeMetadata 可以获得注解的信息。

Condition 接口有一个实现，与 Profile 注解相关，Profile 注解也是用来判断是否要注入 Bean 的，Profile 注解其实就是一个 Conditional 注解，其源码如代码清单 15-53 所示。

代码清单 15-53　Profile 注解源码

```
@Target({ElementType.TYPE, ElementType.METHOD})
@Retention(RetentionPolicy.RUNTIME)
@Documented
@Conditional(ProfileCondition.class)
public @interface Profile {
   String[] value();
}
```

使用 Profile 注解可以实现在不同环境（开发、测试和部署等）使用不同的配置的目标。从 @Conditional(ProfileCondition.class) 可知 Profile 注解仅仅是一个特殊的 Conditional 注解。ProfileCondition 类的实现如代码清单 15-54 所示。

代码清单 15-54　ProfileCondition 类的实现

```
class ProfileCondition implements Condition {
   @Override
    public boolean matches(ConditionContext context, AnnotatedTypeMetadata
metadata) {
      MultiValueMap<String, Object> attrs = metadata.getAllAnnotationAttributes(Prof
ile.class.getName());
      if (attrs != null) {
         for (Object value : attrs.get("value")) {
            if (context.getEnvironment().acceptsProfiles((String[]) value)) {
               return true;
            }
         }
         return false;
      }
      return true;
   }
}
```

从 ProfileCondition 的 mathes 方法可以看出，Spring 从 ConditionContext 中获得激活的 Profile 并与注解上的字符串进行比对，判断是否实例化这个 Bean。

15.5.2　Conditional 注解的使用

接下来通过一个实例来讲解 Conditional 注解的使用。让一个类在不同的系统环境下实例出不同的对象属性值。

首先创建 User 类，如代码清单 15-55 所示。

代码清单 15-55　User 类

```
package com.uifuture.spring.core.bean.condition.entity;
import java.util.StringJoiner;
```

```
public class User {
    private String name;
    private Integer age; // 省略getter/setter，toString方法
}
```

接下来创建 UserConfig 类，配置两个 User 实例注入 Spring 容器，如代码清单 15-56 所示。

代码清单 15-56　UserConfig 类

```
package com.uifuture.spring.core.bean.condition.config;
import com.uifuture.spring.core.bean.condition.condition.LinuxCondition;
import com.uifuture.spring.core.bean.condition.condition.MacConditional;
import com.uifuture.spring.core.bean.condition.condition.WindowsCondition;
import com.uifuture.spring.core.bean.condition.entity.User;
import org.springframework.context.annotation.Bean;
import org.springframework.context.annotation.Conditional;
import org.springframework.context.annotation.Configuration;
@Configuration
public class UserConfig {

    /**
     * 如果WindowsCondition的matches实现方法返回true，则注入这个bean
     */
    @Conditional({WindowsCondition.class})
    @Bean(name = "windowsUser")
    public User windowsUser() {
        User user = new User();
        user.setName("windows");
        user.setAge(10);
        return user;
    }
    /**
     * 如果LinuxCondition的matches实现方法返回true，则注入这个bean
     */
    @Conditional({LinuxCondition.class})
    @Bean("linuxUser")
    public User linuxUser() {
        User user = new User();
        user.setName("linux");
        user.setAge(20);
        return user;
    }
    /**
     * 如果MacCondition的matches实现方法返回true，则注入这个bean
     */
    @Conditional({MacCondition.class})
    @Bean("macUser")
    public User macUser() {
        User user = new User();
        user.setName("mac");
```

```
        user.setAge(30);
        return user;
    }
}
```

接下来先测试两个 Bean 是否已经都成功注入了，测试类如代码清单 15-57 所示。

代码清单 15-57　测试类

```
@Test
public void loadUserBeanTest() {
    AnnotationConfigApplicationContext applicationContext = new AnnotationConfigApp
licationContext(UserConfig.class);
    Map<String, User> map = applicationContext.getBeansOfType(User.class);
    System.out.println(map);
}
```

在 Mac 系统下的演示结果如图 15-2 所示。

```
✔ Tests passed: 1 of 1 test – 702 ms
/Library/Java/JavaVirtualMachines/jdk1.8.0_201.jdk/Contents/Home/bin/java ...
八月 20, 2022 3:22:16 上午 org.springframework.context.support.AbstractApplicati
信息: Refreshing org.springframework.context.annotation.AnnotationConfigApplica
 context hierarchy
{macUser=User[name='mac', age=30]}

Process finished with exit code 0
```

图 15-2　加载 UserBean 的演示结果

如果需要根据操作系统来注入一个 User 实例，例如，在 Windows 系统下注入 windowsUser，在 Linux 系统下注入 linuxUser,这种情况下就可以使用 Conditional 注解了。要实现 Condition 接口，就要重写自定义规则，如代码清单 15-58 所示。

代码清单 15-58　WindowsCondition 类

```
package com.uifuture.spring.core.bean.condition.condition;
import org.springframework.context.annotation.Condition;
import org.springframework.context.annotation.ConditionContext;
import org.springframework.core.env.Environment;
import org.springframework.core.type.AnnotatedTypeMetadata;
import java.util.Objects;
public class WindowsCondition implements Condition {
    /**
     * @param context 判断条件能使用的上下文环境
     * @param metadata 注解所在位置的注释信息
     */
    @Override
     public boolean matches(ConditionContext context, AnnotatedTypeMetadata
metadata) {
        // 获取当前环境信息
```

```
        Environment environment = context.getEnvironment();
        // 获得当前系统名
        String property = environment.getProperty("os.name");
        // 包含Windows则说明是Windows系统，返回true
        if (Objects.requireNonNull(property).contains("Windows")){
            return true;
        }
        return false;
    }
}
package com.uifuture.spring.core.bean.condition.condition;
import org.springframework.context.annotation.Condition;
import org.springframework.context.annotation.ConditionContext;
import org.springframework.core.env.Environment;
import org.springframework.core.type.AnnotatedTypeMetadata;
import java.util.Objects;
public class LinuxCondition implements Condition {
    @Override
     public boolean matches(ConditionContext context, AnnotatedTypeMetadata
metadata) {
        Environment environment = context.getEnvironment();
        String property = environment.getProperty("os.name");
        if (Objects.requireNonNull(property).contains("Linux")){
            return true;
        }
        return false;
    }
}
```

接下来就可以在 Conditional 注解中使用这两个类了，修改 UserConfig 类，修改部分如代码清单 15-59 所示。

代码清单 15-59　修改的 UserConfig 类

```
/** 如果WindowsCondition的matches实现方法返回true，则注入这个bean */
@Conditional({WindowsCondition.class})
@Bean(name = "windowsUser")
public User windowsUser(){
    User user = new User();
    user.setName("windows");
    user.setAge(10);
    return user;
}
/** 如果LinuxCondition的matches实现方法返回true，则注入该bean */
@Conditional({LinuxCondition.class})
@Bean("linuxUser")
public User linuxUser(){
    User user = new User();
    user.setName("linux");
    user.setAge(20);
```

```
        return user;
    }
```

接下来进行测试，便可以获取到对应的 User 实例了。也可以模拟参数，修改运行时参数，增加参数 -Dos.name=Linux，如图 15-3 所示。模拟 Linux 的环境，注入 linuxUser 实例。

本实例将 Conditional 注解写在了方法上，Conditional 注解标注在方法上只能控制一个 Bean 实例的注入。Conditional 注解还可以标注在类上，可以决定一批 Bean 是否进行注入。另外，Conditional 注解是可以传入一个 Class 数组，即在存在多种条件类的情况下，可以增加一个 TestCondition 类来集成 Condition 接口进行测试。至此，可以发现，只有当全部的 matches 方法均返回 true 时，Bean 才会进行注入。

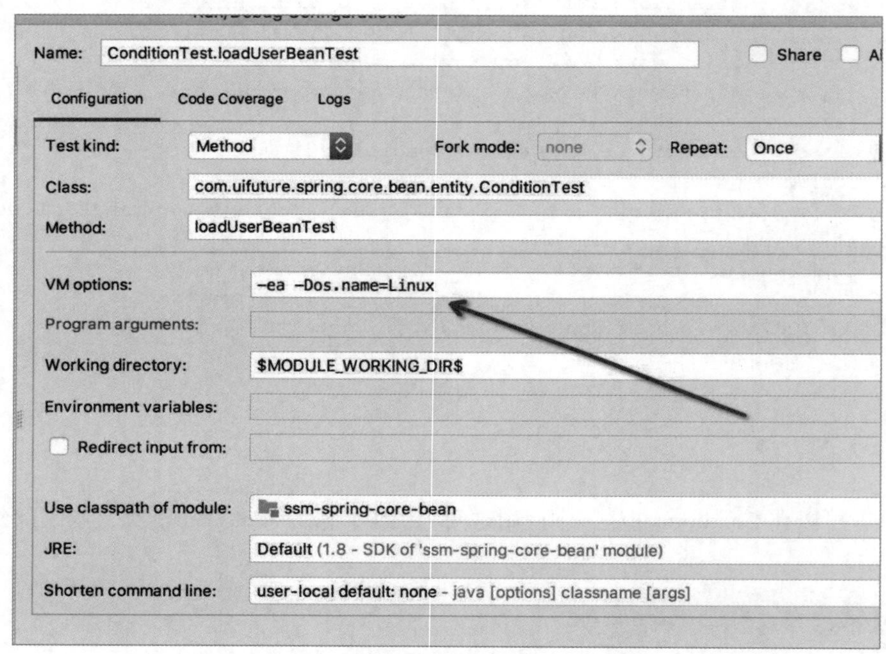

图 15-3　修改运行时 os.name 参数值

15.5.3　ConditionContext 与 AnnotatedTypeMetadata 讲解

matches 有两个参数类型 ConditionContext 与 AnnotatedTypeMetadata，这两个接口的源码如代码清单 15-60 和代码清单 15-61 所示。

代码清单 15-60　ConditionContext 接口

```
public interface ConditionContext {
    /** 返回保存Bean定义的注册表*/
    BeanDefinitionRegistry getRegistry();
    @Nullable
    ConfigurableListableBeanFactory getBeanFactory();
    /** 返回运行当前应用程序的环境*/
```

```
    Environment getEnvironment();
    ResourceLoader getResourceLoader();
    @Nullable
    ClassLoader getClassLoader();
}
```

通过 ConditionContext 的接口，可以获取以下信息：

（1）通过 getRegistry 方法返回的 BeanDefinitionRegistry 检查 Bean 的定义。

（2）通过 getBeanFactory 方法返回的 ConfigurableListableBeanFactory 检查 Bean 是否存在，甚至可以检查 Bean 的属性。

（3）通过 getEnvironment 方法返回的 Environment 可以检查变量的值及环境变量是否存在。

（4）通过 getResourceLoader 方法返回的 ResourceLoader 可以读取并查询所加载的资源。

（5）通过 getClassLoader 方法返回的 ClassLoader 可以加载并检查类是否存在。

代码清单 15-61　AnnotatedTypeMetadata 源码

```
public interface AnnotatedTypeMetadata {
    boolean isAnnotated(String annotationName);
    @Nullable
    Map<String, Object> getAnnotationAttributes(String annotationName);
    @Nullable
    Map<String, Object> getAnnotationAttributes(String annotationName, boolean classValuesAsString);
    @Nullable
    MultiValueMap<String, Object> getAllAnnotationAttributes(String annotationName);
    @Nullable
    MultiValueMap<String, Object> getAllAnnotationAttributes(String annotationName, boolean classValuesAsString);
}
```

AnnotatedTypeMetadata 接口能够让开发者检查带有 @Bean 的方法上还有什么注解。

借助 isAnnotated 方法可以判断带有 @Bean 的方法是不是有其他特定的注解，借助另外的方法，可以检查 @Bean 的方法上其他注解的属性。

从 Spring 4.X 开始，Profile 注解进行了重构，目前 Profile 注解是基于 @Conditional 和 Condition 接口实现的。将 Profile 注解看成一种特殊的 Conditional 注解即可。

15.6　Spring 中的事件机制

事件和回调思想在 GUI（AWT/Swing）相关技术中的应用非常广泛，在 Web 项目中却很少直接使用，但这并不代表面向事件的体系结构在 Web 项目中不够强大。

本节将重点介绍 Spring 框架中的事件处理，首先介绍事件驱动模型，然后介绍 Spring 中事件机制的使用。

15.6.1 事件驱动模型

说到事件驱动，一些开发者会立即联想到观察者模型、发布 / 订阅模型、消息队列、消息驱动、事件、EventSourcing、Nginx 等。下面来详细介绍事件驱动模型。

事件驱动模型其实就是平常说的观察者或者发布 / 订阅模型。事件驱动编程也称为基于事件的编程，是基于对接收到的信号进行反应的编程。这些信号必须通过某种方式来传输信息。简单地举一个例子，在页面上单击按钮，可以将该行为称为事件。也就是说，这些事件可以通过用户操作（如右击、单击按钮、键盘输入等）或者是程序条件执行触发（如某个操作执行结束后可以启动另外一个操作）来产生。

通过以下 3 点，可以更加清晰地理解事件驱动模型。

（1）事件驱动模型中目标和观察者的关系为对象间一对多的关系，如交通信号灯，信号灯是目标，而行人要观察信号灯以便决定走还是停下。

（2）当目标发送改变（发布者），观察者（订阅者）就可以接收信号。

（3）观察者如何处理（如看信号灯的指示，向前走还是向右走，或者停下），目标不需要进行干涉，它们之间的关系是松耦合的。

事件驱动模型如图 15-4 所示。

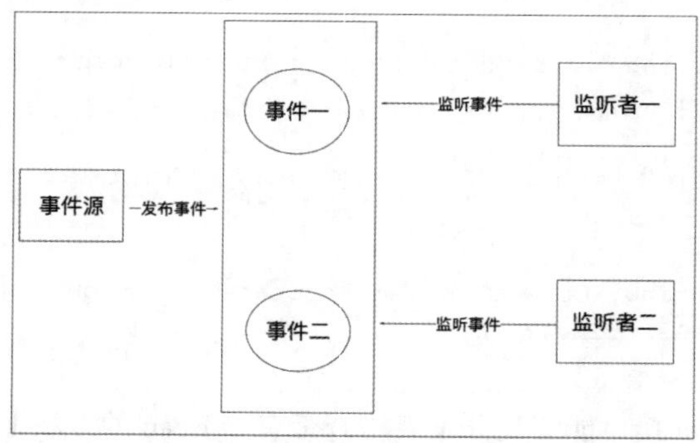

图 15-4 事件驱动模型

Spring 中的事件驱动模型由以下三部分组成：

（1）事件 ApplicationEvent：继承自 JDK 的 EventObject，JDK 要求所有的事件都继承它，并通过 source 得到事件源。

（2）事件发布者：Spring 中的 ApplicationEventPublisher 接口和 ApplicationEventMulticaster 接口，通过这两个接口，服务便能拥有注册和发布事件的能力。

（3）监听器：ApplicationListener，继承自 JDK 的 EventListener，JDK 要求所有的监听器都继承它。下来重点介绍 Spring 中的事件驱动模型。

15.6.2 Spring 中的事件驱动模型

Spring 中的事件驱动模型是一个观察者模式的实现。观察者模式就是一个目标对象管理所有依赖它的观察者对象，并且在它本身的状态改变时主动发出通知。Spring 中通过 ApplicationEvent 类和 ApplicationListener 接口可以实现 ApplicationContext 事件处理。如果容器中有一个 ApplicationListener 的实现类 Bean，每当 ApplicationContext 发布 ApplicationEvent 时，ApplicationListener 实现类 Bean 将自动被触发。

ApplicationListener 接口中只有一个方法 onApplicationEvent，当一个事件发送过来时，该方法会触发相应的处理 ApplicationListener。接口可以通过指定需要接收的事件来实现（方法中有参数 Event）。因此，Spring 会自动筛选可以用来接收给定事件的监听器（Listeners）。

ApplicationListener 接口的源码如代码清单 15-62 所示。

代码清单 15-62　ApplicationListener 接口的源码

```
package org.springframework.context;
import java.util.EventListener;
@FunctionalInterface
public interface ApplicationListener<E extends ApplicationEvent> extends
EventListener {
   /**
    * 处理应用程序事件
    * @param event 事件的响应
    */
   void onApplicationEvent(E event);

}
```

这里的 E 泛型对象通过继承 ApplicationEvent 类进行表示。ApplicationEvent 抽象类又继承并扩展了 EventObject 类，通过使用 EventObject 中的 getSource 方法，开发者可以获得所发生给定事件的对象。

在 Spring 的内置事件中，以父抽象类 ApplicationContextEvent（继承自 ApplicationEvent 类）下的 4 个子事件类为主进行介绍。

ApplicationContextEvent 是 Spring 内置事件中的父抽象类，构造方法传入 Spring 的 Context 容器，同时也有获取 Spring 中上下文容器的方法。

1.ContextStartedEvent 类

当 Spring 容器启动时，即调用 AbstractApplicationContext 类中（上下文）的 start 方法时，会触发此事件。

2.ContextRefreshedEvent 类

当 Spring 容器初始化（此处的初始化是指所有的 Bean 都被 Spring 容器成功装载，后处理 Bean 被检测并且激活，所有单例 Bean 被预实例化，ApplicationContext 容器已就绪可用）或者刷新时，会触发此事件。此事件在开发中常用，用于在 Spring 容器启动时，导入自定义的 Bean 实例到 Spring 容器中。

3.ContextStoppedEvent 类

当 Spring 容器停止时，会触发此事件，即上下文调用 stop 方法时，会触发此事件。此处停止，则容器管理生命周期的 Bean 实例将获得一个指定的停止信号，可以通过再次调用 start 方法重新启动被停止的 Spring 容器。

4.ContextClosedEvent 类

当 Spring 容器关闭时，即当上下文调用 close 方法时，会触发此事件。如 Web 容器关闭时会自动触发 Spring 容器的关闭事件，如果是普通 Java 应用，需要调用 ctx.registerShutdownHook 注册虚拟机关闭时的钩子才会触发。

在 Spring 中，还有一个比较特殊的事件，就是 RequestHandledEvent 事件。Request Handled Event 事件是与 Web 相关的事件，与 request 请求关联，只能应用在使用了 DispatcherServlet 的 Web 应用。例如，当在使用 Spring 作为前端的 MVC 控制器时（SpringMVC），当 Spring 处理完用户请求时，系统便会自动触发该事件。

15.6.3 Spring 中的事件广播器

Spring 是通过事件广播器来将事件分配给专门的监听器的。这里的事件广播器由 ApplicationEventMulticaster 接口的实现表示。ApplicationEventMulticaster 接口的源码如代码清单 15-63 所示。

代码清单 15-63 ApplicationEventMulticaster 接口的源码

```
public interface ApplicationEventMulticaster {
    void addApplicationListener(ApplicationListener<?> listener);
    void addApplicationListenerBean(String listenerBeanName);
    void removeApplicationListener(ApplicationListener<?> listener);
    void removeApplicationListenerBean(String listenerBeanName);
    void removeAllListeners();
    void multicastEvent(ApplicationEvent event);
    void multicastEvent(ApplicationEvent event, @Nullable ResolvableType eventType);
}
```

该接口的方法虽然有 7 个，但是实际上主要定义了以下 3 种方法：

（1）addApplicationListener*：添加新的监听器，定义了两种方法来添加新的监听器 addApplicationListener(ApplicationListener<?> listener) 和 addApplicationListenerBean(String listenerBeanName)。当监听器对象实例已知时，应用第 1 个方法即可。如果使用第 2 个方法，需要通过 Bean 的名称得到 listener 对象（这里应用了依赖查找 DL），然后再将其添加到 listener 列表中。

（2）removeApplication*：删除监听器，与添加方法一样，可以通过传递对象来删除一个监听器 (removeApplicationListener(ApplicationListener<?> listener) 或通过传递 Bean 名称 (removeApplicationListenerBean(String listenerBeanName)，也可以通过 removeAllListeners 方法删除所有已注册的监听器。

（3）multicastEvent：将给定的事件发送到已注册的监听器，它可以用来向所有注册的监听器发送事件（实现可以通过 SimpleApplicationEventMulticaster 类找到）。

15.6.4 演示 Spring 的事件机制

Spring 中的事件机制与其他的事件机制基本相似，都需要事件源、事件及事件监听器组成。只是 Spring 中的事件源是 ApplicationContext，且事件必须由 Java 程序显式触发。接下来演示

Spring 容器中的事件机制。Spring 中内置 ApplicationEvent 类，其对象就是 Spring 容器事件。先用继承 ApplicationEvent 类的子类定义事件，如定义一个注册事件类，如代码清单 15-64 所示。

代码清单 15-64　注册事件类

```
package com.uifuture.spring.core.bean.event;
import org.springframework.context.ApplicationEvent;
/** 注册事件*/
public class RegisterEvent extends ApplicationEvent {
    /** 注册名称*/
    private String username;
    /** 注册邮件*/
    private String email;
    public RegisterEvent(Object source, String username, String email) {
        super(source);
        this.username = username;
        this.email = email;
    }
    public RegisterEvent(Object source) {
        super(source);
    }
    // 省略getter、setter方法
}
```

上面的 RegisterEvent 类继承了 ApplicationEvent 类，除此之外，该类就是一个普通的 Java 类。

容器事件的监听器类必须实现 ApplicationListener 接口，实现该接口就必须实现 onApplicationEvent(ApplicationEvent event) 方法，该方法在容器内发生任何事件时都会被出发。接下来看监听器的实现代码，如代码清单 15-65 所示。

代码清单 15-65　监听器的实现代码

```
package com.uifuture.spring.core.bean.event;
import org.springframework.context.ApplicationEvent;
import org.springframework.context.ApplicationListener;
/**注册监听器*/
public class RegisterListener implements ApplicationListener {
    @Override
    public void onApplicationEvent(ApplicationEvent event) {
        if(event instanceof RegisterEvent){
            RegisterEvent registerEvent = (RegisterEvent) event;
            System.out.println(registerEvent.getUsername());
            System.out.println(registerEvent.getEmail());
        }else {
            System.out.println("其他事件，该事件不在监听范围:"+event);
        }
    }
}
```

要使监听器起作用，还需要将 RegisterListener 注册为 Spring 中的 Bean，如代码清单 15-66 所示。

代码清单 15-66　注册为 Bean

```
<bean id="registerListener" class="com.uifuture.spring.core.bean.event.
RegisterListener"></bean>
```

为 Spring 容器注册监听器，在 Spring 的配置文件中进行简单配置即可。当在 Spring 中配置一个实现 ApplicationListener 接口的 Bean 时，Spring 容器就会把这个 Bean 当成容器事件的监听器。

当系统创建 Spring 容器、加载 Spring 容器时会自动触发容器事件，容器事件监听器可以监听这些事件。另外，程序也可以调用 ApplicationContext 中的 publishEvent 方法来主动触发一个容器事件，如代码清单 15-67 所示。

代码清单 15-67　触发事件

```
public class EventTest {
    /** 测试触发事件 */
    @Test
    public void registerEventTest(){
            ApplicationContext context = new ClassPathXmlApplicationContext("event-
spring-content.xml");
        // 创建一个ApplicationEvent对象
        RegisterEvent registerEvent = new RegisterEvent("任何对象","ssm","abc@**.com");
        // 主动触发事件
        context.publishEvent(registerEvent);
        ((ClassPathXmlApplicationContext) context).stop();
        ((ClassPathXmlApplicationContext) context).close();
    }
}
```

通过运行该代码，可以看到如图 15-5 所示的结果。

```
✔ Tests passed: 1 of 1 test – 546 ms
/Library/Java/JavaVirtualMachines/jdk1.8.0_201.jdk/Contents/Home/bin/java ...
八月 20, 2022 3:39:49 上午 org.springframework.context.support.AbstractApplicationContext prepareRefresh
信息: Refreshing org.springframework.context.support.ClassPathXmlApplicationContext@6ea6d14e: startup date [Sat Aug 20 03:39:49 CST 2022];
 hierarchy
八月 20, 2022 3:39:49 上午 org.springframework.beans.factory.xml.XmlBeanDefinitionReader loadBeanDefinitions
信息: Loading XML bean definitions from class path resource [event-spring-content.xml]
其他事件, 该事件不在监听范围:org.springframework.context.event.ContextRefreshedEvent[source=org.springframework.context.support
 .ClassPathXmlApplicationContext@6ea6d14e: startup date [Sat Aug 20 03:39:49 CST 2022]; root of context hierarchy]
ssm
abc@**.com
其他事件, 该事件不在监听范围:org.springframework.context.event.ContextStoppedEvent[source=org.springframework.context.support
 .ClassPathXmlApplicationContext@6ea6d14e: startup date [Sat Aug 20 03:39:49 CST 2022]; root of context hierarchy]
八月 20, 2022 3:39:49 上午 org.springframework.context.support.AbstractApplicationContext doClose
信息: Closing org.springframework.context.support.ClassPathXmlApplicationContext@6ea6d14e: startup date [Sat Aug 20 03:39:49 CST 2022]; roo
其他事件, 该事件不在监听范围:org.springframework.context.event.ContextClosedEvent[source=org.springframework.context.support
 .ClassPathXmlApplicationContext@6ea6d14e: startup date [Sat Aug 20 03:39:49 CST 2022]; root of context hierarchy]
```

图 15-5　事件测试类运行结果

注意： 如果使用 Spring，并通过 Bean 实例发布事件，则 Bean 必须获得 Spring 上下文容器对象。如果程序中没有直接获取容器的对象，需要让 Bean 实现 ApplicationContextAware 接口或者 BeanFactoryAware 接口，从而获得上下文容器；或者通过工具类，一直保存上下文容器的引用，当使用时进行调用。

15.7 Spring 中的定时器

在项目中，有很多地方需要用到定时器来支持定时任务的运行，如在凌晨 1:00 进行脚本的订正，凌晨 2:00 进行推送消息等。Spring 框架提供了对定时器的支持，通过配置文件或者注解就可以很方便地实现定时器。

在 Spring 中，可以使用 XML 或者注解配置定时器，由于注解使用较多，这里讲解注解的配置，以及动态修改定时任务的调度。

在进行定时器的配置之前，一些 Spring 的基本配置是不可少的，首先需要在 XML 配置文件（项目的 spring-content-task.xml 文件）中引入命名空间（Spring 扫描注解的配置已经配置好），如代码清单 15-68 所示。

代码清单 15-68　引入定时任务命名空间

```
<beans ...
       xmlns:task="http://www.springframework.org/schema/task"
       xsi:schemaLocation="...
       http://www.springframework.org/schema/task
       http://www.springframework.org/schema/task/spring-task.xsd">
</beans>
```

在引入命名空间之后，需要设置注解扫描和自动发现，这里使用 <task:annotation-driven> 标记来完成，如代码清单 15-69 所示。

代码清单 15-69　设置注解扫描定时任务和自动发现

```
<!--设置自动发现-->
<task:annotation-driven/>
<!--设置包扫描-->
<context:component-scan base-package="com.uifuture.spring.core.bean.task">
</context:component-scan>
```

接下来创建一个定时任务类，实现每隔 10s 输出当前时间，如代码清单 15-70 所示。

代码清单 15-70　定时任务类

```
/** 定时任务类*/
@Component
public class PrintTimeTask {
    @Scheduled(fixedDelay=10000)
    public void printTime() {
        SimpleDateFormat formatter=new SimpleDateFormat("yyyy-MM-dd HH:mm:ss");
        Date date = new Date();
        System.out.println(formatter.format(date));
    }
}
```

由于没有使用 Web 项目，那么在测试方法中便需要休眠，以便定时任务有足够的时间执行。测试类如代码清单 15-71 所示。

代码清单 15-71　测试类

```
package com.uifuture.spring.core.bean.entity;
import org.junit.Before;
import org.junit.Test;
import org.springframework.context.ApplicationContext;
import org.springframework.context.support.ClassPathXmlApplicationContext;
public class TaskTest {
    /**
     * 这个before方法在所有的测试方法之前执行，并且只执行一次
     * 所有做Junit单元测试时一些初始化工作可以在这个方法里面进行
   * 如在before方法里面初始化ApplicationContext和userService
     */
    @Before
    public void before() {
        // 使用spring配置文件创建Spring上下文
          ApplicationContext context = new ClassPathXmlApplicationContext("spring-
content-task.xml");
    }
    /** 测试定时任务的执行*/
    @Test
    public void printTime() throws InterruptedException {
        System.out.println("开始执行了...");
        Thread.sleep(1000000);
        System.out.println("结束执行了...");
    }
}
```

演示结果如图 15-6 所示。

```
Tests passed: 0 of 1 test
八月 20, 2022 3:41:02 上午 org.springframework.scheduling.concurrent.ExecutorConfigurati
信息: Initializing ExecutorService  'scheduler'
八月 20, 2022 3:41:02 上午 org.springframework.context.support.PostProcessorRegistration
信息: Bean 'scheduler' of type [org.springframework.scheduling.concurrent.ThreadPoolTas
 BeanPostProcessors (for example: not eligible for auto-proxying)
开始执行了...
2022-08-20 03:41:03
2022-08-20 03:41:13
2022-08-20 03:41:23
2022-08-20 03:41:33
2022-08-20 03:41:43
2022-08-20 03:41:53
2022-08-20 03:42:03
2022-08-20 03:42:13
```

图 15-6　定时任务演示结果

关于使用注解开启定时任务，最重要的是要清楚 Scheduled 注解的使用。下面讲解 Scheduled

注解中常用的属性。

- fixedRate：启动一个固定频率的定时任务，如 @Scheduled(fixedRate=1000)，在这种模式下，每隔 1ms 便会执行定时任务一次。**提示**：下一个任务不会等待上一个任务执行完成。
- fixedDelay：启动一个固定时延的任务，如 @Scheduled(fixedDelay=1000)，在这种配置下，上一个任务的结束时间会影响下一个任务的开始时间，也就是两个任务之间的间隔是固定的。下一个任务需要等待上一个任务运行结束，并且在配置完成之后才会运行。
- initialDelay：配置 Spring 初始启动后有延时的定时任务，如配置 @Scheduled (fixedDelay =1000, initialDelay=1000)，那么任务的第 1 次运行会是在 Spring 启动后的 1000ms 后开始，然后按照 1000ms 的间隔独立运行任务。
- cron：通常情况下，cron 表达式使用得比较多。仅仅有时间间隔和频率配置是远远不够的，需要一个重复执行，复杂的配置控制定时任务，这时候就需要用到 cron 表达式，如 @Scheduled(cron = "0 15 10 15 * ?")，表明在每月的 15 日的上午 10:15 启动任务。

以上内容都是将定时任务的参数编写在代码中的，很不灵活，如果需要改动定时任务的频率或者时间，则需要修改代码并重启项目。有时需要动态地改变定时任务的配置，不用进行重新编译和部署。Spring 提供了从 properties 文件读入参数值，动态配置定时任务的方式。通过使用 properties 文件中的参数值，便可以动态配置定时任务，如代码清单 15-72 所示。

代码清单 15-72　动态配置定时任务

```
//properties文件内容
fixedDelay.milliseconds=1000
fixedRate.milliseconds=1000
cron.expression=0 15 10 15 * ?

//下面是配置在类中方法上的注解使用
//固定时延
//@Scheduled(fixedDelayString = "${fixedDelay.milliseconds}")
//固定频率
//@Scheduled(fixedRateString = "${fixedRate.milliseconds}")
//cron表达式
//@Scheduled(cron = "${cron.expression}")
```

cron 表达式详解如下：

（1）一个 cron 表达式至少有 6 个（最多 7 个）由空格分隔的时间元素。

按顺序依次（括号内的为范围）如下：

秒（0~59）、分钟（0~59）、小时（0~23）、天（0~31）、月（0~11）、星期（1~7，1=SUN 或 SUN，MON，TUE，WED，THU，FRI，SAT）、年份（1970—2099 可选）。

注意：秒、分钟、小时、天、月均是从 0 开始的。其中，每个元素可以是一个值（如 6）、一个连续区间（9 ~ 12）、一个间隔时间（8-18/4，/ 表示每隔 4 小时）、一个列表（1,3,5），通配符为“*”号。由于“月份中的日期”和“星期中的日期”这两个元素是互斥的，必须要对其中一个设置“?”。

（2）有些子表达式能包含一些范围或列表。

例如，子表达式（周）可以表示为“MON-FRI”，“MON，WED，FRI”，“MON-WED,SAT”。

“*”字符代表所有可能的值。

“/”字符用来指定数值的增量。

例如，在子表达式（分钟）里的“0/15”表示从第 0 分钟开始，每 15 分钟触发。在子表达式（分钟）里的“3/20”表示从第 3 分钟开始，每 20 分钟（它和“3，23，43”的含义一样）触发。

“？”字符仅被用于天（月）和天（星期）两个子表达式，表示不指定值。

当 2 个子表达式其中之一被指定了值以后，为了避免冲突，需要将另一个子表达式的值设为“？”。

L 字符仅被用于天（月）和天（星期）两个子表达式，它是单词 last 的缩写。

如果在 L 前有具体的内容，就具有其他的含义了。例如，6L 表示这个月的倒数第 6 天。

注意：在使用 L 参数时，不要指定列表或范围，因为这会出错。

W 字符代表平日 (Mon-Fri)，并且仅能用于日域中。它用来指定离指定日最近的一个平日。大部分的商业处理都是基于工作周的，所以 W 字符非常重要。

例如，日域中的 15W 意味着“离该月 15 日的最近一个平日。”如果 15 日是星期六，那么 trigger 会在 14 日（星期五）触发，因为星期四比星期一离 15 日更近。

C：是 Calendar 的意思，代表计划所关联的日期，如果日期没有被关联，则相当于日历中所有日期。例如，5C 在日期字段中就相当于日历中 5 日以后的第 1 天。1C 在星期字段中相当于星期日后的第 1 天。常见的 cron 表达式及其意义见表 15-1。

表 15-1　常见的 cron 表达式及其意义

cron 表达式	意　义
0 0 12 * * ?	每天中午 12:00 触发
0 */2 * * *	每 2h 触发一次
0 25 10 * * ?	每天上午 10:25 触发
0 25 10 * * ? *	每天上午 10:25 触发
0 25 10 * * ? 2019	2019 年的每天上午 10:25 触发
0 * 14 * * ?	每天下午 2:00（14:00）~ 下午 2:59 每 1min 触发一次
0 0/5 14 * * ?	每天下午 2:00~ 下午 2:55 每 5min 触发
0 0/5 14,18 * * ?	每天下午 2:00~2:55 和下午 6:00~6:55 每 5min 触发
0 0-5 14 * * ?	在每天下午 2:00~ 下午 2:05 每 1min 触发
0 20,44 14 ? 3 WED	每年 3 月的星期三的下午 2:20 和 2:44 触发
0 25 10 ? * MON-FRI	周一至周五的上午 10:25 触发
0 25 10 15 * ?	每月 15 日上午 10:25 触发
0 25 10 L * ?	每月最后一日的上午 10:25 触发
0 25 10 ? * 6L	每月的最后一个星期五上午 10:25 触发
0 25 10 ? * 6L 2019-2029	2019—2029 年的每月的最后一个星期五上午 10:25 触发
0 25 10 ? * 6#3	每月的第 3 个星期五上午 10:25 触发

另外，在多任务异步执行情况下，推荐在 Spring 配置文件中配置定时任务的线程池，如代码清单 15-73 所示。

代码清单 15-73 配置线程池执行定时任务

```
<!-- 配置线程池--配置定时任务-->
<task:executor id="executor" pool-size="5" />
<!--配置线程池-->
<task:scheduler id="scheduler" pool-size="10" />
<!--设置自动发现这里的id为进行异步操作任务是Async异步注解应该指定的属性值-->
<task:annotation-driven executor="executor" scheduler="scheduler" />
<!-- 配置定时任务-->
```

这样可以根据异步执行任务的需要来配置线程池的大小，而不是使用 Spring 默认的线程池大小。

15.8 SpEL

SpEL 全称为 Spring Expression Language，是一种功能强大的表达式语言。SpEL 可以在运行时查询和操作对象图属性、进行对象方法的调用等，并且与 Spring 生态系统中所有的产品无缝对接，如用来配置 Bean 的定义和 Bean 中属性的值等。

SpEL 只是在 Spring 产品中被当作表达式求值的基础模块，本身可以脱离 Spring 单独使用，依赖 core 模块即可。为了体现它的独立性，本节将 SpEL 作为独立的表达式语言来进行讲解。单独讲解需要每次都进行创建一些基础框架类，如解析器。但是在实际项目中，开发者不需要关注这些基础框架类，仅需编写对应的字符串求值表达式。

15.8.1 SpEL 的功能特性

表达式语言一般能够用一些简单的形式来完成主要的工作，减少开发者的工作量。SpEL 支持的用法包括字符表达式，布尔和关系操作符，正则表达式，类型表达式，访问 properties、arrays、lists、maps 等集合，对象方法调用，使用关系操作符，赋值表达式，调用构造器，Bean 对象引用，创建数组，内联 lists、数组、maps，三元操作符，定义使用变量，用户自定义函数，集合投影，集合选择，模板表达式。

注意：在 SpEL 表达式中，关键字不区分大小写。

15.8.2 SpEL 的基础应用

代码清单 15-74 是使用 SpEL API 对字符串解析的示例。

代码清单 15-74 使用 SpEL 对字符串解析

```
package com.uifuture.spring.core.bean.spel;
import org.junit.Assert;
import org.junit.Test;
import org.springframework.expression.EvaluationContext;
import org.springframework.expression.Expression;
import org.springframework.expression.ExpressionParser;
```

```
import org.springframework.expression.spel.standard.SpelExpressionParser;
import org.springframework.expression.spel.support.StandardEvaluationContext;
public class SpELTest {
    @Test
    public void helloWorld1() {
        ExpressionParser parser = new SpelExpressionParser();
        // 连接字符串，并且通过#end赋值变量
        Expression expression = parser.parseExpression("('Hello' + ' World').concat
(#end)");
        EvaluationContext context = new StandardEvaluationContext();
        context.setVariable("end", "!");
        Assert.assertEquals("Hello World!", expression.getValue(context));
    }
}
```

SpEL 求表达式的值的过程一般分为 4 步：

①先构造一个解析器。

②然后使用解析器解析字符串表达式。

③构造上下文。

④最后根据上下文得到表达式运算后的值。

接下来按照这 4 步分析代码清单 15-74。

（1）构造解析器：new SpelExpressionParser();。SpEL 使用 ExpressionParser 接口表示解析器，提供 SpelExpressionParser 的默认实现。

（2）解析表达式：使用 ExpressionParser 的 parseExpression 来解析相应的字符串表达式为 Expression 对象。

（3）构造上下文：准备如变量定义等表达式需要的上下文数据。SpEL 提供 Standard Evaluation Context 来实现。

（4）求值：通过 Expression 接口的 getValue 方法根据上下文获得表达式值。

以上是简单的字符串解析，接下来介绍 SpEl 的原理及接口。

15.8.3 SpEL 的原理及接口

SpEL 提供了一些接口，从而可以便于开发者使用，首先了解以下 4 个概念。

（1）表达式：表达式是表达语言的核心，所有的表达式语言都是围绕着表达式来进行的，表达式语言定义了一些规则，从而可以让开发者方便地使用表达式，简单的理解就是定义了使用该表达式语言要做什么，即告知它需要做什么。

（2）解析器：用于将字符串表达式解析为表达式对象的对象。表达式用于表明做什么，解析器决定了由谁来做。

（3）上下文：表达式对象执行的环境，该环境可能定义变量，定义自定义函数，提供类型转换等，简单的理解就是解析器在哪里执行表达式。

（4）根对象及活动上下文对象：根对象是默认的活动上下文对象，活动上下文对象表示了当前表达式操作的对象，也就是对谁做什么。

理解了前面的概念，下面来了解 SpEL 的工作原理。

首先，定义一个表达式“1+1”；接下来定义解析器 ExpressionParser，SpEL 默认实现 SpelExpressionParser 解析器。SpelExpressionParser 解析器的内部使用 Tokenizer 类进行字符串词法分析，即把字符串流分析为记号流，记号在 SpEL 中使用 Token 类来表示。分析了记号流后，解析器就可以根据记号流生成内部抽象语法树。在 SpEL 中语法树节点由 SpelNode 接口实现类表示。例如，OpPlus 表示加操作节点、IntLiteral 表示 int 型字面量节点。使用 SpelNodel 实现类组成抽象语法树。SpEL 对外提供 Expression 接口简化表示抽象语法树，从而隐藏内部的实现细节，并且提供 getValue 方法来获取表达式值；SpEL 默认实现 SpelExpression。定义表达式上下文对象（可选），SpEL 使用 EvaluationContext 接口表示上下文对象，用于设置根对象、自定义变量、自定义函数、类型转换器等。SpEL 默认实现 StandardEvaluationContext。使用表达式对象根据上下文对象(可选）求值（调用表达式对象的 getValue 方法）获得结果。

以下是 SpEL 中的主要接口。

1.ExpressionParser 接口

ExpressionParser 接口用来解析字符串表达式，该接口的源码如代码清单 15-75 所示。

代码清单 15-75　ExpressionParser 接口的源码

```
package org.springframework.expression;
public interface ExpressionParser {
   Expression parseExpression(String expressionString) throws ParseException;
    Expression parseExpression(String expressionString, ParserContext context)
throws ParseException;
}
```

ExpressionParser 接口的默认实现是 org.springframework.expression.spel.standard 包中的 Spel ExpressionParser 类，使用 parseExpression 方法可以将字符串表达式转换为 Expression 对象，对于 org.springframework.expression.ParserContext 接口而言，用于定义字符串表达式是不是模板，以及定义模板开始字符与结束字符。在 parseExpression 方法中选择传入模块判断，ParserContext 可以定义表达式是模块，且定义模块的开始字符和结束字符。默认只传入字符串而非模板形式。

2.EvaluationContext 接口

EvaluationContext 接口表示上下文环境，在求值表达式中需要解析属性、方法、字段的值以及类型转换时会用到。其默认实现类 StandardEvaluationContext 使用反射机制来操作对象。为获得更好的性能缓存了 java.lang.reflect.Method、java.lang.reflect.Field 和 java.lang.reflect.Constructor 实例。

在 StandardEvaluationContext 中可以使用 setRootObject 方法显式设置根对象，或通过构造器直接传入根对象，还可以通过调用 setVariable 和 registerFunction 方法指定在表达式中用到的变量和函数。使用 StandardEvaluationContext 可以注册自定义的构造器解析器（ConstructorResolvers）、方法解析器（MethodResolvers）和属性存取器（PropertyAccessor）来扩展 SpEL 计算表达式。

使用 StandardEvaluationContext 创建相对比较耗资源，在重复使用的场景下内部会缓存部分中间状态以加快后续的表达式求值效率。因此，建议在使用过程中尽可能被缓存和重用，而不是每次在表达式求值时都重新创建一个对象。

3.Expression 接口

Expression 接口表示表达式对象，默认实现是 org.springframework.expression.spel.standard 包中的SpelExpression，提供的getValue方法用于获取表达式值，提供的setValue方法用于设置对象值。

15.8.4 SpEL 相关语法

下面介绍几种 SpEL 相关语法。

1. 字面量表达式

SpEL 中支持的一些字面量包括字符串、数字类型（int、long、float、double）、布尔类型、null 类型（null 对象）。字面量表达式也是 SpEL 中最简单的表达式。代码清单 15-76 所示为字面量表达式的 double 和 null 类型。

代码清单 15-76 字面量表达式的 double 和 null 类型

```
@Test
public void spelGrammarTest() {
    ExpressionParser parser = new SpelExpressionParser();
    Expression expression = parser.parseExpression("2.5");
    System.out.println(expression.getValue(Double.class));
    expression = parser.parseExpression("null");
    Object obj = expression.getValue();
    // 对象为null，输出true
    System.out.println(obj==null);
    expression = parser.parseExpression("null");
    String str = expression.getValue(String.class);
    // 字段串的值为null
    System.out.println(str);
}
```

2. 算术运算表达式

SpEL 支持加（+）、减（－）、乘（*）、除（/）、求余（%）、幂（^）运算。SpEL 另外提供了求余（MOD）和除（DIV）这两个运算符，与“%”和“/”等价，不区分大小写。代码清单 15-77 所示为算术运算表达式的（DIV）和（MOD）。

代码清单 15-77 算术运算表达式的（DIV）和（MOD）

```
@Test
public void spelArithmeticTest() {
    ExpressionParser parser = new SpelExpressionParser();
    Expression expression = parser.parseExpression("5.0/3.0");
    System.out.println(expression.getValue(Double.class));
    expression = parser.parseExpression("7 MOD 4");
```

```
        System.out.println(expression.getValue(Integer.class));
    }
```

需要注意的是，如果表达式“5.0/3.0”写成了“5/3”，那么计算的值为 int 型，与 Java 一致。

输出结果如下：

```
1.6666666666666667
3
```


3. 关系表达式

等于（==）、不等于（!=）、大于（>）、大于等于（>=）、小于（<）、小于等于（<=），区间（between）运算。下面演示大于（>）、等于（==），以及区间的关系表达式，如代码清单 15-78 所示。

代码清单 15-78　大于（>）、等于（==）和区间（between）的关系表达式

```
@Test
public void spelRelationalTest() {
    ExpressionParser parser = new SpelExpressionParser();
    Expression expression = parser.parseExpression("4>=5");
    System.out.println(expression.getValue(Boolean.class));
    expression = parser.parseExpression("1 between {1, 2}");
    System.out.println(expression.getValue(Boolean.class));
}
```

between 运算符右边的操作数必须是列表类型，且只能包含 2 个元素。第 1 个元素为开始，第 2 个元素为结束，区间运算是包含边界值的，即为左右闭合区间。

SpEL 同样提供了等价的 EQ 、NE、GT、GE、LT、LE 来表示等于、不等于、大于、大于等于、小于、小于等于，同样是不区分大小写的。

演示结果如下：

```
false
true
```

4. 逻辑表达式

逻辑表达式包括且（and）、或 (or)、非 (! 或 NOT)，返回值为布尔类型。

注意：逻辑运算符不支持 Java 中的“&&”和“||”。

5. 字符串连接及截取表达式

使用“+”号可以进行字符串的连接；使用“'String'[index]”可以截取一个字符，index 表示从 0 开始，字符串的字符位。目前只支持截取一个字符，如“'Hello ' + 'World!'”得到“Hello World!”；而“'Hello World!'[1]”将只返回 e。

6. 三目运算及 Elivis 运算表达式

三目运算表达式“表达式 1? 表达式 2: 表达式 3”用于构造三目运算表达式，如“1>2?true:false”将返回 false。

Elivis 运算表达式“表达式 1?: 表达式 2”是从 Groovy 语言引入用于简化三目运算符的，当表达式 1 为非 null 时则返回表达式 1，当表达式 1 为 null 时则返回表达式 2，简化了三目运算表达式。例如，“null?:false”将返回 false，而“true?:false”将返回 true，相当于三目运算符的 null 值判断。

7. 正则表达式

正则表达式 str matches regex，返回值为布尔类型。用于判断是否匹配正则。如“'123' matches '\\d{3}'”将返回 true。

8. 括号优先级表达式

表达式用括号括起来，即使用（表达式）构造，括号里的表达式具有高优先级。

9. 集合

集合（List）可以用大括号“{}”进行直接引用，元素之间使用英文逗号进行分隔，如代码清单 15-79 所示。

代码清单 15-79　集合元素

```
@Test
public void spelListTest() {
    ExpressionParser parser = new SpelExpressionParser();
    EvaluationContext context = new StandardEvaluationContext();
    Expression expression = parser.parseExpression("{1,2,3,4}");
    // 从Expression中获取list
    List numbers = (List) expression.getValue(context);
    System.out.println(numbers);
    // 集合嵌套
     List listOfLists = (List) parser.parseExpression("{{'a','b'},{'c','d'}}").
getValue(context);
    System.out.println(listOfLists);
}
```

“{}”本身代表一个空 list，因为性能的关系，如果列表本身完全由固定的常量值组成，这时会创建一个常量列表来代替表达式，而不是每次在求值时创建新列表。

10.Map

Map 可以直接通过 {key:value} 标记的方式在表达式中使用，如代码清单 15-80 所示。

代码清单 15-80　Map 使用的方式

```
public void spelMapTest() {
    ExpressionParser parser = new SpelExpressionParser();
    EvaluationContext context = new StandardEvaluationContext();
     Map inventorInfo = (Map) parser.parseExpression("{name:'jack',age:22}").
getValue(context);
    Map mapOfMaps = (Map) parser.parseExpression("{name:{first:'tom',last:'ming'},age
:{first:11,last:22}}").getValue(context);
```

```
    System.out.println(inventorInfo);
    System.out.println(mapOfMaps);
}
```

输出结果如下：

```
{name=jack, age=22}
{name={first=tom, last=ming}, age={first=11, last=22}}
```

"{:}"本身代表一个空的 Map。因为性能的原因，如果 Map 本身包含固定的常量或者其他级联的常量结构（lists 或者 maps），则会创建一个常量 Map 来代表表达式，而不是每次求值时都创建一个新的 Map。Map 的 Key 并不一定用引号引用，上面的例子就没有用引号。

11. 方法

方法可以使用典型的 Java 语法来调用,也可以直接在字符串常量上调用。方法也支持可变参数。

12. 类类型

使用 T(Type) 来表示 java.lang.Class 实例，Type 必须是类的全限定名，使用类类型表达式还可以访问类静态方法及类静态字段。

StandardEvaluationContext 使用 TypeLocator 来查找类型，其中 StandardTypeLocator（可以被替换为其他类）默认对 java.lang 包里的类型可见。也就是说 T() 引用 java.lang 包中的类型不需要限定包全名，但是其他类型的引用必须要。具体的使用方法如代码清单 15-81 所示。

代码清单 15-81　类类型的使用方法

```
public void testClassTypeExpressionTest() {
    ExpressionParser parser = new SpelExpressionParser();
    // 其他类的访问
     Class dateClass = parser.parseExpression("T(java.util.Date)").getValue(Class.
class);
    Assert.assertEquals(Date.class, dateClass);
    // ujava.lang包下的类访问
    Class stringClass = parser.parseExpression("T(String)").getValue(Class.class);
    Assert.assertEquals(String.class, stringClass);
    // 类静态字段访问
      int integer = parser.parseExpression("T(Integer).MAX_VALUE").getValue(int.
class);
    Assert.assertEquals(Integer.MAX_VALUE, integer);
    // 类静态方法调用
      int intParse = parser.parseExpression("T(Integer).parseInt('1')").
getValue(int.class);
    Assert.assertEquals(1, intParse);
}
```

13. 类的实例化

类的实例化可以使用 new 操作符。除了元数据类型和 string（如 int、float 等可以直接使用，

java.lang 包内的类型除外）都需要限定类的全名，类的实例化如代码清单 15-82 所示。

代码清单 15-82　类的实例化

```
public void testConstructorExpressionTest() {
    ExpressionParser parser = new SpelExpressionParser();
    String string = parser.parseExpression("new String('类的实例化')").getValue(String.class);
    Assert.assertEquals("类的实例化", string);
    Date date = parser.parseExpression("new java.util.Date()").getValue(Date.class);
    System.out.println(date);
}
```

14. 变量

表达式中的变量可以通过语法“# 变量名”使用。变量可以在 StandardEvaluationContext 中通过方法 setVariable 设置，如代码清单 15-83 所示。

代码清单 15-83　变量的使用

```
public void testVariablesTest() {
    ExpressionParser parser = new SpelExpressionParser();
    EvaluationContext context = new StandardEvaluationContext();
    context.setVariable("variable", "spring");
    context.setVariable(" name", "#variable");
    String name = parser.parseExpression("#variable").getValue(context, String.class);
    System.out.println(name);

    context = new StandardEvaluationContext("mybatis");
    System.out.println(parser.parseExpression("#root").getValue(context, String.class));
    System.out.println(parser.parseExpression("#this").getValue(context, String.class));
}
```

变量定义通过 EvaluationContext 接口的 setVariable(variableName, value) 方法定义，在表达式中使用“#variableName”引用。除了引用自定义变量,SpEL 还允许引用根对象及当前上下文对象。

#this 变量永远指向当前表达式正在求值的对象（这时不需要限定全名），变量 #root 总是指向根上下文对象。#this 在表达式不同部分解析过程中可能会改变，但是 #root 总是指向根。

15.9　<context:annotation-config> 标记

前面讲到关于 <context:component-scan> 的使用，它支持 <context:annotation-config> 标记的全部功能，而且支持指定包下扫描注册 Bean。在本节讲解 <context:annotation-config> 标记的功能。

15.9.1　<context:annotation-config> 标记的作用

当要 Spring 注入类时，需要在配置文件中配置 Bean 标记。如果需要注入的类非常多，对逐个类添加 Bean 标记非常麻烦，而使用注解可以解决这个麻烦。但是若使用 Autowired 注解注入，需要向 Spring 容器注入 Bean 处理器。

（1）使用 @Autowired，必须事先在 Spring 容器中声明 AutowiredAnnotationBeanPostProcessor 的 Bean，在配置文件中进行以下声明。

<bean class="org.springframework.beans.factory.annotation.AutowiredAnnotationBeanPostProcessor "/>

（2）使用 @Required 时，需要在 Spring 容器中声明 RequiredAnnotationBeanPostProcessor 的 Bean。

（3）使用 @PersistenceContext 时，需要在 Spring 容器中声明 PersistenceAnnotationBeanPostProcessor 的 Bean。

（4）使用 @Resource、@PostConstruct 和 @PreDestroy 等注解时，需要在 Spring 容器中声明 CommonAnnotationBeanPostProcessor 的 Bean。

在 Spring 容器中注册这 4 个 BeanPostProcessor（Bean 处理器）的作用是为了 Spring 在项目中能够识别相应的注解。

传统的注册声明方式如下：

```
<bean class="org.springframework.beans.factory.annotation. AutowiredAnnotationBe
anPostProcessor "/>
```

前面讲到 4 个 Bean 处理器处理不同的注解，要注册声明 4 个 Bean 处理器，这种方式太烦琐、不优雅，所以 Spring 提供了一种极为方便注册 BeanPostProcessor 的方式，即使用 <context:annotation-config/> 隐式地向 Spring 容器注册 AutowiredAnnotationBeanPostProcessor、RequiredAnnotationBeanPostProcessor、CommonAnnotationBeanPostProcessor 及 PersistenceAnnotationBeanPostProcessor 这 4 个 BeanPostProcessor。

Spring 会自动完成声明，并且还会自动搜索 @Component、@Controller、@Service、@Repository 等注解标注的类。

注意：在配置文件中使用 context 命名空间之前，必须在 <beans> 标记中声明 context 命名空间。

15.9.2 <context:annotation-config> 标记的源码分析

在 Spring 容器解析配置文件时，会先加载 <beans> 标记中配置的 URL，如代码清单 15-84 所示。

代码清单 15-84 加载 <beans> 标记中配置的 URL

```
<beans xmlns:xsi="http://www.w3.org/2001/XMLSchema-instance"
       xmlns="http://www.springframework.org/schema/beans"
       xmlns:context="http://www.springframework.org/schema/context"
       xsi:schemaLocation="http://www.springframework.org/schema/beans
        http://www.springframework.org/schema/beans/spring-beans.xsd
        http://www.springframework.org/schema/context
        http://www.springframework.org/schema/context/spring-context.xsd"
       >
```

这里介绍 context 命名空间。

（1）根据 <beans> 标记中配置的 xsi:schemaLocation 属性的 URL，到 spring-context 包下的 spring.handlers 文件中，找到对应的 NamespaceHandler，如图 15-7 所示。

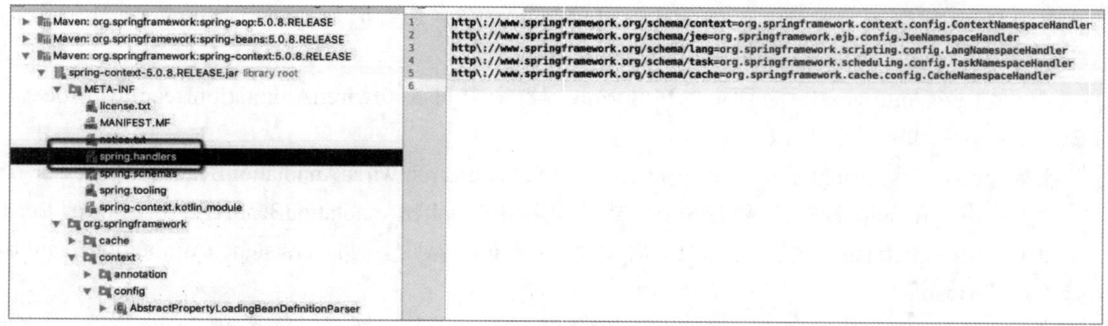

图 15-7 spring.handlers 文件

通过 spring.handlers 文件可以看到 context 对应的 NamespaceHandler 类 http://www.springframework.org/schema/context=org.springframework.context.config.ContextNamespaceHandler。

（2）根据 http://www.springframework.org/schema/context 可以找到 ContextNamespaceHandler 类。ContextNamespaceHandler 类的源码如代码清单 15-85 所示。

代码清单 15-85 ContextNamespaceHandler 类的源码

```
package org.springframework.context.config;
public class ContextNamespaceHandler extends NamespaceHandlerSupport {
    @Override
    public void init() {
        registerBeanDefinitionParser("property-placeholder", new PropertyPlaceholder
BeanDefinitionParser());
        registerBeanDefinitionParser("property-override", new PropertyOverrideBeanDe
finitionParser());
        registerBeanDefinitionParser("annotation-config",new AnnotationConfigBeanDefi
nitionParser());
        registerBeanDefinitionParser("component-scan", new ComponentScanBeanDefinitio
nParser());
        registerBeanDefinitionParser("load-time-weaver", new LoadTimeWeaverBeanDefini
tionParser());
        registerBeanDefinitionParser("spring-configured", new SpringConfiguredBeanDef
initionParser());
        registerBeanDefinitionParser("mbean-export", new MBeanExportBeanDefinitionParser());
        registerBeanDefinitionParser("mbean-server", new MBeanServerBeanDefinitionParser());
    }
}
```

ContextNamespaceHandler类只有一个init方法，该方法的作用是创建这些标记所对应的解析类。

解析类就是所有的自定义命名空间（如 mvc、context 等）下的标记解析都是由 Bean Definition sParser 接口的子类来完成的。

选择一个标记，可以看到 annotation-config 的解析类是 AnnotationConfigBeanDefinitionParser，接下来进入该类看它是如何解析这个标记的。AnnotationConfigBeanDefinitionParser 类的源码如代码清单 15-86 所示。

代码清单 15-86 AnnotationConfigBeanDefinitionParser 类的源码

```
public class AnnotationConfigBeanDefinitionParser implements BeanDefinitionParser {
   @Override
   @Nullable
   public BeanDefinition parse(Element element, ParserContext parserContext) {
      Object source = parserContext.extractSource(element);
      Set<BeanDefinitionHolder> processorDefinitions =
            AnnotationConfigUtils.registerAnnotationConfigProcessors(parserContext.
getRegistry(), source);
      CompositeComponentDefinition compDefinition = new CompositeComponentDefinitio
n(element.getTagName(), source);
      parserContext.pushContainingComponent(compDefinition);
      for (BeanDefinitionHolder processorDefinition : processorDefinitions) {
         parserContext.registerComponent(new BeanComponentDefinition(processorDefinition));
      }
      parserContext.popAndRegisterContainingComponent();
      return null;
   }
}
```

该类也只有一个 parse 方法，也就是解析的方法。其中，代码 AnnotationConfigUtils.registerAnnotationConfigProcessors(parserContext.getRegistry(), source) 将注册获取到所有与 BeanPost Processors 相关的 Bean 定义。解析过程就在这个方法中，registerAnnotationConfigProcessors 方法的源码如代码清单 15-87 所示。

代码清单 15-87 registerAnnotationConfigProcessors 方法的源码

```
public static Set<BeanDefinitionHolder> registerAnnotationConfigProcessors(
        BeanDefinitionRegistry registry, @Nullable Object source) {
    DefaultListableBeanFactory beanFactory = unwrapDefaultListableBeanFactory(registry);
    if (beanFactory != null) {
        /*...省略一些代码
          实例化AutowiredAnnotationBeanPostProcessor */
         if (!registry.containsBeanDefinition(AUTOWIRED_ANNOTATION_PROCESSOR_BEAN_
NAME)) {
            RootBeanDefinition def = new RootBeanDefinition(AutowiredAnnotationBeanP
ostProcessor.class);
            def.setSource(source);
            beanDefs.add(registerPostProcessor(registry, def, AUTOWIRED_ANNOTATION_
PROCESSOR_BEAN_NAME));
        }
        //实例化RequiredAnnotationBeanPostProcessor
        if (!registry.containsBeanDefinition(REQUIRED_ANNOTATION_PROCESSOR_BEAN_NAME)) {
            RootBeanDefinition def = new RootBeanDefinition(RequiredAnnotationBeanPo
stProcessor.class);
            def.setSource(source);
            beanDefs.add(registerPostProcessor(registry, def, REQUIRED_ANNOTATION_
PROCESSOR_BEAN_NAME));
```

```
        }
        // 检查是否支持JSR-250，如果支持，则添加CommonAnnotationBeanPostProcessor
          if (jsr250Present && !registry.containsBeanDefinition(COMMON_ANNOTATION_
PROCESSOR_BEAN_NAME)) {
            // 实例化CommonAnnotationBeanPostProcessor
            RootBeanDefinition def = new RootBeanDefinition(CommonAnnotationBeanPost
Processor.class);
            def.setSource(source);
            beanDefs.add(registerPostProcessor(registry, def, COMMON_ANNOTATION_
PROCESSOR_BEAN_NAME));
        }
        // 检查是否支持JPA，如果支持，则添加 PersistenceAnnotationBeanPostProcessor
        if (jpaPresent && !registry.containsBeanDefinition(PERSISTENCE_ANNOTATION_
PROCESSOR_BEAN_NAME)) {
            RootBeanDefinition def = new RootBeanDefinition();
            try {
            def.setBeanClass(ClassUtils.forName(PERSISTENCE_ANNOTATION_PROCESSOR_
CLASS_NAME,
                    AnnotationConfigUtils.class.getClassLoader()));
            } catch (ClassNotFoundException ex) {
                throw new IllegalStateException(
                        "Cannot load optional framework class: " + PERSISTENCE_
ANNOTATION_PROCESSOR_CLASS_NAME, ex);
            }
            def.setSource(source);
                beanDefs.add(registerPostProcessor(registry, def, PERSISTENCE_
ANNOTATION_PROCESSOR_BEAN_NAME));
        }
        if (!registry.containsBeanDefinition(EVENT_LISTENER_PROCESSOR_BEAN_NAME)) {
            RootBeanDefinition def = new RootBeanDefinition(EventListenerMethodProce
ssor.class);
            def.setSource(source);
                beanDefs.add(registerPostProcessor(registry, def, EVENT_LISTENER_
PROCESSOR_BEAN_NAME));
        }
        if (!registry.containsBeanDefinition(EVENT_LISTENER_FACTORY_BEAN_NAME)) {
            RootBeanDefinition def = new RootBeanDefinition(DefaultEventListenerFactory.class);
            def.setSource(source);
                beanDefs.add(registerPostProcessor(registry, def, EVENT_LISTENER_
FACTORY_BEAN_NAME));
        }
        return beanDefs;
    }
}
```

通过该方法，可以明确看到注册 AutowiredAnnotationBeanPostProcessor、RequiredAnnotation BeanPostProcessor、CommonAnnotationBeanPostProcessor 和 PersistenceAnnotationBeanPostProcessor 这 4 个 BeanPostProcessor 的代码。

第16章

Spring AOP 详解及案例分析

本章要点

1. Spring AOP 相关概念
2. Spring AOP 核心接口及类
3. 使用 XML 配置方式演示 Spring AOP 实例
4. 使用注解方式演示 Spring AOP 实例

16.1　了解 AOP

前面已经介绍了 Spring IoC，如果说 IoC 是 Spring 思想的核心，那么 AOP 便是 Spring 功能的核心。本节先介绍 Spring 的 AOP，通过讲解基础概念进行实例和源码的理解。

在 1.1.3 小节中简单介绍了 AOP（面向切面编程），在 2.3 节中讲解了三种动态代理。有了前面的知识铺垫，学习本章内容相对比较容易。

在介绍 AOP 前，很多书籍都会先介绍 OOP。OOP 是面向对象编程，AOP 就是对 OOP 的补充和完善。OOP 使用继承、封装、多态来处理业务模块和逻辑，而 AOP 则用来封装非业务但是被业务模块频繁调用的功能。在实际的业务场景中，通常会遇到一些公共逻辑的代码，如日志记录、事务管理、权限认证，利用 AOP 可以有效地减少系统中的冗余代码、降低模块之间的耦合和减小维护成本等。简单地理解，就是 AOP 可帮助开发者更好地抽象出代码，进行复用。

AOP 本身并不能解决具体问题，因为 AOP 是一种思想，解决问题的是 AOP 的具体实现，如 AspectJ、JBoss AOP，以及大家最熟悉的 Spring AOP。AOP 代理实现方式分为静态代理和动态代理。静态代理的代表为 AspectJ，动态代理的代表为 Spring AOP。Spring AOP 在 Spring 2.0 之后便开始集成了 AspectJ，从那以后，Spring AOP 便同时支持动态代理和静态代理的 AOP 实现方式。

16.1.1　Spring AOP 相关概念

1. 连接点（JoinPoint）

连接点是程序执行过程中明确的点，通俗地说就是哪些地方可以被拦截，哪些地方就是连接点，如方法的调用或者特定的异常被抛出。连接点又由两个信息确认：方法和相对点。方法表示程序执行点，即目标的方法；相对点表示方位，即目标方法的位置，如调用前、调用后等。可以理解为 Spring 允许使用额外逻辑的地方。根据代码清单 16-1 可以进一步了解。

代码清单 16-1　应用连接点

```
public void doAfter(JoinPoint jp) {//这里的jp参数便可以理解为一个连接点
    System.out.println(jp.getTarget().getClass().getName() + "." +
jp.getSignature().getName()+"方法执行完毕");
}
```

2. 通知（Advice）

通知是在某个特定的连接点上执行的动作，也就是指拦截到连接点之后需要执行的代码。在需要调用时，在连接点被拦截后便会调用。通知分为 5 种类型：

（1）Before：前置通知，在方法被调用之前进行调用通知 。

（2）After：最终通知，在方法完成后进行调用通知，无论方法是否执行成功或者抛出异常。

（3）After-returning：后置通知，在方法成功执行之后进行调用通知。

（4）After-throwing：异常通知，在方法抛出异常后进行调用通知。

（5）Around：环绕通知，包含了被通知的方法，在被通知方法调用之前和调用之后执行自定义行为逻辑。

应用通知的代码如代码清单 16-2 所示。

代码清单 16-2　应用通知

```
<aop:config>
    <aop:aspect id="logAspect" ref="aspectBean">
        <aop:before pointcut-ref="servicePointcut" method="doBefore"/>
   <!-- 这里的aop:before说明这是前置通知，doBefore是要执行的方法，aspectBean便是切面-->
    </aop:aspect>
</aop:config>
<bean id="aspectBean" class="com.uifuture.spring.aop.PrintLog" />
```

3. 切点（Pointcut）

切点就是被拦截的连接点，也就是后续进行添加通知的位置。一般来说，在 Spring 项目中，可以认为所有方法都是连接点，假设一个类有 20 个方法，但是并不需要拦截所有方法，只需要拦截某个或者某几个方法，就需要切点。切点是对连接点的条件定义。切点的作用就是提供一组表达式规则来匹配连接点，给满足规则的连接点添加通知，如代码清单 16-3 所示。

代码清单 16-3　应用切点

```
<aop:config>
    <aop:aspect id="logAspect" ref="aspectBean">
        <!--配置com.spring.service包下所有类或接口的所有方法-->
          <aop:pointcut id="servicePointcut" expression="execution(* com.uifuture.
spring.aop.service.*.*(..))" />
        <!--这里配置的切点便是com.uifuture.spring.aop.service包下的所有类的所有方法 -->
</aop:config>

<bean id="aspectBean" class="com.uifuture.spring.aop.PrintLog" />
```

4. 切面（Aspect）

切面可以理解为通知和切点的结合。切面是对横切关注点的抽象，要对哪些方法进行拦截，拦截后怎么处理，将整个过程定义为切面。连接点其实并不是 Spring AOP 中的概念，连接点这个概念只是为能更好地理解切点才有的。

通知说明了做什么和什么时候做，切点说明了在哪个地方做，而合起来就是一个完整的切面定义。

5. 目标对象（Target）

目标对象指将要被增强功能的对象，也就是包含了业务逻辑的类。简单地理解，就是在业务中需要被实际增强的业务对象。

6. 织入（Weaving）

织入是一个过程，表示将切面连接到其他的应用程序或者对象上，并且创建一个被通知对象的过程。简单来说就是把切面作用到目标对象上，然后产生一个代理对象的过程。织入可以在编译时、类加载运行时完成。根据织入的时机，可以用来区分是静态代理和动态代理。在编译时进行织入便是静态代理方式，在加载运行时进行织入则是动态代理方法。

7. 引入（Introduction）

引入可以让开发者向现有的类（被代理的对象）添加新的方法或者属性。也就是可以给一个类型进行额外的方法或者属性声明。简单地说，就是可以在运行时，将一个对象转换为实现另一个接口的对象，那么就可以使用另一个接口的方法了。通过 @DeclareParents 配置使用，后面会讲到具体例子。

16.1.2 Spring AOP 核心接口和类

前面介绍了一些与 AOP 相关的概念，现在再来了解与 Spring AOP 相关的核心接口和类。

1.InvocationHandler 接口

InvocationHandler 接口是 Java 动态代理机制中非常重要的接口，是实现动态代理的核心。该接口是 proxy 代理实例调用处理程序实现的一个接口，每个 proxy 代理实例都会有一个相关联的调用处理程序；当代理实例调用方法时，方法调用会被分派到调用处理程序的 invoke 方法。InvocationHandler 接口源码如代码清单 16-4 所示。

代码清单 16-4 InvocationHandler 接口源码

```
public interface InvocationHandler
extends Callback
{
     public Object invoke(Object proxy, Method method, Object[] args) throws
Throwable;
}
```

InvocationHandler 接口只有一个 invoke 方法，其中的参数如下：

- proxy：代理类代理的真实对象。
- method：被调用的某个对象真实方法的 Method 对象。
- args：被代理对象方法需传递的参数。

每一个动态代理类的调用处理程序都必须要实现 InvocationHandler 接口，并且每个代理类的实例都关联到实现该接口的动态代理类调用处理程序。当通过动态代理对象调用一个方法时，该方法的调用便会转发到实现 InvocationHandler 接口的 invoke 方法。

2.AopProxy 接口

AopProxy 接口是 Spring AOP 提供的代理类，通过该接口的实现类可以获取代理类。AopProxy 接口的实现类有三个，分别为 CglibAopProxy、JdkDynamicAopProxy、ObjenesisCglibAopProxy。AopProxy 接口的源码如代码清单 16-5 所示。

代码清单 16-5　AopProxy 接口的源码

```
public interface AopProxy {
    /** 获取一个代理对象*/
    Object getProxy();
    /** 根据类加载器获取代理对象 */
    Object getProxy(@Nullable ClassLoader classLoader);
}
```

AopProxy 接口实现类的功能就是生成目标代理类。

3.AopProxyFactory 接口

AopProxyFactory 接口可生成目标代理类的工厂类。该接口只有一个方法，其作用为获取 AopProxy 实现类，其源码如代码清单 16-6 所示。

代码清单 16-6　AopProxyFactory 接口的源码

```
public interface AopProxyFactory {
    AopProxy createAopProxy(AdvisedSupport config) throws AopConfigException;
}
```

在 Spring 5.X 中，AopProxyFactory 接口只有一个实现类 DefaultAopProxyFactory。DefaultAopProxyFactory 实现了接口中的 createAopProxy 方法，定义了在什么情况下使用 JdkDynamicAopProxy 或 ObjenesisCglibAopProxy。DefaultAopProxyFactory 类的源码如代码清单 16-7 所示。

代码清单 16-7　DefaultAopProxyFactory 类的源码

```
public class DefaultAopProxyFactory implements AopProxyFactory, Serializable {
     @Override
     public AopProxy createAopProxy(AdvisedSupport config) throws AopConfigException {
         //启用了优化配置||启用了直接代理目标类模式||没有指定要代理的接口
         if (config.isOptimize() || config.isProxyTargetClass() || hasNoUserSupplied
ProxyInterfaces(config)) {
             Class<?> targetClass = config.getTargetClass();
             if (targetClass == null) {
                 throw new AopConfigException("TargetSource cannot determine target class: " +
                         "Either an interface or a target is required for proxy creation.");
             }
             if (targetClass.isInterface() || Proxy.isProxyClass(targetClass)) {
                 //返回JDK动态代理对象
                 return new JdkDynamicAopProxy(config);
             }
             //返回CGLIB动态代理的对象
             return new ObjenesisCglibAopProxy(config);
         }
         else {
             //返回JDK动态代理对象
             return new JdkDynamicAopProxy(config);
         }
     }
```

```
    private boolean hasNoUserSuppliedProxyInterfaces(AdvisedSupport config) {
        Class<?>[] ifcs = config.getProxiedInterfaces();
            return (ifcs.length == 0 || (ifcs.length == 1 && SpringProxy.class.
isAssignableFrom(ifcs[0])));
    }
}
```

从 DefaultAopProxyFactory 类的源码中可以看出，Spring AOP 默认应用 JDK 动态代理来实现 AOP 功能，如果需要使用 CGLIB 动态代理来实现 AOP 功能，则可以将 proxyTargetClass 的值设置为 true，通过 <aop:aspectj-autoproxy proxy-target-class="true"/> 或者 @EnableAspectJAutoProxy(proxyTargetClass = true) 实现。但根据源码可以看出，即使是强制了 CGLIB 动态代理的实现，当目标类是接口时，Spring 选择的依旧是 JDK 动态代理（Java 原生支持）。

4.JdkDynamicAopProxy 类

JdkDynamicAopProxy 类是一个代理类，实现了 AopProxy 和 InvocationHandler 接口，使用 JDK 自带的动态代理机制代理目标类。通过 AopProxy 接口的 getProxy 方法返回代理对象。JdkDynamicAopProxy 类的 getProxy 方法的源码如代码清单 16-8 所示。

代码清单 16-8　JdkDynamicAopProxy 类的 getProxy 方法的源码

```
@Override
public Object getProxy() {
   return getProxy(ClassUtils.getDefaultClassLoader());
}
@Override
public Object getProxy(@Nullable ClassLoader classLoader) {
   if (logger.isDebugEnabled()) {
      logger.debug("Creating JDK dynamic proxy: target source is " + this.advised.
getTargetSource());
   }
    Class<?>[] proxiedInterfaces = AopProxyUtils.completeProxiedInterfaces(this.
advised, true);
  /*查找被代理的接口中是否包含 equals方法和hashCode方法，如果有，则在代理类中进行相应的标记，
    从而在invoke方法中直接调用相关的方法，避免递归调用导致栈溢出*/
    findDefinedEqualsAndHashCodeMethods(proxiedInterfaces);
   // 生成JDK动态代理对象
    return Proxy.newProxyInstance(classLoader, proxiedInterfaces, this);
}
```

通过 AopProxyUtils.completeProxiedInterfaces 方法，可以获取需要代理的所有接口，该方法的源码如代码清单 16-9 所示。

代码清单 16-9　AopProxyUtils.completeProxiedInterfaces 方法的源码

```
static Class<?>[] completeProxiedInterfaces(AdvisedSupport advised, boolean
decoratingProxy) {
    //获取到所有被代理的接口集合
    Class<?>[] specifiedInterfaces = advised.getProxiedInterfaces();
```

```
    if (specifiedInterfaces.length == 0) {
        //通过AdviceSupport没有获取到目标对象的实现接口时，则通过直接通过target目标对象来获取
        Class<?> targetClass = advised.getTargetClass();
        if (targetClass != null) {
            if (targetClass.isInterface()) {
                advised.setInterfaces(targetClass);
            }
            else if (Proxy.isProxyClass(targetClass)) {
                advised.setInterfaces(targetClass.getInterfaces());
            }
            specifiedInterfaces = advised.getProxiedInterfaces();
        }
    }
    /*是否追加SpringProxy，在AdvisedSupport的isInterfaceProxied方法中会判断传入的接口是否
已经由目标对象实现。此处传入SpringProxy.class判断目标对象是否已经实现该接口，如果没有实现则在代
理对象中需要新增SpringProxy，如果实现了则不必新增 */
    boolean addSpringProxy = !advised.isInterfaceProxied(SpringProxy.class);
    /*是否追加Adviced接口(注意不是Advice通知接口)。ProxyConfig的isOpaque方法用于返回由这个
配置创建的代理对象是否应该避免被强制转换为Advised类型。还有一个条件和上面的方法一样，同理，传入
Advised.class判断目标对象是否已经实现该接口，如果没有实现则在代理对象中需要新增Advised，如果实
现了则不必新增 */
    boolean addAdvised = !advised.isOpaque() && !advised.isInterfaceProxied
(Advised.class);
    /*是否追加DecoratingProxy接口，同样的判断条件有两个，第一个参数decoratingProxy，在调用
completeProxiedInterfaces方法时传入的是true；第二个判断条件和上面一样，判断被代理的目标对象是否
已经实现了DecoratingProxy接口。通常情况下这个接口也会被加入代理对象中，这是Spring 4.3新增的 */
    boolean addDecoratingProxy = (decoratingProxy && !advised.isInterfaceProxied(De
coratingProxy.class));
    int nonUserIfcCount = 0;
    if (addSpringProxy) {
        nonUserIfcCount++;
    }
    if (addAdvised) {
        nonUserIfcCount++;
    }
    if (addDecoratingProxy) {
        nonUserIfcCount++;
    }
    Class<?>[] proxiedInterfaces = new Class<?>[specifiedInterfaces.length +
nonUserIfcCount];

    //复制目标对象的接口
    System.arraycopy(specifiedInterfaces, 0, proxiedInterfaces, 0,
specifiedInterfaces.length);
    //下面便是追加这三个接口到proxiedInterfaces数组中并返回
    int index = specifiedInterfaces.length;
    if (addSpringProxy) {
        proxiedInterfaces[index] = SpringProxy.class;
        index++;
```

```
        }
        if (addAdvised) {
            proxiedInterfaces[index] = Advised.class;
            index++;
        }
        if (addDecoratingProxy) {
            proxiedInterfaces[index] = DecoratingProxy.class;
        }
        return proxiedInterfaces;
    }
```

completeProxiedInterfaces 方法首先获取所有要代理的接口，默认情况下 SpringProxy、Advised、DecoratingProxy 接口会添加到配置的代理接口后面，SpringProxy 仅仅是一个标记接口，它的作用就是标记当前对象是不是由 Spring 生成的一个代理；Advised 接口是为了使代理类可以动态操作其 AOP 通知，用于封装生成代理对象所需要的所有信息；DecoratingProxy 是 Spring 4.3 之后新增的接口，用于返回当前代理对象的最终对象的 Class 类型，便于核心模块自检。

通过前面介绍，基本知道了 JdkDynamicAopProxy 代理类的生成逻辑。接下来介绍该类中最核心的 invoke 方法，如代码清单 16-10 所示。

代码清单 16-10 JdkDynamicAopProxy 代理类的 invoke 方法

```
    @Override
    @Nullable
    public Object invoke(Object proxy, Method method, Object[] args) throws Throwable {
        MethodInvocation invocation;
        Object oldProxy = null;
        boolean setProxyContext = false;
        TargetSource targetSource = this.advised.targetSource;
        Object target = null;
        try {
            if (!this.equalsDefined && AopUtils.isEqualsMethod(method)) {
                // 不代理 equals 方法，通过 JdkDynamicAopProxy 中定义的 equals 方法进行比较
                return equals(args[0]);
            }
            else if (!this.hashCodeDefined && AopUtils.isHashCodeMethod(method)) {
                // 不代理 hashCode 方法，调用 JdkDynamicAopProxy 中定义的 hashCode 方法
                return hashCode();
            }
            else if (method.getDeclaringClass() == DecoratingProxy.class) {
                // DecoratingProxy 接口中只有 getDecoratedClass 方法，通过下面的方法实现其逻辑
                return AopProxyUtils.ultimateTargetClass(this.advised);
            }
            else if (!this.advised.opaque && method.getDeclaringClass().isInterface() &&
                    method.getDeclaringClass().isAssignableFrom(Advised.class)) {
                /*this.advised.opaque为false时才可能代理Advised接口。如果method是在Advised
    中声明的，则直接把method转移到Advised对象上调用。这样就可以通过代理对象直接操作代理的配置，如新
    增Advised。简单的理解就是，Spring AOP不会增强直接实现Advised接口的目标对象，再重复一次，也就是
    说如果目标对象实现的是Advised接口，则不会对其应用切面进行方法的增强*/
```

```
            return AopUtils.invokeJoinpointUsingReflection(this.advised, method, args);
        }
        //方法的返回值
        Object retVal;
        /*是否暴露代理对象，默认false可配置为true，如果暴露就意味着允许在线程内共享代理对象，注
意这是在线程内，也就是说同一线程的任意地方都能通过AopContext获取该代理对象*/
        if (this.advised.exposeProxy) {
            // 如果需要在拦截器中暴露 proxy 对象，则把 proxy 对象添加到 ThreadLocal 中
            oldProxy = AopContext.setCurrentProxy(proxy);
            setProxyContext = true;
        }
        //通过目标源获取目标对象
        target = targetSource.getTarget();
        //获取目标对象Class对象
        Class<?> targetClass = (target != null ? target.getClass() : null);
        // 获取此方法的拦截器链
        List<Object> chain = this.advised.getInterceptorsAndDynamicInterceptionAdvic
e(method, targetClass);
        if (chain.isEmpty()) {
            // 如果拦截器集合为空，说明当前 method 不需要被增强，则通过反射直接调用目标对象上的方法
            Object[] argsToUse = AopProxyUtils.adaptArgumentsIfNecessary(method, args);
            // 没有拦截器链则直接执行目标方法
            retVal = AopUtils.invokeJoinpointUsingReflection(target, method, argsToUse);
        }
        else {
             /* 创建 ReflectiveMethodInvocation，用来管理方法拦截器责任链.通过ReflectiveMethod Invocation.
proceed调用拦截器中的方法和目标对象方法*/
                invocation = new ReflectiveMethodInvocation(proxy, target, method,
args, targetClass, chain);
            //ReflectiveMethodInvocation对象完成对AOP功能实现的封装，并获取到返回值
            retVal = invocation.proceed();
        }
        // 返回 this，需要替换为 proxy 对象
        Class<?> returnType = method.getReturnType();
        if (retVal != null && retVal == target &&
                returnType != Object.class && returnType.isInstance(proxy) &&
                !RawTargetAccess.class.isAssignableFrom(method.getDeclaringClass())) {
            retVal = proxy;
        }
        else if (retVal == null && returnType != Void.TYPE && returnType.isPrimitive()) {
             throw new AopInvocationException("Null return value from advice does not match
primitive return type for: " + method);
        }
        return retVal;
    }
    finally {
        if (target != null && !targetSource.isStatic()) {
            targetSource.releaseTarget(target);
        }
```

```
        if (setProxyContext) {
            // 把第一次调用 setCurrentProxy 返回的对象，重新设置到 ThreadLocal 中
            AopContext.setCurrentProxy(oldProxy);
        }
    }
}
```

在 invoke 方法中，有 3 个比较关键的地方：

（1）this.advised.getInterceptorsAndDynamicInterceptionAdvice(method, targetClass)：获取拦截器链，相当于是获取潜在的增强方法，后续还有匹配的判断机制。

（2）AopUtils.invokeJoinpointUsingReflection(target, method, argsToUse)：没有获取到拦截器链，那么此时相当于直接调用目标对象的方法，invokeJoinpointUsingReflection 方法实际上是对 JDK 反射机制调用的一个封装。

（3）invocation.proceed：调用拦截器链中的增强方法和调用目标对象。代码清单 16-11 演示了 proceed 方法到底执行了什么动作。

代码清单 16-11　proceed 方法

```
/* 该方法的主要逻辑在于通过拦截器链，遍历其中的拦截器，再通过匹配判断是否适用，如果适用则取出拦截器
中的通知器并通过通知器的invoke方法进行调用，如果不适用则继续递归调用*/
@Override
@Nullable
public Object proceed() throws Throwable {
    if (this.currentInterceptorIndex == this.interceptorsAndDynamicMethodMatchers.
size() - 1) {
            /*调用完成所有拦截器链中的拦截器增强方法，则直接调用模板对象的方法并且退出。
invokeJoinpoint方法就直接调用了AopUtils.invokeJoinpointUsingReflection方法*/
        return invokeJoinpoint();
    }
    // 从拦截器链中获取拦截器

    Object interceptorOrInterceptionAdvice =
            this.interceptorsAndDynamicMethodMatchers.get(++this.currentInterceptorIndex);
/*InterceptorAndDynamicMethodMatcher代表运行时继续匹配再次执行match，匹配通过后执行拦截逻
辑，如果是普通的MethodInterceptor直接调用invoke */
    if (interceptorOrInterceptionAdvice instanceof InterceptorAndDynamicMethodMatcher) {
//在此处进行动态匹配，静态部分的匹配已被评估并找到匹配项
        InterceptorAndDynamicMethodMatcher dm =
                (InterceptorAndDynamicMethodMatcher) interceptorOrInterceptionAdvice;
            if (dm.methodMatcher.matches(this.method, this.targetClass, this.
arguments)) {
            // 如果和定义的切点匹配，则该通知便会执行
            return dm.interceptor.invoke(this);
        }
        else {
            // 如果没有使用的拦截进行递归，则继续匹配、判断和调用拦截器
            return proceed();
```

```
            }
        }
        else {
            // 判断出该拦截器是MethodInterceptor，直接进行调用
            return ((MethodInterceptor) interceptorOrInterceptionAdvice).invoke(this);
        }
    }
```

在 this.advised.getInterceptorsAndDynamicInterceptionAdvice 方法执行时，获取拦截器链，通过 DefaultAdvisorChainFactory 类的 getInterceptorsAndDynamicInterceptionAdvice 方法将注册的 Advice 和 Advisor 都转换为 MethodInterceptor 和 InterceptorAndDynamicMethodMatcher。因此，在 proceed 方法中只要不是 InterceptorAndDynamicMethodMatcher 便匹配 MethodInterceptor。

16.2 Spring AOP 实例分析

16.2.1 用 XML 方式解析 Spring AOP 实例

前面讲了这么多概念和源码分析，接下来便通过实例来深入分析 XML 文件、注解的配置方式。先配置 XML 配置文件，如代码清单 16-12 所示。

代码清单 16-12 配置 XML 配置文件

```
<beans xmlns:xsi="http://www.w3.org/2001/XMLSchema-instance"
       xmlns="http://www.springframework.org/schema/beans"
       xmlns:context="http://www.springframework.org/schema/context"
       xmlns:aop="http://www.springframework.org/schema/aop"
       xsi:schemaLocation="http://www.springframework.org/schema/beans
        http://www.springframework.org/schema/beans/spring-beans.xsd
        http://www.springframework.org/schema/context
        http://www.springframework.org/schema/context/spring-context.xsd
         http://www.springframework.org/schema/aop http://www.springframework.org/
schema/aop/spring-aop.xsd
">
</beans>
```

将需要的 AOP 命名空间设置好，接下来创建 Service 接口和 Service 实现类。这里创建的是 UserService 接口和 UserServiceImpl 类，如代码清单 16-13 所示和代码清单 16-14 所示。

代码清单 16-13 UserService 接口

```
package com.uifuture.spring.aop.service;
public interface UserService {
    String say(String name);
}
```

代码清单 16-14　UserServiceImpl 类

```
package com.uifuture.spring.aop.service.impl;
import com.uifuture.spring.aop.service.UserService;
public class UserServiceImpl implements UserService {
    @Override
    public String say(String name) {
        System.out.println("进入UserServiceImpl方法");
        return "say"+name;
    }
}
```

配置需要织入的代码，也就是前面概念所说的切面，如代码清单 16-15 所示。

代码清单 16-15　切面代码

```
package com.uifuture.spring.aop;
import org.aspectj.lang.JoinPoint;
import org.aspectj.lang.ProceedingJoinPoint;

public class PrintLog {
    /** 方法执行后 */
    public void doAfter(JoinPoint jp) {
            System.out.println(jp.getTarget().getClass().getName() + "." +
jp.getSignature().getName()+"方法执行完毕");
    }
    /** 环绕方法 */
    public Object doAround(ProceedingJoinPoint pjp) throws Throwable {
        long time = System.currentTimeMillis();
        Object retVal = pjp.proceed();
        time = System.currentTimeMillis() - time;
           System.out.println(pjp.getTarget().getClass().getName() + "." + pjp.
getSignature().getName()+"方法执行时间: " + time + " ms");
   System.out.println( "返回参数:"+retVal);
           return retVal;
       }
       /** 方法执行前 */
       public void doBefore(JoinPoint jp) {

           System.out.println( jp.getTarget().getClass().getName() + "." +
jp.getSignature().getName() +"方法即将执行");
   StringBuilder stringBuilder = new StringBuilder();
          Object[] objects = jp.getArgs();
          for (int i = 0; i < objects.length; i++) {
                         stringBuilder.append("参数     ").append(i).append(":").
append(objects[i]).append("\n");
          }
          System.out.println( "入参:"+stringBuilder);
       }
```

```
        /** 抛出异常执行 */
        public void doThrowing(JoinPoint jp, Throwable ex) {
                System.out.println(jp.getTarget().getClass().getName() + "." +
jp.getSignature().getName() +"方法抛出异常");
            System.out.println(ex.getMessage());
        }
    }
```

接下来在 Spring 的配置文件中注入 Bean、切面和切点，如代码清单 16-16 所示。

代码清单 16-16　注入 Bean、切面和切点

```
<aop:config>
    <aop:aspect id="logAspect" ref="aspectBean">
        <!--配置com.spring.service包下所有类或接口的所有方法-->
          <aop:pointcut id="servicePointcut" expression="execution(* com.uifuture.
spring.aop.service.*.*(..))" />
        <aop:before pointcut-ref="servicePointcut" method="doBefore"/>
        <aop:after pointcut-ref="servicePointcut" method="doAfter"/>
        <aop:around pointcut-ref="servicePointcut" method="doAround"/>
          <aop:after-throwing pointcut-ref="servicePointcut" method="doThrowing"
throwing="ex"/>
    </aop:aspect>
</aop:config>
<bean id="aspectBean" class="com.uifuture.spring.aop.PrintLog" />
<bean id="userServiceImpl" class="com.uifuture.spring.aop.service.impl.
UserServiceImpl"></bean>
```

配置中 pointcut 便是切入点，也就是需要拦截的方法。在这里配置了 Service 层下的所有类的所有方法。

通过 aop:* 标签，可以配置通知，这里有一点需要注意：如果某个方法配置为通知，那么该方法第 1 个参数便可以定义为 org.aspectj.lang.JoinPoint 类型。JoinPoint 提供了一系列可以获取方法信息、代理对象、返回模板的方法，并且其中的 getSignature 方法还可以强制转换为 MethodSignature 类型，通过该类型可以获取被代理方法的所有信息。

用 XML 配置 AOP 的源码如代码清单 16-17 所示。

代码清单 16-17　用 XML 配置 AOP 的源码

```
package com.uifuture.spring.aop.service.impl;
import com.uifuture.spring.aop.service.UserService;
import org.junit.Test;
import org.springframework.beans.factory.annotation.Autowired;
import org.springframework.context.ApplicationContext;
import org.springframework.context.support.ClassPathXmlApplicationContext;
public class UserServiceImplTest{
    @Test
    public void say() {
        ApplicationContext applicationContext = new ClassPathXmlApplicationContext("
spring-content.xml");
```

```
        UserService userService = applicationContext.getBean("userServiceImpl",
UserService.class);
        userService.say("你好");
    }
}
```

运行测试方法后的输出如下：

```
com.uifuture.spring.aop.service.impl.UserServiceImpl.say方法即将执行
入参为:参数0:你好
进入UserServiceImpl方法
com.uifuture.spring.aop.service.impl.UserServiceImpl.say方法执行时间: 0 ms
返回参数:say你好
com.uifuture.spring.aop.service.impl.UserServiceImpl.say方法执行完毕
```

16.2.2 用注解方式解析 Spring AOP 实例

接下来便解析如何通过注解实现 AOP 功能。相对于 XML，在现在的企业级项目中注解的使用更为频繁，因为其更加方便。接下来通过实例进行分析。

以房东、中介和租客为例，房东比较关注的是签合同和收房租，而带租客看房、讨论租房价格及房屋钥匙保管都可以交给中介。这也可以体现 AOP 的思想，让关注点与业务进行分离。这里的关注点是基于中介而言的关注点，下面用代码来分析。

1. 选择连接点

由于 Spring 是方法级别的 AOP 框架，所以选择某个类的某个方法作为连接点，如代码清单 16-18 所示。

代码清单 16-18　连接点代码与接口

```
public interface RentingService {
    /** 房东的核心业务功能 */
    void service();
}
@Service("rentingServiceImpl")
public class RentingServiceImpl implements RentingService {
    @Override
    public void service(){
        // 输出仅仅代表业务处理
        System.out.println("签合同...");
        System.out.println("收房租...");
    }
}
```

2. 创建切面

创建好连接点之后，就可以创建切面了。可以把切面理解为一个拦截器，当程序运行到连接点时，会被拦截下来，可以在方法调用前后加入自身的处理逻辑。在 Spring 中，使用 @Aspect 可以让 Spring IoC 容器认为这是一个切面，如代码清单 16-19 所示。

代码清单 16-19　创建切面

```
package com.uifuture.spring.aop.annotation.aspect;
import org.aspectj.lang.annotation.After;
import org.aspectj.lang.annotation.Aspect;
import org.aspectj.lang.annotation.Before;
import org.springframework.stereotype.Component;
/** 中介所需要做的事情*/
@Component
@Aspect
public class RentingAspect {
    @Before("execution(* com.uifuture.spring.aop.annotation.impl.RentingServiceImpl.
service())")
    public void before(){
        System.out.println("带租客看房");
        System.out.println("谈价格");
    }
    @After("execution(* com.uifuture.spring.aop.annotation.impl.RentingServiceImpl.
service())")
    public void after(){
        System.out.println("交钥匙");
    }
}
```

被定义为切面的类仍然是一个普通 Bean，需要使用 @Component 将其注册为 Spring 的 Bean。更多 Spring 中的 AspectJ 注解见表 16-1。

表 16-1　AspectJ 注解

注　解	说　明
@Before	前置通知，在连接点方法前调用
@Around	环绕通知，它将覆盖原有方法，但是允许通过反射机制调用原有方法
@After	后置通知，在连接点方法后调用
@AfterReturning	返回通知，在连接点方法处执行并正常返回后调用，要求连接点方法在执行过程中没有发生异常
@AfterThrowing	异常通知，当连接点方法抛出异常时调用

3. 定义切点

表 16-1 的注解汇总定义了 execution 的表达式，Spring 通过该表达式可以判断出具体要拦截哪个类的哪个方法。下面以表达式 @Before("execution(* com.uifuture.spring.aop.annotation.impl.RentingServiceImpl.service())") 为例进行分析。

- execution：表示执行方法时会进行触发。
- *：代表任意返回类型的方法。
- com.uifuture.spring.aop.annotation.impl.RentingServiceImpl：表示类的全限定名。
- service：被拦截的方法名称。

通过上面的表达式可以向 Spring 传达拦截 RentingServiceImpl 的 service 方法的诉求。如果有多个相同的表达式需要切面，那么可以使用 @Pointcut 来定义一个切点，避免重复编写表达式，如代码清单 16-20 所示。

代码清单 16-20　使用 Pointcut 注解避免重复编写表达式

```
package com.uifuture.spring.aop.annotation.aspect;
import org.aspectj.lang.annotation.After;
import org.aspectj.lang.annotation.Aspect;
import org.aspectj.lang.annotation.Before;
import org.aspectj.lang.annotation.Pointcut;
import org.springframework.stereotype.Component;
/** 中介所需要做的事情*/
@Component
@Aspect
public class RentingAspectPointcut {
        @Pointcut("execution(* com.uifuture.spring.aop.annotation.impl.
RentingServiceImpl.service())")
    public void pointcutService(){
    }
    @Before("pointcutService()")
    public void before(){
        System.out.println("带租客看房");
        System.out.println("谈价格");
    }
    @After("pointcutService()")
    public void after(){
        System.out.println("交钥匙");
    }

}
```

注意：不要忘记配置 Spring 配置文件，需要通过配置扫描注解、开启 Aspect 注解，如代码清单 16-21 所示。

代码清单 16-21　Spring 配置文件

```
<beans xmlns:xsi="http://www.w3.org/2001/XMLSchema-instance"
       xmlns:aop="http://www.springframework.org/schema/aop"
        xmlns="http://www.springframework.org/schema/beans" xmlns:context="http://
www.springframework.org/schema/context"
       xsi:schemaLocation="http://www.springframework.org/schema/beans
        http://www.springframework.org/schema/beans/spring-beans.xsd
         http://www.springframework.org/schema/aop http://www.springframework.org/
schema/aop/spring-aop.xsd http://www.springframework.org/schema/context http://
www.springframework.org/schema/context/spring-context.xsd">
    <aop:aspectj-autoproxy/>
    <context:component-scan base-package="com.uifuture">
    </context:component-scan>
</beans>
```

4. 测试

测试代码如代码清单 16-22 所示。

代码清单 16-22　测试代码

```
package com.uifuture.spring.aop.annotation;
import org.junit.Test;
import org.springframework.context.ApplicationContext;
import org.springframework.context.support.ClassPathXmlApplicationContext;
/**测试注解AOP方式 */
public class RentingServiceImplTest  {
    @Test
    public void service() {
        ApplicationContext applicationContext = new ClassPathXmlApplicationContext("
spring-content.xml");
/* RentingServiceImpl rentingServiceImpl = applicationContext.getBean("rentingServi
ceImpl",RentingService.class);
注意：不要使用实现类来进行强转，Spring AOP部分使用JDK或者CGLIB动态代理来为目标对象创建代理。如果
被代理的目标实现了至少一个接口，则会使用JDK动态代理。
 java.lang.ClassCastException: com.sun.proxy.$Proxy17 cannot be cast to com.
uifuture.spring.aop.annotation.impl.RentingServiceImpl */
        RentingService rentingService = applicationContext.getBean("rentingServiceI
mpl",RentingService.class);
        rentingService.service();
    }
}
```

输出结果如图 16-1 所示。

```
带租客看房
谈价格
签合同...
收房租...
交钥匙

Process finished with exit code 0
```

图 16-1　测试结果

Spring AOP 的出现极大地提升了项目的可扩展性和可维护性，这也是 AOP 最重要的价值所在。

第17章 Spring 的数据库事务管理

本章要点

1. 数据库事务的介绍
2. 事务的隔离模式
3. 事务并发问题与类型介绍
4. Spring 事务管理核心接口
5. Spring 事务使用案例演示
6. 编程式事务与声明式事务的演示
7. Transactional 注解讲解与实例演示

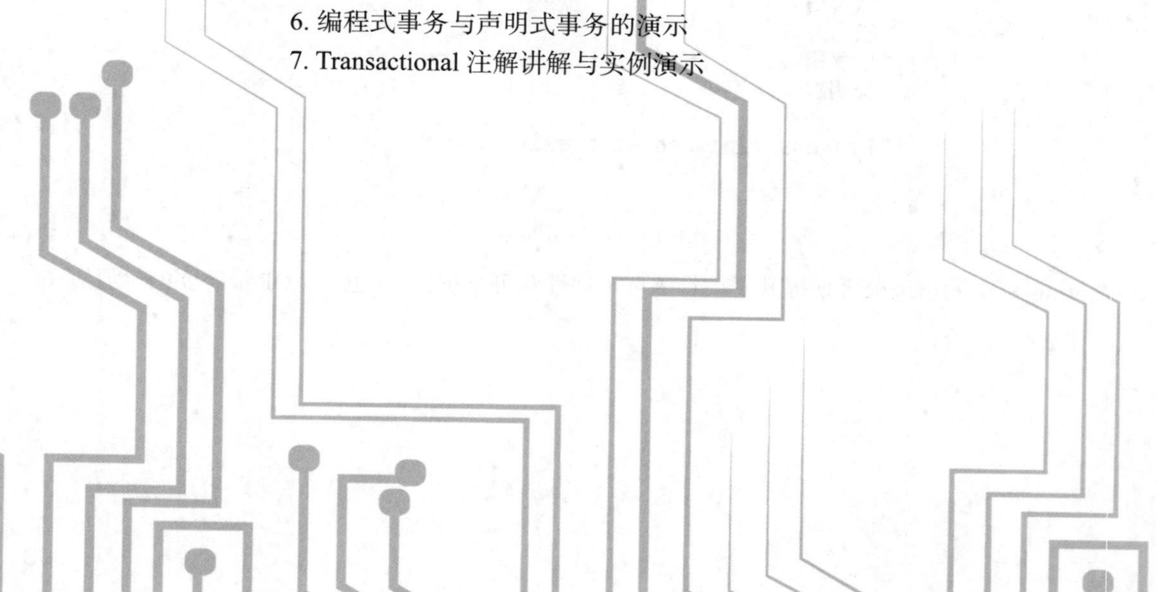

17.1 数据库事务基础

传统的 JDBC 数据库编程连接到数据库分为以下步骤：

（1）加载数据库驱动程序 JAR 文件。

（2）注册驱动器类。

JDBC 4 的驱动程序必须包括自动注册机制，如果驱动程序 JAR 中包含 METAINF/services/java.sql.Driver，则它会自动注册驱动器类。

还有几种手动注册驱动器的方法：在 Java 程序中使用反射 Class.forName("org.postgresql.Driver")；设置 jdbc.drivers 属性 System.setProperty("jdbc.drivers" , "org.postgresql.Driver") 或者在启动命令中使用 java -Djdbc.driver=org.postgresql.Driver ProgramName。

（3）设置数据库 URL。

设置数据库 URL 源码为 jdbc:mysql:xxx。这里 mysql 表示用于连接数据库的具体驱动程序，在这里表示 MySQL 数据库，而 xxx 是与驱动程序相关的。

（4）获取数据库链接，如通过 DriverManager 获取 Connection 对象。

（5）准备 SQL 语句，通过 Connection 对象执行 SQL 语句。

从上面的内容可以看出，如果要执行一条简单的 SQL 语句，步骤比较复杂，更不用说执行事务或者其他复杂的 SQL 操作。而使用 Spring 就不会如此，简单的一行注入，就能方便地操作数据库。

本节首先介绍数据库事务的一些基础知识，然后再讲解 Spring 的数据库事务操作，以及 Spring 事务的实现原理。

17.1.1 什么是事务

事务是一个逻辑工作单元，它可以包括一系列的操作。事务处理可以确保事务性单元内的所有操作都成功完成，否则不会永久更新面向数据的资源。也就是说，在事务处理中，所有的操作要么都执行，要么都不执行。通过将一组相关操作组合为一个要么全部成功要么全部失败的单元，可以简化错误恢复，使应用程序更加可靠。

事务包括 4 个基本特性，也就是常说的 ACID。

（1）Atomic（原子性，这里的“原子”即代表事务中的各个操作不可分割）：事务中包含的操作可看作一个逻辑单元，这个逻辑单元中的操作要么全部成功，要么全部失败。

（2）Consistency（一致性）：意味着只有合法的数据才会被写入数据库，否则会回滚到最初状态。事务确保数据库的状态从一个一致状态转变为另一个一致状态。

（3）Isolation（隔离性）：当多个事务并发执行时，一个事务的执行不应影响其他事务的执行。

（4）Durability（持久性）：已被提交的事务对数据库的修改应该永久保存在数据库中。

正是由于这几个特性，才使得 SQL 的性能并没有 NoSQL 这么高。当然，基于内存操作也是 NoSQL 更快的一个原因，在这里就不展开介绍了。

17.1.2 事务的隔离模式

为了避免并发事务处理下发生意外的情况，标准 SQL 规范中规定了 4 种事务隔离级别。

1. Read Uncommitted（读取未提交内容）

Read Uncommitted 是最低等级的事务隔离，即使另一个并发事务没有提交，当前的事务也可以读取到这个事务中数据的改变，非常不安全。Read Uncommitted 允许脏读（Drity Read），但是不允许更新丢失。如果另外一个事务已经开始写数据，当前事务则不允许同时进行写数据操作，但是允许当前事务进行读数据操作。

注意：除非开发者明确知道自己在做什么，并且有充分的理由选择这样做，否则不建议使用该等级的事务隔离级别。

2. Read Committed（读取已提交内容）

在 Read Committed 级别的事务隔离等级下，当前事务不会读到另一个事务已经修改但是未提交的数据。Read Committed 允许不可重复读取，但是不允许脏读。读数据操作的事务允许其他并发事务继续访问，但是未提交的写数据操作事务将会禁止其他事务访问。

注意：此级别的事务隔离等级是最常用的，并且是大多数数据库的默认隔离级别（不是 MySQL 的默认级别），同时也适用于大多数系统。它满足了隔离的早先定义：一个事务在开始时只能读取已经提交事务所做的改变，一个事务从开始到提交前，所做的任何数据改变都是不可见的，除非已经提交。这种隔离级别也支持所谓的“不可重复读”。这意味着用户运行同一语句两次，看到的结果是不同的。

3. Repeatable Read（可重复读取）

Repeatable Read 隔离级别表示同一事务先后执行相同的查询语句时，得到的结果是一样的。这也意味着一个事务不可能更新已经由另一个事务读取但未提交的数据。Repeatable Read 禁止不可重复读取和脏读，但是有时可能出现幻读。所谓幻读，是指当用户读取某一范围的数据行时，另一个事务又在修改该范围内插入新行，当用户再读取该范围的数据行时，会发现有新的“幻影”。可重复读取对于数据行数的查询结果是不保证的，这也是幻读的原因所在。读取数据的事务将会禁止写事务（但允许读事务），写事务则禁止任何其他事务。

注意：Repeatable Read 隔离级别解决了 Read Uncommitted 隔离导致的问题。它确保当同一事务的多个实例在并发读取数据时，会“看到同样的”数据行，不过这可能导致“幻读”。Repeatable Read 是 MySQL 的默认事务隔离级别（如 MySQL InnoDB 存储引擎）。

4. Serializable（序列化）

Serializable 是最高等级的事务隔离级别，提供了最严格的隔离机制，可以防止脏读，不可重复读取和幻读。Serializable 要求事务序列化执行，即事务只能一个接着一个地执行，不能并发执行。

注意：该级别可能导致大量的超时现象和锁竞争现象，使得数据库执行效率很慢，所以不建议对数据库事务使用该隔离级别。

17.1.3 事务并发的问题

由于事务隔离级别设置的不同，可能出现一些由并发引起的问题。

（1）丢失更新。

当两个或者多个事务选择同一行数据，然后更新该行数据时，会发生丢失更新问题。每个事务都不知道其他事务的存在。最后的更新将重写由其他事务所做的更新，这将导致数据丢失。

例如，事务 A 和事务 B 同时修改某行的值，将执行以下操作。

①事务 A 将值改为 1 并提交。

②事务 B 将值改为 2 并提交。

最后数据的值会为 2，事务 A 所做的更新将会丢失。

解决办法：可以对行加锁，只允许一个更新事务执行。

（2）脏读。

一个事务读取了另外一个事务没有提交的数据，所以可能会读取最后被另一个事务回滚的数据。

例如，值 a 原值为 100，事务 A 将值 a 修改为 200，但是事务 A 未提交，此时事务 B 读取 a 值，读取的值为 200。后续事务 A 发生异常进行了回滚，事务 B 就是产生了脏读。

（3）不可重复读。

在同一个事务中，两次读取同一数据，得到结果数据不同。即当在一个事务中再次读取之前读取过的数据时，发现该数据已经被另一个已提交的事务修改，这两次读取的数据不同。

例如以下过程：

①事务 1 读取了 a 值为 100，事务未提交。

②事务 2 修改 a 值为 200，并提交了事务。

③事务 1 再次读取 a 值时，a 的值变为了 200。

解决办法：如果只有在修改事务完全提交之后才可以读取数据，则可以避免该问题。

（4）幻读。

在同一个事务中，同样的查询语句前后进行查询，由于另外一个事务执行了插入操作，导致后一次得到的查询数据多出了被插入的数据。简单地说，就是一个事务读取了另一个事务提交的插入的数据。

例如，以下过程：

①事务 1 查询表中所有记录。

②事务 2 插入一条或一批记录。

③事务 2 调用 commit 进行提交事务。

④事务 1 再次查询表中所有记录（此时事务 1 两次查询到的记录是不一样的）。

注意：很多人容易搞混不可重复读和幻读，这两者确实有些相似。但不可重复读重点在于修改和删除，而幻读的重点在于插入新数据。

假设使用锁机制来实现这两种隔离级别，在事务第 1 次读取数据后，将其加锁，则其他事务无法修改这些数据，就可以实现可重复读。但这种方法却无法锁住插入的数据，所以当事务 A 先读取了数据，或者修改了全部数据，事务 B 还是可以插入数据提交，这时事务 A 就会发现莫名其妙多了一条之前没有的数据，这就是幻读，不能通过行锁来避免。

幻读需要 Serializable 隔离级别，读用读锁，写用写锁，读锁和写锁互斥，这样可以有效地避免幻读、不可重复读和脏读等问题，但也会极大地降低数据库的并发能力。

17.1.4 事务类型

数据库事务类型分为本地事务和分布式事务。

（1）本地事务：也就是普通事务，可以保证单台机器数据库上操作的 ACID，被限定在一台机器的数据库上。

（2）分布式事务：涉及两个或多个数据库源的事务，即跨越多台同类或异类数据库的事务（由每台数据库的本地事务组成），分布式事务旨在保证这些本地事务的所有操作的 ACID，使事务可以跨越多台数据库。

Java 事务类型分为 JDBC 事务和 JTA 事务。

（1）JDBC 事务：指数据库事务类型中的本地事务，通过 Connection 对象的控制来管理事务。

（2）JTA 事务：Java 事务 API（Java Transaction API），是 Java EE 数据库事务规范，JTA 只提供了事务管理接口，由应用程序服务器厂商（如 WebSphere Application Server）提供实现，JTA 事务比 JDBC 更强大，支持分布式事务。

Java EE 事务类型分为本地事务和全局事务。

（1）本地事务：使用 JDBC 编程实现事务。

（2）全局事务：由应用程序服务器提供，使用 JTA 事务。

Java EE 按是否通过编程实现，事务分为声明式事务和编程式事务两种。

（1）声明式事务：通过注解或 XML 配置文件指定事务信息。在 Spring 的配置文件中配置隔离级别，传播特性，并结合 Spring 的 AOP 功能对所配置的方法做动态代理来做到某方法中的多次数据库操作同时成功或者同时失败。

（2）编程式事务：通过编写代码实现事务。在方法内部对数据库的多次操作采用手工事务包裹的方式做到事务内的数据库操作同时成功或者同时失败。对于编程式事务管理，Spring 推荐使用 TransactionTemplate。

注意：关于声明式事务与编程式事务，简单地说，编程式事务侵入了业务代码，但是提供了更加详细的事务管理；声明式事务由于基于 AOP，可以既起到事务管理的作用，又可以不影响业务代码的具体实现。

17.2 Spring 对事务管理的支持

Spring 为事务的管理提供了一致的编程模板，在高层次建立了统一的事务抽象，不管选择 Spring JDBC、Hibernate、JPA，还是选择 MyBatis、Spring，都可以让开发者用统一的编程模型进行事务管理。

这种统一处理的方式带来的好处是用户可以完全抛开事务管理的问题，专注编写业务程序，只需要在 Spring 中通过配置完成事务的管理工作。

17.2.1 Spring 事务管理核心接口

在 Spring 事务管理 SPI（Service Provider Interface）的抽象层主要包括 3 个接口，分别是

PlatformTransactionManager 接口、TransactionDefinition 接口和 TransactionStatus 接口。接下来分别对这 3 个接口进行介绍。

1. PlatformTransactionManager 接口

Spring 并不直接管理事务，而是提供了多种事务管理器，它们将事务管理的职责委托给 MyBatis 或者 JTA 等持久化机制所提供的相关平台框架的事务来实现。

Spring 事务管理器的接口是 org.springframework.transaction.PlatformTransactionManager，通过该接口，Spring 为如 JDBC、MyBatis、Hibernate 等平台都提供了对应的事务管理器，但是由各个平台自身进行具体实现。其中，DataSourceTransactionManager 实现类便是使用 Spring JDBC 或者 MyBatis 进行持久化数据时使用的事务管理器。

Spring 的事务管理器 PlatformTransactionManager 接口中定义了 3 个方法，如代码清单 17-1 所示。

代码清单 17-1　PlatformTransactionManager 中的方法

```
// 基于事务的传播特性，返回一个已经存在的事务或者创建一个新的事务
TransactionStatus getTransaction(TransactionDefinition definition) throws
TransactionException;
// 提交事务
void commit(TransactionStatus status) throws TransactionException;
// 回滚事务
void rollback(TransactionStatus status) throws TransactionException;
```

通过 getTransaction 方法，再根据 TransactionDefinition 提供的事务属性配置信息创建事务，并用 TransactionStatus 描述该激活事务的状态。

2. TransactionDefinition 接口

事务管理器接口 PlatformTransactionManager 通过 getTransaction(TransactionDefinition definition) 方法来获取事务，参数是 TransactionDefinition 类，该类定义了一些基本的事务属性。TransactionDefinition 接口用于描述事务的隔离级别、超时时间、是否为只读事务和事务的传播规则等事务属性。

DefaultTransactionDefinition 类是 TransactionDefinition 接口的默认实现，它的传播行为是 PROPAGATION_REQUIRED（如果当前没事务，则要创建新事务，否则加入到当前事务中），隔离级别是数据库默认级别。

事务属性可以理解为事务的一些基本配置，描述了事务策略如何应用到方法上，包含 5 个方面：隔离级别、传播行为、回滚规则、是否只读、事务超时。

TransactionDefinition 接口中的方法如代码清单 17-2 所示。

代码清单 17-2　TransactionDefinition 接口方法

```
// 返回事务的传播行为
int getPropagationBehavior();
// 返回事务的隔离级别，事务管理器根据它来控制另外一个事务可以访问到本事务内的哪些数据
int getIsolationLevel();
// 返回事务的名字
String getName();
```

```
// 获取超时时间
int getTimeout();
// 返回是否优化为只读事务
boolean isReadOnly();
```

代码清单 17-2 给出了 TransactionDefinition 接口中的方法，另外在该接口中还有一些表示事务属性的常量，其中的事务隔离级别的常量如代码清单 17-3 所示。

代码清单 17-3　TransactionDefinition 接口中的事务隔离级别常量

```
/**
 * TransactionDefinition.ISOLATION_DEFAULT：使用后端数据库默认的隔离级别，MySQL 默认采用的是 REPEATABLE_READ隔离级别，Oracle 默认采用的是READ_COMMITTED隔离级别
 */
int ISOLATION_DEFAULT = -1;
/**
 * TransactionDefinition.ISOLATION_READ_UNCOMMITTED： 最低的隔离级别，允许读取尚未提交的数据变更，可能会导致脏读、幻读或不可重复读
 */
int ISOLATION_READ_UNCOMMITTED = Connection.TRANSACTION_READ_UNCOMMITTED;
/**
 * TransactionDefinition.ISOLATION_READ_COMMITTED：允许读取并发事务已经提交的数据，可以阻止脏读，但是幻读或不可重复读仍有可能发生
 */
int ISOLATION_READ_COMMITTED = Connection.TRANSACTION_READ_COMMITTED;
/**
 * TransactionDefinition.ISOLATION_REPEATABLE_READ：对同一字段的多次读取结果都是一致的，除非数据是被本身事务自身所修改，可以阻止脏读和不可重复读，但幻读仍有可能发生
 */
int ISOLATION_REPEATABLE_READ = Connection.TRANSACTION_REPEATABLE_READ;
/**
 * TransactionDefinition.ISOLATION_SERIALIZABLE：最高的隔离级别，完全服从ACID的隔离级别。所有的事务依次执行，这样事务就完全不可能互相干扰。也就是说，该级别可以防止脏读、不可重复读以及幻读。但是这将严重影响程序的性能。通常情况下也不会用到该级别
 */
int ISOLATION_SERIALIZABLE = Connection.TRANSACTION_SERIALIZABLE;
```

上面代码列出了 5 个表示隔离级别的常量。当事务方法被另一个事务方法调用时，必须指定事务应该如何传播。常见的情况是，新的方法可能继续在现有事务中运行，也可能开启一个新事务，并在自身的事务中运行。TransactionDefinition 接口中的事务传播行为的常量如代码清单 17-4 所示。

代码清单 17-4　TransactionDefinition 接口中的事务传播行为的常量

```
//支持当前事务的情况
/**
 * TransactionDefinition.PROPAGATION_REQUIRED: 如果当前存在事务，则加入该事务；如果当前没有事务，则创建一个新的事务
 */
int PROPAGATION_REQUIRED = 0;
```

```
/**
 * TransactionDefinition.PROPAGATION_SUPPORTS：如果当前存在事务，则加入该事务；如果当前没有
事务，则以非事务的方式继续运行
 */
int PROPAGATION_SUPPORTS = 1;
/**
 * TransactionDefinition.PROPAGATION_MANDATORY：如果当前存在事务，则加入该事务；如果当前没
有事务，则抛出异常(mandatory：强制性)
 */
int PROPAGATION_MANDATORY = 2;
//不支持当前事务的情况
/**
 * TransactionDefinition.PROPAGATION_REQUIRES_NEW：创建一个新的事务，如果当前存在事务，则
把当前事务挂起
 */
int PROPAGATION_REQUIRES_NEW = 3;
/**
 * TransactionDefinition.PROPAGATION_NOT_SUPPORTED：以非事务方式运行，如果当前存在事务，
则把当前事务挂起
 */
int PROPAGATION_NOT_SUPPORTED = 4;
/**
 * DTransactionDefinition.PROPAGATION_NEVER：以非事务方式运行，如果当前存在事务，则抛出异常
 */
int PROPAGATION_NEVER = 5;

// 其他情况
/**
 * TransactionDefinition.PROPAGATION_NESTED：如果当前存在事务，则创建一个事务作为当前事务的
嵌套事务来运行；如果当前没有事务，则该取值等价于TransactionDefinition.PROPAGATION_REQUIRED
 */
int PROPAGATION_NESTED = 6;
```

在这里需要注意的是，前面支持当前事务的常量与不支持当前事务的常量，都是 Spring 从 EJB 中引入的，它们共享相同的概念。

PROPAGATION_NESTED 是 Spring 特有的。以 PROPAGATION_NESTED 启动的事务内嵌于外部事务中（如果存在外部事务），此时，内嵌事务并不是一个独立的事务，它依赖于外部事务，只有通过外部事务的提交，才能引起内部事务的提交，嵌套的子事务不能单独提交。另外，外部事务的回滚也会导致嵌套子事务的回滚。

3. TransactionStatus 接口

TransactionStatus 接口代表一个事务具体的运行状态，事务管理器可以通过该接口获取事务运行期的状态信息或者间接地回滚事务。

PlatformTransactionManager.getTransaction 方法返回一个 TransactionStatus 对象。返回的 TransactionStatus 对象可能代表一个新的或已经存在的事务（如果在当前调用堆栈有一个符合条件的事务）。

TransactionStatus 接口的方法如代码清单 17-5 所示。

代码清单 17-5　TransactionStatus 接口的方法

```
public interface TransactionStatus{
    /** 是否是新的事务*/
    boolean isNewTransaction();
    /** 事务是否有恢复点*/
    boolean hasSavepoint();
    /** 设置为只回滚 */
    void setRollbackOnly();
    /** 此事务是否为只回滚*/
    boolean isRollbackOnly();
    /** 将基础会话刷新到数据存储区  */
    void flush();
    /** 此事务是否已完成  */
    boolean isCompleted();
}
```

TransactionStatus 接口描述的是处理事务提供简单的控制事务执行和查询事务状态的方法，在回滚或提交时需要应用对应的事务状态。

接下来梳理以上 3 个核心接口之间的关系。

PlatformTransactionManager 根据 TransactionDefinition 定义进行事务管理，管理过程中事务可能存在多种不同的状态，每个事务状态信息都通过 TransactionStatus 进行表示、存储。PlatformTransactionManager 的默认实现类 AbstractPlatformTransactionManager 针对不同的数据库持久化操作技术，又分别有 3 个不同的实现类：DataSourceTransactionManager、HibernateTransactionManager 和 JpaTransactionManager，其他的实现类这里不一一列举。

Spring 事务管理最核心的接口是 org.springframework.transaction.PlatformTransactionManager，通过该接口，Spring 为各个平台提供了对应的事务管理器，至于具体的实现由各个平台负责。通过该接口，Spring 为不同的平台事务 API 提供了一致的编程模型。

17.2.2　Spring 使用事务案例的准备

前面介绍了事务的基础概念，以及讲解了 Spirng 中事务的 3 个核心接口。接下来以转账为例介绍 Spring 的事务。

首先创建一个 users 表，用来记录用户信息。创建表的 SQL 语句如代码清单 17-6 所示。

代码清单 17-6　创建 users 表的 SQL 语句

```
CREATE TABLE `users` (
  `id` int(11) NOT NULL AUTO_INCREMENT COMMENT '自增id',
  `username` varchar(32) NOT NULL COMMENT '用户名',
  `money` int(22) NOT NULL DEFAULT '0' COMMENT '金额，单位：分',
  `create_time` datetime NOT NULL DEFAULT CURRENT_TIMESTAMP COMMENT '创建时间',
  `update_time` datetime NOT NULL DEFAULT CURRENT_TIMESTAMP ON UPDATE CURRENT_
TIMESTAMP COMMENT '修改时间',
```

```
  `delete_time` int(22) NOT NULL DEFAULT '0' COMMENT '删除时间，0-未删除 ',
  PRIMARY KEY (`id`),
  UNIQUE KEY `uniq_username` (`username`) USING BTREE COMMENT '用户名唯一建'
) ENGINE=InnoDB DEFAULT CHARSET=utf8mb4 COLLATE=utf8mb4_0900_ai_ci;
```

这里金额使用了 int(22) 进行存储，单位是分。另外，关于创建时间和修改时间，建议每个表都必须有，并且不要使用代码操作时间，可以在数据结构上设置默认值进行操作。

然后插入两个数据，a 用户和 b 用户，金额都默认为 1000，其 SQL 语句如代码清单 17-7 所示。

代码清单 17-7　插入 a 用户和 b 用户的 SQL 语句

```
INSERT INTO `ssm_test`.`users` (`username`, `money`, `create_time`, `update_time`,
`delete_time`) VALUES ('a', 100000, DEFAULT, DEFAULT, DEFAULT);
INSERT INTO `ssm_test`.`users` (`username`, `money`, `create_time`, `update_time`,
`delete_time`) VALUES ('b', 100000, DEFAULT, DEFAULT, DEFAULT);
```

假设一个简单的业务场景：有两个用户 a 和 b，他们的初始账户余额都为 100 000 分（即 ¥1000.00 元）。此时进行以下业务：a 向 b 转账 ¥100 元。在程序中可以分解为两个步骤：

① a 的账户余额 1000*100 分减少 100*100 分，剩余 90 000 分。

② b 的账户余额 1000*100 分增加 100*100 分，变为 110 000 分。

以上两个步骤要么都执行成功，要么都不执行。可通过 TransactionTemplate 编程式事务来控制。

17.2.3　不使用事务进行转账

接下来使用 MyBatis 进行事务的操作。

先准备好项目，对应的 GitHub 项目的模块是 ssm-mybatis-transaction 模块。构建环境这里不具体讲解。

为了方便开发，这里使用 Mybatis Plus 作为 MyBatis 操作数据库的增强框架。引入 MyBatis Plus 的依赖，如代码清单 17-8 所示。

代码清单 17-8　引入 MyBatis Plus 的依赖

```
<!--MyBatis--Plus的依赖-->
<dependency>
    <groupId>com.baomidou</groupId>
    <artifactId>mybatis-plus</artifactId>
    <version>3.0.7</version>
</dependency>
<dependency>
    <groupId>com.baomidou</groupId>
    <artifactId>mybatis-plus-generator</artifactId>
    <version>3.0.7</version>
</dependency>
<!--freemarker生成文件需要-->
<dependency>
    <groupId>org.freemarker</groupId>
    <artifactId>freemarker</artifactId>
    <version>2.3.28</version>
```

```
        <scope>test</scope>
    </dependency>
```

如果手写 Dao 层 Mapper 接口和 XML 文件，当表非常多时，无疑非常耗时。既然已经使用了 MyBatis Plus，那么可以使用简单的类来生成数据库表的映射接口及对应实体类的代码，如代码清单 17-9 所示。

代码清单 17-9　使用 MyBatis Plus 代码生成工具

```
package com.uifuture.chapter17;
import com.baomidou.mybatisplus.generator.AutoGenerator;
import com.baomidou.mybatisplus.generator.config.DataSourceConfig;
import com.baomidou.mybatisplus.generator.config.GlobalConfig;
import com.baomidou.mybatisplus.generator.config.PackageConfig;
import com.baomidou.mybatisplus.generator.config.StrategyConfig;
import com.baomidou.mybatisplus.generator.config.rules.NamingStrategy;
import com.baomidou.mybatisplus.generator.engine.FreemarkerTemplateEngine;
/** MyBatis Plus代码生成 */
public class CodeGenerator {
    /** 注意，这里的MySQL连接和依赖是MySQL 8.0的配置 */
    public static void main(String[] args) {
        // 代码生成器
        AutoGenerator mpg = new AutoGenerator();
        // 全局配置
        GlobalConfig gc = new GlobalConfig();
        String projectPath = "ssm-mybatis-transaction";
        // 文件生成目录
        gc.setOutputDir(projectPath + "/src/main/java");
        gc.setAuthor("chenhx");
        gc.setOpen(false);
        // 实体类后缀

        gc.setEntityName("%sEntity");
        mpg.setGlobalConfig(gc);
        // 数据源配置
        DataSourceConfig dsc = new DataSourceConfig();
        dsc.setUrl("jdbc:mysql://127.0.0.1:3306/ssm_test?useUnicode=true&characterEn
coding=UTF-8&zeroDateTimeBehavior=convertToNull&serverTimezone=GMT%2B8");
        /** dsc.setDriverName("com.mysql.jdbc.Driver");
        MySQL 8.0+的驱动地址 */
        dsc.setDriverName("com.mysql.cj.jdbc.Driver");
        dsc.setUsername("root");
        dsc.setPassword("12345678");
        mpg.setDataSource(dsc);
        // 包配置
        PackageConfig packageConfig = new PackageConfig();
        // 配置父类包
        packageConfig.setParent("com.uifuture.chapter17");
        // 配置实体包
```

```
        packageConfig.setEntity("domain.entity");
        mpg.setPackageInfo(packageConfig);

        // 策略配置
        StrategyConfig strategy = new StrategyConfig();
        strategy.setNaming(NamingStrategy.underline_to_camel);
        strategy.setColumnNaming(NamingStrategy.underline_to_camel);
        // 设置实体类的父类
        strategy.setSuperEntityClass("com.uifuture.chapter17.domain.base.BaseEntity");
        strategy.setEntityLombokModel(true);
        strategy.setRestControllerStyle(true);
        // 表名
        strategy.setInclude("users");
        // 设置父类公共属性，注意是数据库列名
            strategy.setSuperEntityColumns("id",   "create_time", "update_time",
"delete_time");
        strategy.setEntityColumnConstant(true);
        strategy.setControllerMappingHyphenStyle(true);
        strategy.setTablePrefix(packageConfig.getModuleName() + "_");
        mpg.setStrategy(strategy);
        mpg.setTemplateEngine(new FreemarkerTemplateEngine());
        mpg.execute();
    }
}
```

运行 main 方法即可生成一套 Dao 层的代码。生成的代码结构如图 17-1 所示。

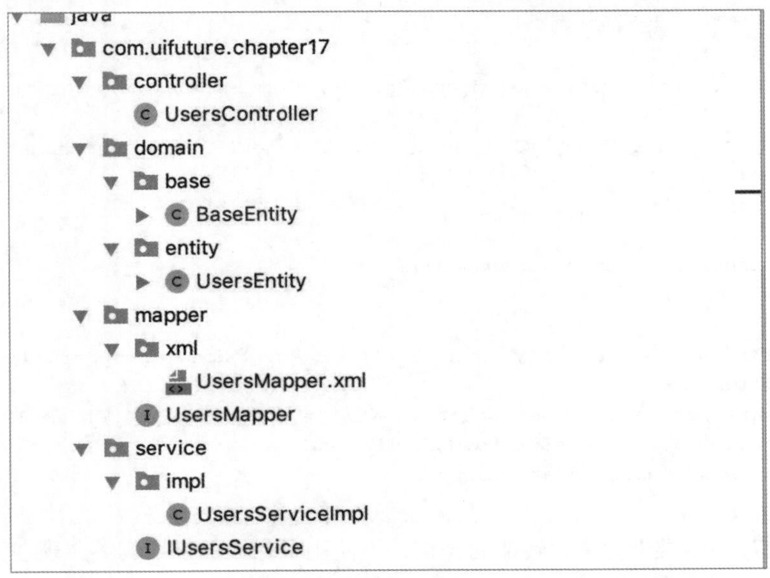

图 17-1　代码结构

接下来便可以使用 MyBatis Plus 进行数据的 CRUD（Create、Retrieve、Update、Delete 的缩写，也称"增删查改"）了。在 Service 层增加一个通过用户名操作数据的接口，如代码清单 17-10 所示。

代码清单 17-10　通过用户名操作数据

```
public interface IUsersService extends IService<UsersEntity> {
    /** 通过用户名修改用户 */
    Boolean updateByUsername(UsersEntity usersEntity);
}
@Service
public class UsersServiceImpl extends ServiceImpl<UsersMapper, UsersEntity>
implements IUsersService {
    @Override
    public Boolean updateByUsername(UsersEntity usersEntity) {
        if(usersEntity==null){
            return false;
        }
        if(StringUtils.isEmpty(usersEntity.getUsername())){
            return false;
        }
        QueryWrapper<UsersEntity> queryWrapper = new QueryWrapper<>();
        queryWrapper.eq(UsersEntity.USERNAME,usersEntity.getUsername());
        return this.update(usersEntity,queryWrapper);
    }
}
```

上面的改动肯定不能用于金额的修改，数值的修改可以直接通过 SQL 语句进行，这样不会出现并发执行的情况。新增一个操作金额的接口，如代码清单 17-11 所示。

代码清单 17-11　新增操作金额的接口

```
/** 通过用户名修改金额 */
Boolean updateMoneyByUsername(Integer money,String username);
@Override
public Boolean updateMoneyByUsername(Integer money, String username) {
    if(money==null){
        return false;
    }
    if(StringUtils.isEmpty(username)){
        return false;
    }
    UpdateWrapper<UsersEntity> updateWrapper = new UpdateWrapper<>();
    // 拼接SQL
    updateWrapper.setSql(UsersEntity.MONEY+"="+UsersEntity.MONEY+"+"+money);
    updateWrapper.eq(UsersEntity.USERNAME,username);
    return this.update(updateWrapper);
}
```

接下来便通过测试类进行模拟转账的操作，不使用事务。但是事实上，现在无法通过测试类调用 service 方法，只能通过前面讲的 MyBatis 知识点来理解和完成，可以在测试类中调用 Mapper 接口的方法。因此，接下来介绍 Spring 与 MyBatis 如何整合。

Spirng 的配置文件 applicationContext.xml 文件内容如代码清单 17-12 所示。

代码清单 17-12　applicationContext.xml 文件内容

```
<beans xmlns:xsi="http://www.w3.org/2001/XMLSchema-instance"
          xmlns:context="http://www.springframework.org/schema/context"
xmlns="http://www.springframework.org/schema/beans"
       xsi:schemaLocation="http://www.springframework.org/schema/beans
        http://www.springframework.org/schema/beans/spring-beans.xsd
              http://www.springframework.org/schema/context http://www.
springframework.org/schema/context/spring-context.xsd">
    <!--扫描并注入使用注入注解的Bean-->
    <context:component-scan base-package="com.uifuture.chapter17">
    </context:component-scan>
    <!--导入application-service.xml配置文件内容-->
    <import resource="classpath*:application-service.xml"></import>
</beans>
```

然后使用 application-service.xml 配置文件将 Spring 与 MyBatis 进行了整合。为什么还要分出 application-service.xml 配置文件？主要是方便管理。如果后续再有缓存或者 AOP 等需要注入，那么可以再导入 application-cache.xml、application-aop.xml 配置文件。application-service.xml 配置文件如代码清单 17-13 所示。

代码清单 17-13　application-service.xml 配置文件

```
<?xml version="1.0" encoding="UTF-8"?>
<beans xmlns:xsi="http://www.w3.org/2001/XMLSchema-instance"
       xmlns:tx="http://www.springframework.org/schema/tx"
       xmlns="http://www.springframework.org/schema/beans"

       xsi:schemaLocation="http://www.springframework.org/schema/beans http://www.
springframework.org/schem
a/beans/spring-beans.xsd
         http://www.springframework.org/schema/tx http://www.springframework.org/
schema/tx/spring-tx.xsd">
    <bean id="propertyConfigurer"
          class="org.springframework.beans.factory.config.PropertyPlaceholderConfigurer">
        <property name="locations">
            <list>
         <!-- 导入数据库的配置-->
                <value>classpath:jdbc.properties</value>
            </list>
        </property>
    </bean>
    <!-- 数据源配置, 使用DriverManagerDataSource -->
      <bean id="dataSource" class="org.springframework.jdbc.datasource.
DriverManagerDataSource">
        <property name="driverClassName" value="${driver}"/>
        <property name="url" value="${url}"/>
        <property name="username" value="${username}"/>
        <property name="password" value="${password}"/>
    </bean>
```

```
    <bean id="sqlSession" class="org.mybatis.spring.SqlSessionTemplate">
        <constructor-arg index="0" ref="sqlSessionFactory"/>
    </bean>
    <!-- MyBatis的SQLSessionFactorybean:SqlSessionFactoryBean -->
    <!-- MyBatis的sqlsession:org.mybatis.spring.SqlSessionFactoryBean -->
    <!--配置MyBatis--Plus的sqlSessionFactory -->
    <bean id="sqlSessionFactory" class="com.baomidou.mybatisplus.extension.spring.
MybatisSqlSessionFactoryBean">
        <property name="dataSource" ref="dataSource"/>
        <!--自动扫描加载指定位置的Mapper和XML配置具体SQL);
        若XML与接口在同一个包下面，则不需要配置该属性-->
        <property name="mapperLocations">
            <list>
                <value>classpath*:com/uifuture/chapter17/mapper/**/*.xml</value>
            </list>
        </property>
    </bean>
    <!-- MyBatis自动扫描加载SQL映射文件/接口:MapperScannerConfigurer -->
    <bean class="org.mybatis.spring.mapper.MapperScannerConfigurer">
        <!-- 可以考虑使用通配符"*" 扫描多个包，Mapper接口上面使用@Repository -->
        <property name="basePackage" value="com.uifuture.chapter17.mapper"/>
        <property name="sqlSessionFactoryBeanName" value="sqlSessionFactory"/>
    </bean>
    <!-- *************事务管理****************** -->
    <bean id="transactionManager"

          class="org.springframework.jdbc.datasource.DataSourceTransactionManager">
        <property name="dataSource" ref="dataSource"/>
    </bean>
    <!-- 注解方式配置事务 -->
    <tx:annotation-driven transaction-manager="transactionManager"/>
</beans>
```

接下来编写测试类。先准备测试基类，这样在后续进行测试类编写时，就不需要再注入配置文件了，编写的 Junit 类测试注解如代码清单 17-14 所示。

代码清单 17-14　Junit 类测试注解

```
package com.uifuture.chapter17;
import org.junit.runner.RunWith;
import org.springframework.test.context.ContextConfiguration;
import org.springframework.test.context.junit4.SpringJUnit4ClassRunner;
import org.springframework.test.context.web.WebAppConfiguration;
/** 测试基类*/
@RunWith(SpringJUnit4ClassRunner.class)
@WebAppConfiguration
@ContextConfiguration({"classpath*:applicationContext.xml"})
public class BaseTest {
}
```

接下来编写 a 对 b 进行转账的操作测试，且不使用事务，如代码清单 17-15 所示。

代码清单 17-15　不使用事务的转账测试

```
package com.uifuture.chapter17;
import com.uifuture.chapter17.service.IUsersService;
import org.junit.Test;
import org.springframework.beans.factory.annotation.Autowired;
import java.io.IOException;
public class UsersServiceImplTest extends BaseTest {
    @Autowired
    private IUsersService usersService;
    /** 不使用事务的测试 */
    @Test
    public void updateMoneyByUsername() throws IOException {
        //a用户给b用户转账100元，不使用事务
        usersService.updateMoneyByUsername(-100*100, "a");
        usersService.updateMoneyByUsername(100*100, "b");
    }
}
```


通过代码可以看出，修改 a 用户的金额和修改 b 用户的金额是分两步进行的。正常情况下，运行后的结果为 a 用户拥有 900 元，b 用户拥有 1100 元。

现在，将代码稍做修改，在两个步骤之间增加一个计算，如代码清单 11-16 所示。

代码清单 17-16　增加错误计算

```
/** 不使用事务的测试，步骤之间增加错误计算 */
@Test
public void updateMoneyByUsername2() throws IOException {
    //a用户给b用户转账100元，不使用事务
    usersService.updateMoneyByUsername(-100*100, "a");
    // 运行分母为0的除法运算
    int size = 100/0;
    usersService.updateMoneyByUsername(100*100, "b");
}
```

运行代码 17-16 的测试方法，毫无疑问，程序会抛出 java.lang.ArithmeticException: /by zero 异常。但是，给 a 用户转账的步骤已经完成了。此时看数据库。可以看到 a 用户的金额为 900 元，但是 b 用户的金额还是 1000 元。也就是说给 b 用户的转账肯定是失败的或没有执行。这个肯定是不符合需求的，通过事务才能解决转账过程中出现的这个问题。

17.2.4　编程式事务处理

编程式事务处理允许开发者在源代码编程的帮助下处理事务，这极大地提高了灵活性，缺点是比较难维护。

上面转账的两步操作中间发生了异常，但是第 1 步 a 用户减少金额操作依然在数据库中执行了。实际应用中不会允许这样的情况发生，所以这里用事务来进行处理。使用事务需要在配置文件中配置事务管理器，如代码清单 17-17 所示。

代码清单 17-17　配置事务管理器

```
<!-- *************事务管理器****************** -->
<bean id="transactionManager"
      class="org.springframework.jdbc.datasource.DataSourceTransactionManager">
    <property name="dataSource" ref="dataSource"/>
</bean>
```

接下来使用事务管理器进行编程式事务的处理，如代码清单 17-18 所示。

代码清单 17-18　编程式事务的处理

```
@Slf4j
public class UsersServiceImplTest extends BaseTest {
    @Autowired
    private IUsersService usersService;
    // 事务管理器
    @Autowired
    private DataSourceTransactionManager transactionManager;
     /** 使用编程式事务的测试，步骤之间增加错误计算 */
    @Test
    public void updateMoneyByUsername3() throws IOException {
        // 开启事务
        TransactionStatus transaction = transactionManager.getTransaction(new Defa
ultTransactionDefinition());
        try {
            //a用户给b用户转账100元
            usersService.updateMoneyByUsername(-100 * 100, "a");
            // 运行分母为0的除法运算
            int size = 100 / 0;
            usersService.updateMoneyByUsername(100 * 100, "b");
            // 事务提交
            transactionManager.commit(transaction);
        } catch (Exception e) {
            log.error("转账异常，事务回滚",e);
            // 回滚
            transactionManager.rollback(transaction);
        }
    }
}
```

由于 a 用户减少 100 元后遇到了异常，所以事务会进行回滚，a 用户的金额不会减少。也就实现了事务的处理。

17.2.5　声明式事务处理

声明式事务处理允许开发者在配置的帮助下（而不是通过编写源代码）处理事务，这意味着开发者可以将事务管理从事务代码中隔离出来，只使用注解或基于配置的 XML 文件来管理事务。

在 Spring 配置文件中以配置注解方式处理事务，如代码清单 17-19 所示。

代码清单 17-19　配置声明式事务

```
<!-- 声明式事务处理 -->
<tx:annotation-driven transaction-manager="transactionManager"/>
```

可以在 <tx:annotation-driven> 标记中通过配置 proxy-target-class="true" 将代理方式强制为 CGLIB 代理。接下来便可以通过 Transactional 注解来进行方法或者类的事务的开启。如果在类上添加注解，则说明类中的所有方法都添加事务；如果在方法上添加注解，则说明只有该方法添加事务。

使用声明式事务处理分为两步走：

①在 applicationContext.xml 中配置事务管理器，将并事务管理器交给 Spring 容器管理。

②在目标类或目标方法中添加 Transactional 注解即可。

下面通过使用声明式事务处理方法进行转账操作，如代码清单 17-20 所示。

代码清单 17-20　声明式事务处理

```
// service增加转账接口
/**
 * 转账接口
 * @param fromName 转账方
 * @param toName 收账方
 * @param money 金额
 */
void transfer(String fromName,String toName,Integer money);


@Override
@Transactional
public void transfer(String fromName, String toName, Integer money) {
    //a用户给b用户转账100元，不使用事务
    updateMoneyByUsername(-1*money, fromName);
    // 运行分母为0的除法运算
    int size = 100 / 0;
    updateMoneyByUsername(money, toName);
}
```

在上面的实现方法中，依然通过分母为 0 模拟转账失败的情况，根据 a 用户的金额来判断事务是否生效。

测试声明式事务处理，如代码清单 17-21 所示。

代码清单 17-21　测试声明式事务处理

```
/**声明式事务处理 */
@Test
public void updateMoneyByUsername4() throws IOException {
    usersService.transfer("a","b",100);
}
```

运行后，a 用户的金额不会改变。

通过使用 Spring 的事务管理器来处理数据库的事务是非常方便的。接下来分析 Transactional 注

解的基本用法与实现原理。

17.3 Transactional 注解

前面介绍了事务的基本知识与如何在 Spring 中使用编程式事务和声明式事务，本节将讲解 Transactional 注解的用法，再从源码角度来分析 Spring 是如何实现事务管理的。

17.3.1 Transactional 注解的用法

Transactional 注解可以作用于接口、接口方法、类及类方法上。当作用于类上时，该类的所有的 public 方法都将具有事务属性。同时，可以在方法级别上使用该注解来覆盖类级别的定义。

注意：在 Spring 中不建议在接口或者接口方法上使用 Transactional 注解，因为它只有在使用基于接口的代理时才会生效。

另外，Transactional 注解只能被应用到 public 方法上，这是由 Spring AOP 的本质决定的。如果在 protected、private 或默认可见性的方法上使用 Transactional 注解，其作用将被忽略，也不会抛出任何异常，即不生效。

默认情况下，只有来自外部的方法调用才会被 AOP 代理捕获，也就是类内部方法调用本类内部的其他方法并不会引起事务行为，即使被调用方法使用 Transactional 注解进行修饰。

Transactional 注解的属性见表 17-1。

表 17-1　Transactional 注解的属性

属　性	类　型	描　述
value	String	可选的限定描述符，指定使用的事务管理器
propagation	enum: Propagation	可选的事务传播行为设置
isolation	enum: Isolation	可选的事务隔离级别设置
readOnly	boolean	读写或只读事务，默认为读写
timeout	int (in seconds granularity)	事务超时时间设置
rollbackFor	Class 对象数组，必须继承自 Throwable	导致事务回滚的异常类数组
rollbackForClassName	类名数组，必须继承自 Throwable	导致事务回滚的异常类名字数组
noRollbackFor	Class 对象数组，必须继承自 Throwable	不会导致事务回滚的异常类数组
noRollbackForClassName	类名数组，必须继承自 Throwable	不会导致事务回滚的异常类名字数组

17.3.2 Transactional 注解的实现原理

在讲解 Transactional 注解的实现原理之前，先来了解事务的基本原理。Spring 事务在本质上就是数据库对事务的支持，如果没有数据库的事务支持，Spring 肯定是无法提供事务功能的。当只使用 JDBC 操作数据库时，如果要用到事务，则可以按照下面的步骤进行。

①获取连接 Connection con=DriverManager.getConnection()。

②开启事务 con.setAutoCommit(true/false)。

③执行增、删、改、查操作。

④正常逻辑块提交事务要使用 con.commit()，异常逻辑块中使用 con.rollback()，finally 逻辑块中使用 con.close()。

⑤关闭连接使用 conn.close()。

使用 Spring 的事务管理功能后，可以省略步骤②和步骤④的代码，这两步代码由 Spring 自动完成。

接下来便从整体上理解 Spring 是如何在增、删、改、查之前和之后开启事务和关闭事务的。以 Transactional 注解方式为例，有以下 3 个步骤：

①在 Spring 配置文件中开启注解驱动，在需要开启事务的类或者方法上加入 Transactional 注解。

②当 Spring 启动时，会解析生成相关注入的 Bean，这时 Spring 会查看拥有相关注解的类和方法，并且为这些类和方法生成代理，根据 Transaction 注解的相关参数进行配置后注入。这样 Spring 在代理中将相关的事务进行了处理，也就是开启事务，正常时提交事务，异常时进行事务回滚。

③实际上，数据库事务的提交和回滚是通过 binlog 或者 redolog 进行实现的。

Transactional 注解便是用来把事务开启、提交或者回滚的操作，在 Spring 中通过 AOP 的方式进行管理。通过 Transactional 注解就能让 Spring 管理事务，免去了重复的事务管理逻辑，减少了对业务代码的侵入，使开发人员能够专注于业务层面。

Transactional 注解的原理是基于 Spring AOP，AOP 是动态代理模式的实现，通过对源码的阅读，可以了解在 Spring 中是如何利用 AOP 来实现 Transactional 注解功能的。

通过对 AOP 的了解可知，如果要对一个方法进行代理，则必须定义切点。那么可以猜想，在实现 Transactional 注解的过程中 Spring 同样定义了以 Transactional 注解为织入点的切点，Spring 便知道 Transactional 注解标注的方法或类需要被代理。

有了切面定义后，在 Spring Bean 的初始化过程中，就需要对标注了 Transactional 注解的 Bean 或者方法进行代理，生成代理对象。

在生成代理对象的代理逻辑中，当调用方法时，需要先获取切面的逻辑，Transactional 注解的切面逻辑类似于 Around 注解，在 Spring 中是一种类似的代理实现逻辑。

17.3.3 声明式事务实现大体分析

先来了解 <tx> 标记的解析。Spring 解析标记时会使用 spring.handles 指定标记的解析类，如图 17-2 所示。

spring.handles 文件中的内容如下：

```
http\://www.springframework.org/schema/tx=org.springframework.transaction.config.
TxNamespaceHandler
```

从这里便可以知道 <tx> 标记的解析类是 org.springframework.transaction.config.TxNamespace-Handler。TxNamespaceHandler 类的源码如代码清单 17-22 所示。

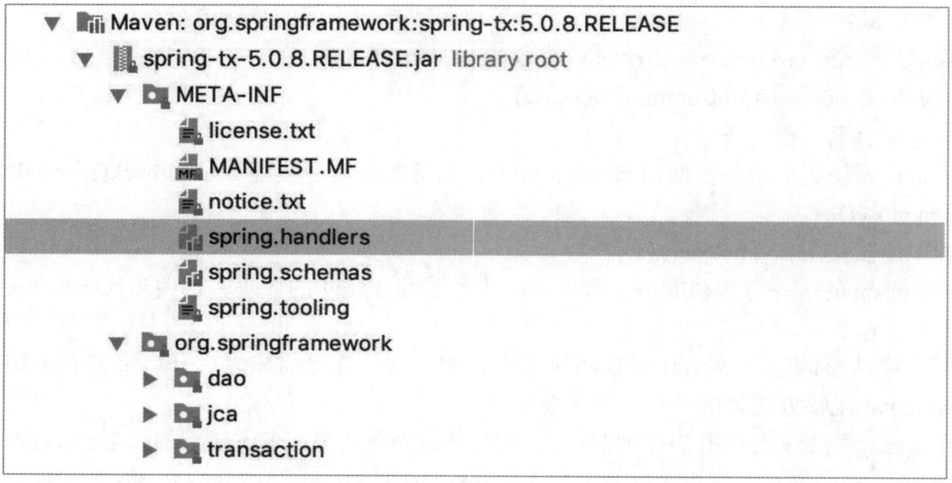

图 17-2 Spring 指定标记解析类配置文件

代码清单 17-22 TxNamespaceHandler 源码

```
public class TxNamespaceHandler extends NamespaceHandlerSupport {
    static final String TRANSACTION_MANAGER_ATTRIBUTE = "transaction-manager";
    static final String DEFAULT_TRANSACTION_MANAGER_BEAN_NAME = "transactionManager";
    static String getTransactionManagerName(Element element) {
      return (element.hasAttribute(TRANSACTION_MANAGER_ATTRIBUTE) ?
                element.getAttribute(TRANSACTION_MANAGER_ATTRIBUTE) : DEFAULT_
TRANSACTION_MANAGER_BEAN_NAME);
    }
    @Override
    public void init() {
      registerBeanDefinitionParser("advice", new TxAdviceBeanDefinitionParser());
      registerBeanDefinitionParser("annotation-driven", new AnnotationDrivenBeanDef
initionParser());
      registerBeanDefinitionParser("jta-transaction-manager", new JtaTransactionMan
agerBeanDefinitionParser());
    }
}
```

由代码清单 17-22 的 init 方法可知，<tx:annotation-driven> 标记的解析器是 AnnotationDriven-BeanDefinitionParser 类。从源码可知，AnnotationDrivenBeanDefinitionParser 类解析方法就是 parse，该方法在注释中有说明，也可通过英文翻译来学习。AnnotationDrivenBeanDefinitionParser 类如代码清单 17-23 所示。

代码清单 17-23 AnnotationDrivenBeanDefinitionParser 类

```
@Override
@Nullable
public BeanDefinition parse(Element element, ParserContext parserContext) {
    registerTransactionalEventListenerFactory(parserContext);
```

```
        String mode = element.getAttribute("mode");
        if ("aspectj".equals(mode)) {
            // mode="aspectj"
            registerTransactionAspect(element, parserContext);
        }
        else {
            // mode="proxy"
            AopAutoProxyConfigurer.configureAutoProxyCreator(element, parserContext);
        }
        return null;
    }
```

在方法上很清楚地说明了如何解析 <tx> 标记。首先获取标记上的 mode 的属性，默认情况下，mode=proxy。从图 17-3 可知默认值是 proxy。

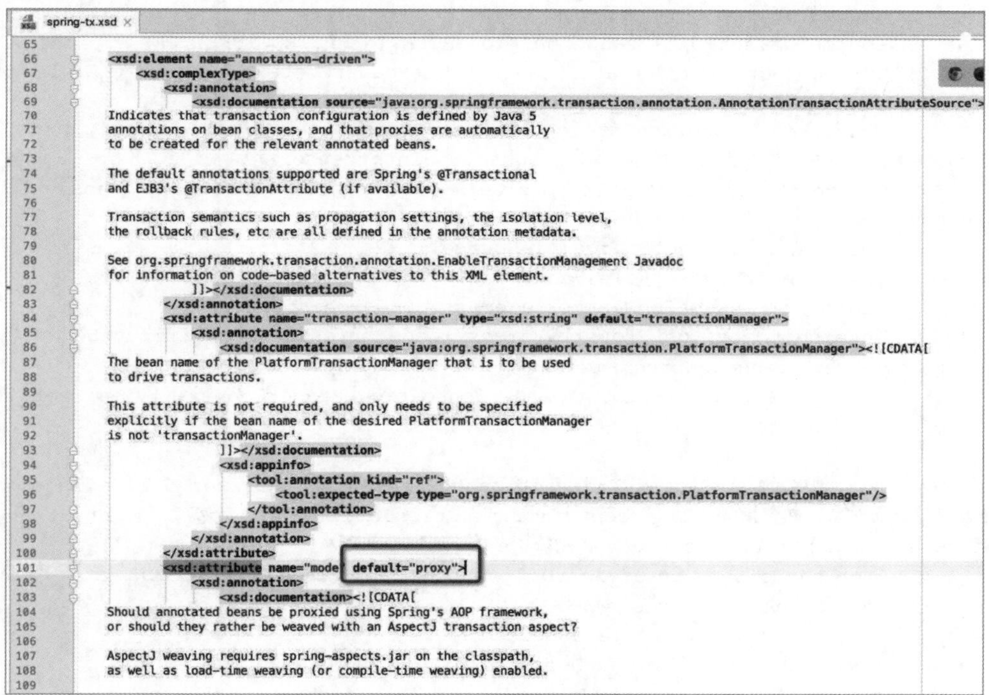

```
spring-tx.xsd
    <xsd:element name="annotation-driven">
        <xsd:complexType>
            <xsd:annotation>
                <xsd:documentation source="java:org.springframework.transaction.annotation.AnnotationTransactionAttributeSource">
    Indicates that transaction configuration is defined by Java 5
    annotations on bean classes, and that proxies are automatically
    to be created for the relevant annotated beans.

    The default annotations supported are Spring's @Transactional
    and EJB3's @TransactionAttribute (if available).

    Transaction semantics such as propagation settings, the isolation level,
    the rollback rules, etc are all defined in the annotation metadata.

    See org.springframework.transaction.annotation.EnableTransactionManagement Javadoc
    for information on code-based alternatives to this XML element.
                ]]></xsd:documentation>
            </xsd:annotation>
            <xsd:attribute name="transaction-manager" type="xsd:string" default="transactionManager">
                <xsd:annotation>
                    <xsd:documentation source="java:org.springframework.transaction.PlatformTransactionManager"><![CDATA[
    The bean name of the PlatformTransactionManager that is to be used
    to drive transactions.

    This attribute is not required, and only needs to be specified
    explicitly if the bean name of the desired PlatformTransactionManager
    is not 'transactionManager'.
                    ]]></xsd:documentation>
                    <xsd:appinfo>
                        <tool:annotation kind="ref">
                            <tool:expected-type type="org.springframework.transaction.PlatformTransactionManager"/>
                        </tool:annotation>
                    </xsd:appinfo>
                </xsd:annotation>
            </xsd:attribute>
            <xsd:attribute name="mode" default="proxy">
                <xsd:annotation>
                    <xsd:documentation><![CDATA[
    Should annotated beans be proxied using Spring's AOP framework,
    or should they rather be weaved with an AspectJ transaction aspect?

    AspectJ weaving requires spring-aspects.jar on the classpath,
    as well as load-time weaving (or compile-time weaving) enabled.
```

图 17-3　mode 属性的默认值

因此，根据代码 17-23，可知接下来将执行 AopAutoProxyConfigurer.configureAutoProxyCreator (element, parserContext)，如代码清单 17-24 所示。

代码清单 17-24　AopAutoProxyConfigurer.configureAutoProxyCreator 方法

```
/** 内部类只是在实际处于代理模式时引入AOP框架依赖项*/
private static class AopAutoProxyConfigurer {
    public static void configureAutoProxyCreator(Element element, ParserContext parserContext) {
```

```
// 注册AutoProxyCreator，用来生成代理
        AopNamespaceUtils.registerAutoProxyCreatorIfNecessary(parserContext,
element);
        String txAdvisorBeanName = TransactionManagementConfigUtils.TRANSACTION_
ADVISOR_BEAN_NAME;
 // 如果没有注册事务的通知(advice)
      if (!parserContext.getRegistry().containsBeanDefinition(txAdvisorBeanName)) {
         Object eleSource = parserContext.extractSource(element);
          /*创建一个AnnotationTransactionAttributeSource BeanDefinition，它的作用类似于
一个pointcut切入点，主要是判断方法或者类是否含有@Transactional。设置事务属性源为
AnnotationTransactionAttributeSource*/
         RootBeanDefinition sourceDef = new RootBeanDefinition(
               "org.springframework.transaction.annotation.AnnotationTransactionAtt
ributeSource");
         sourceDef.setSource(eleSource);
         sourceDef.setRole(BeanDefinition.ROLE_INFRASTRUCTURE);
         String sourceName = parserContext.getReaderContext().registerWithGenerated
Name(sourceDef);
          /* 创建了TransactionInterceptor BeanDefinition，作用主要是拦截@Transactional，
进行后续事务的一些处理。设置事务拦截器为TransactionInterceptor */
         RootBeanDefinition interceptorDef = new RootBeanDefinition(TransactionInter
ceptor.class);
         interceptorDef.setSource(eleSource);
         interceptorDef.setRole(BeanDefinition.ROLE_INFRASTRUCTURE);
         registerTransactionManager(element, interceptorDef);
         interceptorDef.getPropertyValues().add("transactionAttributeSource", new
 RuntimeBeanReference(sourceName));
         String interceptorName = parserContext.getReaderContext().registerWithGene
ratedName(interceptorDef);

          /* 创建了BeanFactoryTransactionAttributeSourceAdvisor BeanDefinition，是一
个真正的切面对象，包含拦截器Interceptor(TransactionInterceptor)、切入点pointcut(Annotati
onTransactionAttributeSource)，提供给Spring AOP使用 */
         RootBeanDefinition advisorDef = new RootBeanDefinition(BeanFactoryTransacti
onAttributeSourceAdvisor.class);
         advisorDef.setSource(eleSource);
         advisorDef.setRole(BeanDefinition.ROLE_INFRASTRUCTURE);
         advisorDef.getPropertyValues().add("transactionAttributeSource", new Runt
imeBeanReference(sourceName));
         advisorDef.getPropertyValues().add("adviceBeanName", interceptorName);
      // 设置通知的order
   if (element.hasAttribute("order")) {
               advisorDef.getPropertyValues().add("order", element.
getAttribute("order"));
         }
    // 注册通知
           parserContext.getRegistry().registerBeanDefinition(txAdvisorBeanName,
advisorDef);
    // 注册组件
```

```
        CompositeComponentDefinition compositeDef = new CompositeComponentDefiniti
on(element.getTagN ame(), eleSource);

            compositeDef.addNestedComponent(new BeanComponentDefinition(sourceDef,
sourceName));
          compositeDef.addNestedComponent(new BeanComponentDefinition(interceptorDef,
interceptorName));
           compositeDef.addNestedComponent(new BeanComponentDefinition(advisorDef,
txAdvisorBeanName));
        parserContext.registerComponent(compositeDef);
      }
    }
}
```

接下来分析注册 AutoProxyCreator 的源码（AopNamespaceUtils.registerAutoProxyCreatorIfNecessary(ParserContext parserContext, Element sourceElement)），如代码清单 17-25 所示。

代码清单 17-25　AopNamespaceUtils.registerAutoProxyCreatorIfNecessary 源码

```
public static void registerAutoProxyCreatorIfNecessary(
        ParserContext parserContext, Element sourceElement) {
    BeanDefinition beanDefinition = AopConfigUtils.registerAutoProxyCreatorIfNecessary(
            parserContext.getRegistry(), parserContext.extractSource(sourceElement));
    useClassProxyingIfNecessary(parserContext.getRegistry(), sourceElement);
    registerComponentIfNecessary(beanDefinition, parserContext);
}
```

AopConfigUtils.registerAutoProxyCreatorIfNecessary(parserContext.getRegistry(), parserContext.extractSource(sourceElement)) 方法源码如代码清单 17-26 所示。

代码清单 17-26　AopConfigUtils.registerAutoProxyCreatorIfNecessary() 方法源码

```
@Nullable
public static BeanDefinition registerAutoProxyCreatorIfNecessary(BeanDefinitionRegis
try registry,
        @Nullable Object source) {
    return registerOrEscalateApcAsRequired(InfrastructureAdvisorAutoProxyCreator.
class, registry, source);
}
```

从代码 17-26 中可以得知，此处代码向 Spring 容器中注册了一个 InfrastructureAdvisorAutoProxyCreator 类。InfrastructureAdvisorAutoProxyCreator 类的作用是利用后置处理器机制，在对象创建后包装对象，返回一个代理对象（增强器），代理对象执行方法利用拦截器链进行调用。接下来继续分析 AopConfigUtils.registerOrEscalateApcAsRequired 方法，如代码清单 17-27 所示。

代码清单 17-27　AopConfigUtils.registerOrEscalateApcAsRequired 方法源码

```
@Nullable
private static BeanDefinition registerOrEscalateApcAsRequired(Class<?> cls,
BeanDefinitionRegistry registry,
```

```
        @Nullable Object source) {
    Assert.notNull(registry, "BeanDefinitionRegistry must not be null");
// 如果已存在AutoProxyCreator
    if (registry.containsBeanDefinition(AUTO_PROXY_CREATOR_BEAN_NAME)) {
        BeanDefinition apcDefinition = registry.getBeanDefinition(AUTO_PROXY_
CREATOR_BEAN_NAME);
        if (!cls.getName().equals(apcDefinition.getBeanClassName())) {
            int currentPriority = findPriorityForClass(apcDefinition.
getBeanClassName());

            int requiredPriority = findPriorityForClass(cls);---
  // 取优先级大的
            if (currentPriority < requiredPriority) {
                apcDefinition.setBeanClassName(cls.getName());
            }
        }
        return null;
    }
    // 如果不存在AutoProxyCreator，则将InfrastructureAdvisorAutoProxyCreator设置为AutoProxyCreator
    RootBeanDefinition beanDefinition = new RootBeanDefinition(cls);
    beanDefinition.setSource(source);
    beanDefinition.getPropertyValues().add("order", Ordered.HIGHEST_PRECEDENCE);
    beanDefinition.setRole(BeanDefinition.ROLE_INFRASTRUCTURE);
    registry.registerBeanDefinition(AUTO_PROXY_CREATOR_BEAN_NAME, beanDefinition);
    return beanDefinition;
}
```

接着分析 AopNamespaceUtils 类中的 AopNamespaceUtils.useClassProxyingIfNecessary 方法，如代码清单 17-28 所示。

代码清单 17-28　AopNamespaceUtils.useClassProxyingIfNecessary 方法源码

```
private static void useClassProxyingIfNecessary(BeanDefinitionRegistry registry, @
Nullable
Element sourceElement) {
    if (sourceElement != null) {
        boolean proxyTargetClass = Boolean.parseBoolean(sourceElement.
getAttribute(PROXY_TARGET_CLASS_ATTRIBUTE));
        if (proxyTargetClass) {
            AopConfigUtils.forceAutoProxyCreatorToUseClassProxying(registry);
        }
        boolean exposeProxy = Boolean.parseBoolean(sourceElement.getAttribute(EXPOSE_
PROXY_ATTRIBUTE));
        if (exposeProxy) {
            AopConfigUtils.forceAutoProxyCreatorToExposeProxy(registry);
        }
    }
}
```

该方法用来设置是否强制使用 CGLIB 来生成代理。最后的 registerComponentIfNecessary 是用来注册 AutoProxyCreator 组件的。

将代码清单 17-24 进行简单总结，其共做了 4 件事：

①注册 AutoProxyCreator（主要是用来创建代理的。此处若已经存在 AutoProxyCreator，如 AOP 的，则使用优先级高的那个）。

②创建一个切入点 pointcut，即 AnnotationTransactionAttributeSource 类，用于判断方法或者类是否含有 Transactional 注解。

③创建一个拦截器 Interceptor，即 TransactionInterceptor 类，主要拦截 Transactional 注解标识的方法或者类，然后进行一些事务处理。

④创建一个切面 Aspect，即 BeanFactoryTransactionAttributeSourceAdvisor 类，用来组装切入点和拦截器，提供给 Spring AOP 生成代理。

因此，从第②步可知，Spring 对事务的操作是由 TransactionInterceptor 进行处理的。接下来分析 TransactionInterceptor 类，如代码清单 17-29 所示。

代码清单 17-29　TransactionInterceptor 类源码

```
public class TransactionInterceptor extends TransactionAspectSupport implements 
MethodInterceptor, Serializable {
...//其他代码省略
   @Override
   @Nullable
   public Object invoke(MethodInvocation invocation) throws Throwable {
         Class<?> targetClass = (invocation.getThis() != null ? AopUtils.
getTargetClass(invocation.getThis()) : null);
         return invokeWithinTransaction(invocation.getMethod(), targetClass, 
invocation::proceed);
   }
...//其他代码省略
}
```

从 Spring AOP 的原理实现可知，DynamicAdvisedInterceptor 类的 intercept 方法调用了 TransactionInterceptor 类的 invoke 方法。接下来可知核心代码是 TransactionInterceptor.invokeWithinTransaction 方法（在 TransactionInterceptor 的父类 TransactionAspectSupport 中），如代码清单 17-30 所示。

代码清单 17-30　TransactionAspectSupport.invokeWithinTransaction 方法源码

```
@Nullable
protected Object invokeWithinTransaction(Method method, @Nullable Class<?> 
targetClass,final InvocationCallback invocation) throws Throwable {
     /* 获取对应事务属性.如果事务属性为空(则目标方法不存在事务)。如果项目使用@Transactional，那
么这里就对应Transactional的值。如果项目使用了<tx:attributes>，就是标签配置的属性的值 */
    TransactionAttributeSource tas = getTransactionAttributeSource();
     /* 根据事务的属性获取beanFactory中的PlatformTransactionManager(Spring事务管理器的顶
级接口)，一般这里是DataSourceTransactiuonManager*/
    final TransactionAttribute txAttr = (tas != null ? tas.getTransactionAttribute(method, 
targetClass) : null);
```

```
    final PlatformTransactionManager tm = determineTransactionManager(txAttr);
      final String joinpointIdentification = methodIdentification(method,
targetClass, txAttr);
    //如果事务属性为空或者tm变量不是编程式事务管理器
    if (txAttr == null || !(tm instanceof CallbackPreferringPlatformTransactionManager)) {
            /*开启事务。如果事先没有添加指定任何transactionmanger最终会从容器中按照类型获取一个
PlatformTransactionManager*/
            TransactionInfo txInfo = createTransactionIfNecessary(tm, txAttr,
joinpointIdentification);
        Object retVal = null;
        try {
            /*invocation.proceedWithInvocation()方法是一个around advice,会进行链式调用。
被代理的方法就在这里执行的，执行完后，真正事务提交在commitTransactionAfterReturning中 */
            retVal = invocation.proceedWithInvocation();
        }

        catch (Throwable ex) {
            // 出现异常，进行事务回滚
            completeTransactionAfterThrowing(txInfo, ex);
            throw ex;
        }
        finally {
            //重置TransactionInfo ThreadLocal
            cleanupTransactionInfo(txInfo);
        }
        //提交事务
        commitTransactionAfterReturning(txInfo);
        return retVal;
    }
    else {
        //下面是编程式事务处理，暂不重点讨论
        final ThrowableHolder throwableHolder = new ThrowableHolder();
        try {
              Object result = ((CallbackPreferringPlatformTransactionManager) tm).
execute(txAttr, status -> {
                    TransactionInfo txInfo = prepareTransactionInfo(tm, txAttr,
joinpointIdentification, status);
                try {
                    /*invocation.proceedWithInvocation()方法是一个around advice，会进行链
式调用。被代理的方法就是在这里执行的，执行完后，真正事务提交在commitTransactionAfterReturning中*/
                    return invocation.proceedWithInvocation();
                }
                catch (Throwable ex) {
                    if (txAttr.rollbackOn(ex)) {
                        if (ex instanceof RuntimeException) {
                            throw (RuntimeException) ex;
                        }
```

```
                    else {
                        throw new ThrowableHolderException(ex);
                    }
                }
                else {
                    throwableHolder.throwable = ex;
                    return null;
                }
            }
            finally {
                cleanupTransactionInfo(txInfo);
            }
        });
        if (throwableHolder.throwable != null) {
            throw throwableHolder.throwable;
        }
        return result;
    }
    catch (ThrowableHolderException ex) {
        throw ex.getCause();
    }
    catch (TransactionSystemException ex2) {
        if (throwableHolder.throwable != null) {
                logger.error("Application exception overridden by commit
exception", throwableHolder.throwable);
            ex2.initApplicationException(throwableHolder.throwable);
        }
        throw ex2;
    }
    catch (Throwable ex2) {
        if (throwableHolder.throwable != null) {
                logger.error("Application exception overridden by commit
exception", throwableHolder.throwable);
        }
        throw ex2;
    }
  }
}
```

上面代码的核心是 TransactionInfo txInfo = createTransactionIfNecessary(tm, txAttr, joinpoint Identification)。接下来重点分析 createTransactionIfNecessary 方法，它用来判断是否存在事务，并根据传播行为做出相应的判断，主要是通过 TransactionStatus 对象进行包装。TransactionAspect Support 类的 createTransactionIfNecessary 方法源码如代码清单 17-31 所示。

代码清单 17-31　TransactionAspectSupport 类的 createTransactionIfNecessary 方法源码

```
protected TransactionInfo createTransactionIfNecessary(@Nullable PlatformTransactionManager
tm, @Nullable TransactionAttribute txAttr, final String joinpointIdentification) {
```

```
    // 如果未指定名称，则将方法标识应用为事务名称
    if (txAttr != null && txAttr.getName() == null) {
       txAttr = new DelegatingTransactionAttribute(txAttr) {
          @Override
          public String getName() {
             return joinpointIdentification;
          }
       };
    }
    TransactionStatus status = null;
    if (txAttr != null) {
       if (tm != null) {
          status = tm.getTransaction(txAttr);
       }
       else {
          if (logger.isDebugEnabled()) {
             logger.debug("Skipping transactional joinpoint [" + joinpointIdentification +
                   "] because no transaction manager has been configured");
          }
       }
    }
    return prepareTransactionInfo(tm, txAttr, joinpointIdentification, status);
 }
```

TransactionAspectSupport 类的 createTransactionIfNecessary 方法实际上调用了 getTransaction 方法来判断 TransactionAttribute 是否具有事务，并且根据传播行为去做相应的处理。getTransaction 方法源码如代码清单 17-32 所示。

代码清单 17-32　getTransaction 方法源码

```
public final TransactionStatus getTransaction(@Nullable TransactionDefinition
definition) throws TransactionException {
    // 这里主要是调用了PlatformTransactionManager的doGetTransaction方法生成一个事务
    Object transaction = doGetTransaction();
    // 缓存调试标志以避免重复检查
    boolean debugEnabled = logger.isDebugEnabled();
    if (definition == null) {
        // 如果没有给出事务定义，则使用默认值
        definition = new DefaultTransactionDefinition();
    }
    // 判断是否已经存在事务
    if (isExistingTransaction(transaction)) {
        //为现有事务创建TransactionStatus。
        return handleExistingTransaction(definition, transaction, debugEnabled);
    }

    // 检查新事务的定义设置
    if (definition.getTimeout() < TransactionDefinition.TIMEOUT_DEFAULT) {
          throw new InvalidTimeoutException("Invalid transaction timeout",
```

```
definition.getTimeout());
    }

    // 如果找不到现有的事务，则检查传播行为
    if (definition.getPropagationBehavior() == TransactionDefinition.PROPAGATION_
MANDATORY) {
        throw new IllegalTransactionStateException(
                "No existing transaction found for transaction marked with
propagation 'mandatory'");
    }
    /*如果是PROPAGATION_REQUIRED、PROPAGATION_REQUIRES_NEW、PROPAGATION_NESTED这三种
类型，则开启一个新的事务*/
    else if (definition.getPropagationBehavior() == TransactionDefinition.
PROPAGATION_REQUIRED ||
            definition.getPropagationBehavior() == TransactionDefinition.
PROPAGATION_REQUIRES_NEW ||
            definition.getPropagationBehavior() == TransactionDefinition.
PROPAGATION_NESTED) {
        // 因为是全新的事务，不需要挂起事务，所以调用suspend方法传入null
        SuspendedResourcesHolder suspendedResources = suspend(null);
        if (debugEnabled) {
            logger.debug("Creating new transaction with name [" + definition.
getName() + "]: " + definition);
        }
        try {
            boolean newSynchronization = (getTransactionSynchronization() !=
SYNCHRONIZATION_NEVER);
            //创建一个TransactionStatus对象，后面介绍这个对象的作用
            DefaultTransactionStatus status = newTransactionStatus(
                    definition, transaction, true, newSynchronization,
debugEnabled, suspendedResources);
            // 开启一个事务
            doBegin(transaction, definition);
            prepareSynchronization(status, definition);
            return status;
        }
        catch (RuntimeException | Error ex) {
            resume(null, suspendedResources);
            throw ex;
        }
    }
    else {
        if (definition.getIsolationLevel() != TransactionDefinition.ISOLATION_
DEFAULT && logger.isWarnEnabled()) {
            logger.warn("Custom isolation level specified but no actual transaction
initiated; " +
                    "isolation level will effectively be ignored: " + definition);
        }
```

```
            boolean newSynchronization = (getTransactionSynchronization() ==
    SYNCHRONIZATION_ALWAYS);
            return prepareTransactionStatus(definition, null, true, newSynchronization,
    debugEnabled, null);
        }
    }
```

先对 PlatformTransactionManager 中 getTransaction 方法逻辑进行总结，后面再依次进行分析。

（1）调用 PlatformTransactionManager 类中的 doGetTransaction 方法，创建 Transaction 事务对象，该对象封装了数据库连接对象 Connection。

（2）调用 DataSourceTransactionManager 的 isExistingTransaction 方法判断之前是否开启了事务，如果已经开启事务，那么调用 AbstractPlatformTransactionManager 中的 handleExistingTransaction 方法为现有事务创建 TransactionStatus。

（3）如果没有开启事务，那么获取事务的传播行为，再进行相应的处理。如果传播行为是 TransactionDefinition.PROPAGATION_MANDATORY，那么抛出 IllegalTransactionStateException 异常。如果是 TransactionDefinition.PROPAGATION_REQUIRED、TransactionDefinition.PROPAGATION_REQUIRES_NEW、TransactiSonDefinition.PROPAGATION_NESTED 三个中的一个，那么调用 doBegin 方法开启一个新的事务。

17.3.4 声明式事务实现具体分析

当调用 PlatformTransactionManager 类中的 doGetTransaction 方法时，事务管理器默认使用的是 DataSourceTransactionManager。DataSourceTransactionManager 的 doGetTransaction 方法源码如代码清单 17-33 所示。

代码清单 17-33 doGetTransaction 方法源码

```
@Override
protected Object doGetTransaction() {
// 创建一个DataSourceTransactionObject对象
// 该对象主要用于存储数据库连接对象Connection和一些Connection的状态标识
   DataSourceTransactionObject txObject = new DataSourceTransactionObject();
   txObject.setSavepointAllowed(isNestedTransactionAllowed());
/* TransactionSynchronizationManager对象有一个ThreadLocal变量，用于存储当前线程的
ConnectionHolder，并且该对象具有Connection的创建、删除等功能。这里实际上是从ThreadLocal取出
ConnectionHolder。所以当线程第1次进来时，这里取出的ConnectionHolder为空*/
   ConnectionHolder conHolder =
   (ConnectionHolder) TransactionSynchronizationManager.getResource(obtainDataSource());
/* 把ConnectionHolder放入DataSourceTransactionObject对象中，并且传入false，标记该
ConnectionHolder不是一个新的连接 */
   txObject.setConnectionHolder(conHolder, false);
   return txObject;
}
```

从代码 17-33 的注释可以看出，核心是 TransactionSynchronizationManager.getResource(obtainDataSource()) 方法，TransactionSynchronizationManager.getResource 方法源码如代码清单 17-34 所示。

代码清单 17-34　TransactionSynchronizationManager.getResource 方法源码

```
// key是dataSource，也就是数据源
@Nullable
public static Object getResource(Object key) {
   Object actualKey = TransactionSynchronizationUtils.unwrapResourceIfNecessary(key);
// 把dataSource作为key，并在ThreadLocal中取出ConnectionHolder
   Object value = doGetResource(actualKey);
   if (value != null && logger.isTraceEnabled()) {
      logger.trace("Retrieved value [" + value + "] for key [" + actualKey + "]
bound to thread [" +
            Thread.currentThread().getName() + "]");
   }
// 返回ConnectionHolder
   return value;
}
```

接下来分析 TransactionSynchronizationManager 的 doGetResource 方法，其源码如代码清单 17-35 所示。

代码清单 17-35　TransactionSynchronizationManager.doGetResource 方法源码

```
@Nullable
private static Object doGetResource(Object actualKey) {
  // 从ThreadLocal取出一个Map
   Map<Object, Object> map = resources.get();
   if (map == null) {
  // 如果该MAP为null，那么直接返回null
      return null;
   }
   Object value = map.get(actualKey);
...// 删除标记为void的ResourceHolder
   if (value instanceof ResourceHolder && ((ResourceHolder) value).isVoid()) {
      map.remove(actualKey);
...// 如果为空，那么删除整个ThreadLocal
      if (map.isEmpty()) {
         resources.remove();
      }
      value = null;
   }
   return value;
}
```

resources 变量的定义为 private static final ThreadLocal<Map<Object, Object>> resources=new NamedThreadLocal<>("Transactional resources")。

Spring 通过 TransactionSynchronizationManager.getResource() 方法把 Connection 对象放入 ThreadLocal 中进行存储。

接下来分析 isExistingTransaction 方法是如何判断事务是否存在的。

DataSourceTransactionManager.isExistingTransaction 方法的源码如代码清单 17-36 所示。**提示：**

这里使用的是 PlatformTransactionManager 接口的实现类 DataSourceTransactionManager。

代码清单 17-36　DataSourceTransactionManager.isExistingTransaction 方法源码

```
@Override
protected boolean isExistingTransaction(Object transaction) {
   DataSourceTransactionObject txObject = (DataSourceTransactionObject) transaction;
/* 判断ConnectionHolder是否为空以及判断isTransaction是否为true*/
    return (txObject.hasConnectionHolder() && txObject.getConnectionHolder().
isTransactionActive());
}
```

接下来根据 isExistingTransaction 方法的返回结果，进行不同的处理。因为之前没有开启事务，所以会执行 DataSourceTransactionManager.doBegin 方法去开启新事务。

DataSourceTransactionManager.doBegin 方法的源码如代码清单 17-37 所示。

代码清单 17-37　DataSourceTransactionManager.doBegin 方法源码

```
//此方法实现设置隔离级别，但会忽略超时
@Override
protected void doBegin(Object transaction, TransactionDefinition definition) {
    DataSourceTransactionObject txObject = (DataSourceTransactionObject) transaction;
    Connection con = null;
    try {
        // 这里判断Transaction的ConnectionHolder为null
        if (!txObject.hasConnectionHolder() ||
                txObject.getConnectionHolder().isSynchronizedWithTransaction()) {
            // 从数据源DataSource获取一个数据库的连接Connection
            Connection newCon = obtainDataSource().getConnection();
            if (logger.isDebugEnabled()) {
                logger.debug("Acquired Connection [" + newCon + "] for JDBC transaction");
            }
            // 把新的ConnectionHolder连接放入Transaction对象，并且传入true，标记为全新的Connection
            txObject.setConnectionHolder(new ConnectionHolder(newCon), true);
        }
        txObject.getConnectionHolder().setSynchronizedWithTransaction(true);
        // 从ConnectionHolder取出一个数据库连接
        con = txObject.getConnectionHolder().getConnection();
        Integer previousIsolationLevel = DataSourceUtils.prepareConnectionForTransa
ction(con, definition);
        txObject.setPreviousIsolationLevel(previousIsolationLevel);
        // 必要时可以设置手动提交，默认从数据源取出的连接Connection是自动提交
        if (con.getAutoCommit()) {
            txObject.setMustRestoreAutoCommit(true);
            if (logger.isDebugEnabled()) {
                logger.debug("Switching JDBC Connection [" + con + "] to manual commit");
            }
            // 这里把Connection设置为手动提交，关闭自动提交
            con.setAutoCommit(false);
```

```
        }
        prepareTransactionalConnection(con, definition);
        // 把transaction标记为active，之前判断transaction是否存在时，就是通过这个标识
        txObject.getConnectionHolder().setTransactionActive(true);
        int timeout = determineTimeout(definition);
        if (timeout != TransactionDefinition.TIMEOUT_DEFAULT) {
            txObject.getConnectionHolder().setTimeoutInSeconds(timeout);
        }
        // 如果是全新的Connection，将连接绑定到线程
        if (txObject.isNewConnectionHolder()) {
                TransactionSynchronizationManager.bindResource(obtainDataSource(),
txObject.getConnectionHolder());
        }
    } catch (Throwable ex) { //...
    }
}
```


当同一个线程调用时，事务已经存在，那么调用 AbstractPlatformTransactionManager 中 handleExistingTransaction(definition, transaction, debugEnabled) 方法，根据传播行为的设置，进行相应的处理。handleExistingTransaction 方法源码如代码清单 17-38 所示。

代码清单 17-38 handleExistingTransaction 方法源码

```
private TransactionStatus handleExistingTransaction(
        TransactionDefinition definition, Object transaction, boolean debugEnabled)
        throws TransactionException {
    // 如果是TransactionDefinition.PROPAGATION_NEVER
    if (definition.getPropagationBehavior() == TransactionDefinition.PROPAGATION_NEVER) {
        throw new IllegalTransactionStateException(
                "Existing transaction found for transaction marked with propagation 'never'");
    }
    // 如果是TransactionDefinition.PROPAGATION_NOT_SUPPORTED
    if (definition.getPropagationBehavior() == TransactionDefinition.PROPAGATION_
NOT_SUPPORTED) {
        if (debugEnabled) {
            logger.debug("Suspending current transaction");
        }
        /* 把之前的Transaction直接挂起，返回一个SuspendedResourcesHolder，里面包装了挂起的
数据库连接*/
        Object suspendedResources = suspend(transaction);
            boolean newSynchronization = (getTransactionSynchronization() ==
SYNCHRONIZATION_ALWAYS);
        /* 创建一个TransactionStatus，把当前Transaction置为null；包装挂起的事务Suspended-
ResourcesHolder，用于后面恢复事务；把newTransaction标记置为false*/
        return prepareTransactionStatus(
                definition, null, false, newSynchronization, debugEnabled,
suspendedResources);
    }
```

```
    // 如果是TransactionDefinition.PROPAGATION_REQUIRES_NEW
    if (definition.getPropagationBehavior() == TransactionDefinition.PROPAGATION_
REQUIRES_NEW) {
        if (debugEnabled) {
            logger.debug("Suspending current transaction, creating new transaction with name [" +
                    definition.getName() + "]");
        }
        /* 把之前的Transaction直接挂起，返回一个SuspendedResourcesHolder，里面包装了挂起的
数据库连接*/
        SuspendedResourcesHolder suspendedResources = suspend(transaction);
        try {
                boolean newSynchronization = (getTransactionSynchronization() !=
SYNCHRONIZATION_NEVER);
                /* 创建一个TransactionStatus，把当前Transaction置为null，包装挂起的事务
SuspendedResourcesHolder，用于后面恢复事务；把newTransaction标记置为true*/
            DefaultTransactionStatus status = newTransactionStatus(
                        definition, transaction, true, newSynchronization,
debugEnabled, suspendedResources);
            //之前已经讲过doBegin方法了，用于为transaction绑定一个全新的数据库连接Connection
            doBegin(transaction, definition);
            prepareSynchronization(status, definition);
            return status;
        }
        catch (RuntimeException | Error beginEx) {

            resumeAfterBeginException(transaction, suspendedResources, beginEx);
            throw beginEx;
        }
    }
    // 如果是TransactionDefinition.PROPAGATION_NESTED
    if (definition.getPropagationBehavior() == TransactionDefinition.PROPAGATION_
NESTED) {
        if (!isNestedTransactionAllowed()) {
            throw new NestedTransactionNotSupportedException(
                    "Transaction manager does not allow nested transactions by
default - " +
                        "specify 'nestedTransactionAllowed' property with value 'true'");
        }
        if (debugEnabled) {
            logger.debug("Creating nested transaction with name [" + definition.
getName() + "]");
        }
        if (useSavepointForNestedTransaction()) {
                /*在现有的Spring管理的事务中创建保存点，通过TransactionStatus实现的
SavepointManager API。通常使用JDBC 3.0保存点。永远不会激活Spring同步。如果之前已经开启了事务，
那么就继续使用当前的事务，但是要把newTransaction的标记置为false*/
            DefaultTransactionStatus status =
                    prepareTransactionStatus(definition, transaction, false, false,
debugEnabled, null);
```

```
            // 创建一个savepoint
            status.createAndHoldSavepoint();
            return status;
        }
        else {
            /*通过嵌套的begin和commit/rollback调用嵌套事务。通常用于JTA；在存在JTA事务的情况下，Spring同步会在此处激活。*/
                boolean newSynchronization = (getTransactionSynchronization() != SYNCHRONIZATION_NEVER);
            // 如果之前没有开启事务，那么就重新开启一个事务，与PROPAGATION_REQUIRES相似
            DefaultTransactionStatus status = newTransactionStatus(
                            definition, transaction, true, newSynchronization, debugEnabled, null);
            doBegin(transaction, definition);
            prepareSynchronization(status, definition);
            return status;
        }
    }
    // 可能是PROPAGATION_SUPPORTS 或者 PROPAGATION_REQUIRED
    if (debugEnabled) {
        logger.debug("Participating in existing transaction");
    }

    if (isValidateExistingTransaction()) {
        if (definition.getIsolationLevel() != TransactionDefinition.ISOLATION_DEFAULT) {
            Integer currentIsolationLevel = TransactionSynchronizationManager.getCurrentTransactionIsolationLevel();
                if (currentIsolationLevel == null || currentIsolationLevel != definition.getIsolationLevel()) {
                Constants isoConstants = DefaultTransactionDefinition.constants;
                        throw new IllegalTransactionStateException("Participating transaction with definition [" +
                                definition + "] specifies isolation level which is incompatible with existing transaction: " +
                        (currentIsolationLevel != null ?
                                isoConstants.toCode(currentIsolationLevel, DefaultTransactionDefinition.PREFIX_ISOLATION) :
                                "(unknown)"));
            }
        }
        if (!definition.isReadOnly()) {
            if (TransactionSynchronizationManager.isCurrentTransactionReadOnly()) {
                        throw new IllegalTransactionStateException("Participating transaction with definition [" +
                            definition + "] is not marked as read-only but existing transaction is");
            }
        }
    }
```

```
    boolean newSynchronization = (getTransactionSynchronization() != SYNCHRONIZATION_
NEVER);
      /* 如果是PROPAGATION_SUPPORTS or PROPAGATION_REQUIRED，那么直接使用当前的
Transaction，但是需要把newTransaction的标记置为false */
      return prepareTransactionStatus(definition, transaction, false,
newSynchronization, debugEnabled, null);
}
```

AbstractPlatformTransactionManager 中 handleExistingTransaction 方法就是根据不同的事务传播行为进行一些不同处理。

在代码清单 17-31 中，已经调用了 getTransaction，且返回了 TransactionStatus 对象，该对象主要包含当前的 Transaction 和挂起的 Transaction，并且该对象有一个很重要的标识对象 new Transaction，该标识表示当前的事务是否为有权限的，后面提交事务时会用到。另外，还有一个 savepoint。

接着是调用 prepareTransactionInfo 方法，其源码如代码清单 17-39 所示。

代码清单 17-39　prepareTransactionInfo 方法源码

```
// 为给定的属性和状态对象准备TransactionInfo
protected TransactionInfo prepareTransactionInfo(@Nullable PlatformTransactionManagertm,
        @Nullable TransactionAttribute txAttr, String joinpointIdentification,
        @Nullable TransactionStatus status) {
// 创建一个新的TransactionInfo
    TransactionInfo txInfo = new TransactionInfo(tm, txAttr, joinpointIdentification);
    if (txAttr != null) {

        // 需要一个事务方法
        if (logger.isTraceEnabled()) {
            logger.trace("Getting transaction for [" + txInfo.getJoinpointIdentification()
+ "]");
        }
        // 如果已存在不兼容的tx，则将事务管理器标记为错误
        txInfo.newTransactionStatus(status);
    }
    else {
        // TransactionInfo.hasTransaction()方法返回false，保留此类中维护的ThreadLocal堆栈
的完整性
        if (logger.isTraceEnabled())
          logger.trace("Don't need to create transaction for [" + joinpointIdentification +
                  "]: This method isn't transactional.");
    }
  //  将TransactionInfo绑定到线程上，也就是ThreadLocal
    txInfo.bindToThread();
    return txInfo;
}
```

最后便是事务的提交和回滚了。通过调用 createTransactionIfNecessary 方法，返回一个

TransactionInfo 对象，并且把这个对象绑定到 ThreadLocal，然后开始执行下一个拦截器。当所有拦截器执行完，并且调用了原始方法后都没有出现异常时，就会执行 commitTransactionAfterReturning 方法进行事务的提交。如果出现了异常，则最终通过调用 AbstractPlatformTransactionManager 中的 doRollback 方法进行当前线程事务的回滚。

接下来通过图 17-4 概括 Transactional 注解的实现流程。

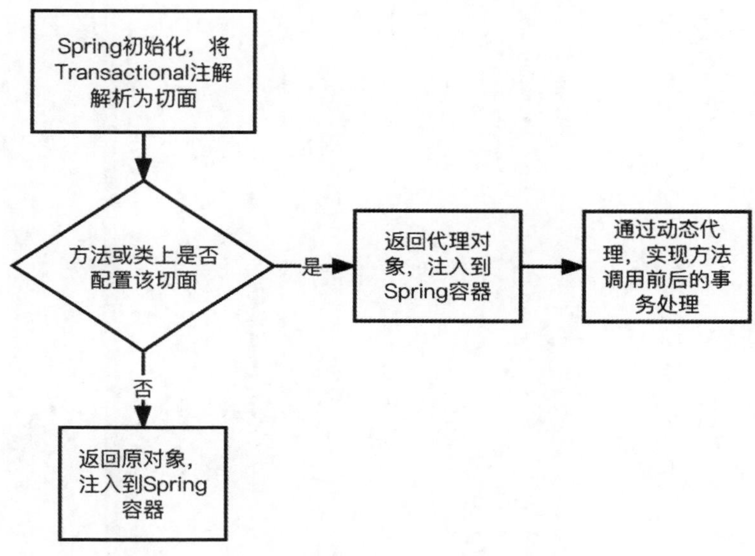

图 17-4　Transactional 注解的实现流程

本书讲到了很多源码，希望读者可以打开 IDE，配合本书进度，一个方法一个方法地阅读、实践。建议读者到 GitHub 网站将 Spring 的源码 fork 到自己的仓库，在阅读过程中，将自己的理解和注释同步到源码上，这样的阅读效果会更好。

Spring 最核心的还是 AOP 和 IoC 的思想是，希望读者能够自行体会。限于篇幅，本书无法讲解 Spring 的所有源码，其他源码希望读者能够按照 Spring 的各个功能进行查看；也可以在 Spring 源码中添加断点，在项目启动时，查看 Spring 的初始化过程。

第V篇　项目实战

本篇的案例参考的是一个实际案例。前面学习的Spring、Spring MVC、MyBatis相关知识，在本篇就会派上用场。在本篇读者可以了解到企业级项目研发的流程，从项目设计分析，到数据库设计，再到框架搭建，均有讲解。本篇结合Spring、Spring MVC、MyBatis框架，集成MyBatis-Plus快速操作Dao层，通过代码自动生成工具，自动生成非业务代码，集成Maven、快速管理项目依赖，集成log4j日志输出以及异常日志自动发送邮件；集成Redis框架，进行缓存和实现分布式锁，通过Redis实现Web层的Session数据共享，实现多端统一登录。

本案例实现了系统登录注册、全局异常处理、异常日志发送邮件、自动登录、多端统一自动登录、图片上传下载、资源发表、登录拦截、集成OSS图片存储、用户登录轨迹记录、用户关注收藏功能。

通过本篇的学习，希望读者可以完整掌握一个企业级项目的研发流程，对于项目功能有自己的思考。在企业的业务项目开发中，功能实现在业务成功与否的重要度上占比较小，前期的需求调研、项目评审，到功能上线后的运营策略、活动策略才是重中之重。希望读者在工作岗位上不仅是代码写得好，在思维上还要比产品经理想得远。

项目设计

本章要点

1. 企业级项目的简介与分析
2. 项目模块完整的项目设计流程
3. 项目依赖的技术分析
4. 项目的数据库设计

本章将完整地阐述如何实现一个企业级项目，但这里要提前说明以下 3 点：

（1）内容不会涉及对前端页面的开发，只讲解后端技术。

（2）在此项目中，前后端是不分离的。对于前后端分离问题，本书建议：如果有专门的前端团队和后端团队，那么可以选择前后端分离，否则前后端不分离更为便捷。

（3）该项目是边开发边记录的，目的是把开发过程真实完整地展现出来。因此，本章在叙述时可能会有设计不合理或者功能遗漏的地方，以此体现没有任何人从一开始就能把项目的所有内容设计得完美无缺。只有初学者真正感受了完整的过程并参与其中，才能获得成就感和宝贵的经验。当然，本书配套的 GitHub 仓库中的项目是完善的。

18.1 项目简介与分析

本项目主要实现的功能是资源互换。例如，用户 A 需要资源 a，却有资源 b；用户 B 需要资源 b，却有资源 c；用户 C 需要资源 c，却有资源 a，他们可以上架自己的资源，也可以购买自己需要的资源。所以该项目称为“资源平台”。资源平台作为中间商，为用户提供存储、交易的功能，并提供一种虚拟货币 U 币，用于购买资源。

用户 A 通过资源平台出售资源 b，当有人购买资源 b 时，用户 A 能获得相应数量的 U 币，用户可用获得的 U 币在平台上购买所需的资源 a。这里仅实现资源平台的两个核心需求：出售（分享）资源与购买（下载）资源。

下面分析该项目需要的功能。

18.2 模块与需求分析

本节介绍项目包含的各个模块，以及进行模块的需求分析。

18.2.1 模块划分

项目有两大模块：前台模块与管理模块。前台模块分为资源、用户、活动和用户中心；管理模块分为用户管理、资源管理、活动管理和权限管理。

前台模块与管理模块的划分如图 18-1 所示。

根据图 18-1 的功能划分，将项目代码模块分为三大块：Admin-Web（管理后台项目）、Web（前台功能项目）和 Service（后台服务项目）。

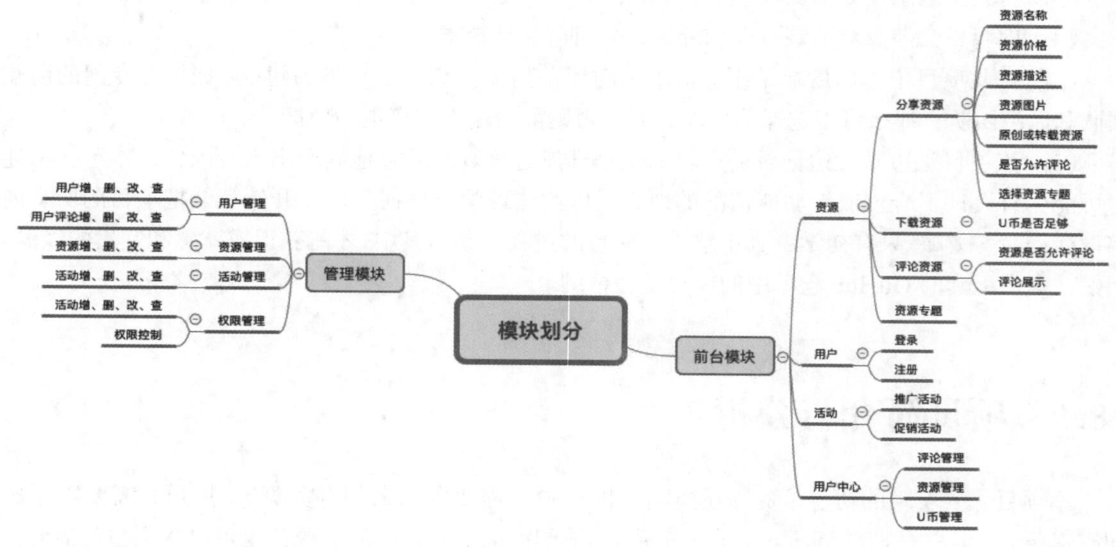

图 18-1　模块划分

18.2.2　前台模块

前台模块是面向用户的，要求分析如下：

- 资源：重点实现资源的分享、评论和下载功能。分享资源只针对已注册的用户，且分享时需要定义下载价格。对于资源下载功能，设置下载有效时间为 24h。对于评论功能，要进行敏感词过滤，实现自动审核。
- 用户：这里使用邮箱激活与手机号注册的方式实现用户的登录注册功能。登录注册功能需记录用户的 IP 地址、登录时间等登录信息。
- 活动：这是一个扩展功能，开启活动功能后，可在某些节假日对资源进行打折、抽奖、兑换等操作，这里暂时只实现打折功能。
- 用户中心：用户可以在用户中心编辑个人资料、管理评论、查看资源，以及 U 币的充值和获取记录查看等。

18.2.3　管理模块

管理模块就是实现后台的管理功能，用来让系统管理员对前台模块的用户、资源、活动和权限进行管理，需求分析如下：

- 用户管理：对用户的增、删、改、查，主要实现用户列表、用户编辑、禁用用户（不建议删除用户，保留原始数据非常重要）等功能。此外，还要实现用户评论管理功能，主要用来删除评论，如果自动审核未过滤，则还可以人工删除。
- 资源管理：对资源的增、删、改、查。

- 活动管理：对活动的增、删、改、查。
- 权限管理：权限管理中的“权限”，指购买之后方可操作的功能权限。例如，赋予注册用户用 U 币购买资源、下载资源的权限，控制某个资源的下载权限等（在项目中，除了管理员、普通注册用户及游客，不再对具体用户进行权限细分）。

18.3 技术及依赖分析

接下来介绍实现各种功能要用到的技术。对于开发者来说，通过需求就可以很快判断出哪些技术可以快速方便地实现。

该项目并没有什么难点，都是一些非常简单易懂的操作。关于技术框架的选择，笔者建议用 Spring Boot 来进行开发，因为 Spring Boot 足够快、足够方便，而且有非常多脚手架开源项目，可以一键生成从 Dao 层到 Controller 层的代码，生成数据库表，以及对应表单的增、删、改、查操作。

但是，本书选择基于 Spring MVC、Spring 和 MyBatis 框架进行开发，因为如果要更好地运用 Spring Boot，则一定要熟悉 Spring，最好是用 Spring 开发过项目。Spring Boot 可以看作 Spring 的注解增强版，中间省去了非常多的操作，开发者如果不明白这些原理，那么使用 Spring Boot 后会出现问题。

Web 容器的话，作为 Java 开发者，毫无疑问会选择 Tomcat。

数据库这里选用 MySQL，因为它开源，也是目前开发者使用最多的数据库。同时，为了能够提升系统的性能，需要选择缓存中间件，这里选择 Redis，因为它操作简单、性能高、开发效率高。

邮件的发送，使用专门的 Jar 包；短信的发送，可选择阿里大鱼短信平台；资源的存放，选择阿里云的 OSS。

至此，需要的技术依赖列举完毕。

18.4 数据库设计

数据库设计是比较重要的环节。前面讲过，由于并不是将项目全部完成后才讲解本章内容，所以在开发前期，数据库设计存在不合理的地方，后面会介绍相应的表结构改动的 SQL 语句。

在设计之前，对数据应该具有敏感性，即要清楚数据至少要有哪几列：创建时间、删除时间，以及更新时间，id 也是需要有的。由于数据量不多，这里选择自增的整型类型作为 id。

先用思维导图来进行数据库表设计，再进行表结构操作。用户相关表设计如图 18-2 所示。

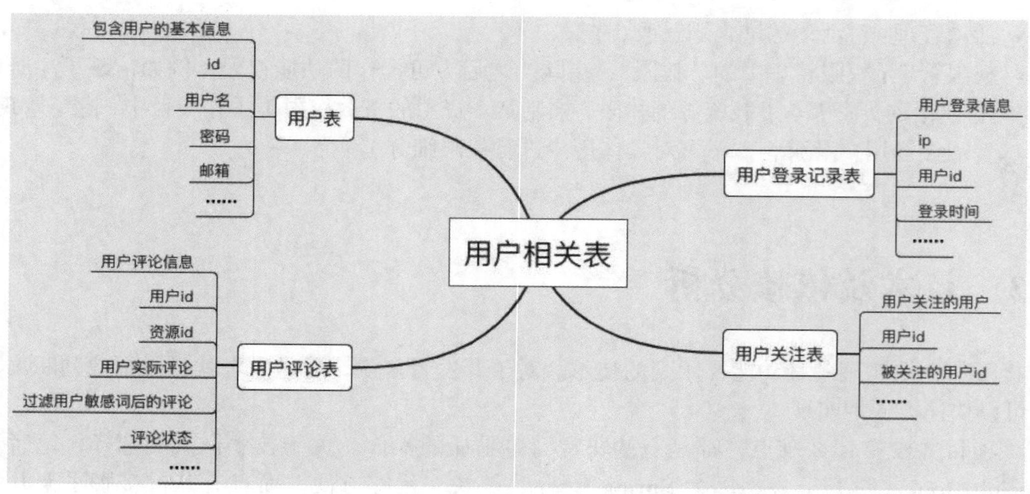

图 18-2 用户相关表设计

对管理员以及菜单的操作相关表可以进行菜单的权限控制，以及用户的权限控制，另外也集成了部门的管理，如图 18-3 所示。

既然系统定义了一种虚拟货币，那么肯定也需要围绕该虚拟货币进行数据库的定义，如图 18-4 所示。

接下来设计资源相关表，也是本项目的核心，如图 18-5 所示。

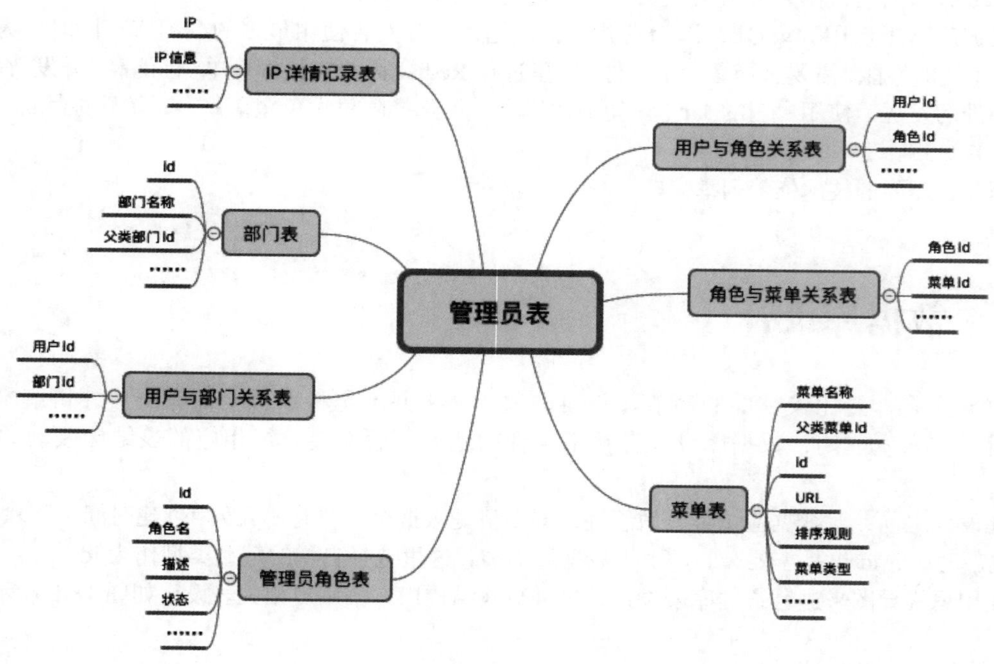

图 18-3 管理员表

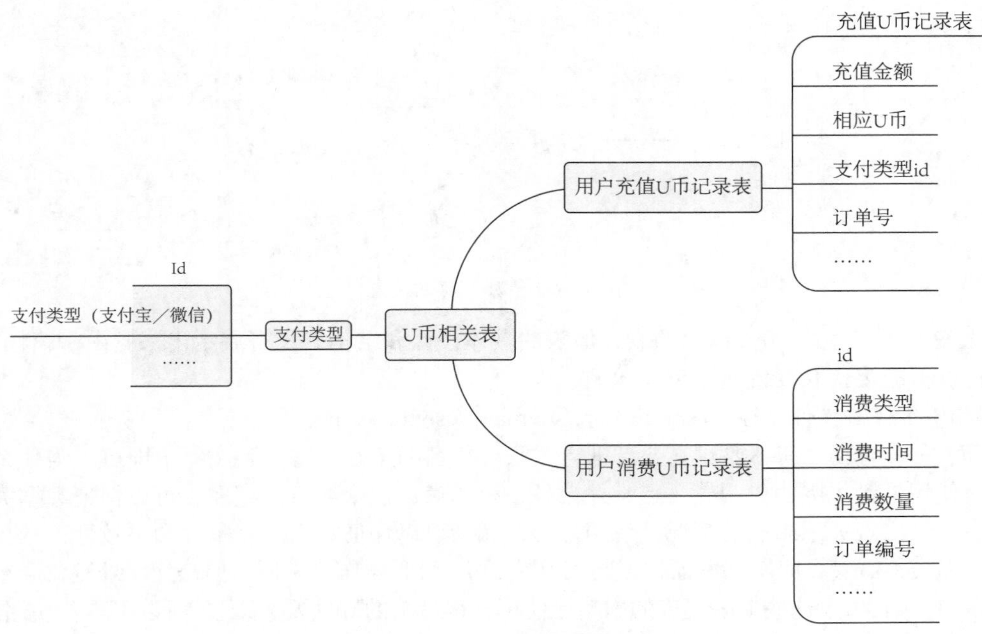

图 18-4　U 币相关表

资源有专题之分，方便在前端页面进行分类展示，以及便于用户自己开设专题，如图 18-6 所示。

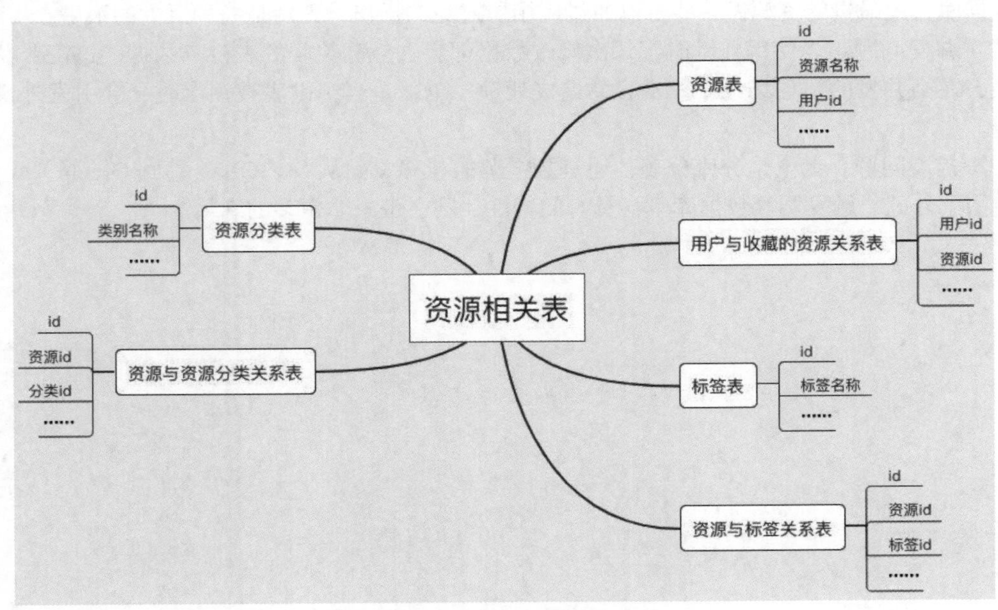

图 18-5　资源相关表

18

项目设计

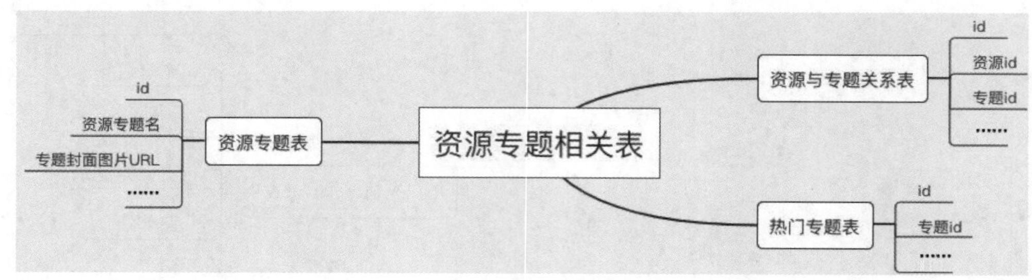

图 18-6　资源专题相关表

以上囊括了本系统 90% 以上的表，但表的具体设计和结构并没有展示出来。由于篇幅有限，可以通过以下链接获取表结构的 SQL 文件。

表结构的 SQL 文件路径：/ssm-uifuture/src/main/resources/sql/。

注意：在表与表之间不要建立外键进行关联，关系约定好，编写到注释中即可。为什么现在都不推荐使用外键了呢？先明确一点，外键是一种约束，这个约束的存在，可以保证数据关系的完整性，不会出现关系断裂的情况。使用外键，无疑可以保证数据的完整性和一致性，减少代码工作量（因为不需要在应用层保证数据的完整性了），级联操作更方便。但是使用外键，会给系统带来很多的缺陷，由于这些缺陷造成的影响比使用外键带来的好处要大得多，所以不推荐使用外键。

使用外键带来的问题可以简单地概括为以下 3 种：

（1）性能问题。当两张表直接有外键关联时，对其中一张表进行增、删、改操作，数据库会对另一张表的相关操作也进行检查，会带来额外的检查消耗。现在的互联网平台的瓶颈一般都体现在数据库，用昂贵的资源做一些可以通过应用层避免的操作，无疑是一种巨大的浪费。

（2）并发问题。当使用外键时，每次修改数据都会去外键关联的表检查数据，这需要获取额外的锁，在高并发的事务场景中非常容易造成死锁。也就是当主键表被锁定时，会引发外键表也被锁定。

（3）扩展问题。无论是分库分表，还是数据源的迁移（如从 MySQL 迁移到 Oracle）或扩展，如果使用了外键、触发器等数据库本身具有的特性，则会带来非常多的麻烦。

第19章

初步开发——框架集成

本章要点

1. 项目集成 Spring、Spring MVC、MyBatis 框架
2. 集成 log4j 日志与发送日志邮件
3. JDBC 集成以及数据库账户密码加密
4. 集成 MyBatis-Plus，快速生成数据库实体类
5. IDEA 配置 Tomcat Web 项目

19.1 框架集成简介

在前面4篇中，对Spring MVC、Spring、MyBatis这三大框架的知识点已经有了比较深入的介绍。可以将Spring理解成整合项目的大容器，在Spring中使用注解或者XML配置文件去实例化对象，其实就是IoC/DI，即控制反转/依赖注入（对IoC和DI在第14章中有深入介绍）。也就是说，不用再显式地新建一个对象，Spring框架已经完成了实例化操作。

Spring MVC在项目中可以简单理解为是与用户打交道的。在Spring MVC中使用配置拦截用户的请求，用户的请求会通过HandlerMapping匹配Controller。Controller就是具体请求所对应执行的操作，可以理解为具体的类或方法。学过Struts的读者对此会比较容易理解。

MyBatis是在项目中对数据库进行操作的，它让数据库操作变得简单且透明。MyBatis的操作都是通过操作sqlSessionFactory实例来展开的。通过配置文件关联到每个实体类对应的Mapper文件，在每个Mapper文件中配置了每个类对应数据库所需要进行的SQL语句映射；在Dao层，每个接口中的方法对应Mapper文件中的每个增、删、改、查操作。在学习MyBatis前，如果能手写JDBC，连接数据库并进行操作（第14章有实例介绍），会更容易理解MyBatis。在MyBatis项目中，用户每次与数据库进行的交互都是通过sqlSessionFactory实例拿到一个sqlSession，再执行SQL语句。

本章主要进行Spring、Spring MVC与MyBatis框架的整合，以及解决和总结在整合过程中遇到的问题。

19.2 搭建项目框架

本书主要的项目开发环境、工具有Mac、IntelliJ IDEA 2018.3、Tomcat 7.0.30、Java 1.8.0_31(JDK8)、Spring 5.0.8.RELEASE、MyBatis 3.4.6、Maven 3.3.9和MySQL 8等，其他的开源工具类版本可以在代码中看到。

19.2.1 创建Maven的Web项目

通过IntelliJ IDEA（以下简称为IDEA）创建Maven项目，选择File → New → Project...命令，在打开的界面中进行相应的设置，如图19-1所示。

单击Next按钮进行下一步，为项目起唯一的标识，如图19-2所示。

- GroupId：项目的唯一标识（最好是分为多个段）。一般来说，选择用户的域名进行命名。在这里，域名是uifuture.com，又因为是一个SSM整合项目，所以使用com.uifuture.ssm来进行命名。
- ArtifactId：对应项目的实际名称，也就是项目根目录的名称。

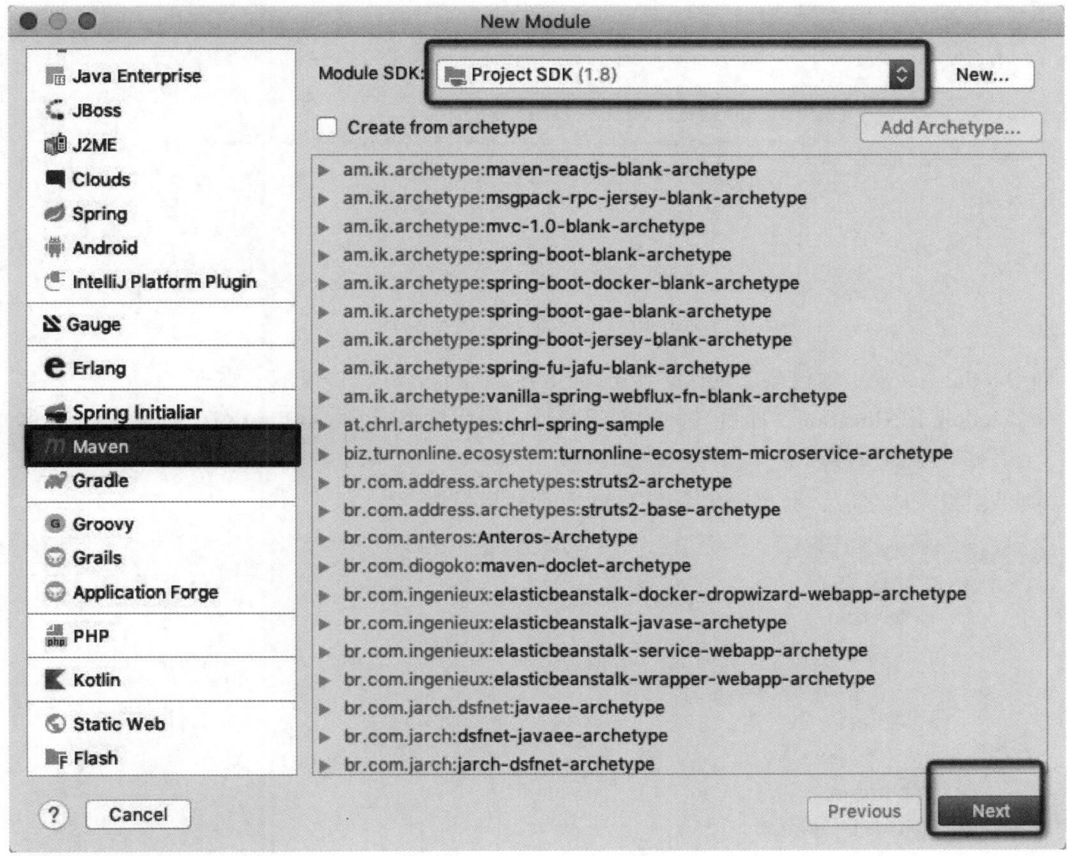

图 19-1　创建 Maven 项目

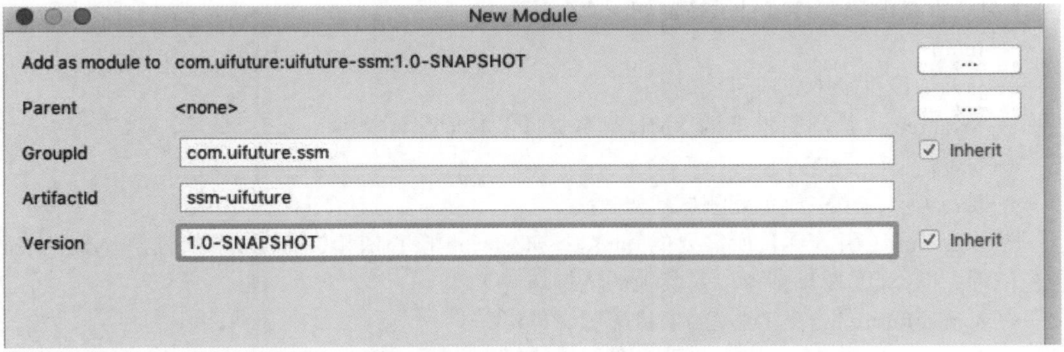

New Module

Add as module to　com.uifuture:uifuture-ssm:1.0-SNAPSHOT　...

Parent　<none>　...

GroupId　com.uifuture.ssm　Inherit

ArtifactId　ssm-uifuture

Version　1.0-SNAPSHOT　Inherit

图 19-2　为项目起唯一的标识

GroupId 和 ArtifactId 统称为“坐标”，这是为了保证项目唯一性而提出的。如果要把项目存储到 Maven 本地仓库或中央仓库，可以根据这两个 ID 导入该 .jar 包。

填写完成后继续下一步，进入图 19-3 所示的界面。

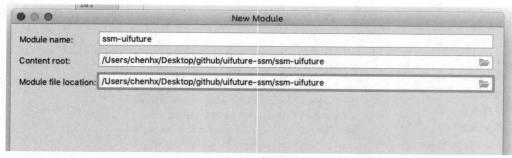

图 19-3　项目名及存储路径

- Module name：项目名。默认是 ArtifactId 中填写的名称，建议不要修改。
- Module file location：项目文件的创建路径。最后的目录 ssm-uifuture 在 SSM 中默认为项目名称，不需要修改。

项目创建完成之后，会自动生成图 19-4 所示的目录树。

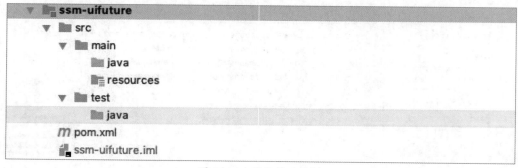

图 19-4　项目的目录树

src 目录下是源码，具体目录结构如下：

——main；

——java（该目录中放项目代码）；

——resources（放项目框架的 XML 配置文件及其他配置文件）；

——test；

——java（该目录中放单元测试代码）；

——pom.xml（引入项目所需要的 .jar 包，如 spring 核心依赖、MyBatis 依赖、mybatis-spring 整合包依赖、MySQL 驱动依赖以及其他的依赖包等）；

——ssm-uifuture.iml（IDEA 的项目配置文件）；

——External Libraries（项目中导入的 jar 包都在这里）。

19.2.2　配置项目依赖

配置 pom.xml 文件来引入所需依赖的 .jar 包，如代码清单 19-1 所示。

代码清单 19-1　pom.xml 配置文件

```
<?xml version="1.0" encoding="UTF-8"?>
<project xmlns="http://maven.apache.org/POM/4.0.0"
         xmlns:xsi="http://www.w3.org/2001/XMLSchema-instance"
           xsi:schemaLocation="http://maven.apache.org/POM/4.0.0 http://maven.
apache.org/xsd/maven-4.0.0.xsd">
    <modelVersion>4.0.0</modelVersion>
    <groupId>com.uifuture.ssm</groupId>
    <artifactId>ssm-uifuture</artifactId>
    <version>1.0-SNAPSHOT</version>
    <properties>
        <java.version>1.8</java.version>
        <!-- Spring -->
        <spring-framework.version>5.0.8.RELEASE</spring-framework.version>
        <!--MyBatis -->
        <mybatis.version>3.4.6</mybatis.version>
        <!-- MyBatis与Spring整合的核心包 -->
        <mybatis-spring.version>2.0.2</mybatis-spring.version>
        <!-- MySQL -->
        <mysql.version>8.0.15</mysql.version>
        <!-- Junit -->
        <junit.version>4.12</junit.version>
        <alibaba.druid.version>1.1.10</alibaba.druid.version>
        <jackson-mapper-asl.version>1.9.13</jackson-mapper-asl.version>
        <aspectjweaver.version>1.9.0</aspectjweaver.version>
    </properties>
    <dependencies>
        <!-- 添加spring-core包 -->
        <dependency>
            <groupId>org.springframework</groupId>
            <artifactId>spring-core</artifactId>
            <version>${spring-framework.version}</version>
        </dependency>
        <!-- 添加spring-context包 -->
        <dependency>
            <groupId>org.springframework</groupId>
            <artifactId>spring-context</artifactId>
            <version>${spring-framework.version}</version>
        </dependency>
        <dependency>
            <groupId>org.springframework</groupId>
            <artifactId>spring-context-support</artifactId>
            <version>${spring-framework.version}</version>
        </dependency>

        <!-- 添加spring-tx包 -->
        <dependency>
            <groupId>org.springframework</groupId>
            <artifactId>spring-tx</artifactId>
```

```
        <version>${spring-framework.version}</version>
    </dependency>
    <!-- 添加spring-jdbc包 -->
    <dependency>
        <groupId>org.springframework</groupId>
        <artifactId>spring-jdbc</artifactId>
        <version>${spring-framework.version}</version>
    </dependency>
    <!-- 为了方便进行单元测试，添加spring-test包 -->
    <dependency>
        <groupId>org.springframework</groupId>
        <artifactId>spring-test</artifactId>
        <version>${spring-framework.version}</version>
        <scope>test</scope>
    </dependency>
    <!-- 添加Spring MVC包 -->
    <dependency>
        <groupId>org.springframework</groupId>
        <artifactId>spring-webmvc</artifactId>
        <version>${spring-framework.version}</version>
    </dependency>
    <!-- 添加MyBatis的核心包 -->
    <dependency>
        <groupId>org.mybatis</groupId>
        <artifactId>mybatis</artifactId>
        <version>${mybatis.version}</version>
    </dependency>
    <!-- 添加MyBatis与Spring整合的核心包 -->
    <dependency>
        <groupId>org.mybatis</groupId>
        <artifactId>mybatis-spring</artifactId>
        <version>${mybatis-spring.version}</version>
    </dependency>
    <!-- 添加mysql驱动包 -->
    <dependency>
        <groupId>mysql</groupId>
        <artifactId>mysql-connector-java</artifactId>
        <version>${mysql.version}</version>
        <scope>runtime</scope>

    </dependency>
    <!-- 添加junit单元测试包 -->
    <dependency>
        <groupId>junit</groupId>
        <artifactId>junit</artifactId>
        <version>${junit.version}</version>
        <scope>test</scope>
    </dependency>
    <!-- 添加druid连接池包 -->
```

```
<dependency>
    <groupId>com.alibaba</groupId>
    <artifactId>druid</artifactId>
    <version>${alibaba.druid.version}</version>
</dependency>
<!-- spring-aop支持-->
<dependency>
    <groupId>org.springframework</groupId>
    <artifactId>spring-aop</artifactId>
    <version>${spring-framework.version}</version>
</dependency>
<!-- aspectj支持 -->
<!-- https://mvnrepository.com/artifact/org.aspectj/aspectjrt -->
<dependency>
    <groupId>org.aspectj</groupId>
    <artifactId>aspectjrt</artifactId>
    <version>${aspectjweaver.version}</version>
</dependency>
<!--添加aspectjweaver包 -->
<dependency>
    <groupId>org.aspectj</groupId>
    <artifactId>aspectjweaver</artifactId>
    <version>${aspectjweaver.version}</version>
</dependency>

<!--导入Jackson包，在SpringMVC返回JSON数据时用,防止IE出问题-->
<dependency>
    <groupId>org.codehaus.jackson</groupId>
    <artifactId>jackson-mapper-asl</artifactId>
    <version>${jackson-mapper-asl.version}</version>
</dependency>
<dependency>
    <groupId>org.codehaus.jackson</groupId>

    <artifactId>jackson-core-asl</artifactId>
    <version>${jackson-mapper-asl.version}</version>
</dependency>

<!--文件上传依赖的.jar包-->
<dependency>
    <groupId>commons-fileupload</groupId>
    <artifactId>commons-fileupload</artifactId>
    <version>1.3.2</version>
</dependency>

<!--J2EE相关包Servlet、JSP、JSTL-->
<!--javax.servlet HttpServletRequest必须要有的包-->
<dependency>
    <groupId>javax.servlet</groupId>
```

```
            <artifactId>javax.servlet-api</artifactId>
            <version>3.0.1</version>
        </dependency>
        <!-- jstl -->
        <dependency>
            <groupId>javax.servlet</groupId>
            <artifactId>jstl</artifactId>
            <version>1.2</version>
        </dependency>
        <dependency>
            <groupId>javax.servlet.jsp</groupId>
            <artifactId>jsp-api</artifactId>
            <version>2.2.1-b03</version>
        </dependency>

        <!--阿里巴巴的JSON解析工具-->
        <dependency>
            <groupId>com.alibaba</groupId>
            <artifactId>fastjson</artifactId>
            <version>1.2.60</version>
        </dependency>

        <dependency>
            <groupId>org.apache.commons</groupId>
            <artifactId>commons-lang3</artifactId>
            <version>3.4</version>
        </dependency>
        <dependency>

            <groupId>commons-lang</groupId>
            <artifactId>commons-lang</artifactId>
            <version>2.6</version>
        </dependency>
        <dependency>
            <groupId>commons-codec</groupId>
            <artifactId>commons-codec</artifactId>
            <version>1.10</version>
        </dependency>
        <!-- lombok依赖，实体类自动添加setter、getter、toString等方法 -->
        <dependency>
            <groupId>org.projectlombok</groupId>
            <artifactId>lombok</artifactId>
            <version>1.16.18</version>
        </dependency>
    </dependencies>
    <build>
        <!--为了解决IDEA不将src/main/java下的XML文件也编译成.class文件的问题，而添加下面的
resources-->
        <resources>
          <!--表示把java目录下有关XML文件、properties文件在编译/打包时放在resource目录下-->
```

```
            <resource>
                <directory>${basedir}/src/main/java</directory>
                <includes>
                    <include>**/*.properties</include>
                    <include>**/*.xml</include>
                </includes>
            </resource>
            <resource>
                <directory>${basedir}/src/main/resources</directory>
            </resource>
        </resources>
        <plugins>
            <!-- 指定Maven编译的JDK版本。如果不指定，则Maven 3默认用JDK 1.5，Maven 2默认用
JDK 1.3。在用Maven打包的过程中用到JDK高版本新特性时会出现错误 -->
            <plugin>
                <groupId>org.apache.maven.plugins</groupId>
                <artifactId>maven-compiler-plugin</artifactId>
                <configuration>
                    <source>1.8</source>
                    <target>1.8</target>

                </configuration>
            </plugin>
        </plugins>

    </build>
</project>
```

19.2.3 Spring 和 MyBatis 整合配置

配置 MySQL 的 properties 配置文件（JDBC 属性配置文件），命名为 config.properties。JDBC 属性配置文件如代码清单 19-2 所示。

代码清单 19-2　JDBC 属性配置文件

```
driverClassName=com.mysql.jdbc.Driver
jdbc_url=jdbc:mysql://localhost:3306/uifuture?useUnicode=true&characterEncoding=UTF-
8&zeroDateTimeBehavior=convertToNull
jdbc_username=4B1F1099FE3DCDEE0C4B5ED079322488
jdbc_password=52F6086FF5DF8415A9986AAE0C6FD56D
#初始化连接大小
jdbc_init=50
#连接池最小空闲连接数
jdbc_minIdle=20
#获取连接最大等待时间(单位为ms)
jdbc_maxActive=60000
```

19.3 节中会讲到如何对此处数据库的账号和密码进行加密，以让账号和密码的明文不直接泄露。这里对 jdbc_url 中的 zeroDateTimeBehavior 参数进行介绍。

Java 连接 MySQL 数据库后，对于将 TIMESTAMP 赋值为“0000-00-00 00:00:00”（或者其他不能正常解析的类型）的情况，会抛出“java.sql.SQLException: Cannot convert value '******' from column 1 to TIMESTAMP”的 SQLException 异常。

在 JDBC 连接中有一个 zeroDateTimeBehavior 属性，用来在连接中配置出现赋值解析异常的处理策略。该属性有以下 3 种属性值：

（1）exception（默认值）：不指定情况下的默认值，抛出 SQLException 异常。

（2）convertToNull：将错误的类型值转换为 NULL。

（3）round：替换成离现在最近的日期（即 ××××-01-01）。

没有特殊要求的情况下，选择 convertToNull 即可。**提示**：应根据个人实际项目来进行选择。

在 src/main/java 目录中创建包 com.uifuture.ssm，在 resources 目录下创建 spring-mybatis.xml 和 spring.xml 的配置文件。

spring.xml 文件内容如代码清单 19-3 所示。

代码清单 19-3　spring.xml 文件内容

```
<?xml version="1.0" encoding="UTF-8"?>
<beans xmlns="http://www.springframework.org/schema/beans"
       xmlns:xsi="http://www.w3.org/2001/XMLSchema-instance"
       xmlns:p="http://www.springframework.org/schema/p"
       xmlns:tx="http://www.springframework.org/schema/tx"
       xmlns:aop="http://www.springframework.org/schema/aop"
       xmlns:context="http://www.springframework.org/schema/context"
       xsi:schemaLocation="
http://www.springframework.org/schema/beans
http://www.springframework.org/schema/beans/spring-beans.xsd
http://www.springframework.org/schema/tx
http://www.springframework.org/schema/tx/spring-tx.xsd
http://www.springframework.org/schema/aop
http://www.springframework.org/schema/aop/spring-aop.xsd
http://www.springframework.org/schema/context
http://www.springframework.org/schema/context/spring-context.xsd
">
    <!-- 引入属性文件classpath就是在resource下的文件-->
     <context:property-placeholder location="classpath:config.properties" ignore-
unresolvable="true"/>
    <!-- 包自动扫描，不扫描Controller注解-->
    <context:component-scan base-package="com.uifuture.ssm">
        <context:exclude-filter type="annotation" expression="org.springframework.
stereotype.Controller"/>
    </context:component-scan>
    <!--配置sqlSessionFactory -->
    <bean id="sqlSessionFactory" class="org.mybatis.spring.SqlSessionFactoryBean">
        <property name="configLocation" value="classpath:spring-mybatis.xml"/>
        <property name="dataSource" ref="dataSource"/>
        <property name="mapperLocations">
            <list>
```

```
            <value>classpath*:com/uifuture/ssm/*/mapping/*.xml</value>
        </list>
    </property>
</bean>
<!-- 自动扫描mapper接口-->
<bean class="org.mybatis.spring.mapper.MapperScannerConfigurer"
      p:basePackage="com.uifuture.ssm.**.dao"
      p:sqlSessionFactoryBeanName="sqlSessionFactory"/>
<!-- 配置数据源 -->
<bean id="dataSource" class="com.alibaba.druid.pool.DruidDataSource"
      p:driverClassName="${driverClassName}"
      p:url="${jdbc_url}"
      p:username="${jdbc_username}"
      p:password="${jdbc_password}"
      p:initialSize="${jdbc_init}"
      p:minIdle="${jdbc_minIdle}"
      p:maxActive="${jdbc_maxActive}"
      p:filters="stat,wall"
>
</bean>
<!-- 配置事务管理类 -->
  <bean id="transactionManager" class="org.springframework.jdbc.datasource.
DataSourceTransactionManager"
      p:dataSource-ref="dataSource">
</bean>
<!-- 开启注解式事务扫描 -->
<tx:annotation-driven transaction-manager="transactionManager"/>
<!--使用基于AspectJ的自动代理-->
<aop:aspectj-autoproxy proxy-target-class="true"/>
 <!--proxy-target-class="true" , Spring建议通过 proxy-target-class 属性值来控制是基
于接口的还是基于类的代理被创建:
如果 proxy-target-class属性值被设置为true，那么基于类的代理将起作用(这时需要CGLIB库cglib.
jar在CLASSPATH中)。
如果 proxy-target-class 属性值被设置为false或者这个属性被省略，那么标准的JDK基于接口的代理将
起作用。接口上的注解不能继承 -->
</beans>
```

在 spring.xml 中，主要进行了以下配置：

（1）引入属性文件。

（2）包自动扫描，将标注 Spring 注解的类自动转换为 Bean，同时完成 Bean 的注入。

（3）配置 sqlSessionFactory。

（4）自动扫描 Mapper 接口。

（5）开启注解式事务扫描。

（6）配置数据源。

（7）配置事务管理类。

（8）使用基于 AspectJ 的自动代理。

以上是最基本的 spring-mybatis 的整合配置信息，还有一些配置信息，笔者会在后面的讲解中逐渐加入。

MyBatis 已经在 spring-mybatis.xml 文件中配置好了，如代码清单 19-4 所示。

代码清单 19-4　spring-mybatis.xml 文件

```
<?xml version="1.0" encoding="UTF-8" ?>
<!DOCTYPE configuration
        PUBLIC "-//mybatis.org//DTD Config 3.0//EN"
        "http://mybatis.org/dtd/mybatis-3-config.dtd">
<configuration>
    <!-- -->
    <settings>
        <!-- 指定 LOG4J为MyBatis 所用日志的具体实现，未指定时将自动查找。
         需要设置 log4j.properties /log4j.xml 文件-->
        <setting name="logImpl" value="LOG4J"/>
         <!-- 启用下划线到驼峰式命名规则的映射(如first_name => firstName)。开启后，mapper配
置不需要写字段与属性的配置，会自动映射-->
        <setting name="mapUnderscoreToCamelCase" value="true"/>
    </settings>
    <!-- 别名配置 -->
    <!--<typeAliases>-->
    <!--进行Dao实体类的一些别名配置-->
    <!--</typeAliases>-->
</configuration>
```

logImpl 配置可以指定 MyBatis 所用日志的具体实现（SLF4J | LOG4J | LOG4J2 | JDK_LOGGING | COMMONS_LOGGING | STDOUT_LOGGING | NO_LOGGING），未指定时将自动查找。在这里需要在 resources 目录下设置 log4j.properties 或 log4j.xml 文件。

mapUnderscoreToCamelCase 可以设置启用下划线到驼峰式命名规则的映射（如 user_name => userName）。开启后，Mapper 配置不需要编写字段与属性的配置，将自动映射，这样当编写实体类的映射文件时会节省很多时间。不过，因为使用的是 MyBatis-Generator（逆向代码生成器），它也能自动生成映射文件。关于 Spring MVC 的集成，将在第 20 章进行详细配置讲解。

19.2.4　配置 log4j 与发送日志邮件

接下来进行 log4j 日志的配置。先在 pom.xml 中添加 log4j 的依赖。当然，如果使用的是 slf4j，还需要另外两个依赖，如代码清单 19-5 所示。

代码清单 19-5　log4j 依赖

```
<!--log4j的jar包-->
<dependency>
    <groupId>log4j</groupId>
    <artifactId>log4j</artifactId>
    <version>1.2.17</version>
</dependency>
```

```
<dependency>
    <groupId>org.slf4j</groupId>
    <artifactId>slf4j-api</artifactId>
    <version>1.7.7</version>
</dependency>
<dependency>
    <groupId>org.slf4j</groupId>
    <artifactId>slf4j-log4j12</artifactId>
    <version>1.7.25</version>
</dependency>
```

在 resources 目录下创建 log4j.properties 文件。要对日志进行配置，如代码清单 19-6 所示。

代码清单 19-6 日志配置

```
#####配置根元素
#调试模式，写成DEBUG——设置日志级别，存储DEBUG以及以上级别的记录
log4j.rootLogger=DEBUG,console,file,error,Mail
#生产级项目,写成INFO即可
#INFO级别比DEBUG级别高——console,file,error都是开发者编写的，可以自定义名称
####以下配置为输出到控制台的配置####
log4j.appender.console=org.apache.log4j.ConsoleAppender
log4j.appender.console.layout=org.apache.log4j.PatternLayout
log4j.appender.console.Target=System.out
log4j.appender.console.layout.ConversionPattern=[%p] %d{yyyy-MM-dd HH:mm:ss} [%l]
-> %m [%t] [uif]%n
### 输出到日志文件 ###
log4j.appender.file=org.apache.log4j.RollingFileAppender
#RollingFileAppender按log文件最大长度限度生成新文件
#以下是具体的目录
log4j.appender.file.File=../logger/info/info.log
#每个文件的最大尺寸
log4j.appender.file.MaxFileSize=30MB
#最多可以是多少个文件
log4j.appender.file.MaxBackupIndex=30
##输出INFO级别以上的日志——Threshold是一个全局的过滤器，它将低于所设置INFO级别的信息过滤掉
log4j.appender.file.Threshold=DEBUG
#服务器启动后默认是追加日志，false:服务器启动后会生成日志文件，并把旧的覆盖掉
log4j.appender.file.Append=true
#信息的布局格式，按指定的格式输出
log4j.appender.file.layout=org.apache.log4j.PatternLayout
#具体的布局格式 ，%d为时间
#log4j.appender.file.layout.ConversionPattern=%d %p [%c] %m %l %n
log4j.appender.file.layout.ConversionPattern=[%p] %d{yyyy-MM-dd HH:mm:ss} [%l] ->
%m [%t] [uif]%n
#设置输出日志文件编码(可以控制乱码情况)
log4j.appender.file.encoding=UTF-8
### 保存异常信息到单独文件 ###
log4j.appender.error=org.apache.log4j.DailyRollingFileAppender
#DailyRollingFileAppender按日期生成新文件
##异常日志文件名
```

```
log4j.appender.error.File=../logger/error/error.log
#在每天产生的文件后面追加
log4j.appender.error.DatePattern='.'yyyyMMdd
log4j.appender.error.Append=true
##只输出ERROR级别以上的日志
log4j.appender.error.Threshold=error
log4j.appender.error.layout=org.apache.log4j.PatternLayout
log4j.appender.error.layout.ConversionPattern=[%p] %d{yyyy-MM-dd HH:mm:ss} [%l] ->
%m [%t] [uif]%n
log4j.appender.error.encoding=UTF-8
#发送邮件
#########################
#邮件发送配置
########################
log4j.appender.Mail=org.apache.log4j.net.SMTPAppender
#日志的错误级别
log4j.appender.Mail.Threshold=ERROR
#缓存文件大小
log4j.appender.Mail.BufferSize=512
#发送邮件的邮箱
log4j.appender.Mail.From=*****
log4j.appender.Mail.SMTPProtocol=smtp
#发送邮件的服务器
log4j.appender.Mail.SMTPHost=smtp.*****
#发送邮件的服务器端口号
log4j.appender.Mail.SMTPPort=25
#发送邮件的用户
log4j.appender.Mail.SMTPUsername=*****
#发送邮件的邮箱密码
log4j.appender.Mail.SMTPPassword=*****
#邮件主题
log4j.appender.Mail.Subject=Log4J Message
#接收邮件的邮箱
log4j.appender.Mail.To=*****
#发送邮件的格式
#log4j.appender.Mail.layout=org.apache.log4j.HTMLLayout
log4j.appender.Mail.layout=org.apache.log4j.PatternLayout
log4j.appender.Mail.layout.ConversionPattern=[%p] %d{yyyy-MM-dd HH:mm:ss} [%l] ->
%m [%t] [uif]%n
```

导入邮件发送包，如代码清单 19-7 所示。

代码清单 19-7　导入邮件发送包

```
<!-- 发送邮件相关表 -->
<dependency>
    <groupId>javax.mail</groupId>
    <artifactId>mail</artifactId>
    <version>1.4.7</version>
</dependency>
<dependency>
```

```
        <groupId>javax.activation</groupId>
        <artifactId>activation</artifactId>
        <version>1.1.1</version>
    </dependency>
```

对于 log4j 日志的输出来说，要对其配置进行一一讲解是比较麻烦的。在这里，读者可以看代码中的注解。对于 ConversionPattern 输出格式，请参见表 19-1 列出的详细说明及示例。

表 19–1　ConversionPattern 输出格式详细说明及示例

参　数	说　明	示　例	
		log4j 配置文件参数举例	举例输出说明（假设当前 logger 类全名是 "a.b.c"）
%c	列出 logger 名称空间的全称（用于输出的记录事件的类别全称），后面加上 %d{<层数>} 表示列出从最后面那层算起的指定层数的名称空间（不理解名称空间全称的，可以直接将其当作类名全称来理解）	%c	a.b.c（输出日志信息所属的类目，通常是所在类的全称）
		%c{2}	b.c（以“.”为分隔符，选择后面两个值）
		%20c	指定输出类目的名称，最小的宽度是 20。如果类目的名称小于 20，默认情况下右对齐，左边用空格填充补全
		%-20c	指定输出类目的名称，最小的宽度是 20。如果类目的名称小于 20，"-" 指定左对齐。右边用空格填充补全
		%20.30c	如果类目的名称小于 20 就补空格，并且右对齐（也就是用空格补在左边，注意这里与单独的 %20c 有点不同）。如果类目的名称多于 30 个字符，就截去右边超出的部分
		%-20.30c	如果类目的名称小于 20 个字符就补空格，并且左对齐（也就是空格补在右边，注意这里与单独的 %-20c 有点不同）。如果类目的名称多于 30 字符，就截去右边超出的部分
%C	列出调用 logger 的类的全称（包含包路径）	假设当前类是 org.apache.xyz.SomeClass	
		%C	org.apache.xyz.SomeClass
		%C{1}	SomeClass
		%C{2}	xyz.SomeClass
%d	用于输出记录事件的日期。默认为 ISO8601 格式，也可以在其后指定格式	%d{yyyy-MM-dd HH:mm:ss SSS}	2022-06-30 18:53:46 119
		%d{ABSOLUTE}	18:53:46,119
		%d{DATE}	30 六月 2022 18:53:46,119 （在英文系统环境下月份输出的是英文）
		%d{ISO8601}	2022-06-30 18:55:13,900
%F	显示调用 logger 类的源文件名	%F	TestLog.java （即输出调用的文件名）
%l	输出日志事件的发生位置，其中包括类目名、发生的线程，以及在代码中的行数（相当于 %C.%M(%F:%L) 的组合）	%l	com.uifuture.TestLog.main(TestLog.java:22)

续表

参　数	说　明	示　例	
		log4j 配置文件参数举例	举例输出说明（假设当前 logger 类全名是 "a.b.c"）
%L	显示调用输出日志的代码行	%L	22
%m	输出代码中指定的消息，产生的日志具体信息	%m	自定义填写的输出字符串
%M	显示调用 logger 类的方法名	%M	main
%n	当前系统平台下的换行符	%n	Windows 平台下表示 /r/n UNIX 平台下表示 /n
%p	输出日志信息优先级，即 DEBUG、INFO、WARN、ERROR、FATAL	%p	INFO
%r	显示从程序启动时到记录该条日志时已经经过的毫秒数	%r	1250
%t	输出产生该日志事件的线程名	%t	main （也就是主线程）
%x	输出和当前线程相关联的 NDC（嵌套诊断环境）	%x	可以使用 org.apache.log4j.NDC 类进行测试
%%	显示出一个百分号	%%	%

这里对 log4j 的其他配置不进行过多讲解。在输出格式上，常用的、需要记住的是 %d、%l、%m、%p、%n，其他的了解就行。特别是像日志输出的配置文件，第 1 次自行配置之后，基本上就有了一个初步的配置框架。对于以后的项目来说，复用和查看文档优化即可。

19.3　加密数据库账号密码

也许在某些时刻需要对数据库的配置文件进行加密，如至少要让数据库的账号密码不会在配置文件中以明文的方式出现。下面不只是讲解如何进行文件配置，重点讲解了当遇到问题时，该如何主动地去学习、去调试。

19.3.1　调试与查看源码

有时需要对 JDBC 配置文件中的账号密码进行加密，以达到不泄露明文的效果，当然这种加密是可以解密的。解密的过程就是继承 Spring 中的 PropertySourcesPlaceholderConfigurer 类的过程，并重写它的 processProperties 方法（也可以使用 BeanFactoryPostProcessor 接口，重写其 postProcessBeanFactory 方法实现，还可以继承 Properties 类来实现）。下面主要讲解如何使用 PropertySourcesPlaceholderConfigurer 类来实现。

直接将代码粘贴出来意义不大，重点在于怎么知道 PropertySourcesPlaceholderConfigurer 类就是需要继承的类，为什么重写 postProcessBeanFactory 方法就可以达到解密的效果。

在 Spring 的配置文件中配置数据源，其实就是配置 Spring 管理的 Bean。这个 Bean 和手动编写的 Bean 是一样的，在这个项目中用的是阿里的 Druid 进行数据库连接池的管理，而这个 Bean 的参数值是由配置文件注入的。引入配置文件，如代码清单 19-8 所示。

代码清单 19-8　引入配置文件

```
<!-- 引入配置文件，其classpath在resource目录下-->
<context:property-placeholder location="classpath:config.properties" ignore-
unresolvable="true"/>
```

引入配置文件后在 IDEA 下按住 Ctrl 键并单击 property-placeholder，即可跳转到如代码清单 19-9 所示的 spring-context.xsd 源码处。

代码清单 19-9　spring-context.xsd 源码

```
<xsd:element name="property-placeholder">
   <xsd:annotation>
      <xsd:documentation><![CDATA[
Activates replacement of ${...} placeholders by registering a PropertySourcesPlaceh
olderConfigurer within the application context. Properties willbe resolved against
the specified properties file or Properties object -- so called"local properties",
if any, and against the Spring Environment's current set of PropertySources.

Note that as of Spring 3.1 the system-properties-mode attribute has been removed
infavor of the more flexible PropertySources mechanism. However, applications
maycontinue to use the 3.0 (and older) versions of the spring-context schema in
orderto preserve system-properties-mode behavior. In this case, the traditionalProp
ertyPlaceholderConfigurer component will be registered instead of the newer
PropertySourcesPlaceholderConfigurer.

See ConfigurableEnvironment javadoc for more information on usage.
      ]]></xsd:documentation>
      <xsd:appinfo>
         <tool:annotation>
            <tool:exports type="org.springframework.context.support.PropertySources
PlaceholderConfigurer"/>
         </tool:annotation>
      </xsd:appinfo>
   </xsd:annotation>
```

对以上代码进行简单解释：PropertySourcesPlaceholderConfigurer 类用于解决“${}”的占位符问题，如本地变量、系统变量或环境变量等。从 Spring 3.1 开始，应该在实现中优先使用新的 PropertySourcesPlaceholderConfigurer 类，但是为了兼容旧版本的 Spring，开发者也可以使用

PropertyPlaceholderConfigurer 类。

19.3.2 继承 PropertySourcesPlaceholderConfigurer 类进行解密处理

从 19.2.1 小节可知，PropertySourcesPlaceholderConfigurer 类和 PropertyPlaceholderConfigurer 类都是可以解决加密问题的。因为需要管理 propertyConfigurer，所以这里将由 <context:property-placeholder> 配置改为手动配置 Bean。Spring 引入属性文件解密数据库账号密码，如代码清单 19-10 所示。

代码清单 19-10 Spring 引入属性文件解密数据库账号密码

```
<!-- 引入属性文件，classpath在resource目录下-->
<!--<context:property-placeholder local-override="true" location=
"classpath:config.properties" ignore-unresolvable="true"/>-->
<bean id="propertyConfigurer" class="com.uifuture.ssm.config.EncryptPropertyPlaceho
lderConfigurer">
    <property name="locations">
        <list>
            <value>classpath:config.properties</value>
        </list>
    </property>
</bean>
```

EncryptPropertyPlaceholderConfigurer 该类继承 PropertySourcesPlaceholderConfigurer 类，进入该类，添加断点，用 Debug 工具运行代码，如图 19-5 所示。

这里的 mergeProperties 方法是关键，其运行的是父类的方法，如图 19-6 所示。

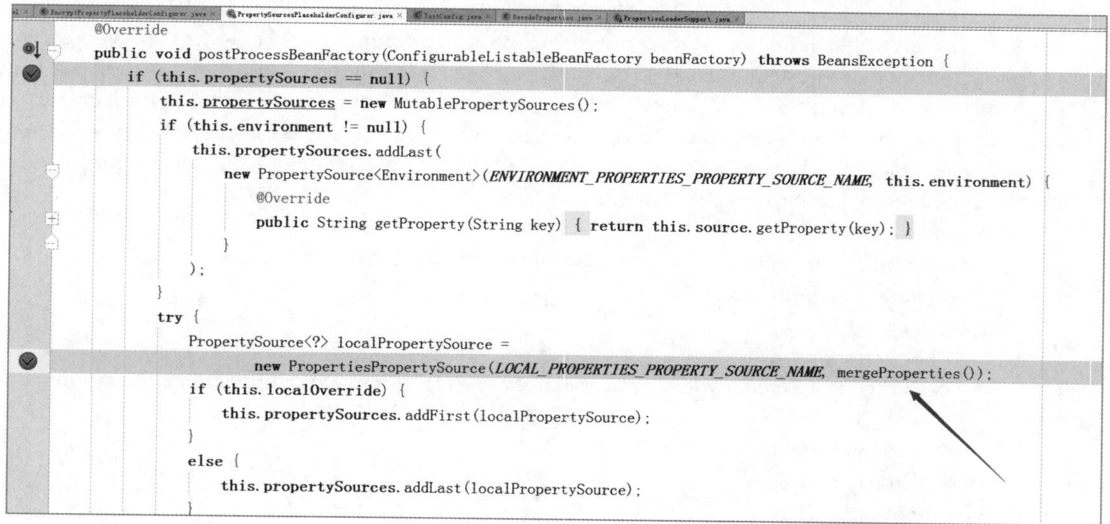

图 19-5 PropertySourcesPlaceholderConfigurer 类的 postProcessBeanFactory 方法

```java
 * Return a merged Properties instance containing both the
 * loaded properties and properties set on this FactoryBean.
 */
protected Properties mergeProperties() throws IOException {
    Properties_result = new Properties();

    if (this.localOverride) {
        // Load properties from file upfront, to let local properties override.
        loadProperties(result);
    }

    if (this.localProperties != null) {
        for (Properties localProp : this.localProperties) {
            CollectionUtils.mergePropertiesIntoMap(localProp, result);
        }
    }

    if (!this.localOverride) {
        // Load properties from file afterwards, to let those properties override.
        loadProperties(result);
    }
```

图 19-6　mergeProperties 方法

mergeProperties 方法调用的是 loadProperties 方法，进入 loadProperties 方法，可以看到 loadProperties 方法中就是对配置文件的操作。运行下一步，可以看到配置属性已经被读取出来，如图 19-7 所示。

配置属性 result 已被读取出来，并已返回。继续从断点往下运行，将执行 addLast 方法，如图 19-8 所示。

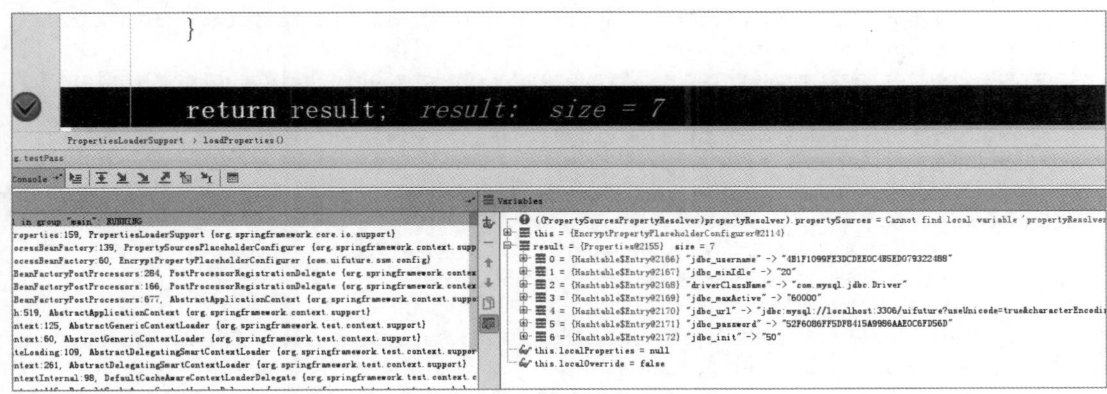

图 19-7　读出配置属性 result

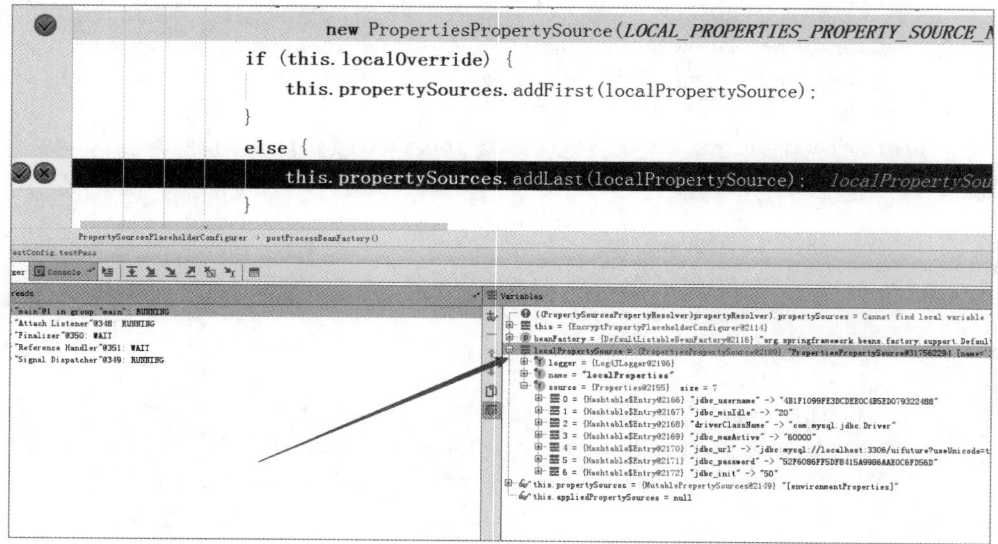

图 19-8 addLast 方法

现在已经将值全部放在了 propertySourceList 中。此时 Spring 是还没有注入 Bean 的，需要找到一个在 Spring 中注入 Bean 之前运行的方法进行重写。从调试过程和源码可以发现，重写 postProcessBeanFactory 方法、processProperties 方法或者 doProcessProperties 方法都是可以的。EncryptPropertyPlaceholderConfigurer 类源码如代码清单 19-11 所示。

代码清单 19-11 EncryptPropertyPlaceholderConfigurer 类源码

```
package com.uifuture.ssm.config;

import com.uifuture.ssm.util.AES;
import org.springframework.beans.BeansException;
import org.springframework.beans.MutablePropertyValues;
import org.springframework.beans.PropertyValue;
import org.springframework.beans.factory.config.BeanDefinition;
import org.springframework.beans.factory.config.ConfigurableListableBeanFactory;
import org.springframework.beans.factory.config.TypedStringValue;
import org.springframework.context.support.PropertySourcesPlaceholderConfigurer;
import org.springframework.core.env.ConfigurablePropertyResolver;
import org.springframework.util.StringValueResolver;
/** 加密解密账号密码 */
public class EncryptPropertyPlaceholderConfigurer extends PropertySourcesPlaceholde
rConfigurer {
    /** 注意！这里的值是真正需要注入的Bean的属性名，而不是配置文件中的属性名 */
    private String[] encryptPropNames = {"username", "password"};
    @Override
    protected void processProperties(ConfigurableListableBeanFactory beanFactoryToProcess,
ConfigurablePropertyResolver propertyResolver) throws BeansException {
        super.processProperties(beanFactoryToProcess, propertyResolver);
        BeanDefinition bd = beanFactoryToProcess.getBeanDefinition("dataSource");
```

```
        MutablePropertyValues pv = bd.getPropertyValues();
        for (String s : encryptPropNames) {
            PropertyValue p = pv.getPropertyValue(s);
            //先解密处理再覆盖原来的值
            assert p != null;
            String value = p.getValue().toString();
            if(p.getValue() instanceof TypedStringValue){
                        TypedStringValue typedStringValue = (TypedStringValue)
p.getValue();
                value = typedStringValue.getValue();
            }
            String v = AES.decrypt(value);
            pv.add(s, v);
        }
    }
}
```

注意：AES类为自行编写的加密解密类，可以在项目的源码链接中找到。

另外，也可以重写doProcessProperties方法或者postProcessBeanFactory方法，如代码清单19-12所示。

代码清单 19-12　另外一种加密解密方式

```
@Override
protected void doProcessProperties(ConfigurableListableBeanFactory
beanFactoryToProcess, StringValueResolver valueResolver) {
    super.doProcessProperties(beanFactoryToProcess, valueResolver);
    BeanDefinition bd = beanFactoryToProcess.getBeanDefinition("dataSource");
    MutablePropertyValues pv =  bd.getPropertyValues();
    for(String s:encryptPropNames){
        PropertyValue p = pv.getPropertyValue(s);
        //先解密处理再覆盖原来的值
        assert p != null;
        String value = p.getValue().toString();
        if(p.getValue() instanceof TypedStringValue){
            TypedStringValue typedStringValue = (TypedStringValue) p.getValue();
            value = typedStringValue.getValue();
        }
        String v = AES.decrypt(value);
        pv.add(s, v);
    }
}
/** 重写postProcessBeanFactory方法 */
@Override
public void postProcessBeanFactory(ConfigurableListableBeanFactory beanFactory)
throws BeansException {
    super.postProcessBeanFactory(beanFactory);
    BeanDefinition bd = beanFactory.getBeanDefinition("dataSource");
    MutablePropertyValues pv =  bd.getPropertyValues();
    for(String s:encryptPropNames){
```

```
            PropertyValue p = pv.getPropertyValue(s);
            //先解密处理再覆盖原来的值
            assert p != null;
            String value = p.getValue().toString();
            if(p.getValue() instanceof TypedStringValue){
                TypedStringValue typedStringValue = (TypedStringValue) p.getValue();
                value = typedStringValue.getValue();
            }
            String v = AES.decrypt(value);
            pv.add(s, v);
        }
    }
```

19.3.3 继承 Properties 类进行解密处理

还有一种方法可以进行解密处理，即继承 Properties 类并重写 mergeProperties 方法进行解密，如图 19-9 所示。

由注释 Load properties from file upfront, to let local properties override（先从文件加载属性，后面让本地属性覆盖）可以看出，localOverride 属性是用来控制是否覆盖 Spring 读取的属性配置的。下面紧接着判断 this.localProperties!=null，如果 localProperties 不为 null，则会将这些配置信息读取到 Spring 容器中，并覆盖 Spring 中已存在的属性。

清楚以上内容后，下面来编写一个类继承 Properties 类并重写构造方法，在构造方法中对加密的值进行解密，再设置值。

```
protected Properties mergeProperties() throws IOException {
    Properties result = new Properties();

    if (this.localOverride) {
        // Load properties from file upfront, to let local properties override.
        loadProperties(result);
    }

    if (this.localProperties != null) {
        for (Properties localProp : this.localProperties) {
            CollectionUtils.mergePropertiesIntoMap(localProp, result);
        }
    }

    if (!this.localOverride) {
        // Load properties from file afterwards, to let those properties override.
        loadProperties(result);
    }

    return result;
}
```

图 19-9 mergeProperties 方法

首先为 Spring 配置加密、解密信息，如代码清单 19-13 所示。

代码清单 19-13　为 Spring 配置加密、解密信息

```
<!-- 引入属性文件，classpath在resource目录下-->
<context:property-placeholder local-override="true" location="classpath:config.
properties"
file-encoding="UTF-8" properties-ref="decodeProperties" ignore-
unresolvable="true"/>
<!-- 用来解密jdbc.properties中的属性，然后存放到Properties类中 -->
<bean id="decodeProperties" class="com.uifuture.ssm.config.DecodeProperties">
<constructor-arg value="jdbc_username,jdbc_password"/>
</bean>
```

接着实现 DecodeProperties 类，如代码清单 19-14 所示。

代码清单 19-14　实现 DecodeProperties 类

```
package com.uifuture.ssm.config;
import com.uifuture.ssm.util.AES;
import java.io.IOException;
import java.util.Properties;
public class DecodeProperties extends Properties {
    /**
     * 构造方法
     * @param properties属性配置的键值对
     */
    public DecodeProperties(Properties properties) {
        try {
this.load(DecodeProperties.class.getResourceAsStream("/config.properties"));
            for (String keys : properties.stringPropertyNames()) {
                for (String key : keys.split(",")) {
                    // 进行解密并设置值
                    this.setProperty(key, AES.decrypt(this.getProperty(key)));
                }
            }
        } catch (IOException e) {
            e.printStackTrace();
        }
    }
}
```

config.properties 中的加密属性如下：

```
jdbc_username=4B1F1099FE3DCDEE0C4B5ED079322488
jdbc_password=52F6086FF5DF8415A9986AAE0C6FD56D
```

这样就实现了数据库账号密码的明文加密。对 properties 文件中的属性加密，当 Spring 读取时会先由自行实现的 DecodeProperties 类来进行解密并完成存储和替换，然后 Spring 会把解密后的属性存放在容器中，并且根据 Spring 配置文件中的 EL 表达式注入数据源中。

对前面的 4 种方法，读者都可以研究一下，以便理解 Spring，并且有助于加强读者的调试能力。

该加密解密问题的解决方案不止上面介绍的 4 种，其他解决方案读者可以自行在项目中调试。

注意：最后实际的代码可能与书中的代码稍微有出入，请以 GitHub 仓库中的代码为准。

19.4 快速生成数据库实体类

前面已经将项目框架搭建好并对数据库的账号、密码进行了加密，接下来根据创建的数据库表生成 Java 类。由数据库表可知，每个实体类都有 id、创建时间、修改时间、删除时间字段，所以将这 4 个属性抽取到公共类中，命名为 BaseEntity，如代码清单 19-15 所示。

代码清单 19-15 BaseEntity 类

```
package com.uifuture.ssm.entity;
import java.io.Serializable;
import java.util.Date;
/** 所有实体类的超类 */
public class BaseEntity implements Serializable {
    private static final long serialVersionUID = -1549634521453074321L;
    /** 主键，唯一标识符 */
    protected Integer id;
    /** 创建时间 */
    protected Date createTime;
    /** 更新时间 */
    protected Date updateTime;
    /** 删除时间，0代表为未删除，单位为s */
    protected Integer deleteTime;
…//省略getter()方法、setter()方法
}
```

19.4.1 集成 MyBatis-Plus 工具

MyBatis-Plus（MP）是一款 MyBatis 的增强工具，它在 MyBatis 的基础上只做增强，不做改变，为简化开发过程、提高效率而诞生。

MyBatis-Plus 的特性如下：

（1）无侵入：MyBatis-Plus 是在 MyBatis 的基础上进行的扩展，只做增强，不做改变；引入 MyBatis-Plus 不会对用户现有的 MyBatis 构架产生任何影响，而且 MyBatis-Plus 支持所有 MyBatis 原生的特性。

（2）依赖少：仅仅依赖 Mybatis 及 MyBatis-Spring。

（3）损耗小：启动便会自动注入基本 CURD 配置，性能基本无损耗，直接面向对象操作。

（4）预防 SQL 注入：内置 SQL 注入剥离器，有效预防 SQL 注入攻击。

（5）通用 CRUD 操作：内置通用 Mapper、通用 Service，仅仅通过少量配置即可实现单表大部分的 CRUD 操作；更有强大的条件构造器，可满足各类使用需求。

（6）多种主键策略：支持 4 种主键策略（内含分布式唯一 ID 生成器），可自由配置，完美解决主键问题。

（7）支持热加载：Mapper 对应的 XML 支持热加载，对简单的 CRUD 操作甚至可以无 XML 启动。

（8）支持 ActiveRecord：支持 ActiveRecord 形式调用，实体类只需继承 Model 类即可实现基本 CRUD 操作。

（9）支持代码生成：采用代码或者 Maven 插件可以快速生成 Mapper、Model、Service、Controller 层代码，支持模板引擎，更有超多自定义配置可供选择（比 MyBatis 的 Generator 更加强大）。

（10）支持自定义全局通用操作：支持全局通用方法注入（践行"一次编写，到处运行"理念）。

（11）支持关键词自动转义：支持数据库关键词（order、key）自动转义，还可以自定义关键词。

（12）内置分页插件：基于 MyBatis 物理分页，开发者无须关心具体操作，配置好插件之后，编写分页等同于普通的列表查询。

（13）内置性能分析插件：可以输出 SQL 语句及其执行时间。建议在开发、测试时启用该功能，能有效解决慢查询问题。

（14）内置全局拦截插件：提供全表 Delete、Update 操作智能分析阻断，预防误操作。

当然，在本项目中不可能用到所有的特性，只需要代码生成、通用 CRUD 操作特性。

下面来集成 MyBatis-Plus，首先在项目中导入相关依赖，如代码清单 19-16 所示。

代码清单 19-16 集成 MyBatis-Plus

```
<!--MyBatis-plus的依赖-->
<dependency>
    <groupId>com.baomidou</groupId>
    <artifactId>mybatis-plus</artifactId>
    <version>3.1.0</version>
</dependency>
```

接下来使用 AutoGenerator。AutoGenerator 是 MyBatis-Plus 的代码生成器，通过 AutoGenerator 可以快速生成 Entity、Mapper、Mapper XML、Service、Controller 等各个模块的代码，极大提升开发效率。

下面进行演示。MyBatis-Plus 3.0.3 之后的版本移除了代码生成器与模板引擎的默认依赖，开发者需要手动添加相关依赖。AutoGenerator 依赖添加如代码清单 19-17 所示。

代码清单 19-17 AutoGenerator 依赖添加

```
<!--AutoGenerator依赖-->
<dependency>
    <groupId>com.baomidou</groupId>
    <artifactId>mybatis-plus-generator</artifactId>
    <version>3.1.0</version>
</dependency>
```

由于代码生成会使用到 freemarker 模板引擎，所以还需要加入 freemarker 模块引擎的依赖，如代码清单 19-18 所示。

代码清单 19-18　freemarker 模块引擎依赖

```
<!--freemarker模板引擎-->
<dependency>
    <groupId>org.freemarker</groupId>
    <artifactId>freemarker</artifactId>
    <version>2.3.29</version>
</dependency>
```

关于 AutoGenerator 便不进行更加详细的参数讲解了，有兴趣的读者可以通过 MyBatis 官网访问更多关于 MyBatis-Plus 的解析。接下来直接通过代码生成器配置代码，如代码清单 19-19 所示。

代码清单 19-19　代码生成器配置代码

```
/**
 * 自动生成模块代码
 *执行main方法，在控制台输入模块表名，按回车键，自动在目录中生成对应项目 */
public class CodeGenerator {
    /** 读取控制台内容 */
    public static String scanner(String tip) {
        Scanner scanner = new Scanner(System.in);
        StringBuilder help = new StringBuilder();
        help.append("请输入" + tip + ":");
        System.out.println(help.toString());
        if (scanner.hasNext()) {
            String ipt = scanner.next();
            if (StringUtils.isNotEmpty(ipt)) {
                return ipt;
            }
        }
        throw new MybatisPlusException("请输入正确的" + tip + "！");
    }
    public static void main(String[ ] args) {
        // 代码生成器
        AutoGenerator mpg = new AutoGenerator();

        String projectPath = System.getProperty("user.dir");
        // 子模块
        String model = "/ssm-uifuture";
        projectPath = projectPath + model;
        // ①全局配置
        GlobalConfig config = new GlobalConfig();
        // 生成的Service接口名称首字母是否为“I”，这样设置就没I；配置%s为占位符
        config.setServiceName("%sService");
        // 设置生成的Service接口名称首字母是否为“I”，即生成IEmployeeService
        config.setServiceImplName("%sServiceImpl");
        config.setEntityName("%sEntity");
        // 是否打开输出目录
```

```
        config.setOpen(false);
        // 是否支持AR模式
        config.setActiveRecord(false);
        // 设置作者名字
        config.setAuthor("chenhx");
        // 设置文件生成的路径
        config.setOutputDir(projectPath + "/src/main/java");
        // 是否将文件覆盖
        config.setFileOverride(true);
        // 主键策略
        config.setIdType(IdType.AUTO);
        // 是否生成基础的映射结果
        config.setBaseResultMap(true);
        // 生成基础的SQL列，即SQL片段
        config.setBaseColumnList(true);

        //②数据源配置
        DataSourceConfig dsConfig = new DataSourceConfig();
        dsConfig.setDbType(DbType.MYSQL)//设置数据库的类型
                .setDriverName("com.mysql.jdbc.Driver")
                .setUrl("jdbc:mysql://localhost:3306/uifuture?useUnicode=true&charac
terEncoding=UTF-8&useSSL=false&serverTimezone=GMT%2B8&remarks=true")
                .setUsername("root")
                .setPassword("12345678");

        // ③包名策略
        PackageConfig pkConfig = new PackageConfig();
        pkConfig.setParent("com.uifuture.ssm")      //父包
                .setMapper("mapper")
                .setService("service")
                .setController("controller")
                .setEntity("entity");
        // 自定义配置
        InjectionConfig cfg = new InjectionConfig() {
            @Override
            public void initMap() {
                …// to do nothing
            }
        };
        // 如果模板引擎是freemarker
        String templatePath = "/templates/mapper.xml.ftl";
        /* 如果模板引擎是velocity
         String templatePath = "/templates/mapper.xml.vm";
         自定义输出配置*/
        List<FileOutConfig> focList = new ArrayList<>();
        // 自定义配置会被优先输出
        String path = projectPath;
        focList.add(new FileOutConfig(templatePath) {
```

19

初步开发——框架集成

```
            @Override
            public String outputFile(TableInfo tableInfo) {
 // 自定义输出文件名，如果Entity设置了前后缀，此处注意XML的名称会跟着发生变化
                if (StringUtils.isEmpty(pkConfig.getModuleName())) {
                        return path + "/src/main/resources/mapper/" + tableInfo.
getMapperName() + StringPool.DOT_XML;
                }
                    return path + "/src/main/resources/mapper/" + pkConfig.
getModuleName()
                      + "/" + tableInfo.getMapperName() + StringPool.DOT_XML;
            }
        });
        /*
        cfg.setFileCreate(new IFileCreate() {
            @Override
            public boolean isCreate(ConfigBuilder configBuilder, FileType fileType,
String filePath) {
                // 判断自定义文件夹是否需要创建
                checkDir("调用默认方法创建的目录");
                return false;
            }
        });
        */
        cfg.setFileOutConfigList(focList);
        // 策略配置
        StrategyConfig strategy = new StrategyConfig();
        strategy.setNaming(NamingStrategy.underline_to_camel);
        strategy.setColumnNaming(NamingStrategy.underline_to_camel);
        strategy.setSuperEntityClass("com.uifuture.ssm.base.BaseEntity");
        strategy.setEntityLombokModel(true);
        strategy.setRestControllerStyle(true);
        // 公共父类
        strategy.setSuperControllerClass("com.uifuture.ssm.base.BaseController");
        // 写于父类中的公共字段
        strategy.setSuperEntityColumns("id",
                "create_time",
                "update_time",
                "delete_time");
        // 生成的表
        strategy.setInclude(scanner("表名，可用多个英文逗号分隔").split(","));
        strategy.setControllerMappingHyphenStyle(true);
        // 为表添加前缀
        strategy.setTablePrefix(pkConfig.getModuleName() + "_");

        // 配置模板
        TemplateConfig templateConfig = new TemplateConfig();

        /*配置自定义输出模板
```

```
        指定自定义模板路径，注意不要带.ftl/.vm，会根据使用的模板引擎自动识别
        templateConfig.setEntity("templates/entity2.java");
        templateConfig.setService();
        templateConfig.setController();*/
        templateConfig.setXml(null);

        //④整合配置
        mpg.setGlobalConfig(config)
                .setDataSource(dsConfig)
                .setStrategy(strategy)
                .setPackageInfo(pkConfig);
        mpg.setTemplate(templateConfig);
        mpg.setCfg(cfg);
        mpg.setTemplateEngine(new FreemarkerTemplateEngine());
        mpg.execute();
    }
}
```


这里的 BaseController 是为了在 Controller 层更好地操作请求，如获取 IP 等公共方法。当然，在运行之前，记得引入 slf4j 的依赖，否则运行时会报 java.lang.ClassNotFoundException: org.slf4j.LoggerFactory 异常。添加 slf4j 依赖，如代码清单 19-20 所示。

代码清单 19-20 添加 slf4j 依赖

```
<dependency>
    <groupId>org.slf4j</groupId>
    <artifactId>slf4j-api</artifactId>
    <version>1.7.7</version>
</dependency>
```

数据库此时已经是准备好。如果读者没有准备好数据库或相关数据库表，请通过 GitHub 仓库查看相关 SQL 文件。

19.4.2 自动生成 Dao 层和 Service 层代码

首先来生成 users 表的代码。运行 CodeGenerator 类的 main 方法，然后输入表名（此处输入 users），如图 19-10 所示。

```
CodeGenerator
/Library/Java/JavaVirtualMachines/jdk1.8.0_201.jdk/Contents/Home/bin/java ...
SLF4J: Failed to load class "org.slf4j.impl.StaticLoggerBinder".
SLF4J: Defaulting to no-operation (NOP) logger implementation
SLF4J: See http://www.slf4j.org/codes.html#StaticLoggerBinder for further details.
请输入表名，多个英文逗号分割：
users
```

图 19-10 运行 CodeGenerator 类的 main 方法并输入表名

此时，可以看到生成代码后的项目结构，如图 19-11 所示。

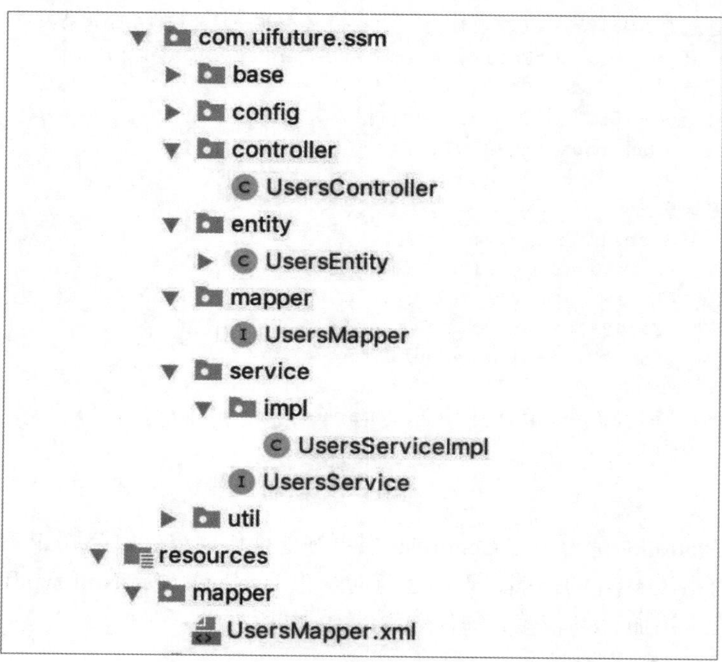

图 19-11　生成代码后的项目结构

在这里并没有一次性生成所有的表。由于没有外键关联，因此完全可以根据后续的功能来逐渐增加相关类的生成。

前面在 Spring 中注入的 sqlSessionFactory 是 org.mybatis.spring.SqlSessionFactoryBean，现在使用 MyBatis-Plus，需要将其修改为 com.baomidou.mybatisplus.extension.spring.MybatisSqlSession Factory Bean。修改后的代码如代码清单 19-21 所示。

代码清单 19-21　修改后的 sqlSessionFactory

```
<!-- MyBatis的sqlSessionFactoryBean:org.mybatis.spring.SqlSessionFactoryBean-->
<!--配置MyBatis-Plus的sqlSessionFactory -->
<bean id="sqlSessionFactory" class="com.baomidou.mybatisplus.extension.spring.
MybatisSqlSessionFactoryBean">
    <property name="dataSource" ref="dataSource"/>
    <property name="mapperLocations">
        <list>
            <value>classpath*:/mapper/*.xml</value>
        </list>
    </property>
</bean>
```

接下来便通过测试类验证是否可以连接数据库。**提示**：记得先在数据库的 users 表中插入数据。

在测试之前，为了更好地分层，以及对每个配置文件的配置信息一目了然，要将配置文件重新写出来。由于 Spring 的配置文件将 MyBatis 需要做的事情都做了，因此不需要 spring-mybatis.xml 文件。最新的配置文件结构如图 19-12 所示。

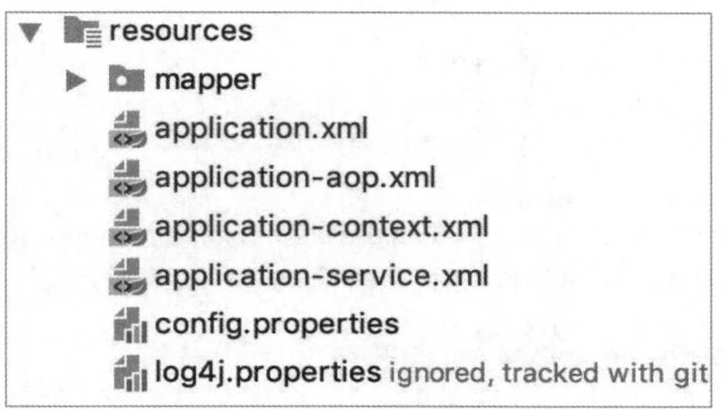

图 19-12　最新的配置文件结构

后面因为集成 Spring MVC 要配置 web.xml 文件，那时要指定 application-context.xml 作为配置文件的入口。

首先看 application-context.xml 中的内容，这是配置文件的入口，如代码清单 19-22 所示。

代码清单 19-22　application-context.xml 中的内容

```
<beans xmlns:xsi="http://www.w3.org/2001/XMLSchema-instance"
       xmlns:aop="http://www.springframework.org/schema/aop"
          xmlns:context="http://www.springframework.org/schema/context"
xmlns="http://www.springframework.org/schema/beans" xsi:schemaLocation="http://
www.springframework.org/schema/beans
http://www.springframework.org/schema/beans/spring-beans.xsd
 http://www.springframework.org/schema/aop
http://www.springframework.org/schema/aop/spring-aop.xsd http://www.
springframework.org/schema/context http://www.springframework.org/schema/context/
spring-context.xsd">
    <aop:aspectj-autoproxy/>
    <context:component-scan base-package="com.uifuture.ssm">
    </context:component-scan>
    <import resource="classpath*:application.xml"></import>
</beans>
```

接下来在 application.xml 中继续加载子配置文件，如代码清单 19-23 所示。

代码清单 19-23　application.xml 中的内容

```
<?xml version="1.0" encoding="UTF-8"?>
<beans xmlns:xsi="http://www.w3.org/2001/XMLSchema-instance"
       xmlns:context="http://www.springframework.org/schema/context"
       xmlns="http://www.springframework.org/schema/beans"
       xsi:schemaLocation="http://www.springframework.org/schema/beans http://www.
springframework.org/schema/beans/spring-beans.xsd
http://www.springframework.org/schema/context http://www.springframework.org/
schema/context/spring-context.xsd">
    <context:component-scan base-package="com.uifuture.**">
          <context:exclude-filter type="regex" expression="com\.uifuture\.ssm\.
swagger\.view\..*"/>
```

```
    </context:component-scan>
    <!--导入需要的配置，可按功能或者模块来分-->
    <import resource="application-service.xml"></import>
    <import resource="application-aop.xml"></import>
</beans>
```

然后依次配置 application-service.xml 和 application-aop.xml。application-service.xml 是 Service 层的配置文件，集成了 MyBatis，其内容如代码清单 19-24 所示。

代码清单 19-24　application-service.xml 中的内容

```
<?xml version="1.0" encoding="UTF-8"?>
<beans xmlns:xsi="http://www.w3.org/2001/XMLSchema-instance"
       xmlns:p="http://www.springframework.org/schema/p"
       xmlns:tx="http://www.springframework.org/schema/tx"
       xmlns:aop="http://www.springframework.org/schema/aop"
       xmlns:context="http://www.springframework.org/schema/context"
       xmlns="http://www.springframework.org/schema/beans"
       xsi:schemaLocation="
          http://www.springframework.org/schema/beans
          http://www.springframework.org/schema/beans/spring-beans-4.2.xsd
          http://www.springframework.org/schema/tx
          http://www.springframework.org/schema/tx/spring-tx-4.2.xsd
          http://www.springframework.org/schema/aop
          http://www.springframework.org/schema/aop/spring-aop-4.2.xsd
          http://www.springframework.org/schema/context
          http://www.springframework.org/schema/context/spring-context-4.2.xsd">
    <!-- 引入属性文件，classpath在resource目录下-->
    <!--<context:property-placeholder local-override="true" location="classpath:config.
properties"  ignore-unresolvable="true"/>-->
    <bean id="propertyConfigurer" class="com.uifuture.ssm.config.EncryptPropertyPla
ceholderConfigurer">
        <property name="locations">
            <list>
                <value>classpath:config.properties</value>
            </list>
        </property>
    </bean>
    <!--spring解密数据库账号密码的第1种方式-->
    <!-- 引入属性文件，classpath在resource目录下-->
    <!--<context:property-placeholder local-override="true" location="classpath:config.
properties"-->
    <!--file-encoding="UTF-8" properties-ref="decodeProperties"
ignore-unresolvable="true"/>-->
    <!--解密jdbc.properties中的属性，然后存储到Properties类中-->
    <!--<bean id="decodeProperties" class="com.uifuture.ssm.config.DecodeProperties">-->
    <!--<constructor-arg value="jdbc_username,jdbc_password"/>-->
    <!--</bean>-->
    <!-- 包自动扫描，不扫描Controller注解-->
```

```
    <context:component-scan base-package="com.uifuture.ssm">
        <context:exclude-filter type="annotation" expression="org.springframework.
stereotype.Controller"/>
    </context:component-scan>
    <!-- MyBatis的SQLSession的工厂:SqlSessionFactoryBean-->
    <!-- MyBatis的SqlSessionFactoryBean:org.mybatis.spring.SqlSessionFactoryBean-->
    <!--配置MyBatis-Plus的sqlSessionFactory -->
    <bean id="sqlSessionFactory" class="com.baomidou.mybatisplus.extension.spring.
MybatisSqlSessionFactoryBean">
        <property name="dataSource" ref="dataSource"/>
        <property name="mapperLocations">
            <list>
                <value>classpath*:/mapper/*.xml</value>
            </list>
        </property>
    </bean>
    <!-- 自动扫描Mapper接口-->
    <bean class="org.mybatis.spring.mapper.MapperScannerConfigurer"
          p:basePackage="com.uifuture.ssm.**.mapper"
          p:sqlSessionFactoryBeanName="sqlSessionFactory"/>

    <bean id="sqlSession" class="org.mybatis.spring.SqlSessionTemplate">
        <constructor-arg index="0" ref="sqlSessionFactory"/>
    </bean>
    <!-- 数据源配置，使用DriverManagerDataSource-->
    <!--<bean id="dataSource" class="org.springframework.jdbc.datasource.
DriverManagerDataSource">-->
    <!--<bean id="dataSource" class="com.alibaba.druid.pool.DruidDataSource">-->
    <!--<property name="driverClassName" value="${driverClassName}"/>-->
    <!--<property name="url" value="${jdbc_url}"/>-->
    <!--<property name="username" value="${jdbc_username}"/>-->
    <!--<property name="password" value="${jdbc_password}"/>-->
    <!--</bean>-->
    <!-- 配置数据源 -->
    <bean id="dataSource" class="com.alibaba.druid.pool.DruidDataSource">
        <!-- 基本属性 url、username、password -->
        <property name="url" value="${jdbc_url}"/>
        <property name="username" value="${jdbc_username}"/>
        <property name="password" value="${jdbc_password}"/>
        <property name="driverClassName" value="${driverClassName}"/>

        <!-- 配置初始化尺寸、最小值和最大值 -->
        <property name="initialSize" value="${jdbc_init}"/>
        <property name="minIdle" value="${jdbc_minIdle}"/>
        <property name="maxActive" value="${jdbc_maxActive}"/>

        <!-- 配置获取连接等待超时的时间 -->
        <property name="maxWait" value="60000"/>

        <!-- 配置检测的间隔时间，检测需要关闭的空闲连接，单位为ms -->
```

```
        <property name="timeBetweenEvictionRunsMillis" value="2000"/>

        <!-- 配置一个连接在池中最小生存时间，单位为ms -->
        <property name="minEvictableIdleTimeMillis" value="600000"/>
        <property name="maxEvictableIdleTimeMillis" value="900000"/>

        <property name="validationQuery" value="select 1"/>
        <property name="testWhileIdle" value="true"/>
        <property name="testOnBorrow" value="false"/>
        <property name="testOnReturn" value="false"/>

        <property name="keepAlive" value="true"/>

        <!-- 配置监控统计拦截的filters -->
        <property name="filters" value="stat,wall"/>
    </bean>

    <!-- 配置druid用来监控spring jdbc -->
    <bean id="druid-stat-interceptor"
          class="com.alibaba.druid.support.spring.stat.DruidStatInterceptor">
    </bean>
     <bean id="druid-stat-pointcut" class="org.springframework.aop.support.
JdkRegexpMethodPointcut"
          scope="prototype">
        <property name="patterns">
            <list>
                <value>com.uifuture.ssm.service.*</value>
            </list>
        </property>
    </bean>

    <!-- 配置事务管理类 -->
     <bean id="transactionManager" class="org.springframework.jdbc.datasource.
DataSourceTransactionManager"
          p:dataSource-ref="dataSource">
    </bean>
    <!-- 开启注解式事务扫描 -->
    <tx:annotation-driven transaction-manager="transactionManager"/>

</beans>
```

对 application-aop.xml 文件进行 AOP 配置，如代码清单 19-25 所示。

代码清单 19-25　application-aop.xml 中的内容

```
<?xml version="1.0" encoding="UTF-8"?>
<beans xmlns:xsi="http://www.w3.org/2001/XMLSchema-instance" xmlns:aop="http://
www.springframework.org/schema/aop"
       xmlns="http://www.springframework.org/schema/beans"
       xsi:schemaLocation="http://www.springframework.org/schema/beans
```

```
                 http://www.springframework.org/schema/beans/spring-beans-3.0.xsd
                 http://www.springframework.org/schema/aop
                 http://www.springframework.org/schema/aop/spring-aop-3.0.xsd">
    <!—使用基于AspectJ的自动代理-->
    <aop:aspectj-autoproxy proxy-target-class="true"/>
     <!--proxy-target-class="true" ，Spring建议通过 proxy-target-class 属性值来控制是
基于接口的还是基于类的代理被创建：如果 proxy-target-class 属性值被设置为 true，那么基于类的代
理将起作用(这时需要CGLIB库cglib.jar在CLASSPATH中)；如果 proxy-target-class 属性值被设置为
false，或者这个属性被省略，那么标准的JDK基于接口的代理将起作用
     接口上的注解不能继承-->
</beans>
```

通过以上配置，就可以直接操作数据了。为了方便测试，先创建一个测试基类——BaseTest 类，如代码清单 19-26 所示。

代码清单 19-26　BaseTest 类代码

```
package com.uifuture.ssm;
import org.junit.runner.RunWith;
import org.springframework.test.context.ContextConfiguration;
import org.springframework.test.context.junit4.SpringJUnit4ClassRunner;
@RunWith(SpringJUnit4ClassRunner.class)
@ContextConfiguration(locations = {"classpath*:application-context.xml"})
public class BaseTest {
}
```

接下来可以创建测试类，如代码清单 19-27 所示。

代码清单 19-27　创建测试类

```
package com.uifuture.ssm.service.impl;
import com.uifuture.ssm.BaseTest;
import com.uifuture.ssm.service.UsersService;
import lombok.extern.slf4j.Slf4j;
import org.junit.Test;
import org.springframework.beans.factory.annotation.Autowired;
@Slf4j
public class UsersServiceImplTest extends BaseTest {
    @Autowired
    private UsersService usersService;
    @Test
    public void list() {
        log.info("获取的数据:" + usersService.list());
    }
}
```

由于这里只插入了一条数据，所以查询的结果也只有一条，测试结果如图 19-13 所示。

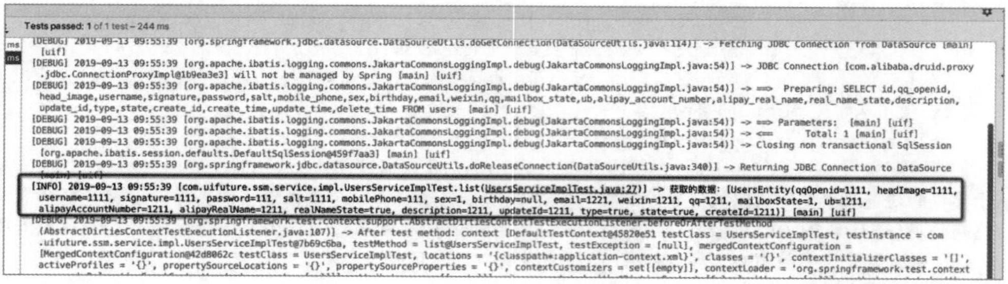

图 19-13　测试结果

19.5　集成 Spring MVC

前面已经集成了 MyBatis，并且通过 MyBatis-Plus 可以非常快速地创建数据库表对应的实体类、Dao 层、Service 层的代码。接下来便通过集成 Spring MVC 进行页面数据的访问。

在这里，选择 freemarker 模板引擎来输出动态页面。如果没有引入 freemarker 依赖，则需要先引入，如代码清单 19-28 所示。

代码清单 19-28　引入 freemarker 依赖

```
<!--freemarker模板引擎-->
<dependency>
    <groupId>org.freemarker</groupId>
    <artifactId>freemarker</artifactId>
    <version>2.3.29</version>
</dependency>
```

接下来便配置 Spring MVC 的配置文件，application-mvc.xml 文件中的内容如代码清单 19-29 所示。

代码清单 19-29　application-mvc.xml 文件中的内容

```
<?xml version="1.0" encoding="UTF-8"?>
<beans xmlns:mvc="http://www.springframework.org/schema/mvc"
       xmlns:xsi="http://www.w3.org/2001/XMLSchema-instance"
       xmlns:context="http://www.springframework.org/schema/context"
       xmlns="http://www.springframework.org/schema/beans"
       xsi:schemaLocation="http://www.springframework.org/schema/beans
http://www.springframework.org/schema/beans/spring-beans.xsd
http://www.springframework.org/schema/context
http://www.springframework.org/schema/context/spring-context.xsd
http://www.springframework.org/schema/mvc
http://www.springframework.org/schema/mvc/spring-mvc.xsd">
    <!-- 配置自定义扫描的包 -->
    <context:component-scan base-package="com.uifuture.ssm.controller" />
    <!-- 整合freemarker -->
```

```
    <bean id="freemarkerConfig" class="org.springframework.web.servlet.view.
freemarker.FreeMarkerConfigurer">
        <property name="templateLoaderPath" value="/WEB-INF/templates"/>
        <property name="freemarkerSettings">
            <props>
                <!--模板更新延迟，0为开发时使用，正式使用时应大于3600 -->
                <prop key="template_update_delay">3600</prop>
                <!--设置编码-->
                <prop key="defaultEncoding">UTF-8</prop>
                <prop key="locale">zh_CN</prop>
            </props>
        </property>
    </bean>
    <!-- 配置视图解析器 -->
    <bean id="viewResolver" class="org.springframework.web.servlet.view.
freemarker.FreeMarkerViewResolver">
        <property name="prefix" value=""/>
        <property name="suffix" value=".ftl"/>
        <property name="contentType" value="text/html; charset=UTF-8"/>
  </bean>
    <!-- 配置注解驱动 -->
    <mvc:annotation-driven />
     <!--   配置直接跳转的页面，无须经过Controller层，输入http://localhost:8080/
freemarker01/index，会跳转到WEB-INF/templates/index.html页面-->
    <mvc:view-controller path="/" view-name="index"/>
    <!-- 静态资源配置 -->
    <mvc:resources location="/static/" mapping="/static/**"></mvc:resources>
</beans>
```

其中，设置 defaultEncoding 是为了防止页面出现乱码的情况，将编码统一为 UTF-8。接下来配置 web.xml 文件，如代码清单 19-30 所示。

代码清单 19-30　web.xml 文件中的内容

```
<?xml version="1.0" encoding="UTF-8"?>
<web-app xmlns:xsi="http://www.w3.org/2001/XMLSchema-instance"
         xmlns="http://xmlns.jcp.org/xml/ns/javaee"
         xsi:schemaLocation="http://xmlns.jcp.org0/xml/ns/javaee http://xmlns.jcp.
org/xml/ns/javaee/web-app_4_0.xsd"
         version="4.0">
    <!--Spring的配置-->
    <context-param>
        <!-- param的name必须为contextConfigLocation,Spring内部会解析的 -->
        <param-name>contextConfigLocation</param-name>
        <!-- contextConfigLocation参数的值可配置多个，用英文逗号隔开 -->
        <param-value>
            classpath:application-context.xml
        </param-value>
    </context-param>
      <listener> <listener-class>org.springframework.web.context.
ContextLoaderListener </listener-class>
```

```
    </listener>
    <!—Spring MVC 入口，配置文件位置 -->
    <servlet>
            <servlet-name>springDispatcherServlet</servlet-name> <servlet-
class>org.springframework.web.servlet.DispatcherServlet</servlet-class>
        <init-param>
            <param-name>contextConfigLocation</param-name>
            <param-value>classpath:application-mvc.xml</param-value>
        </init-param>
        <load-on-startup>1</load-on-startup>
    </servlet>
    <!-- 拦截所有请求 -->
    <servlet-mapping>
        <servlet-name>springDispatcherServlet</servlet-name>
        <url-pattern>/</url-pattern>
    </servlet-mapping>
    <!-- 编码过滤器 -->
    <filter>
        <filter-name>encodingFilter</filter-name>
        <filter-class>
            org.springframework.web.filter.CharacterEncodingFilter
        </filter-class>
        <init-param>
            <param-name>encoding</param-name>
            <param-value>UTF-8</param-value>
        </init-param>
    </filter>
    <filter-mapping>
        <filter-name>encodingFilter</filter-name>
        <url-pattern>/*</url-pattern>
    </filter-mapping>
</web-app>
```

这里 web.xml 文件做了以下 5 件事情：

（1）在 org.springframework.web.context.ContextLoaderListener 类中 ContextLoaderListener 的作用就是当启动 Web 容器时，自动装配 ApplicationContext 的配置信息。因为它实现了 ServletContextListener 接口，在 web.xml 中配置该监听器，启动容器时会默认执行该监听器实现的方法。ContextLoaderListener 中关联了 ContextLoader 类，所以整个加载配置过程由 ContextLoader 来完成。ContextLoader 可以由 ContextLoaderListener 和 ContextLoaderServlet 生成。查看 ContextLoader Servlet 的 API 就可以看到 ContextLoaderServlet 也关联了 ContextLoader 类且它实现了 HttpServlet 接口。ContextLoader 创建的是 XmlWebApplicationContext 这样的类，它实现的接口是 WebApplicationContext → ConfigurableWebApplicationContext → ApplicationContext → BeanFactory，这样一来 Spring 中的所有 Bean 都是由该类来创建。

（2）如果在 web.xml 中不加入任何参数配置信息，默认的路径是 /WEB-INF/applicationContext.xml，也就是说在 WEB-INF 目录下创建的 .xml 文件的名称必须是 applicationContext.xml。如果要自定义文件名，则可以在 web.xml 中加入 contextConfigLocation 这个 context 参数，在 <param-value></param-value> 标记中指定相应的 .xml 文件名。当有多个 .xml 文件时，可以将它们写在一起并以“,”分隔。另外，也可以使用 application-*.xml 通配符形式，当该目录下有 application-context.xml、application-mvc.xml 等文件时，都会一同被载入。

（3）使用 Spring MVC，在 web.xml 中配置 DispatcherServlet（前置控制器）是第 1 步。DispatcherServlet 是一个 Servlet，所以可以配置多个 DispatcherServlet。在这里配置的是名称为 springDispatcherServlet 的 DispatcherServlet。DispatcherServlet 被配置在 web.xml 文件中，当拦截匹配的请求时，要自定义 Servlet 拦截匹配规则，并把拦截下来的请求依据设置的规则分发到目标 Controller 处理。

（4）配置 DispatcherServlet 的拦截规则，这里配置的是拦截所有。

（5）编码的设置。Spring 提供了过滤器 CharacterEncodingFilter，该过滤器就是针对每次浏览器请求进行过滤的，然后在其上添加了父类没有的功能，即处理字符编码。其中，encoding 用来设置编码格式；forceEncoding 用来设置是否调用 request.getCharacterEncoding 方法，将其设置为 true 则强制覆盖之前的编码格式。

接下来，在 web/WEB-INF/templates 下创建 index.ftl（.ftl 是 freemarker 文件的后缀名），index.ftl 文件中的内容如代码清单 19-31 所示。

代码清单 19-31　index.ftl 文件中的内容

```
<!DOCTYPE html>
<html lang="en">
<head>
    <meta charset="UTF-8">
    <meta http-equiv="X-UA-Compatible" content="IE=edge">
    <meta name="viewport" content="width=device-width, initial-scale=1">
    <meta name="keywords" content="spring"/>
    <title>spring mvc</title>
</head>
<body>
你好，世界
</body>
</html>
```

运行项目，在这里可以直接通过域名端口 +“/”访问页面。在 Controller 中通过配置方法来跳转页面，如代码清单 19-32 所示。

代码清单 19-32　通过配置方法跳转页面

```
@RequestMapping("/testController")
@Slf4j
@Controller
public class TestController extends BaseController {
    @RequestMapping("index")
    public ModelAndView index() {
        log.info("访问testController/index");
        return new ModelAndView("index");
    }


}
```

此时，项目还不能运行起来，需要进行一些 IDEA 的配置。

19.6 在 IDEA 中配置 Tomcat Web 项目

首先配置 Web 的路径，选择 Project Structure → Facets 命令，单击“+”按钮，选择 Web 选项，如图 19-14 所示。

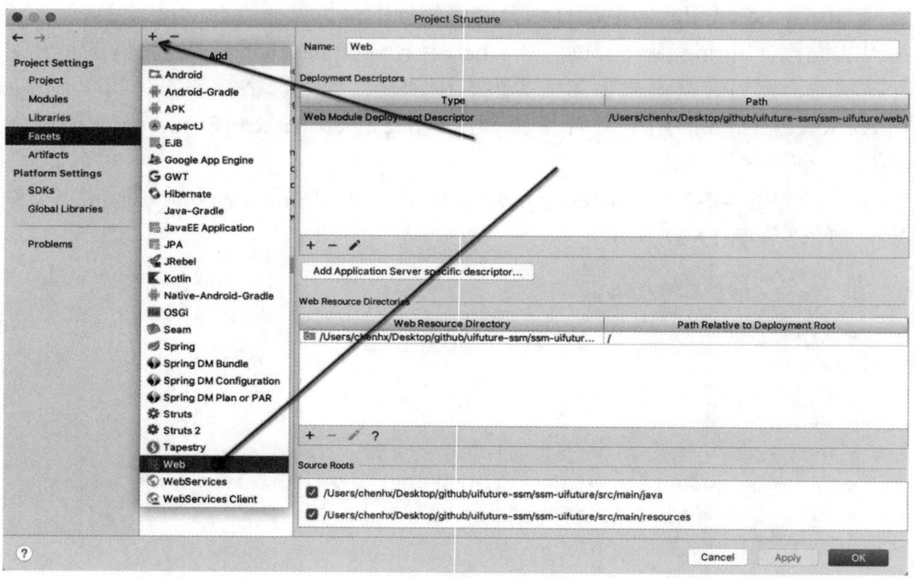

图 19-14　配置 Web 的路径

接下来设置 Web 资源的路径，如图 19-15 所示。

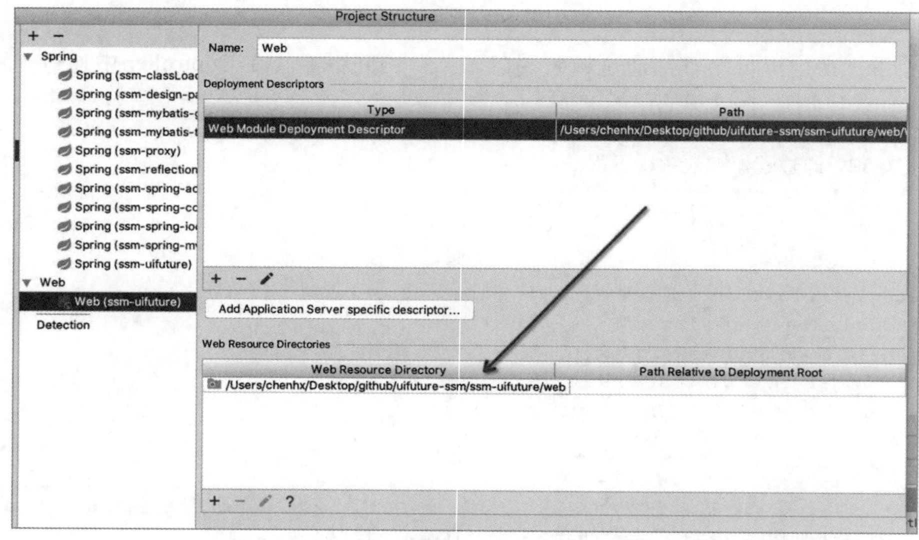

图 19-15　设置 Web 资源的路径

由于 Tomcat 容器需要 war 包才能运行，因此，在 IDEA 中可以通过 Project Structure → Artifacts 命令快速生成热更新 war 包。生成 war 包的过程如图 19-16 所示。

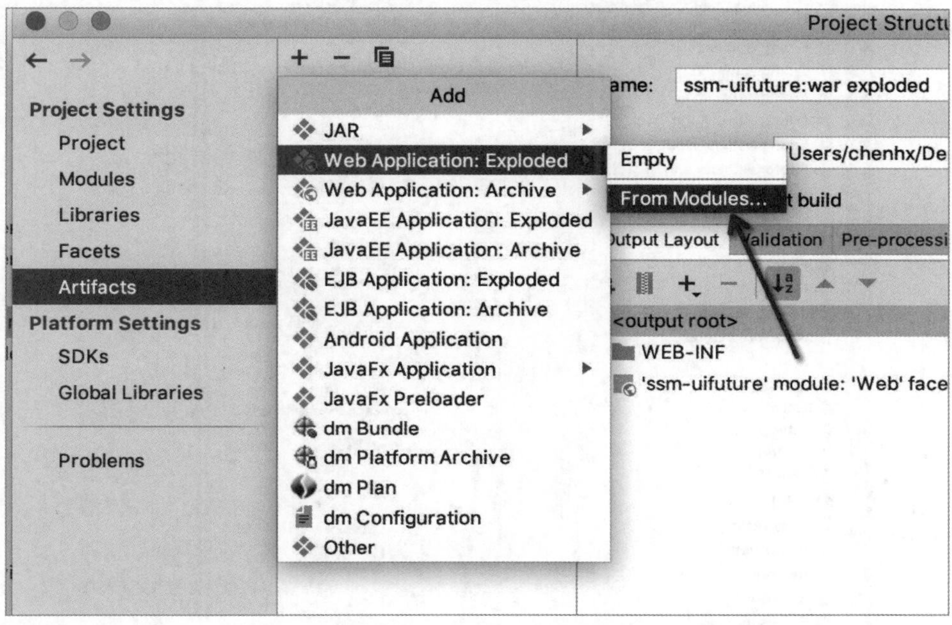

图 19-16　生成 war 包的过程

接下来便会出现列表，让开发者选择含有 Web 资源的项目，如图 19-17 所示。

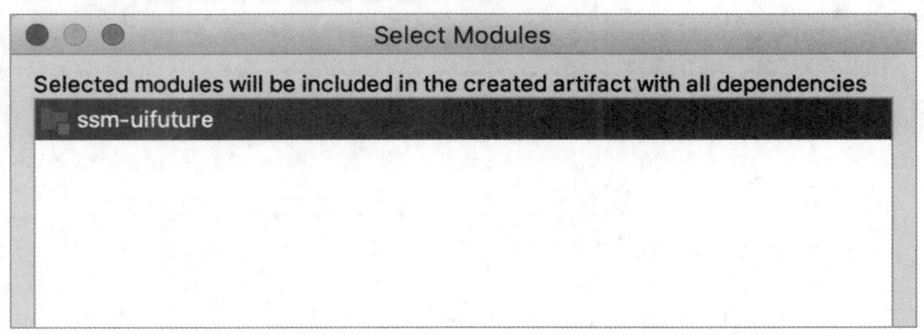

图 19-17　选择含有 Web 资源的项目

war 包配置好后，接下来便可以配置 Tomcat 容器了。在 Run/Debug Configurations 界面中，单击“+”按钮，然后选择 Tomcat Server → Local（本地的容器），如图 19-18 所示。

下面配置 Tomcat 容器的 war 包。在左侧 Tomcat Server 选项下单击 Tomcat 9.0.24，选择 Deployment 选项卡，单击“+”按钮，选择 Artifact... 选项，如图 19-19 所示。

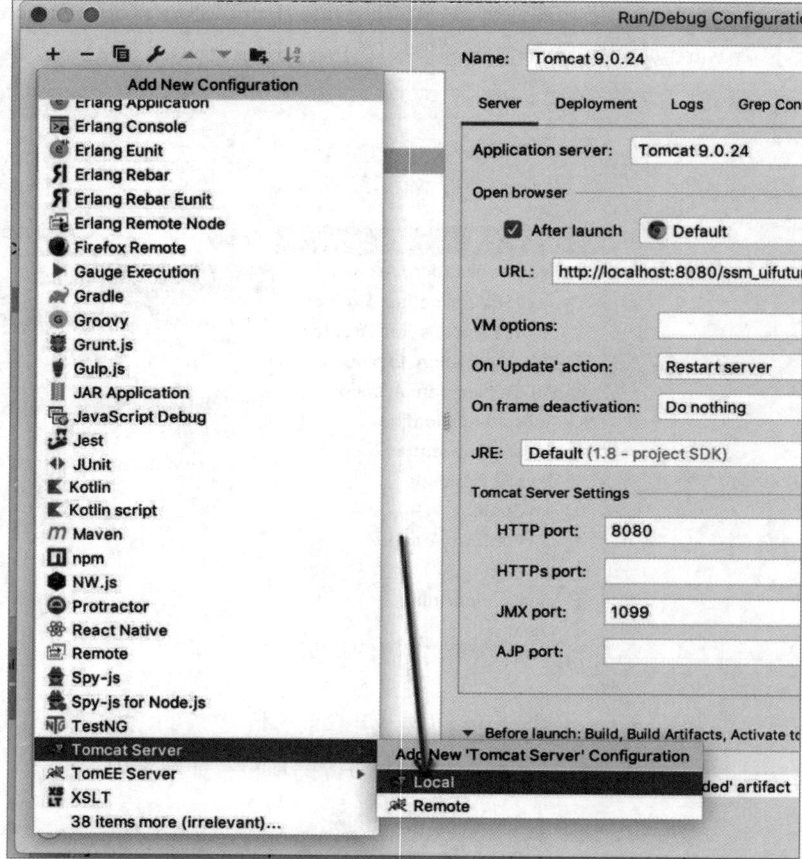

图 19-18　添加 Tomcat 容器

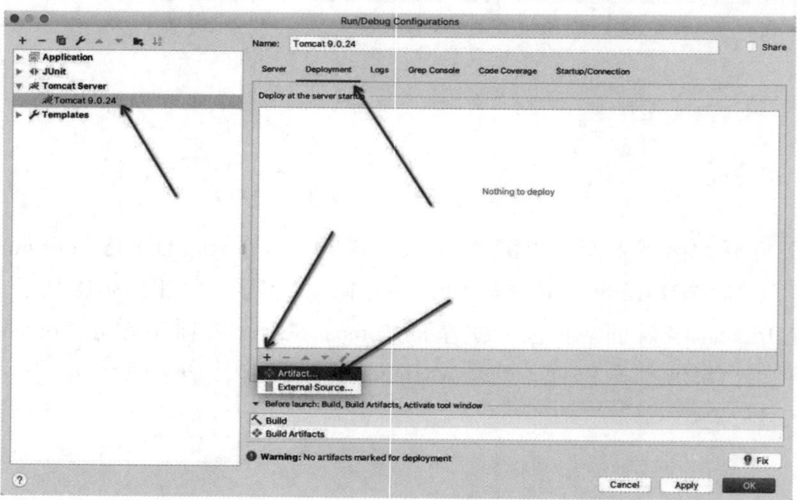

图 19-19　配置 Tomcat 容器的 war 包

最后会出现图 19-20 所示的界面。

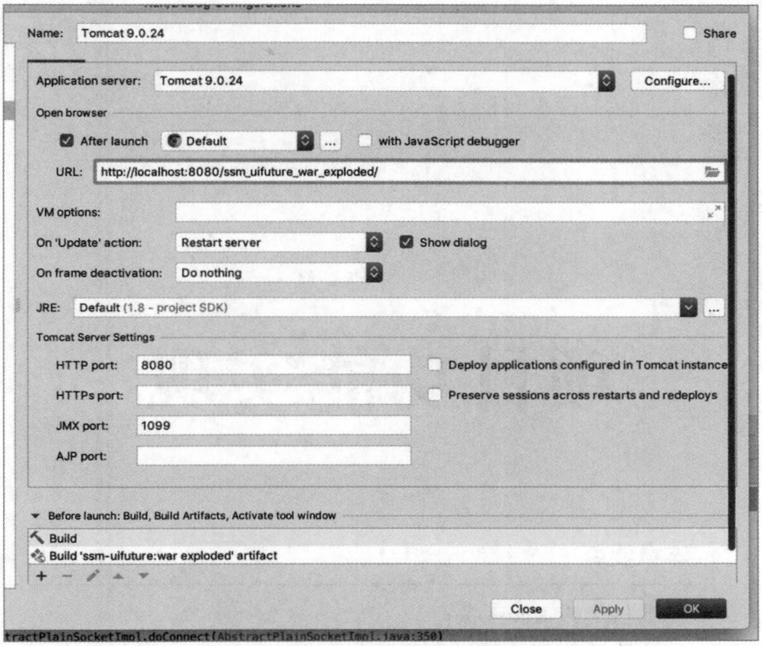

图 19-20　配置好的 Tomcat

运行 Tomcat，可以查看首页及配置的 Controller 页面的跳转情况，如图 19-21 和图 19-22 所示。

图 19-21　查看首页

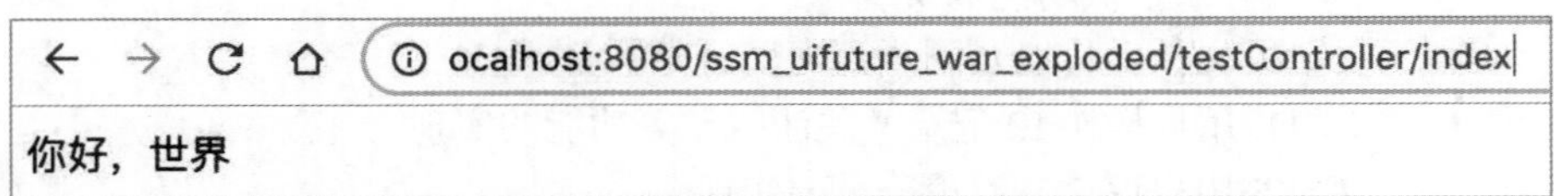

图 19-22　查看 Controller 页面跳转情况

至此，基本的 Spring、Spring MVC 与 MyBatis 的整合已经完成。接下来，随着项目功能的完善，读者便可以逐渐了解更多的框架知识和项目知识。

第 20 章

项目功能实现

本章要点

1. 项目集成 Redis 缓存
2. 注册功能邮件发送的集成与配置
3. 利用 Redis 实现数字的原子性自增
4. 项目登录、注册、退出等功能的开发
5. 使用 MapStruct 快速实现 Bean 复制
6. 使用 Redis 实现 Session 共享
7. 自动登录功能的实现
8. 全局异常与日志输出的实现
9. 登录拦截功能的实现
10. 文件上传以及同步到阿里云 OSS 的实现
11. 用户操作 IP 记录、关注收藏等功能的实现
12. 用户评论、评论分页的实现
13. 敏感词过滤、防范 XSS 攻击的实现
14. 资源数据、交易功能的结构设计与实现

20.1 注册功能

登录、注册是任何一个面向用户的系统需要具备的功能。用户可以选择使用手机号码、邮箱等方式进行注册，本项目使用邮箱注册。

20.1.1 接入 Redis

邮箱注册在便利性方面是不如用手机号码注册的，但是邮箱注册的优势是通用且免费。

邮箱注册的步骤类似于手机号码注册，系统会向用户填写的邮箱地址发送验证码，该验证码用于验证邮箱的所有权。

验证码具有时效性，接入 Redis 无疑是最好的选择。下面便将 Redis 整合到项目中，先引入与依赖相关的 Jar 包，如代码清单 20-1 所示。

代码清单 20-1 添加与依赖相关的 Jar 包

```
<!--Redis依赖，注意jedis与spring-data-redis之间有版本关联-->
<!-- https://mvnrepository.com/artifact/redis.clients/jedis -->
<dependency>
    <groupId>redis.clients</groupId>
    <artifactId>jedis</artifactId>
    <version>2.9.3</version>
    <exclusions>
        <exclusion>
            <artifactId>slf4j-api</artifactId>
            <groupId>org.slf4j</groupId>
        </exclusion>
    </exclusions>
</dependency>
<!-- https://mvnrepository.com/artifact/org.springframework.data/spring-data-
redis -->
<dependency>
    <groupId>org.springframework.data</groupId>
    <artifactId>spring-data-redis</artifactId>
    <version>2.0.9.RELEASE</version>
    <exclusions>
        <exclusion>
            <artifactId>slf4j-api</artifactId>
            <groupId>org.slf4j</groupId>
        </exclusion>
    </exclusions>
</dependency>
<dependency>
    <groupId>com.fasterxml.jackson.core</groupId>
    <artifactId>jackson-databind</artifactId>
    <version>2.9.9.3</version>
```

```
        <scope>compile</scope>
    </dependency>
```

配置 Redis 的参数，在 resource 目录下创建 redis.properties 文件，如代码清单 20-2 所示。

代码清单 20-2　redis.properties 文件内容

```
## 访问地址
redis.host=127.0.0.1
## 访问端口
redis.port=6379
## 注意，如果没有password，此处不设置值，但这一项要保留
redis.password=
# Redis库
redis.database=0
```

接下来将 Redis 的配置及 Bean 注入 Spring 中，新建 application-redis.xml 文件，如代码清单 20-3 所示。

代码清单 20-3　application-redis.xml 文件内容

```
<?xml version="1.0" encoding="UTF-8"?>
<beans xmlns:xsi="http://www.w3.org/2001/XMLSchema-instance"
       xmlns="http://www.springframework.org/schema/beans"
       xsi:schemaLocation="http://www.springframework.org/schema/beans
                           http://www.springframework.org/schema/beans/spring-beans-
4.2.xsd">

    <!-- Redis 配置-->
    <!-- scanner redis properties -->
    <!--<context:property-placeholder location="classpath:redis.properties" ignore-
unresolvable="true"/>-->
     <!--如果有多个数据源需要通过<context:property-placeholder>标记管理，且不愿意放在一个配
置文件里，那么一定要加上ignore-unresolvable= "true"(每一个都需要加这个属性)-->

    <!-- Redis Standalone单节点配置 -->
    <bean id="redisStandaloneConfiguration"
          class="org.springframework.data.redis.connection.RedisStandaloneConfiguration">
        <property name="hostName" value="${redis.host}"/>
        <property name="port" value="${redis.port}"/>
        <property name="database" value="${redis.database}"/>
        <property name="password">
            <bean class="org.springframework.data.redis.connection.RedisPassword">
  <!-- Redis数据库索引(默认为0)-->
                <constructor-arg index="0" value="${redis.password}"/>
            </bean>
        </property>
    </bean>

    <!-- Redis连接配置 -->
```

```
    <bean id="jedisConnectionFactory" class="org.springframework.data.redis.
connection.jedis.JedisConnectionFactory">
        <constructor-arg name="standaloneConfig" ref="redisStandaloneConfiguration"/>

    </bean>
    <!-- Redis序列化 -->
     <bean id="stringRedisSerializer" class="org.springframework.data.redis.
serializer.StringRedisSerializer"/>
     <bean id="jsonRedisSerializer" class="org.springframework.data.redis.
serializer.GenericJackson2JsonRedisSerializer">
        <!-- 使用默认的ObjectMapper，移除JSON格式化后的"@Class"节点 -->
        <constructor-arg name="mapper">
            <bean class="com.fasterxml.jackson.databind.ObjectMapper"/>
        </constructor-arg>
    </bean>
    <!-- Redis持久化模板 -->
     <bean id="redisTemplate" class="org.springframework.data.redis.core.
RedisTemplate">
        <property name="connectionFactory" ref="jedisConnectionFactory"/>
        <!-- 如果不配置Serializer，那么存储时只能使用String；如果用对象类型存储，那么会提示错误
"can't cast to String！！！"-->
        <property name="keySerializer">
            <!--对key的默认序列化器，默认值是String Redis Serializer-->
                  <bean class="org.springframework.data.redis.serializer.
StringRedisSerializer"/>
        </property>
        <!--对value的默认序列化器，默认值是JdkSerializationRedisSerializer-->
        <property name="valueSerializer">
            <bean class="org.springframework.data.redis.serializer.JdkSerialization
RedisSerializer"/>
        </property>
        <!--存储Map时key需要的序列化配置-->
        <property name="hashKeySerializer">
                  <bean class="org.springframework.data.redis.serializer.
StringRedisSerializer"/>
        </property>
        <!--存储Map时value需要的序列化配置-->
        <property name="hashValueSerializer">
            <bean class="org.springframework.data.redis.serializer.JdkSerialization
RedisSerializer"/>
        </property>
          <!--开启事务，系统自动获得了事务中绑定的连接。在一个方法的多次对Redis增、删、改、查中，
可以始终使用同一个连接
        -->
        <property name="enableTransactionSupport" value="true"/>
        <!--在Spring中@Transactional也是可以进行事务控制的-->
    </bean>

</beans>
```

注意：在这里并没有加载 Redis 的配置文件，这是由于前面使用了自定义的 EncryptPropertyPlaceholderConfigurer 类作为配置文件的加载类，所以多个配置文件只能配置在一起。新增 application-properties.xml 文件，将原来加载配置文件的代码删除。application-properties.xml 文件内容如代码清单 20-4 所示。

代码清单 20-4　application-properties.xml 文件内容

```
<?xml version="1.0" encoding="UTF-8"?>
<beans xmlns:xsi="http://www.w3.org/2001/XMLSchema-instance"
      xmlns="http://www.springframework.org/schema/beans"
      xsi:schemaLocation="http://www.springframework.org/schema/beans
                    http://www.springframework.org/schema/beans/spring-beans-
4.2.xsd">
    <!-- 引入属性文件，classpath就是在resource目录下-->
     <!--在这里使用了EncryptPropertyPlaceholderConfigurer类进行配置文件加载，注入Bean，无
法分开写配置文件-->
    <bean id="propertyConfigurer" class="com.uifuture.ssm.config.EncryptPropertyPla
ceholderConfigurer">
        <property name="locations">
            <list>
                <value>classpath:config.properties</value>
                <value>classpath:redis.properties</value>
            </list>
        </property>
    </bean>
</beans>
```

为了方便用 Redis 持久化模板操作数据，新增了很多类。由于其代码过多，读者可以通过 GitHub 访问（路径为 ssm-uifuture/src/main/java/com/uifuture/ssm/redis）。

还有一处需要配置，就是新增的类。因为没有通过注解形式注入，所以需要在 application-redis.xml 中新增并注入 Spring 中，如代码清单 20-5 所示。

代码清单 20-5　application-redis.xml 文件内容

```
<!--配置-->
<bean id="connConfig" class="com.uifuture.ssm.redis.config.ConnConfig">
    <property name="host" value="${redis.host}"></property>
    <property name="port" value="${redis.port}"></property>
    <property name="password" value="${redis.password}"></property>
    <property name="app" value="${redis.app}"></property>
</bean>
<!--Redis常规操作-->
<bean id="redisClient" class="com.uifuture.ssm.redis.RedisClient">
    <property name="connConfig" ref="connConfig"></property>
</bean>
<!--Redis分布式锁操作-->
<bean id="redisLock" class="com.uifuture.ssm.redis.RedisLock">
    <property name="client" ref="redisClient"/>
</bean>
```

20.1.2 发送邮件配置

接下来进入功能开发,使用邮箱注册账号。用户注册时,需要填写基本的信息,UsersReq 类(使用 Req 结尾的实体类来代表前端请求的参数)如代码清单 20-6 所示。

代码清单 20-6 UsersReq 类

```
@Data
public class UsersReq implements Serializable {
    /** 用户名 */
    @NotEmpty(message="用户名不能为空")
    @Length(min=6, max=32, message="用户名长度应在6~32之间")
    private String username;
    /** 密码(使用MD5+盐值加密) */
    @NotEmpty(message="密码不能为空")
    @Length(min=6, max=32, message="密码长度应在6~32之间")
    private String password;
    /** 邮箱 */
    @NotEmpty(message="邮箱不能为空")
    @Length(max=100, message="邮箱最长为100位")
    private String email;
    /** 验证码 */
    @NotEmpty(message="邮箱验证码不能为空")
    @Length(min=6, max=6, message="邮箱验证码为6位数字")
    private String emailCode;
}
```

在这里集成 hibernate-validator 依赖进行参数校验,hibernate-validator 所需要的依赖如代码清单 20-7 所示。

代码清单 20-7 hibernate-validator 依赖

```
<!--增加参数校验依赖 -->
<dependency>
    <groupId>org.hibernate</groupId>
    <artifactId>hibernate-validator</artifactId>
    <version>6.0.9.Final</version>
</dependency>
<!-- https://mvnrepository.com/artifact/javax.validation/validation-api -->
<dependency>
    <groupId>javax.validation</groupId>
    <artifactId>validation-api</artifactId>
    <version>2.0.1.Final</version>
</dependency>
<!--end-->
```

相较于添加注解,笔者更倾向于编写校验工具类,如代码清单 20-8 所示。

20 项目功能实现

代码清单 20-8　校验工具类

```
/** 参数校验(根据参数上的注解) */
public class ValidateUtils {
    private static final Validator VALIDATOR = Validation.buildDefaultValidatorFactory().
getValidator();
    public static <T> void validate(T object) {
        if (object==null) {
            throw new CommonException(ResultCodeEnum.PARAMETER_ERROR);
        }
            Set<ConstraintViolation<T>> constraintViolations=VALIDATOR.
validate(object);
        // 如果有验证信息，则抛出异常
            ConstraintViolation<T> constraintViolation= IteratorsUtils.
getFirst(constraintViolations, null);
        if (constraintViolation != null) {
            throw new CommonException(ResultCodeEnum.PARAMETER_ERROR.getValue(),
constraintViolation.getMessage());
        }
    }
}
```

在代码中使用时，可以直接通过 ValidateUtils.validate(变量名) 来进行实体类的校验。

通过邮箱注册需要发送邮件，接下来将引入与邮箱相关的依赖，如代码清单 20-9 所示。

代码清单 20-9　引入与邮箱相关的依赖

```
<dependency>
    <groupId>javax.mail</groupId>
    <artifactId>mail</artifactId>
    <version>1.4.7</version>
    <exclusions>
        <exclusion>
            <artifactId>activation</artifactId>
            <groupId>javax.activation</groupId>
        </exclusion>
    </exclusions>
</dependency>
<dependency>
    <groupId>javax.activation</groupId>
    <artifactId>activation</artifactId>
    <version>1.1.1</version>
</dependency>
```

然后增加读取邮件信息的配置文件，这里新增邮件配置文件 email.properties，如代码清单 20-10 所示。

代码清单 20-10　邮件配置文件 email.properties

```
# 邮箱地址
email.mail.add=uifuture@uifuture.com
```

```
# 发送人名称
email.mail.name=uifuture
# 邮箱密码
email.mail.password=******
email.mail.port=465
# 服务器主机名
email.mail.host=smtp.uifuture.com
# 协议
email.mail.protocol=smtp
# 主题
email.subject=uifuture官方邮件
email.index.mail.add=http://127.0.0.1:8080/
# 客服邮箱
email.service.mail=uifuture@uifuture.com
```

在 application-properties.xml 中增加读取邮件信息的功能，如代码清单 20-11 所示。

代码清单 20-11　增加读取邮件信息的功能

```
<?xml version="1.0" encoding="UTF-8"?>
<beans xmlns:xsi="http://www.w3.org/2001/XMLSchema-instance"
       xmlns="http://www.springframework.org/schema/beans"
       xsi:schemaLocation="http://www.springframework.org/schema/beans
                    http://www.springframework.org/schema/beans/spring-beans-4.2.xsd">
    <bean id="propertyConfigurer" class="com.uifuture.ssm.config.EncryptPropertyPla
ceholderConfigurer">
        <property name="locations">
            <list>
                <value>classpath:config.properties</value>
                <value>classpath:redis.properties</value>
                <value>classpath:email.properties</value>
            </list>
        </property>
    </bean>
</beans>
```

以上操作也就是给 locations 集合增加了一个 value。接下来编写配置类，将配置文件的值设置到 Java 类中。为了扩展，可以先创建一个 EmailConfig 接口，如代码清单 20-12 所示。

代码清单 20-12　EmailConfig 接口

```
public interface EmailConfig {
    /**
     * @return 发件邮箱地址
     */
    String getMainAdd();
    /**
     * @return 发件人的名称
     */
    String getMainName();
    /**
```

```
     * @return 发件邮箱密码(也叫授权码)
     */
    String getPassword();
    /**
     * @return 发件邮箱端口
     */
    String getPort();
    /**
     * @return 设置邮件服务器主机名
     */
    String getMailHost();
    /**
     * @return 发送邮件协议名称
     */
    String getMailProtocol();
    /**
     * @return 邮件：主题
     */
    String getSubject();
    /** 网站地址 */
    String getIndexAdd();
    /** 客户邮件 */
    String getServiceMail();
}
```

接下来编写实现类，也就是实现对邮件信息的读取，如代码清单 20-13 所示。

代码清单 20-13　读取邮件信息

```
@Component @Data
public class EmailConfigImpl implements EmailConfig {
    /** 发件邮箱地址 */
    @Value("${email.mail.add}")
    private String mainAdd;
    /** 发件人的名称 */
    @Value("${email.mail.name}")
    private String mainName;
    /** 发件邮箱授权码 */
    @Value("${email.mail.password}")
    private String password;
    /** 发件邮箱端口 */
    @Value("${email.mail.port}")
    private String port;
    /** 设置邮件服务器主机名 */
    @Value("${email.mail.host}")
    private String mailHost;
    /** 发送邮件协议名称 */
    @Value("${email.mail.protocol}")
    private String mailProtocol;
    /** 邮件：主题 */
```

```
    @Value("${email.subject}")
    private String subject;
    /** 网站地址 */
    @Value("${email.index.mail.add}")
    private String indexAdd;
    /** 客服邮箱 */
    @Value("${email.service.mail}")
    private String serviceMail;
}
```

至此，基本的邮件信息便配置好了。下面介绍核心的邮件发送类。

首先新建一个发送邮件接口，用于获取邮件接收人、验证码等动态信息，如代码清单 20-14 所示。

代码清单 20-14　新建发送邮件接口

```
public interface SendEmail {
    /** 验证码 */
    String getCode();
    /** 用户的昵称 */
    String getName();
    /** 用户的邮箱，也就是收件邮箱 */
    String getEmail();
}
```

通过实现该发送类，可由 getCode、getName、getEmail 三个方法返回动态信息。接下来是实际的发送邮件类，这里要通过异步发送邮件，所以实现了 Callable 接口。实现 Callable 接口的好处是可以选择是否需要返回值来确定邮件是否发送成功。当然，也可以通过消息队列的方式确定，即邮件发送失败后，放入延时消息队列，并通知开发、运营，再通过消费消息队列的消息内容来发送邮件。发送邮件的实现类如代码清单 20-15 所示。

代码清单 20-15　发送邮件的实现类

```
public class SendEmailCallable implements Callable {
    private static Logger logger=LoggerFactory.getLogger(SendEmailCallable.class);
    private EmailConfig emailConfig;
    private SendEmail sendEmail;
    public SendEmailCallable() {
    }
    public SendEmailCallable(EmailConfig emailConfig, SendEmail sendEmail) {
        this.emailConfig=emailConfig;
        this.sendEmail=sendEmail;
    }
    @Override
    public Object call() throws Exception {
        // 与SMTP服务器建立一个连接
        Properties p=new Properties();
        // 设置邮件服务器主机名，指定邮件服务器，默认端口为25
        p.setProperty("mail.host", emailConfig.getMailHost());
```

```
        // 发送服务器需要身份验证，要采用指定用户名密码的方式去认证
        p.setProperty("mail.smtp.auth", "true");
        // 发送邮件协议名称
        p.setProperty("mail.transport.protocol", emailConfig.getMailProtocol());
        p.setProperty("mail.smtp.port", emailConfig.getPort());
        // 开启SSL加密，否则会失败
        MailSSLSocketFactory sf=null;
        try {
            sf=new MailSSLSocketFactory();
        } catch (GeneralSecurityException e1) {
                logger.error("开启SSL加密失败，出现GeneralSecurityException异常！给" +
sendEmail.getName() + "," + sendEmail.getEmail() + "发送邮件失败！ ");
            logger.error("", e1);
            return null;
        }
        // 开启Debug调试，以便在控制台查看p.setProperty("mail.debug", "true");
        sf.setTrustAllHosts(true);
        p.put("mail.smtp.ssl.enable", "true");
        p.put("mail.smtp.ssl.socketFactory", sf);
        Session session=Session.getDefaultInstance(p, new Authenticator() {
            @Override
            protected PasswordAuthentication getPasswordAuthentication() {
                // 用户名可以用邮箱的别名，后面的字符是授权码
                return new PasswordAuthentication(
                        emailConfig.getMainAdd(), emailConfig.getPassword());
            }
        });
        //session.setDebug(true); 设置打开调试状态
        try {
            // 声明一个Message对象(代表一封邮件),从session中创建
            MimeMessage msg=new MimeMessage(session);
            // 邮件信息封装
            // 1.发件人
            msg.setFrom(new InternetAddress(emailConfig.getMainAdd(), emailConfig.
getMainName(), "UTF-8"));
            // 2.收件人
                    msg.setRecipient(Message.RecipientType.TO, new
InternetAddress(sendEmail.getEmail()));
            // 3.邮件：主题、内容
            msg.setSubject(emailConfig.getSubject());
/* StringBuilder是线程不安全的，但是速度快，因为这里只有这个线程来访问，所以可以用这个线程发送
.html格式的文本 */
            String sbd = ("<!DOCTYPE html><html><head><meta charset='UTF-8'>" +
                    "<title>邮箱激活</title>" +
                    "</head><body>" +
                        "<table style='background: #fff; border-collapse: collapse;
border-spacing: 0; color: #222; font-size: 16px; height: 100%; margin: 0; padding:
0; width: 100%'bgcolor='#fff'>" +
```

```
                    "<tbody><tr><td style='-moz-hyphens: auto; -webkit-hyphens: auto; border-collapse: collapse !important; color: #222;  font-size: 16px; hyphens: auto; margin: 0; padding: 0; text-align: center; word-break: break-word'valign='top'align='center'>" +
                      "<center><table style='border-collapse: collapse; border-spacing: 0; font-size: 16px; line-height: 1.5; margin: 0 auto; max-width: 680px; min-width: 300px; width: 95%'>" +
                    "<tbody><tr><td style='-moz-hyphens: auto; -webkit-hyphens: auto; border-collapse: collapse !important; color: #222; font-family: font-size: 16px; hyphens: auto; margin: 0; padding: 0; word-break: break-word'>" +
                      "<hr style='background: #ddd; border: none; color: #ddd; height: 1px; margin: 20px 0 30px'>" +
                      "<table style='border-collapse: collapse; border-spacing: 0'width='100%'>" +
                    "<tbody><tr><td style='-moz-hyphens: auto; -webkit-hyphens: auto; border-collapse: collapse !important; color: #222; font-size: 16px; hyphens: auto; margin: 0; padding: 0; word-break: break-word'>" +
                "<div><p style='color: #222;  font-size: 16px; margin: 0 0 10px; padding: 0'>" +
                sendEmail.getName() + ",你好! " +
                 "</p><p style='color: #222;  font-size: 16px; margin: 0 0 10px; padding: 0'>" +
                "感谢注册uifuture,您的验证码为" +
                   "</p><p style='color: #222; font-size: 16px; margin: 24px 0; padding: 0'>" +
                 "<a target='_blank'style='background: #83a198; border-radius: 4px; color: #fff; padding: 8px 16px; text-decoration: none; word-break: break-all'>" +
                sendEmail.getCode() +
                 "</a></p><p style='color: #999999;  font-size: 12px; margin: 0 0 10px; padding: 0'>" +
                "验证码10分钟内有效。如果你没有使用过" + emailConfig.getMainName() + ",请忽略此邮件。" +
                "</p></div></td></tr></tbody></table>" +
                      "<hr style='background: #ddd; border: none; color: #ddd; height: 1px; margin: 20px 0'>" +
                      "<table style='border-collapse: collapse; border-spacing: 0'width='100%'>" +
                 "<tbody><tr><td colspan='2'style='-moz-hyphens: auto; -webkit-hyphens: auto; border-collapse: collapse !important; color: #222;font-size: 16px; hyphens: auto; margin: 0; padding: 0; word-break: break-word'align='center'>" +
                  "<p style='color: #999999; font-size: 12px; margin: 0 0 10px; padding: 0'>" +
                "如有疑问请联系我们:" +
                "<a href='mailto:" +
                emailConfig.getServiceMail() +
                "'style='color: #999999; text-decoration: underline'target='_blank'>" +
                emailConfig.getServiceMail() +
                  "</a>" + "</p></td></tr></tbody></table></td></tr></tbody></table></center></td></tr></tbody></table></body></html>");
```

20 项目功能实现

```
            msg.setContent(sbd, "text/html;charset=utf-8");
            // 发送动作
            Transport.send(msg);
                logger.info("用 户 为:" + sendEmail.getName() + ",给" + sendEmail.
getEmail() + "发送邮件成功......");
            return 1;
        } catch (MessagingException | UnsupportedEncodingException e) {
                logger.error("用 户 为:" + sendEmail.getName() + ",给" + sendEmail.
getEmail() + "发送邮件失败，异常信息:" + e.getMessage());
             return "用户为:" + sendEmail.getName() + ",给" + sendEmail.getEmail() +
"发送邮件失败，异常信息:" + e.getMessage();
        }
    }
}
```

接下来通过测试类验证是否发送成功，测试类内容如代码清单 20-16 所示。

代码清单 20-16　测试类

```
public class SendEmailCallableTest extends BaseTest {
    @Autowired
    private EmailConfig emailConfig;
    @Test
    public void sendEmail() throws InterruptedException {
        // 发送邮件
        SendEmail sendEmail=new SendEmail() {
            @Override
            public String getCode() {
                return "123456";
            }
            @Override
            public String getName() {
                return "test";
            }
            @Override
            public String getEmail() {
                return "接收邮件的邮箱地址";
            }
        };
        // 异步发送邮件
           SendEmailCallable sendEmailCallable=new SendEmailCallable(emailConfig,
sendEmail);
        ExecutorService executorService=Executors.newSingleThreadExecutor();
        executorService.submit(sendEmailCallable);
        // 因为是异步发送，主线程如果退出，子线程也会直接退出
        Thread.sleep(60000);
    }
}
```

接收的邮件信息如图 20-1 所示。

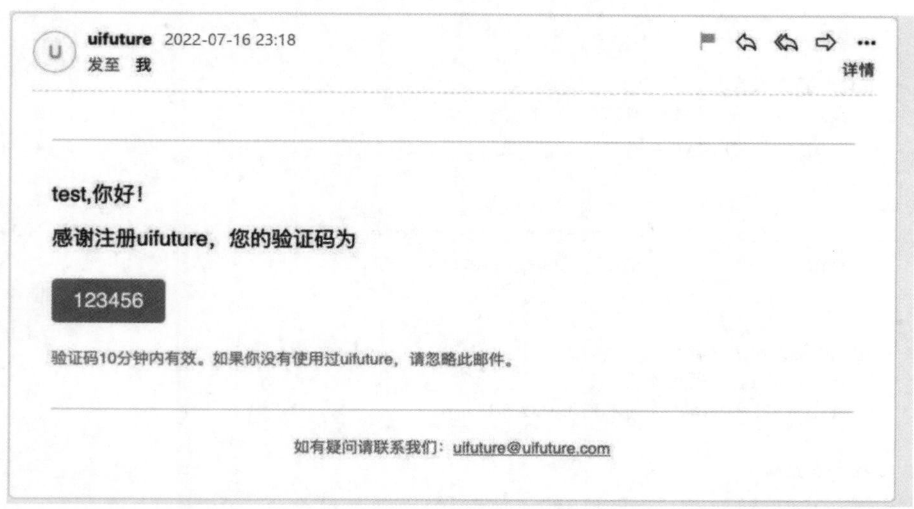

图 20-1　接收的邮件信息

20.1.3　实现 Server 层与 Controller 层

前面已经将 Redis 集成并加入了邮件的发送类，下面直接编写业务代码。首先进行 Server 层的实现。由于用户名及邮箱都是唯一的，因此要先通过用户名或者邮箱来查询其是否已经存在，可以创建 UsersServiceImpl 类来实现，如代码清单 20-17 所示。

代码清单 20-17　UsersServiceImpl 类

```
/** 用户表，服务实现类 */
@Service
public class UsersServiceImpl extends ServiceImpl<UsersMapper, UsersEntity>
implements UsersService {
    /**
     * 查询用户名是否存在
     * @param username 查询用户名
     * @return 0表示不存在
     */
    @Override
    public Integer selectCountByUsername(String username) {
        return getCount(username, UsersEntity.USERNAME);
    }
    /**
     * 通过用户名查询用户数据
     * @param username 代表用户名
     * @return 用户数据，没有该用户，则返回NULL
     */
    @Override
    public UsersEntity selectByUsername(String username) {
        return getUsersEntity(username, UsersEntity.USERNAME);
```

```
    }
    /**
     * 查询邮箱是否存在
     * @param email 代表邮箱
     * @return 0表示不存在
     */
    @Override
    public Integer selectCountByEmail(String email) {
        return getCount(email, UsersEntity.EMAIL);
    }
    /**
     * 通过邮箱查询用户数据
     * @param email 代表邮箱
     * @return 用户数据，查询不到用户返回NULL
     */
    @Override
    public UsersEntity selectByEmail(String email) {
        return getUsersEntity(email, UsersEntity.EMAIL);
    }
    /** 通过唯一键获取数据是否存在 */
    private Integer getCount(String value, String column) {
        QueryWrapper<UsersEntity> queryWrapper=new QueryWrapper<>();
        queryWrapper.eq(column, value);
        return this.count(queryWrapper);
    }
    /** 通过唯一键查询数据  */
    private UsersEntity getUsersEntity(String value, String column) {
        QueryWrapper<UsersEntity> queryWrapper=new QueryWrapper<>();
        queryWrapper.eq(column, value);
        return this.getOne(queryWrapper);
    }
}
```

注册时需要记录用户的 IP，由于获取用户 IP 这个方法比较通用，所以将其编写到 Controller 层类的父类 BaseController 中，如代码清单 20-18 所示。

代码清单 20-18　BaseController 中获取 IP 方法

```
/** 获取IP */
protected static String getIpAddress(HttpServletRequest request) {
    String ip=null;
    // 通过请求头中的X-Forwarded-For:Squid 服务代理
    String ipAddresses=request.getHeader("X-Forwarded-For");
     if (ipAddresses==null || ipAddresses.length()==0 || "unknown".
equalsIgnoreCase(ipAddresses)) {
        // 通过请求头中的Roxy-Client-IP:Apache服务代理
        ipAddresses=request.getHeader("Proxy-Client-IP");
    }
     if (ipAddresses==null || ipAddresses.length()==0 || "unknown".
equalsIgnoreCase(ipAddresses)) {
```

```
        // 通过请求头中的WL-Proxy-Client-IP:weblogic服务代理
        ipAddresses=request.getHeader("WL-Proxy-Client-IP");
    }
    if (ipAddresses==null || ipAddresses.length()==0 || "unknown".
equalsIgnoreCase(ipAddresses)) {
        // 通过请求头中的HTTP_CLIENT_IP: 有些代理服务器
        ipAddresses=request.getHeader("HTTP_CLIENT_IP");
    }
    if (ipAddresses==null || ipAddresses.length()==0 || "unknown".
equalsIgnoreCase(ipAddresses)) {
        // 通过请求头中的X-Real-IP:nginx服务代理
        ipAddresses=request.getHeader("X-Real-IP");
    }
    /*某些网络通过多层代理可能获取到的ip就会有多个; 多个iP一般都是通过逗号(,)分隔开, 并且第1个
ip值为客户端的真实iP*/
    if (ipAddresses!=null && ipAddresses.length()!=0) {
        ip=ipAddresses.split(",")[0];
    }
    //如果还不能获取, 最后可通过request.getRemoteAddr()获取
    if (ip==null || ip.length()==0 || "unknown".equalsIgnoreCase(ipAddresses)) {
        ip=request.getRemoteAddr();
    }
    return ip;
}
```

因为用户名和邮箱需要校验，所以增加了正则工具类，如代码清单 20-19 所示。

代码清单 20-19　正则工具类

```
public class RegexUtils {
    /**
     * 验证账号
     * @param username 查询用户名注册匹配
     * @return 验证成功返回true, 验证失败返回false
     */
    public static boolean checkUsername(String username) {
        if (StringUtils.isEmpty(username)) {
            return false;
        }
        String regex="[a-zA-Z_][a-zA-Z0-9_]{5,31}";
        return Pattern.matches(regex, username);
    }
    /**
     * 验证Email
     * @param email 查询Email地址, 格式为zhangsan@xxx.com.cn, xxx代表邮件服务商
     * @return 验证成功返回true, 验证失败返回false
     */
    public static boolean checkEmail(String email) {
        if (StringUtils.isEmpty(email)) {
            return false;
```

```
        }
        String regex = "\\w+@\\w+\\.[a-z]+(\\.[a-z]+)?";
        return Pattern.matches(regex, email);
    }
}
```

验证用户名和邮箱是否符合规范方法也比较通用，所以将其抽取到 BaseController 中，如代码清单 20-20 所示。

代码清单 20-20　验证用户名和邮箱

```
/**
 * 校验用户名和邮箱
 * @param username 查询用户名
 * @param email 查询邮箱
 */
protected static void checkParam(String username, String email) {
    if (!RegexUtils.checkEmail(email)) {
        throw new CommonException(ResultCodeEnum.INCORRECT_MAILBOX_FORMAT);
    }
    if (!RegexUtils.checkUsername(username)) {
        throw new CommonException(ResultCodeEnum.PARAMETER_ERROR);
    }
}
```

至于这里直接抛出异常的原因，后面介绍到的一个全局异常捕获的类可以解释。接下来介绍实际的注册方法。关于 HTTP 的请求方法，在这里选择使用默认的 GET 和 POST 方法。查询的请求使用 GET 方法，其余全部使用 POST 方法。

然后创建 UsersRestController 类来使用邮箱进行注册。**提示**：要限制单个邮箱接收邮件的频次，这里限制为同一 IP 下，10min 内仅能接收 10 次邮件验证码。当然，这样会出现副作用，即在共用 IP 的内部网中，10min 内最多只能有 10 个人能够注册成功。UsersRestController 类中注册方法如代码清单 20-21 所示。

代码清单 20-21　UsersRestController 类中注册方法

```
/** 用户注册 */
@RequestMapping(value="/registered", method=RequestMethod.POST )
public ResultModel registered(UsersReq usersReq, HttpServletRequest request) {
    // 校验参数
    ValidateUtils.validate(usersReq);
    // 对用户名中数字、字母进行判断，注意要以字母开头
    checkParam(usersReq.getUsername(), usersReq.getEmail());
    // 校验用户名
    Integer num=usersService.selectCountByUsername(usersReq.getUsername());
    if (num>0) {
        return ResultModel.fail(ResultCodeEnum.USERNAME_ALREADY_EXISTS);
    }
    // 校验邮箱
```

```
num=usersService.selectCountByEmail(usersReq.getEmail());
    if (num>0) {
        return ResultModel.fail(ResultCodeEnum.EMAIL_ALREADY_EXISTS);
    }
    // 同一个IP下，10min最多请求10次
     long times=redisClient.incr(RedisConstants.getRegTimesKey(getIpAddress(reque
st)), RedisConstants.REG_MAX_TIME );
    if (times>RedisConstants.REG_MAX_TIMES ) {
        // 请求次数过多，提示用户稍后再试
        return ResultModel.fail(ResultCodeEnum.ALL_TOO_OFTEN);
    }
    // 通过Redis确定验证码是否正确，获取验证码
     String realCode= redisClient.get(RedisConstants.getRegEmailKey(usersReq.
getEmail())).getObject(String.class);
    if (realCode==null) {
        // 验证码过期
        return ResultModel.fail(ResultCodeEnum.VERIFICATION_CODE_HAS_EXPIRED);
    }
    if (!realCode.equals(usersReq.getEmailCode())) {
        // 验证码错误
        return ResultModel.fail(ResultCodeEnum.VERIFICATION_CODE_ERROR);
    }
    // 构建用户数据，包括密码加密、增加密码被破解的难度
    String salt=PasswordUtils.getSalt();
    String password=PasswordUtils.getPassword(usersReq.getPassword(), salt);
    UsersEntity usersEntity=new UsersEntity();
    usersEntity.setUsername(usersReq.getUsername());
    usersEntity.setPassword(password);
    usersEntity.setSalt(salt);
    usersEntity.setEmail(usersReq.getEmail());
    usersEntity.setCreateId(0);
    usersEntity.setMailboxState(1);
    // 数据入库
    usersService.save(usersEntity);
    return ResultModel.success();
}
/** 发送邮箱验证码 */
@RequestMapping(value="/sendEmailCode", method=RequestMethod.POST)
public ResultModel sendEmailCode(String email, String username, HttpServletRequest
request) {
    // 校验参数
    if (StringUtils.isEmpty(email) || StringUtils.isEmpty(username)) {
        return ResultModel.fail(ResultCodeEnum.PARAMETER_ERROR);
    }
    // 对用户名中数字、字母进行判断，注意要以字母开头
    checkParam(username, email);
// 校验用户名
    Integer num=usersService.selectCountByUsername(username);
    if (num>0) {
        return ResultModel.fail(ResultCodeEnum.USERNAME_ALREADY_EXISTS);
```

```
    }
    // 校验邮箱
    num=usersService.selectCountByEmail(email);
    if (num>0) {
        return ResultModel.fail(ResultCodeEnum.EMAIL_ALREADY_EXISTS);
    }
    /*判断是否已经发送，10min内同一IP最多发送10次*/
    // 获取发送次数
    long times= redisClient.incr(RedisConstants.getSendEmailCodeTimesKey(getIpAddres
s(request)), RedisConstants.REG_MAX_TIME);
    if (times > RedisConstants.SEND_CODE_MAX_TIMES ) {
        //请求次数过多，提示用户稍后再试
        return ResultModel.fail(ResultCodeEnum.ALL_TOO_OFTEN );
    }
    // 判断是否已经发送过，激活码使用原来的
    String code;
    // 获取原来的code
     String realCod = redisClient.get(RedisConstants.getRegEmailKey(email)).
getObject(String.class);
    if (StringUtils.isNotEmpty(realCode)) {
        code=realCode;
    } else {
        code=PasswordUtils.randomNumberLower(6);
    }
    // 发送邮件，用户进行激活
    SendEmail sendEmail=new SendEmail() {
        @Override
        public String getCode() {
            return code;
        }
        @Override
        public String getName() {
            return username;
        }
        @Override
        public String getEmail() {
            return email;
        }
    };
    // 异步发送邮件
     SendEmailCallable sendEmailCallable=new SendEmailCallable(emailConfig,
sendEmail);
ExecutorService executorService=Executors.newSingleThreadExecutor();
    executorService.submit(sendEmailCallable);
    //将code记录到Redis，时效为10min
     redisClient.set(RedisConstants.getRegEmailKey(email), code, RedisConstants.
REG_MAX_TIME);
    return ResultModel.success();
}
```

注意：关于 MyBatis-Plus 有一个知识点，即由于该项目中的数据库表 ID 是自增的，所以在 BaseEntity 类中的 ID 属性需要增加策略声明。BaseEntity 类修改后如代码清单 20-22 所示。

代码清单 20-22　BaseEntity 类修改后

```
@Data
public class BaseEntity implements Serializable {
    private static final long serialVersionUID=-1549634521453074321L;
    /**
     * 主键，唯一标识符
     * 主键策略，数据库ID自增
     * IdType.AUTO 数据库ID自增
     * IdType.INPUT 用户输入ID
     * IdType.ID_WORKER 全局唯一ID，内容为空时自动填充(默认配置)
     * IdType.UUID 全局唯一ID，内容为空时自动填充
     */
    @TableId(value="id",type=IdType.AUTO)
    protected Integer id;
    /** 创建时间 */
    protected Date createTime;
    /** 更新时间 */
    protected Date updateTime;
    /** 删除时间，0代表未删除，单位为s */
    protected Integer deleteTime;
}
```

至此，可以通过前面的两个接口进行账号的注册了。

20.1.4　使用 Redis 实现数字的原子性自增

大家都知道，Redis 是单线程，不会出现并发问题，所以前面那个在单位时间内限制用户接收邮件次数的功能，就可以使用 RedisAtomicInteger 类实现。另外，使用 RedisAtomicLong 也是一个生成分布式全局 ID 的解决方案。

在 Redis 客户端中操作数据的方法如代码清单 20-23 所示。

代码清单 20-23　在 Redis 客户端中操作数据的方法

```
/**
 * 以原子方式将当前值增加1
 * @param key
 * @param liveTime 只有第1次才会设置过期时间
 * @return 更新前的值，也就是+1前的值
 */
public int incrInt(String key, long liveTime) {
    RedisAtomicInteger entityIdCounter=new RedisAtomicInteger(key, redisTemplate.
getConnectionFactory());
    int increment=entityIdCounter.getAndIncrement();
    // 设置过期时间
    if (increment==0 && liveTime>0) {
```

```
            entityIdCounter.expire(liveTime, TimeUnit.SECONDS );
        }
        return increment;
    }
```

注意：这里有一个“小坑”。RedisAtomicInteger 的构造函数如代码清单 20-24 所示。

代码清单 20-24　RedisAtomicInteger 构造函数

```
public RedisAtomicInteger(String redisCounter, RedisConnectionFactory factory) {
    this(redisCounter, factory, null);
}
private RedisAtomicInteger(String redisCounter, RedisConnectionFactory factory, @
Nullable Integer initialValue) {
    RedisTemplate<String, Integer> redisTemplate=new RedisTemplate<>();
    redisTemplate.setKeySerializer(new StringRedisSerializer());
     redisTemplate.setValueSerializer(new GenericToStringSerializer<>(Integer.
class));
    redisTemplate.setExposeConnection(true);
    redisTemplate.setConnectionFactory(factory);
    redisTemplate.afterPropertiesSet();
    this.key=redisCounter;
    this.generalOps=redisTemplate;
    this.operations=generalOps.opsForValue();
    if (initialValue==null) {
        if (this.operations.get(redisCounter)==null) {
            set(0);
        }
    } else {
        set(initialValue);
    }
}
```

当第 1 次设置值时，执行的代码是 set(0)，也就是说并没有设置超时时间，如代码清单 20-25 所示。

代码清单 20-25　set(0) 的代码

```
public void set(int newValue) {
    operations.set(key, newValue);
}
```

由于该系统要求 10min 内最多发送 10 次，所以在初始化时需要设置超时时间为 10min，但不能在每次设置值时都设置一个超时时间，故有了如代码清单 20-23 所示的代码设置。if (increment==0 && liveTime>0) {entityIdCounter.expire(liveTime, TimeUnit.SECONDS);} 通过判断 increment 是否等于 0，以及是否已经设置了超时时间，来很好地防止单位时间内计数并发的情况发生。

20.2 系统登录功能

前面讲完了注册功能，接下来开始编写项目的登录功能。用户可以通过用户名、邮箱来登录，也可以使用自动登录功能。自动登录利用了 Cookie 技术。

20.2.1 用户名或邮箱登录

由于用户名只能是字母与数字，且邮箱中会有“@”字符，所以开发时可以通过“@”字符区分填写的是用户名还是邮箱。登录方法如代码清单 20-26 所示。

代码清单 20-26 登录方法

```
@RequestMapping(value="/login", method=RequestMethod.POST )
    public ResultModel login(UsersReq usersReq, HttpServletRequest request,
HttpServletResponse response) {
        // 校验参数
        if (StringUtils.isEmpty(usersReq.getPassword())) {
            return ResultModel.fail(ResultCodeEnum.PARAMETER_ERROR );
        }
        if (StringUtils.isEmpty(usersReq.getUsername())) {
            return ResultModel.fail(ResultCodeEnum.PARAMETER_ERROR );
        }
        UsersEntity usersEntity;
        // 邮箱登录
        if (RegexUtils.checkEmail(usersReq.getUsername())) {
            usersEntity=usersService.selectByEmail(usersReq.getUsername());
        } else {
            usersEntity=usersService.selectByUsername(usersReq.getUsername());
        }
        if (usersEntity==null) {
            return ResultModel.fail(ResultCodeEnum.WRONG_PASSWORD_USERNAME_EMAIL);
        }
            String password=PasswordUtils.getPassword(usersReq.getPassword(),
usersEntity.getSalt());
        if (!password.equals(usersEntity.getPassword())) {
            return ResultModel.fail(ResultCodeEnum.WRONG_PASSWORD_USERNAME_EMAIL);
        }

        // 选择“记住我”选项，自动登录
        if (usersReq.getRememberMe()) {
            UsersCookieDTO usersCookieDTO= UsersConvert.INSTANCE.entityToDTO(usersEntity);
            /*不推荐使用Spring的复制功能，其速度还较慢，推荐使用Java实体映射工具MapStruct
            BeanUtils.copyProperties(usersEntity, usersCookieDTO); */
            usersCookieDTO.setTime(DateUtils.getLongDateTimeMS());
             usersCookieDTO.setToken(PasswordUtils.getToken(usersEntity.getSalt(),
usersEntity.getPassword(), usersCookieDTO.getTime()));
```

```
            // 增加Cookie设置
                CookieUtils.setCookie(response, UsersConstants.COOKIE_USERS_LOGIN_
INFO, JSON.toJSONString(usersCookieDTO), UsersConstants.EXPIRATION_DATE_30);
        }
        // 登录成功
        setLoginInfo(request, usersEntity);
        return ResultModel.success("登录成功");
    }
```

这里有几个新增的知识点：

（1）使用 Cookie 进行用户信息记录，方便用户在 Session 过期后能够自动登录。

（2）为了防止密码泄露，客户端存储的 Cookie 信息不包含密码，使用盐值加密 + 加密密码 + 时间戳的方式进行 MD5 加密存储令牌。

（3）前面已经用过了 Redis，所以建议基于 Redis 的生命周期来派发一个无业务含义的身份令牌，通过令牌和服务器端的用户标签挂钩，效率更高，也更可控。**提示**：完全基于客户端来维护用户登录的生命周期容易失控。

另外，还有一个复制 Bean 的方式，这里选择使用 MapStruct。使用 MapStruct 复制 Bean 比在 Spring 中使用反射机制复制 Bean 快百倍以上。测试复制 Bean 速度的方法如代码清单 20-27 所示。

代码清单 20-27　测试复制 Bean 速度的方法

```
public static void main(String[] args) {
    int size= 5000000;
    UsersEntity usersEntity=new UsersEntity();
    usersEntity.setQqOpenid("");
    usersEntity.setHeadImage("");
    usersEntity.setUsername("");
    usersEntity.setSignature("");
    usersEntity.setPassword("");
    usersEntity.setSalt("");
    usersEntity.setMobilePhone("");
    usersEntity.setSex(0);
    usersEntity.setBirthday(LocalDate.now());
    usersEntity.setEmail("");
    usersEntity.setWeixin("");
    usersEntity.setQq("");
    usersEntity.setUb(0);
    usersEntity.setAlipayAccountNumber("");
    usersEntity.setAlipayRealName("");
    usersEntity.setRealNameState(0);
    usersEntity.setDescription("");
    usersEntity.setUpdateId(0);
    usersEntity.setType(0);
    usersEntity.setState(0);
    usersEntity.setCreateId(0);
    usersEntity.setMailboxState(0);
    usersEntity.setId(0);
```

```java
        usersEntity.setCreateTime(new Date());
        usersEntity.setUpdateTime(new Date());
        usersEntity.setDeleteTime(0);
        long s=System.currentTimeMillis();
        for(int i=0;i<size;i++){
            UsersCookieDTO usersCookieDTO=new UsersCookieDTO();
            BeanUtils.copyProperties(usersEntity,usersCookieDTO);
        }
        long e=System.currentTimeMillis();
        System.out.println("SpringBean复制"+size+"次，共消耗时间:"+(e-s)+"ms");
        s=System.currentTimeMillis();
        for(int i=0;i<size;i++){
            UsersCookieDTO usersCookieDTO=UsersConvert.INSTANCE.entityToDTO(usersEntity);
        }
        e=System.currentTimeMillis();
        System.out.println("mapstruct复制"+size+"次，共消耗时间:"+(e-s)+"ms");
    }
```


以 100 万次、500 万次、1000 万次进行测试，结果如图 20-2 所示。

```
UsersRestControllerTest ×
/Library/Java/JavaVirtualMachines/jdk1.8.0_201.
SpringBean复制1000000次，共消耗时间:699ms
mapstruct复制1000000次，共消耗时间:19ms
SpringBean复制5000000次，共消耗时间:806ms
mapstruct复制5000000次，共消耗时间:5ms
SpringBean复制10000000次，共消耗时间:1376ms
mapstruct复制10000000次，共消耗时间:5ms
```

图 20-2　复制 Bean 测试结果

虽然单次请求可能消耗的实际时间相差不大，但积累的总时间差距还是很大的。

20.2.2　使用 MapStruct 复制 Bean

MapStruct 是一个代码生成器，它基于约定优于配置，极大地简化了 JavaBean 类型之间映射的实现。

本小节只是简单地介绍 MapStruct 的使用。引入依赖，如代码清单 20-28 所示。

代码清单 20-28　引入依赖

```xml
<!--复制Bean-->
<org.mapstruct.version>1.3.0.Final</org.mapstruct.version>
<!-- https://mvnrepository.com/artifact/org.mapstruct/mapstruct-jdk8 -->
<dependency>
    <groupId>org.mapstruct</groupId>
    <artifactId>mapstruct-jdk8</artifactId>
    <version>${org.mapstruct.version}</version>
```

```
</dependency>


<dependency>
    <groupId>org.mapstruct</groupId>
    <artifactId>mapstruct-processor</artifactId>
    <version>${org.mapstruct.version}</version>
    <optional>true</optional>
    <scope>provided</scope>
</dependency>
```

MapStruct 中有一些注解，这里讲解以下两个比较常用的。

（1）Mapper：类注解，标记接口作为一个映射接口，并且是编译时 MapStruct 处理器的入口。其中还有一些可以定义的枚举属性，这里不一一展开介绍。

（2）Mappings：方法注解，可以处理源对象和目标对象中属性名称不同的情况。属性名称相同，则可以省略该注解。

在本系统中将 UsersEntity 中的属性复制给 UsersCookieDTO，需新增 UsersConvert 类，如代码清单 20-29 所示。

代码清单 20-29　UsersConvert 类

```
/** 类复制 */
@Mapper
public interface UsersConvert {
    UsersConvert INSTANCE=Mappers.getMapper(UsersConvert.class);
    /** Entity -> DTO */
    UsersCookieDTO entityToDTO(UsersEntity entity);
    /** DTO -> Entity */
    UsersEntity dtoToEntity(UsersCookieDTO entity);
    /** Entity集合 -> DTO集合 */
    List<UsersCookieDTO> entityToDTOList(List<UsersEntity> entity);
    /** DTO集合 -> Entity集合 */
    List<UsersEntity> dtoToEntityList(List<UsersCookieDTO> entity);
}
```

接下来通过该类复制 Bean，也就是 UsersCookieDTOusersCookieDTO=UsersConvert.INSTANCE.entityToDTO(usersEntity)。

20.2.3　退出登录功能

退出登录功能比较简单，只需删除 Session 与 Cookie 信息。退出登录功能如代码清单 20-30 所示。

代码清单 20-30　退出功能

```
/** 退出登录 */
@RequestMapping(value="/logout", method=RequestMethod.POST)
public ResultModel logout(HttpServletRequest request, HttpServletResponse
response) {
    // 获取用户
```

```
        UsersEntity users=SessionUtils.getAttribute(request, UsersConstants.SESSION_
USERS_LOGIN_INFO);
        if (users==null) {
            return ResultModel.resultModel(ResultCodeEnum.USER_NOT_LOGGED);
        }
        // 删除Session
        SessionUtils.removeAttribute(request, UsersConstants.SESSION_USERS_LOGIN_
INFO);
        SessionUtils.removesession(request);
        // 删除Cookie
        CookieUtils.delCookie(response, UsersConstants.COOKIE_USERS_LOGIN_INFO);
        return ResultModel.success("成功退出");
    }
```

20.2.4 使用 Redis 实现 Session 共享

如果选择使用 Redis 进行 Session 存储，这样可以实现多服务器之间的共享。虽然本系统是单机项目，为了扩展，这里还是使用 Redis 存储 Session。

关于 Session 共享的实现的方式有很多种，如以下 5 种方式：

（1）通过 Nginx 的 ip_hash，根据 ID 将请求分配到对应的服务器。

（2）通过关系型数据库存储。

（3）通过 Cookie 存储 。

（4）通过服务器内置的 Session 复制域 。

（5）通过 NoSQL 存储 。

比较常用的就是（1）和（5），在这里将方式（5）进行讲解。其实现原理并不复杂，只是在所有的请求之前配置一个过滤器，在请求之前操作 Session。spring-session 中真正起作用的 Session 过滤器是 SessionRepositoryFilter，其集成了 Redis 与 Mongodb。

首先添加依赖，如代码清单 20-31 所示。

代码清单 20-31　spring-session 依赖

```
<!-- https://mvnrepository.com/artifact/org.springframework.session/spring-
session-data-redis -->
<dependency>
    <groupId>org.springframework.session</groupId>
    <artifactId>spring-session-data-redis</artifactId>
    <version>2.0.5.RELEASE</version>
</dependency>
```

注意：在 IDEA 启用 Tomcat 的情况下，有可能在 Tomcat 的 lib 中并没有刷新该包的依赖，所以会出现类没有找到的情况。此时可以选择 Project Settings → Artifacts 命令，将 Jar 包添加到容器中，如图 20-3 所示。

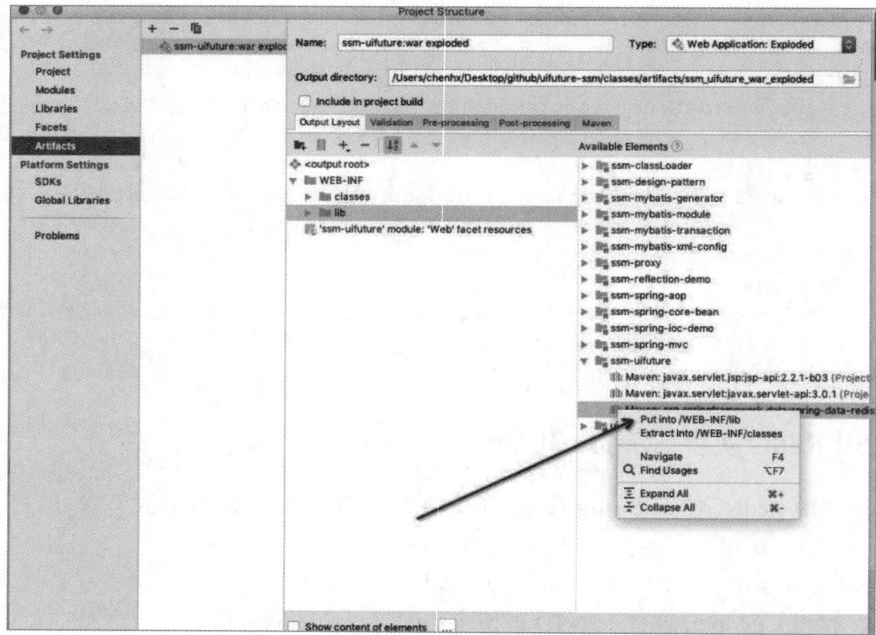

图 20-3　添加 .jar 包到容器

接下来在 web.xml 中配置过滤器 DelegatingFilterProxy 来拦截所有链接，如代码清单 20-32 所示。

代码清单 20-32　配置 DelegatingFilterProxy 过滤器

```
<!--配置Redis存储Session信息-->
<filter>
    <filter-name>springSessionRepositoryFilter</filter-name>
     <filter-class>org.springframework.web.filter.DelegatingFilterProxy</filter-
class>
</filter>
<filter-mapping>
    <filter-name>springSessionRepositoryFilter</filter-name>
    <url-pattern>/*</url-pattern>
</filter-mapping>
```

将 RedisHttpSessionConfiguration 注入 Spring 容器中，在 application-redis.xml 中添加配置，如代码清单 20-33 所示。

代码清单 20-33　将 RedisHttpSessionConfiguration 类注入 Spring 容器

```
<bean id="redisHttpSessionConfiguration"
   class="org.springframework.session.data.redis.config.annotation.web.http.
RedisHttpSessionConfiguration">
    <!--最大空闲时间，单位为s-->
    <property name="maxInactiveIntervalInSeconds" value="1800"/>
</bean>
```

然后通过登录接口验证 Session 是否已经通过 Redis 进行存储。登录后，可查看 Redis 中的存储情况，如图 20-4 所示。

图 20-4　查看 Redis 中的存储情况

此时 Redis 已经将登录用户的 Session 信息进行了存储，且过期时间与设置的 1800s 一致。

20.2.5　自动登录功能

既然通过 Cookie 记录了用户信息，那么当用户登录过期后，如何利用 Cookie 来实现自动登录呢？在这里，可以选择拦截所有的链接，若有没有登录的用户，再查看 Cookie 是否存在该用户信息，存在则自动登录。当然，也可以选择只拦截固定的页面链接，毕竟拦截过多对网站速度有一定影响。

接下来创建拦截器类拦截所有请求，如代码清单 20-34 所示。

代码清单 20-34　拦截器类

```
/** 自动登录拦截器 */
@Slf4j
@Configuration
public class AutoLoginInterceptor extends BaseController implements
HandlerInterceptor {
    @Autowired
    private UsersService usersService;
    @Override
     public boolean preHandle(HttpServletRequest request, HttpServletResponse
response, Object handler) throws Exception {
        // 判断是否登录
           UsersEntity users=SessionUtils.getAttribute(request, UsersConstants.
SESSION_USERS_LOGIN_INFO);
        if (users!=null) {
            // 已经登录
            return true;
        }
        // 从Cookie中获取
```

20
项目功能实现

```
            String usersStr=CookieUtils.getCookie(request, UsersConstants.COOKIE_
USERS_LOGIN_INFO);
        if (StringUtils.isEmpty(usersStr)) {
            return true;
        }
        try {
                UsersCookieDTO usersCookieDTO=JSON.parseObject(usersStr,
UsersCookieDTO.class);
            // 获取用户信息
              UsersEntity realUsers= usersService.selectByUsername(usersCookieDTO.
getUsername());
            if (realUsers==null) {
                return true;
            }
            // 判断时间是否在30天内
            if (DateUtils.getLongDateTimeMS()-usersCookieDTO.getTime()>DAY30_MS) {
                // 删Cookie
                CookieUtils.delCookie(response, UsersConstants.COOKIE_USERS_LOGIN_INFO);
                return true;
            }
               String token=PasswordUtils.getToken(realUsers.getSalt(), realUsers.
getPassword(), usersCookieDTO.getTime());
            if (token.equals(usersCookieDTO.getToken())) {
                // 进行登录
                  SessionUtils.setAttribute(request, UsersConstants.SESSION_USERS_
LOGIN_INFO, realUsers);
            }
        } catch (Exception e) {
            log.error("转换成UsersCookieDTO出现异常,usersStr=" + usersStr, e);
        }
        return true;
    }
}
```

最后还需要到 Spring MVC 的配置文件中配置拦截器的作用范围，即在 application-mvc.xml 中增加拦截器配置，如代码清单 20-35 所示。

代码清单 20-35　在 application-mvc.xml 中增加拦截器配置

```
<!--配置拦截器-->
<mvc:interceptors>
    <mvc:interceptor>
        <mvc:mapping path="/**"/>
        <mvc:exclude-mapping path="/usersRest/login"/>
        <mvc:exclude-mapping path="/usersRest/logout"/>
        <bean class="com.uifuture.ssm.interceptors.AutoLoginInterceptor"/>
    </mvc:interceptor>
</mvc:interceptors>
```

接下来进行验证，先登录，再到 Redis 中删除 Session，然后访问除登录退出外的其他接口，此时可以看到 Redis 中又有了 Session，说明自动登录生效了。

另外，比较重要的功能就是用户的删除和禁用，即不允许用户继续登录，如代码清单 20-36 所示。

代码清单 20-36　用户被删除及被禁用判断

```
//用户是否被禁用、被删除
if(UsersStateEnum.FORBIDDEN.getValue().equals(usersEntity.getState())){
    return ResultModel.fail(ResultCodeEnum.USER_VIOLATIONS_ARE_BANNED);
}
if(!DeleteEnum.NO_DELETE.getValue().equals(usersEntity.getDeleteTime())){
    return ResultModel.fail(ResultCodeEnum.THE_USER_HAS_BEEN_DELETED);
}
```

在用户登录方法及拦截器方法中均加入该判断即可。用户状态枚举类如代码清单 20-37 所示。

代码清单 20-37　用户状态枚举类

```
public enum UsersStateEnum {
    /** 用户状态枚举 */
    NORMAL("正常", 1),
    FORBIDDEN("禁用", 0);
    private String name;
    private Integer value;
    UsersStateEnum(String name, Integer value) {
        this.name=name;
        this.value=value;
    }
    public static UsersStateEnum getByValue(Integer value) {
        UsersStateEnum[] valueList=UsersStateEnum.values();
        for (UsersStateEnum v : valueList) {
            if (v.getValue().equals(value)) {
                return v;
            }
        }
        return null;
    }
    public String getName() {
        return name;
    }
    public Integer getValue() {
        return value;
    }
}
public enum DeleteEnum {
    /** 是否被删的枚举 */
    NO_DELETE("未被删除", 0);
    private String name;
    private Integer value;
...//省略一些方法
}
```

登录注册功能就介绍到这里，接下来介绍项目的其他功能。

20.3 资源发表功能

将数据库表生成实体类，运行 MyBatis-Plus 的生成代码功能。

由于资源的正文文本可能会很长，也可能会包含图片，因此，这里将资源的描述信息与资源的正文分为两个表。正文使用数据库的 mediumtext 类型进行存储，且文本使用 Markdown 格式进行编写，在前端再转换为 HTML 进行展示。

另外，还有资源的专题、资源的分类、资源的标签及相关的表数据实体的生成。

20.3.1 全局异常捕获与日志输出

由于系统可能会出现异常，因此在这里对全局异常的捕获类进行配置，以便排查问题，以及对用户进行友好提示。全局异常捕获如代码清单 20-38 所示。

代码清单 20-38 全局异常捕获

```
@Slf4j
@Configuration
public class ExceptionHandler extends BaseController implements HandlerExceptionResolver {
    @Override
      public ModelAndView resolveException(HttpServletRequest request,
HttpServletResponse response, Object handler, Exception ex) {
        ResultModel result;
        if (ex instanceof ServiceException) {
            // 业务失败的异常
            log.error("{}{}", "[业务异常]", ex.getMessage());
            ServiceException exception=(ServiceException) ex;
            result=ResultModel.resultModel(exception.getCode(), ex.getMessage());
        } else if (ex instanceof NoHandlerFoundException) {
             result = ResultModel.resultModel(ResultCodeEnum.NOT_FOUND.getValue(),
"接口 [" + request.getRequestURI() + "] 不存在");
        } else if (ex instanceof CommonException) {
            log.error("{}{}", "[公共异常]", ex.getMessage());
            CommonException exception=(CommonException) ex;
                 result=ResultModel.resultModel(exception.getCode(), exception.
getMessage());
        } else if (ex instanceof MaxUploadSizeExceededException) {
            log.error("{}{}", "[上传文件过大]", ex.getMessage());
            MaxUploadSizeExceededException exception=(MaxUploadSizeExceededException) ex;
            result=ResultModel.resultModel(500, exception.getMessage());
        } else {
              result=ResultModel.resultModel(ResultCodeEnum.INTERNAL_SERVER_ERROR.
getValue(), "接口 [" + request.getRequestURI() + "] 内部错误，请联系客服，邮箱:uifuture@
uifuture.com");
```

```
            String message;
            if (handler instanceof HandlerMethod) {
                HandlerMethod handlerMethod=(HandlerMethod) handler;
                message = String.format("接口 [%s] 出现异常，方法:%s.%s，异常信息:%s",
                        request.getRequestURI(),
                        handlerMethod.getBean().getClass().getName(),
                        handlerMethod.getMethod().getName(),
                        ex.getMessage());
            } else {
                message=ex.getMessage();
            }
            // 增加url异常输出
            String url=request.getRequestURI();
            Map map=request.getParameterMap();
                log.error("异常链接URL:" + url + ", 请求参数:" + map + ", 请求IP:" +
getIpAddress(request) + ", 异常信息:" + message, ex);
        }
        responseResult(response, result);
        return null;
    }
}
```


在Service层中使用Spring AOP功能对每一个Service层的类方法进行拦截，在拦截方法中进行入参的输出，对于执行慢的方法，输出一个警告日志，如代码清单20-39所示。

代码清单 20-39 Service 层日志输出

```
@Aspect
@Component
@Slf4j
public class ServiceLogAspect {
    /** 通过@Before声明一个建言，此建言直接使用拦截规则作为参数，拦截dubbo接口实现 */
    @Around("execution(* com.uifuture.ssm.service.impl.*.*(*))")
    public Object around(ProceedingJoinPoint joinPoint) {
            MethodSignature methodSignature=(MethodSignature) joinPoint.
getSignature();
        String methodName=methodSignature.getName();
        String className=methodSignature.getDeclaringTypeName();
        Object[] args=joinPoint.getArgs();
        Object result=null;
        try {
                log.info("方 法 名:{}#{}, 入 参:{}", className, methodName, JSON.
toJSONString(args));
            long s=System.currentTimeMillis();
            result=joinPoint.proceed();
            long e=System.currentTimeMillis();
            if (e-s>1000) {
                log.warn("方法名:{}#{}, 消耗时间:{}ms", className, methodName, (e - s));
            }
        } catch (Throwable throwable) {
```

```
            log.error("方法名:" + className + "#" + methodName + ", 入参:" + JSON.
toJSONString(args) + ", 出现异常", throwable);
            if (throwable instanceof BaseException) {
                throw (BaseException) throwable;
            }
            if (throwable instanceof RuntimeException) {
                throw (RuntimeException) throwable;
            }
        }
        return result;
    }
}
```

注意：<aop:aspectj-autoproxy proxy-target-class="true"/> 配置需要加在 application-aop.xml 文件中。

这样，全局的异常输出及 Service 层的日志输出就已经完成了。

20.3.2 登录拦截功能

由于发表资源及上传资源文件和图片需要登录才能进行的，如果在每个方法中都进行判断，无疑会显得臃肿。在这里，将使用拦截器对某些链接进行拦截。登录拦截器类如代码清单 20-40 所示。

代码清单 20-40　登录拦截器类

```
@Slf4j
@Configuration
public class LoginInterceptor extends BaseController implements HandlerInterceptor
{
    @Override
     public boolean preHandle(HttpServletRequest request, HttpServletResponse
response, Object handler) throws Exception {
        // 判断是否登录
        UsersEntity users=getLoginInfo(request);
        //已经登录
        if (users==null) {
            // 写入提醒
            ResultModel resultModel=new ResultModel(500, "登录后方可操作");
            responseResult(response, resultModel);
            return false;
        }
        return true;
    }
}
```

其中，responseResult 方法在父类 BaseController 中，如代码清单 20-41 所示。

代码清单 20-41　responseResult 方法

```
/** 设置编码、响应头和JSON字符串 */
```

```java
    protected void responseResult(HttpServletResponse response, ResultModel result) {
        response.setHeader("Content-type", "application/json;charset=UTF-8");
      response.setHeader("Access-Control-Allow-Origin", "*");
        response.setCharacterEncoding("UTF-8");
        response.setStatus(HttpStatus.OK.value());
        try {
            response.getWriter().print(JSON.toJSONString(result));
        } catch (IOException ex) {
            log.error(ex.getMessage());
        }
    }
```

写好了拦截器，接下来在 Spring MVC 的配置文件中配置要拦截的链接，如代码清单 20-42 所示。

代码清单 20-42　配置要拦截的链接

```xml
<!--配置拦截器-->
<mvc:interceptors>
    <!--登录拦截器-->
    <mvc:interceptor>
        <mvc:mapping path="/resource/submit"/>
        <mvc:mapping path="/resource/uploadImages"/>
        <mvc:mapping path="/resource/uploadResources"/>
        <bean class="com.uifuture.ssm.interceptors.LoginInterceptor"/>
    </mvc:interceptor>
</mvc:interceptors>
```

这样，当用户访问配置的拦截链接时，就需要先登录。如果没有登录，那么将会返回让用户登录后再操作的提醒。

20.3.3　上传资源文件到本地

如果上传文件，那么首先需要在 Spring MVC 配置文件中配置文件上传解析器，并设置文件编码以及单个文件的最大值，如代码清单 20-43 所示。

代码清单 20-43　设置文件上传配置

```xml
<!-- 文件上传配置，配置文件上传解析器MultipartResovler -->
<bean id="multipartResolver"
      class="org.springframework.web.multipart.commons.CommonsMultipartResolver">
    <!--设置默认编码-->
    <property name="defaultEncoding" value="utf-8"></property>
    <!--设置单个文件最大值为50MB-->
    <property name="maxUploadSize" value="52428800"></property>
</bean>
```

接下来便是文件上传方法的编写，这里仅提供 .zip 后缀文件的上传代码，如代码清单 20-44 所示。

代码清单 20-44　将资源文件上传到本地

```
/** 用户上传资源文件的路径 */
private static final String FILE_RESOURCES_UPLOAD_PATH = File.separator +
"resources" + File.separator;

/** 上传资源文件  */
@RequestMapping(value="/uploadResources", method=RequestMethod.POST)
public ResultModel uploadResources(HttpServletRequest request, HttpServletResponse
response,
                                    @RequestParam("uploadFile") MultipartFile
uploadFile) throws IOException {
    if (uploadFile==null) {
        return ResultModel.failNoData("请选择文件再上传");
    }
    UsersEntity users=getLoginInfo(request);
    if (users==null) {
        return ResultModel.fail(ResultCodeEnum.USER_NOT_LOGGED);
    }
    // 单个用户一天最多上传100次资源
    int times=redisClient.incrInt(RedisConstants.getUploadFileTimesKey(users.
getUsername()), RedisConstants.REG_MAX_TIME_1_DAY);
    if (times>UsersConstants.UPLOAD_TIMES ) {
        return ResultModel.fail(ResultCodeEnum.ALL_TOO_OFTEN );
    }
    List<FileInfoDTO> fileOssUrlDTOList=new ArrayList<>();
    Date date=new Date();
    String dateStr=DateUtils.getDateString(date, "yyyyMM") + "/" + DateUtils.
getDateString(date, "dd");
    // 原文件名称，需要带后缀
    String fileName=uploadFile.getOriginalFilename();
    if (StringUtils.isEmpty(fileName)) {
        return ResultModel.fail("文件名为空。原文件名为:" + fileName);
    }
    String fileType=fileName.substring(fileName.lastIndexOf("."));
    if (StringUtils.isEmpty(fileType)) {
        return ResultModel.fail("文件后缀名称错误。原文件名为:" + fileName + ",后缀名为:"
+ fileType);
    }
    // 限制压缩文件后缀名，只有.zip的后缀名文件能通过
    if (!UsersConstants.UPLOAD_SUFFIX.equals(fileType.toLowerCase())) {
        return ResultModel.fail("文件后缀名错误，只能上传.zip压缩文件");
    }
    // 保存文件到本地
    uploadFile(uploadFile, fileOssUrlDTOList, dateStr, fileName, fileType,
request.getSession().getServletContext().getRealPath("user") + FILE_RESOURCES_
UPLOAD_PATH);
    // 返回文件的存储信息
    return ResultModel.resultModel(200, "上传成功", fileOssUrlDTOList);
}
```

```
/** 保存文件到本地 */
private void uploadFile(MultipartFile uploadFile, List<FileInfoDTO>
fileOssUrlDTOList, String dateStr, String fileName, String fileType, String
fileResourcesUploadPath) throws IOException {

InputStream inputStream=uploadFile.getInputStream();
    // 文件上传到OSS，要通过oss 路径上传
    String newFileName=PasswordUtils.getToken()+fileType;
    String path=fileResourcesUploadPath+dateStr+File.separator;
    FileUtils.writeToLocal(path, newFileName, inputStream);
    FileInfoDTO fileInfoDTO=new FileInfoDTO();
    fileInfoDTO.setOldFileName(fileName);
    fileInfoDTO.setNewFileName(newFileName);
    fileInfoDTO.setPath(path);
    fileOssUrlDTOList.add(fileInfoDTO);
}
```

在这里涉及了 FileUtils 工具类的 writeToLocal 方法，该方法将文件写入本地，如代码清单 20-45 所示。

代码清单 20-45　将文件写入本地

```
/** 创建目录 */
public static boolean createDir(String destFilePath) {
    File file=new File(destFilePath);
    if (!file.exists()) {
        if (!file.mkdirs()) {
            return false;
        }
    }
    return true;
}
/**
 * 将InputStream 写入本地文件
 * @param path  写入本地目录
 * @param fileName 写入本地文件名
 * @param input 输入流
 * @throws IOException
 */
public static void writeToLocal(String path, String fileName, InputStream input)
        throws IOException {
    createDir(path);
    int index;
    byte[] bytes=new byte[1024];
    FileOutputStream downloadFile=new FileOutputStream(path + fileName);
    while ((index=input.read(bytes))!=-1) {
        downloadFile.write(bytes, 0, index);
        downloadFile.flush();
    }
    downloadFile.close();
```

```
        input.close();
    }
```

通过该方法便可以将文件上传到项目的运行路径下。

接下来测试上传文件方法是否可用。在 web/WEB-INF/templates/test 目录下新建 login.ftl 文件，用来模拟登录。由于此时已经有了登录拦截，需要先登录才能上传文件，所以登录不可少。当然，也可以临时将登录校验去掉，这里可以使用 login.ftl 文件，如代码清单 20-46 所示。

代码清单 20-46　login.ftl 文件内容

```
<!DOCTYPE html>
<html lang="en">
<head>
    <meta charset="UTF-8">
    <meta http-equiv="X-UA-Compatible" content="IE=edge">
    <meta name="viewport" content="width=device-width, initial-scale=1">
    <meta name="keywords" content="spring"/>
    <title>登录测试</title>
</head>
<body>
<#if SESSION_USERS_LOGIN_INFO??>
    ${SESSION_USERS_LOGIN_INFO.username!''}
</#if>
<form action="/usersRest/login" method="POST">
    用户名：<input type="text" name="username"/>
    密码：<input type="password" name="password"/>
    <input type="submit" value="Submit"/>
</form>
</body>
</html>
```

到该目录下创建 uploadResources.ftl 文件，如代码清单 20-47 所示。

代码清单 20-47　uploadResources.ftl 文件内容

```
<!DOCTYPE html>
<html lang="en">
<head>
    <meta charset="UTF-8"> <meta http-equiv="X-UA-Compatible" content="IE=edge">
    <meta name="viewport" content="width=device-width, initial-scale=1">
    <meta name="keywords" content="spring"/>
    <title>上传资源文件测试</title>
</head>
<body>
资源文件上传测试最大为50MB
<form action="/resource/uploadResources" method="POST" enctype="multipart/form-
data">
    File: <input type="file" name="uploadFile"/>
    <input type="submit" value="Submit"/>
</form>
```

```
</body>
</html>
```

然后还需配置页面的跳转。由于这两个页面没有业务处理，只是单纯地跳转，所以可在 application-mvc.xml 文件中进行如代码清单 20-48 所示的配置。

代码清单 20-48　配置无业务处理的动态页面跳转

```
<mvc:view-controller path="/test/login" view-name="test/login"/>
<mvc:view-controller path="/test/uploadResources" view-name="test/
uploadResources"/>
```

其中，path 是 URL 的访问路径；view-name 是逻辑视图的名称，也就是文件的名称。通过上面的配置，可以运行项目进行校验。验证步骤为：运行项目后访问 test/login 路径，输入注册的用户名和密码登录；登录成功后访问 test/uploadResources 路径，选择文件进行上传；上传成功后可以在项目的 classes 路径下看到 user/resources 路径，然后根据日期继续划分路径，最后可以找到上传的文件。

20.3.4　上传图片文件到阿里云 OSS

20.3.3 小节增加了上传资源文件到本地的功能。本小节使用阿里云的 OSS 存储图片文件，需要提前准备阿里云账号和阿里云 OSS 空间。

在阿里云管理后台创建 Bucket，如图 20-5 所示。

图 20-5　创建 Bucket

关于具体的创建过程不再详解，注意给 OSS 配置一个访问域名，或者直接使用阿里云分配的默认域名。接下来进行接入阿里云的 OSS 操作，首先添加依赖，如代码清单 20-49 所示。

代码清单 20-49　阿里云 OSS 依赖

```
<!--阿里云OSS依赖-->
<dependency>
    <groupId>com.aliyun.oss</groupId>
    <artifactId>aliyun-sdk-oss</artifactId>
    <version>2.8.3</version>
</dependency>
```

此外，还需要阿里云的一些 OSS 配置，如代码清单 20-50 所示。

代码清单 20-50　OSS 配置

```
# 环境配置
application.env=dev
# cdn图片域名
cdn.images.href=OSS绑定的域名
# 阿里云OSS 图片资源配置
aliyun.oss.endpoint=oss-cn-hangzhou.aliyuncs.com
# 阿里云账号的key
aliyun.oss.access.key.id=*****
# 阿里云账号的secret
aliyun.oss.access.key.secret=*****
# bucket名称
aliyun.oss.bucket.img.name=ssm-uifuture
```

接下来创建 AliyunOssConfig 类来读取配置，如代码清单 20-51 所示。

代码清单 20-51　读取配置

```
@Configuration
@Data
public class AliyunOssConfig {
    /** 阿里云API的内或外网域名 */
    @Value("${aliyun.oss.endpoint}")
    private String endpoint;
    /** 阿里云API的密钥Access Key ID */
    @Value("${aliyun.oss.access.key.id}")
    private String accessKeyId;
    /** 阿里云API的密钥Access Key Secret */
    @Value("${aliyun.oss.access.key.secret}")
    private String accessKeySecret;
    /** 阿里云API的bucket名称 */
    @Value("${aliyun.oss.bucket.img.name}")
    private String bucketImgName;
}
@Configuration
@Data
public class SysConfig {
```

```
    /** 系统当前运行环境 */
    @Value("${application.env}")
    private String applicationEnv;
    /** 图片cdn */
    @Value("${cdn.images.href}")
    private String cdnImagesHref;
}
```

为了更加方便地操作 OSS 中的文件，可以通过 OSS 通用操作类来进行，如阿里云 OSS 操作类，如代码清单 20-52 所示。

代码清单 20-52　阿里云 OSS 操作类

```
@Configuration
@Slf4j
public class AliyunOssHandle {
    private static OSSClient client;
    @Autowired
    private AliyunOssConfig aliyunOssConfig;
    /**
     * 修改文件名
     * @param fileName 文件名
     * @return 文件的新名称
     */
    public static String getfileName(String fileName, String userName) {
        String fileType=fileName.substring(fileName.lastIndexOf("."));
        return userName + "-" + PasswordUtils.randomStringLower(10) + fileType;
    }
    public OSSClient getClient() {
        if (client==null) {
            client=new OSSClient(aliyunOssConfig.getEndpoint(), aliyunOssConfig.
getAccessKeyId(), aliyunOssConfig.getAccessKeySecret());
        }
        return client;
    }
    public Boolean folderExist(OSS client, String folderName) {
        return client.doesObjectExist(aliyunOssConfig.getBucketImgName(),
folderName);
    }
    /** 判断文件夹是否存在 */
    public Boolean folderExist(String folderName) {
        return folderExist(getClient(), folderName);
    }
    public String createFolder(OSS client, String folderName) {
        try {
            // 判断文件夹是否存在，不存在则创建
            if (!client.doesObjectExist(aliyunOssConfig.getBucketImgName(), folderName)) {
                // 创建文件夹
                client.putObject(aliyunOssConfig.getBucketImgName(), folderName,
new ByteArrayInputStream(new byte[0]));
```

```
                log.info("创建文件夹成功:{}", folderName);
                // 得到文件夹名
                OSSObject object=client.getObject(aliyunOssConfig.getBucketImgName(),
folderName);
                return object.getKey();
            }
            return folderName;
        } catch (Exception ce) {
            log.error("创建空文件夹到OSS出现异常", ce);
        }
        return "";
    }
    /**
     * 创建文件夹到OSS
     * @param folderName 模拟文件夹名，如"qj_nanjing/"
     * @return 文件夹名
     */
    public String createFolder(String folderName) {
        return createFolder(getClient(), folderName);
    }
    public void deleteFile(OSS ossClient, String bucketName, String folder, String
key) {
        try {
            ossClient.deleteObject(bucketName, folder + key);
        } catch (Exception e) {
            log.error("根据key删除OSS服务器上的文件出现异常,folder=" + folder + ",key="
+ key, e);
        }
        log.info("删除" + bucketName + "下的文件" + folder + key + "成功");
    }
    /**
     * 根据key删除OSS服务器上的文件
     * @param folder 模拟文件夹名，如"qj_nanjing/"
     * @param key Bucket下的文件路径名+文件名，如"upload/cake.jpg"
     */
    public void deleteFile(String folder, String key) {
        OSS client = getClient();
        deleteFile(client, aliyunOssConfig.getBucketImgName(), folder, key);
    }
    /**
     * 上传图片至OSS
     * @param file 上传文件(文件全路径，如:D:\\image\\cake.jpg)
     * @param folder 模拟文件夹名，如"qj_nanjing/" ；前面不能有分隔符，最后需要有分隔符
     * @return String 返回的唯一MD5数字签名
     */
    public String uploadObject2OSS(File file, String folder) {
        return uploadObject2OSS(getClient(), file, aliyunOssConfig.
getBucketImgName(), folder);
```

```
    }
    /**
     * 通过文件名判断并获取OSS服务文件上传时文件的contentType
     * @param fileName 文件名
     * @return 文件的contentType
     */
    public String getContentType(String fileName) {
        // 文件的后缀名
        String fileExtension=fileName.substring(fileName.lastIndexOf("."));
        if (".bmp".equalsIgnoreCase(fileExtension)) {
            return "image/bmp";
        }
        if (".gif".equalsIgnoreCase(fileExtension)) {
            return "image/gif";
        }
          if  (".jpeg".equalsIgnoreCase(fileExtension)  ||  ".jpg".
equalsIgnoreCase(fileExtension) || ".png".equalsIgnoreCase(fileExtension)) {
            return "image/jpeg";
        }
        if (".html".equalsIgnoreCase(fileExtension)) {
            return "text/html";
        }
        if (".txt".equalsIgnoreCase(fileExtension)) {
            return "text/plain";
        }
        if (".vsd".equalsIgnoreCase(fileExtension)) {
            return "application/vnd.visio";
        }
        if (".ppt".equalsIgnoreCase(fileExtension) || "pptx".equalsIgnoreCase(fileE
xtension)) {
            return "application/vnd.ms-powerpoint";
        }
        if (".doc".equalsIgnoreCase(fileExtension) || "docx".equalsIgnoreCase(fileE
xtension)) {
            return "application/msword";
        }
        if (".xml".equalsIgnoreCase(fileExtension)) {
            return "text/xml";
        }
        // 默认返回类型
        return "image/jpeg";
    }
     public String uploadObject2OSS(OSS ossClient, File file, String bucketName,
String folder) {
        String resultStr=null;
        try {
            // 以输入流的形式上传文件
            InputStream inputStream=new FileInputStream(file);
            // 文件名
```

```
            String fileName=file.getName();
            // 文件大小
            long fileSize=file.length();
            // 创建上传Object的Metadata
            ObjectMetadata metadata=new ObjectMetadata();
            // 上传文件的长度
            metadata.setContentLength(fileSize);
            // 指定该Object被下载时网页的缓存行为 一年
            metadata.setCacheControl("max-age=31104000");
            // 指定该Object下设置Header
            metadata.setHeader("Pragma", "no-cache");
            // 指定该Object被下载时的内容编码格式
            metadata.setContentEncoding("utf-8");
            /*文件的MIME，定义文件的类型及网页编码，决定浏览器将以什么形式、什么编码读取文件。如
果用户没有指定则根据Key或文件名的扩展名生成，如果没有扩展名则填默认值application/octet-stream */
            metadata.setContentType(getContentType(fileName));
            //指定该Object被下载时的名称
            metadata.setContentDisposition("filename/filesize=" + fileName + "/" +
fileSize + "Byte.");
            // 上传文件(上传文件流的形式)
            PutObjectResult putResult=ossClient.putObject(bucketName, folder +
fileName, inputStream, metadata);
            // 解析结果
            resultStr=putResult.getETag();
            log.info("上传文件至阿里云成功，文件名称:{},路径:{}", fileName, folder);
        } catch (Exception e) {
                log.error("上传阿里云OSS服务器异常。" + e.getMessage() + ",folder=" +
folder + ",fileName=" + file.getName(), e);
        }
        return resultStr;
    }
    /**
     * 上传图片至OSS
     * @param in 文件流
     * @param name 获取文件后缀时需要
     * @param fileSize 文件大小
     * @param folder OSS的路径
     * @return 图片的全路径
     */
     public String uploadObject2OSS(InputStream in, String name, Long fileSize,
String folder, String fileName, String userName) {
        OSSClient ossClient=getClient();
        String resultStr=null;
        try {
            // 判断folder
            createFolder(folder);
            // 文件名，如果出现重复，则重新生成名称，再上传
                if (ossClient.doesObjectExist(aliyunOssConfig.getBucketImgName(),
folder + fileName)) {
```

```
                fileName=getfileName(name, userName);
            }
            // 创建上传Object的Metadata
            ObjectMetadata metadata = new ObjectMetadata();
            // 上传文件的长度
            metadata.setContentLength(fileSize);
            //指定该Object被下载时网页的缓存行为一年
            metadata.setCacheControl("max-age=31104000");
            // 指定该Object下设置Header
              metadata.setHeader("Pragma", "no-cache");
            // 指定该Object被下载时的内容编码格式
            metadata.setContentEncoding("utf-8");
            /*文件的MIME，定义文件的类型及网页编码，决定浏览器将以什么形式、什么编码读取文件。如
果用户没有指定则根据Key或文件名的扩展名生成，如果没有扩展名则填默认值 */
            metadata.setContentType(getContentType(fileName));
            /*指定该Object被下载时的名称 */
             metadata.setContentDisposition("filename/filesize=" + fileName + "/" +
fileSize + "Byte.");
            // 上传文件(上传文件流的形式)
            PutObjectResult putResult= ossClient.putObject(aliyunOssConfig.
getBucketImgName(), folder + fileName, in, metadata);
            resultStr=putResult.getETag();
            log.info("上传阿里云OSS服务器成功。" + resultStr);
            // 解析结果
        } catch (Exception e) {
            e.printStackTrace();
            log.error("上传阿里云OSS服务器异常。" + e.getMessage(), e);
        }
        return folder + fileName;
    }
}
```

接下来实现上传图片至 OSS 的方法，如代码清单 20-53 所示。

代码清单 20-53　上传图片的请求方法

```
/** 上传图片至阿里云OSS */
    @RequestMapping(value="/uploadImages", method=RequestMethod.POST)
    public ResultModel uploadImages(HttpServletRequest request, HttpServletResponse
response, @RequestParam("uploadFile") MultipartFile[] uploadFile) throws
IoException {
        if (uploadFile.length==0) {
            return ResultModel.failNoData("请选择文件再上传");
        }
        UsersEntity users=getLoginInfo(request);
        if (users==null) {
            return ResultModel.fail(ResultCodeEnum.USER_NOT_LOGGED);
      }
        // 单个用户一天最多上传100张图片
```

```
        int times= redisClient.incrInt(RedisConstants.getUploadFileTimesKey(users.
getUsername()), RedisConstants.REG_MAX_TIME_1_DAY);
        if (times>UsersConstants.UPLOAD_TIMES ) {
            return ResultModel.fail(ResultCodeEnum.ALL_TOO_OFTEN);
        }
        List<FileOssUrlDTO> fileOssUrlDTOList=new ArrayList<>();
        Date date = new Date();
         String dateStr=DateUtils.getDateString(date, "yyyyMM") + "/" + DateUtils.
getDateString(date, "dd");
        for (MultipartFile multipartFile : uploadFile) {
            // 原文件名称，需要带后缀
            String fileName=multipartFile.getOriginalFilename();
            if (StringUtils.isEmpty(fileName)) {
                return ResultModel.fail("文件名为空。原文件名为:" + fileName);
            }
            String fileType=fileName.substring(fileName.lastIndexOf("."));
            if (StringUtils.isEmpty(fileType)) {
                 return ResultModel.fail("文件后缀名称错误。原文件名为:" + fileName + ",
后缀名为:" + fileType);
            }
            InputStream inputStream=multipartFile.getInputStream();
            // 文件上传到OSS，通过oss 路径上传
            String ossFileName=PasswordUtils.getToken() + fileType;
                  String ossUrl=aliyunOssHandle.uploadObject2OSS(inputStream,
fileName, multipartFile.getSize(),"user/" + dateStr + "/", ossFileName, users.
getUsername());
            FileOssUrlDTO fileOssUrlDTO=new FileOssUrlDTO();
            fileOssUrlDTO.setFileName(fileName);
            fileOssUrlDTO.setOssUrl(sysConfig.getCdnImagesHref() + "/" + ossUrl);
            fileOssUrlDTOList.add(fileOssUrlDTO);

        }
        // 返回文件的存储信息
        return ResultModel.resultModel(200, "上传成功", fileOssUrlDTOList);
    }
```

接下来验证功能是否可用。在 web/WEB-INF/templates/test 目录下新建 uploadFile.ftl 文件，注意不要忘记在 Spring MVC 配置文件中配置动态页面的跳转。uploadFile.ftl 文件内容如代码清单 20-54 所示。

代码清单 20-54　uploadFile.ftl 文件内容

```
<!DOCTYPE html>
<html lang="en">
<head>
    <meta charset="UTF-8">
    <meta http-equiv="X-UA-Compatible" content="IE=edge">
    <meta name="viewport" content="width=device-width, initial-scale=1">
    <meta name="keywords" content="spring"/>
    <title>上传图片文件测试</title>
```

```
</head>
<body>
图片文件上传测试最大为50MB
<form action="/resource/uploadImages" method="POST" enctype="multipart/form-data">
    File: <input type="file" name="uploadFile"/>
    <input type="submit" value="Submit"/>
</form>
</body>
</html>
```

运行项目，先访问 /test/login 页面进行登录，再访问 /test/uploadFile 页面，在该页面中上传图片文件。返回结果如代码清单 20-55 所示。

代码清单 20-55　返回结果

```
{"code":200,"message":"上传成功","success":false,"data":[{"fileName":"cxbczl.
jpg","ossUrl":"//ssm.uifuture.com/user/202208/19/20220819221502979wmcnvxs2eymitqr.
jpg"}]
```


查看阿里云 OSS 后端是否已经有了该文件，并通过 http://ssm.uifuture.com/user/202208/19/20220819221502979wmcnvxs2eymitqr.jpg 的链接确认能否正常访问图片，如图 20-6 所示。

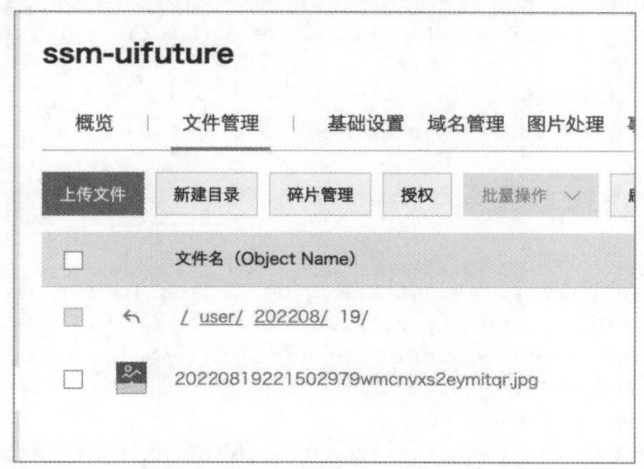

图 20-6　图片上传至 OSS

20.3.5　资源发表

通过数据库表，可以知道资源发表需要的具体信息，如用户登录、标签、分类、专题、资源标题、资源描述、资源内容和资源文件等。资源发表的接口如代码清单 20-56 所示。

代码清单 20-56　资源发表的接口

```
@RequestMapping(value="/submit", method=RequestMethod.POST )
public ResultModel submit(ResourceReq resourceReq, HttpServletRequest request,
HttpServletResponse response) {
    // 校验
ValidateUtils.validate(resourceReq);
```

```
    // 获取当前用户
    UsersEntity usersEntity=getLoginInfo(request);
    if (usersEntity==null) {
        return ResultModel.fail(ResultCodeEnum.USER_NOT_LOGGED);
    }
    // 判断资源文件是否存在
    if (!FileUtils.exists(resourceReq.getPath()+resourceReq.getNewName())) {
        return ResultModel.fail(ResultCodeEnum.PLEASE_UPLOAD_THE_RESOURCE_FILE_FIRST);
    }
    // 资源信息
    ResourceEntity resourceEntity=ResourceConvert.INSTANCE.toEntity(resourceReq);
    // 资源内容
    ResourceContentEntity resourceContentEntity=new ResourceContentEntity();
    resourceContentEntity.setContent(resourceReq.getContent());
    // 查询类型
    List<Integer> typeIds=resourceReq.getTypeIds();
    if (!CollectionUtils.isEmpty(typeIds)) {
            Collection<ResourceTypeEntity> resourceTypeEntityList =
resourceTypeService.listByIds(typeIds);
        typeIds=new ArrayList<>();
        for (ResourceTypeEntity resourceTypeEntity : resourceTypeEntityList) {
                if(DeleteEnum.NO_DELETE.getValue().equals(resourceTypeEntity.
getDeleteTime())) {
                // 未被删除
                typeIds.add(resourceTypeEntity.getId());
            }
        }
    }
    // 查询专题
    List<Integer> subjectIds=resourceReq.getSubjectIds();
    if (!CollectionUtils.isEmpty(subjectIds)) {
            Collection<ResourceSubjectEntity> resourceSubjectEntities =
resourceSubjectService.listByIds(subjectIds);
        subjectIds=new ArrayList<>();
        for (ResourceSubjectEntity resourceSubjectEntity : resourceSubjectEntities) {
                if(DeleteEnum.NO_DELETE.getValue().equals(resourceSubjectEntity.
getDeleteTime())) {
               //未被删除
               subjectIds.add(resourceSubjectEntity.getId());
            }
        }
    }
// 资源发表
     resourceService.saveResource(resourceEntity, resourceContentEntity,
usersEntity, typeIds, subjectIds, com.uifuture.ssm.util.CollectionUtils.
listToSet(resourceReq.getTagsNames()));
    return ResultModel.success("成功");
}
```

这里比较核心的是 resourceService.saveResource 方法，如代码清单 20-57 所示。

代码清单 20-57　resourceService.saveResource 方法实现

```
@Override
public Integer saveResource(ResourceEntity resourceEntity, ResourceContentEntity
resourceContentEntity, UsersEntity usersEntity, List<Integer> typeIds, List<Integer>
subjectIds, Set<String> tagsNames) {
    TransactionStatus transaction=null;
    try {
        // 开启事务
        transaction=transactionManager.getTransaction(new DefaultTransactionDefinit
ion());

        // 增加资源信息
        resourceEntity.setCreateId(usersEntity.getId());
        String token=PasswordUtils.getToken();
        resourceEntity.setToken(token);
        resourceEntity.setState(ResourceStateEnum.IN_THE_REVIEW.getValue());
        resourceEntity.setNewName(resourceEntity.getToken());
        this.save(resourceEntity);

        // 增加资源详情
        resourceContentEntity.setResourceToken(token);
        resourceContentMapper.insert(resourceContentEntity);

        // 增加类型关联
        List<RResourceTypeEntity> rResourceTypeEntities=new ArrayList<>();
        for (Integer typeId : typeIds) {
            RResourceTypeEntity rResourceTypeEntity=new RResourceTypeEntity();
            rResourceTypeEntity.setResourceId(resourceEntity.getId());
            rResourceTypeEntity.setResourceTypeId(typeId);
            rResourceTypeEntities.add(rResourceTypeEntity);
        }
        rResourceTypeService.saveBatch(rResourceTypeEntities);

        // 增加专题关联
        List<RResourceSubjectEntity> rResourceSubjectEntities=new ArrayList<>();
        for (Integer subjectId : subjectIds) {
        RResourceSubjectEntity rResourceSubjectEntity=new RResourceSubjectEntity();
            rResourceSubjectEntity.setResourceId(resourceEntity.getId());
            rResourceSubjectEntity.setSubjectId(subjectId);
            rResourceSubjectEntities.add(rResourceSubjectEntity);
        }
        rResourceSubjectService.saveBatch(rResourceSubjectEntities);

        log.info("增加资源, resourceEntity={},", resourceEntity);
        // 提交事务
        transactionManager.commit(transaction);

    } catch (Exception e) {
        if (transaction != null) {
```

```
                // 回滚事务
                transactionManager.rollback(transaction);
            }
            log.error("发表资源失败", e);
            throw new ServiceException(ResultCodeEnum.BUSINESS_PROCESS_FAILED);
        }
        // 增加标签，标签插入成功对数据实际的影响并不大，可放在事务之外进行
        List<TagsEntity> oldTags=tagsService.listByNameList(tagsNames);
        // 将name抽取出来作为集合
          List<String>oldTagsNames=oldTags.stream().map(TagsEntity::getName).
collect(Collectors.toList());
        Set<String> oldTagsNameSet=CollectionUtils.listToSet(oldTagsNames);

        List<RResourcesTagsEntity> rResourcesTagsEntities = new ArrayList<>();
        for (String tagsName : tagsNames) {
            RResourcesTagsEntity rResourcesTagsEntity = new RResourcesTagsEntity();
            rResourcesTagsEntity.setResourceId(resourceEntity.getId());
            if (oldTagsNameSet.contains(tagsName)) {
                // 原来存在的数据
                rResourcesTagsEntity.setTagsId(0);
            } else {
                // 需要新增的标签
                TagsEntity tagsEntity=new TagsEntity();
                tagsEntity.setName(tagsName);
                  /*这里有可能出现并发的情况，由于数据库存在唯一键约束，可能会抛出异常。如果出现异常，
则重新查询数据库看是否存在该名称的标签，进行处理*/
                try {
                    tagsService.save(tagsEntity);
                } catch (Exception e) {
                    TagsEntity tagsEntity1=tagsService.getByName(tagsName);
                    if (tagsEntity1!=null) {
                        rResourcesTagsEntity.setTagsId(tagsEntity1.getId());
                    } else {
                        // 否则抛弃该标签名，不插入数据库
                        continue;
                    }
                }
            }
            rResourcesTagsEntities.add(rResourcesTagsEntity);
        }
        rResourcesTagsService.saveBatch(rResourcesTagsEntities);
        return 1;
    }
```

这里使用了事务，以确保操作的原子性。另外，由于标签的保存无关资源的发表，所以将其放在事务外进行操作。其实，标签的保存操作是可以通过 Redis 来锁住标签 name，以防止并发操作的，但是在这里应用了补偿机制，外加数据库有唯一键约束，出现二次问题的概率非常小，所以就没有再引入 Redis 分布式锁来进行锁标签操作。当然，锁的粒度很小，影响并不大。在后面会讲到 Redis 分布式锁，会有对应的工具类操作。

前面 3 个小节，仅仅作为项目的初始操作，相当于是搭建了项目的一个“小骨架”。

20.4 用户相关功能

前面基于用户表进行了用户的注册和登录设置，但那些并不完善，如 IP 没有被记录，当用户完全忘记密码时，邮箱也丢失了，则无法找回账号。因此，可以记录用户注册时和登录时的 IP 地址，将来用户申诉时便可以用上。

20.4.1 增加用户 IP 记录

目前只需要在用户注册、登录时进行 IP 的记录。解析 IP 地址有两种方式：一是在记录 IP 时顺便将 IP 的信息进行记录；另一种就是先记录 IP，只有在使用时才去查询 IP 的信息，再记录到数据库。

首先将 ip 和 ip_details 表使用 MyBatis-Plus 自动生成三层代码及实体类。接下来修改 com.uifuture.ssm.controller.UsersRestController 的 registered 方法，在返回注册成功结果时进行以下操作，如代码清单 20-58 所示。

代码清单 20-58 在注册方法中增加用户 IP 记录

```
// 增加IP记录
IpEntity ipEntity=new IpEntity();
ipEntity.setUserId(usersEntity.getId());
ipEntity.setIp(getIpAddress(request));
ipService.save(ipEntity);
```

在注册方法中已经增加了 IP 记录，登录时也需要记录，如代码清单 20-59 所示。

代码清单 20-59 在登录方法中增加 IP 记录

```
// 增加IP记录
IpEntity ipEntity=new IpEntity();
ipEntity.setUserId(usersEntity.getId());
ipEntity.setIp(getIpAddress(request));
ipService.save(ipEntity)
```

此时可以看出，记录 IP 是一个比较通用的方法，可以在 Service 层操作，由于对结果没有影响也可异步操作。在 IpServiceImpl 中异步增加 IP 记录方法，如代码清单 20-60 所示。

代码清单 20-60 异步增加 IP 记录

```
/**
 * 异步增加IP记录
 * @param ipAddress IP地址
 * @param id 用户ID
 */
```

```
@Async
@Override
public void saveIp(String ipAddress, Integer id) {
    // 增加IP记录
    IpEntity ipEntity=new IpEntity();
    ipEntity.setUserId(id);
    ipEntity.setIp(ipAddress);
    this.save(ipEntity);
}
```

通过记录 IP，便可记录用户的登录轨迹。当用户在不是常登录地址进行登录时，还可进行友好提醒，防止用户账户被盗用。

20.4.2 用户关注功能

用户关注功能，在这里是指用户关注用户。当然，如果后面需要扩展功能，也可以关注资源专题、资源分类、资源等。不过，关注资源在本书中体现为用户对资源的收藏，后面会讲到。

既然是关注，那么就是多对多的关系。新建一个表 users_focus，记录关注用户与被关注用户之间的关系。

关注用户的请求方法如代码清单 20-61 所示。

代码清单 20-61　关注用户的请求方法

```
/**
 * 关注用户
 */
@RequestMapping(value="/focusOnTheUser", method=RequestMethod.POST)
public ResultModel focusOnTheUser(Integer userId,HttpServletRequest request,
HttpServletResponse response) {
    UsersEntity usersEntity=checkUserFocus(userId, request);
    // 数据库已经有唯一键约束，不进行查询，直接插入数据
    UsersFocusEntity usersFocusEntity=new UsersFocusEntity();
    usersFocusEntity.setUserId(usersEntity.getId());
    usersFocusEntity.setFocusedUserId(userId);
    usersFocusService.save(usersFocusEntity);
    return ResultModel.success();
}
```

checkUserFocus 方法的代码如代码清单 20-62 所示。

代码清单 20-62　checkUserFocus 方法

```
/**
 * 校验用户
 * @param userId 用户id
 * @param request 请求
 * @return 当前登录的用户
 */
```

```
private UsersEntity checkUserFocus(Integer userId, HttpServletRequest request) {
    if (userId==null || userId<0) {
        throw new CheckoutException(ResultCodeEnum.PARAMETER_ERROR);
    }
    // 获取当前用户
    UsersEntity usersEntity=getLoginInfo(request);
    if (usersEntity==null) {
        throw new CheckoutException(ResultCodeEnum.USER_NOT_LOGGED);
    }
    // 确保userId存在
    UsersEntity users=usersService.getById(userId);
    if (users==null) {
        throw new CheckoutException(ResultCodeEnum.PARAMETER_ERROR);
    }
    return usersEntity;
}
```

注意：该方法需要用户登录才能访问。在 application-mvc.xml 文件中配置登录的拦截链接，将 <mvc:mapping path="/usersFocus/focusOnTheUser"/> 添加到登录拦截器中。

取消关注用户的方法如代码清单 20-63 所示。

代码清单 20-63　取消关注用户的方法

```
/**
 * 取消关注用户
 */
@RequestMapping(value="/cancelFocusOnTheUser", method=RequestMethod.POST)
public ResultModel cancelFocusOnTheUser(Integer userId,HttpServletRequest request,
HttpServletResponse response) {
    UsersEntity usersEntity=checkUserFocus(userId, request);
    usersFocusService.removeByUserIdAndFocusId(usersEntity.getId(),userId);
    return ResultModel.success();
}
```

同样，将该请求链接 cancelFocusOnTheUser 添加到登录拦截器中。

移除关注用户的方法如代码清单 20-64 所示。

代码清单 20-64　移除关注用户的方法

```
public boolean removeByUserIdAndFocusId(Integer userId, Integer focusId) {
    QueryWrapper<UsersFocusEntity> queryWrapper=new QueryWrapper<>();
    queryWrapper.eq(UsersFocusEntity.USER_ID, userId);
    queryWrapper.eq(UsersFocusEntity.FOCUSED_USER_ID, focusId);
    return this.remove(queryWrapper);
}
```

在前端向用户展示是否关注该用户时，还需要一个判断用户是否已经关注该用户的接口，如代码清单 20-65 所示。

代码清单 20-65　判断用户是否已经关注该用户

```
/** 用户是否已经关注该用户  */
@RequestMapping(value="/isFocus", method=RequestMethod.POST)
public ResultModel isFocus(Integer userId,HttpServletRequest request,
HttpServletResponse response) {
    UsersEntity usersEntity=checkUserFocus(userId, request);
    UsersFocusEntity usersFocusEntity= usersFocusService.getByUserIdAndFocusedId(us
ersEntity.getId(),userId);
    if(usersFocusEntity==null){
        return ResultModel.success(false);
    }
    return ResultModel.success(true);
}
```

该方法的访问链接也需要加入登录拦截器中，可使用 getByUserIdAndFocusedId 方法，如代码清单 20-66 所示。

代码清单 20-66　getByUserIdAndFocusedId 方法

```
public UsersFocusEntity getByUserIdAndFocusedId(Integer userId, Integer focusId) {
    QueryWrapper<UsersFocusEntity> queryWrapper=new QueryWrapper<>();
    queryWrapper.eq(UsersFocusEntity.USER_ID, userId);
    queryWrapper.eq(UsersFocusEntity.FOCUSED_USER_ID, focusId);
    return this.getOne(queryWrapper);
}
```

通过设置以上方法，基本的用户关注功能就实现了。但是还缺少一个细节，就是当用户查看自己的关注时需要有一个关注的分页列表来展示用户的头像、昵称等信息。下面便完成该方法，如代码清单 20-67 所示。

代码清单 20-67　获取关注用户的列表

```
/** 获取关注用户的列表*/
@RequestMapping(value="/pageList", method=RequestMethod.POST )
public ResultModel pageList(Integer userId, Integer pageNum, Integer pageSize,
HttpServletRequest request, HttpServletResponse response) {
    if (pageNum==null || pageNum<1) {
        pageNum=1;
    }
    if (pageSize==null || pageSize<1) {
        pageSize=20;
    }
    UsersFocusQueryBo usersFocusQueryBo=new UsersFocusQueryBo();
    usersFocusQueryBo.setUserId(userId);
    usersFocusQueryBo.buildQuery();
        IPage<UsersFocusEntity> entityIPage = usersFocusService.
getPageByFocusId(pageNum, pageSize, usersFocusQueryBo);
    Page<UsersFocusPageDTO> usersFocusPageDTOPage=new Page<>();
    usersFocusPageDTOPage.setPageSize((int) entityIPage.getSize());
```

```
    usersFocusPageDTOPage.setCurrentIndex((int) entityIPage.getCurrent());
    usersFocusPageDTOPage.setTotalNumber((int) entityIPage.getTotal());
    List<UsersFocusPageDTO> usersFocusPageDTOS = new ArrayList<>();
    // 批量查询用户信息
    List<Integer> usersId=new ArrayList<>();
    for (UsersFocusEntity record : entityIPage.getRecords()) {
        usersId.add(record.getUserId());
    }
    if (!CollectionUtils.isEmpty(usersId)) {
        Collection<UsersEntity> usersEntities=usersService.listByIds(usersId);
        usersFocusPageDTOS= UsersConvert.INSTANCE.entityToUsersFocusPageDtoList(user
sEntities);
    }
    usersFocusPageDTOPage.setItems(usersFocusPageDTOS);
    return ResultModel.success(usersFocusPageDTOPage);
}
```

getPageByFocusId 方法如代码清单 20-68 所示。

代码清单 20-68　getPageByFocusId 方法

```
public IPage<UsersFocusEntity> getPageByFocusId(int pageNo, int pageSize,
UsersFocusQueryBo fieldQueryBo) {
    Page<UsersFocusEntity> page=new Page<>();
    page.setSize(pageSize);
    page.setCurrent(pageNo);
    QueryWrapper<UsersFocusEntity> queryWrapper=fieldQueryBo.buildQuery();
    return this.page(page, queryWrapper);
}
```

其中，buildQuery 方法如代码清单 20-69 所示。在这里，为分页查询的一些条件设置了一个条件类以进行统一处理，这样更加方便抽取通用代码，减少代码量。

代码清单 20-69　buildQuery 方法

```
@Data
public class UsersFocusQueryBo extends BaseQueryBo {
    /** 用户id*/
    private Integer userId;
    /** 被关注者的id */
    private Integer focusedUserId;
    public QueryWrapper<UsersFocusEntity> buildQuery() {
        QueryWrapper<UsersFocusEntity> queryWrapper = new QueryWrapper<>();
        if (userId!=null) {
            queryWrapper.eq(UsersFocusEntity.USER_ID, userId);
        }
        if (focusedUserId!=null) {
            queryWrapper.eq(UsersFocusEntity.FOCUSED_USER_ID, focusedUserId);
        }
        addSortQuery(queryWrapper);
```

```
        addIncludeDeleted(queryWrapper);
        return queryWrapper;
    }
}
```

UsersFocusQueryBo 类继承了 BaseQueryBo 类，BaseQueryBo 类源码如代码清单 20-70 所示。

代码清单 20-70　BaseQueryBo 类

```
@Data
public class BaseQueryBo {
    /** 主键ID */
    protected Integer id;
    /** 排序 */
    protected List<SortQueryBo> sorts;
    /** 是否包含被删除的数据，默认不包含被删除的数据 */
    protected Boolean includeDeleted=false;
    /** 查询的数据增加排序规则 */
    protected void addSortQuery(QueryWrapper<?> query) {
        // 默认更新时间倒序
        if (CollectionUtils.isEmpty(sorts)) {
            sorts = new ArrayList<>();
            SortQueryBo sortQueryBo = new SortQueryBo();
            sortQueryBo.setFieldName(BaseEntity.CREATE_TIME);
            sortQueryBo.setSortTypeEnum(SortQueryBo.SortTypeEnum.DESC);
            sorts.add(sortQueryBo);
        }
        sorts.forEach(sortReq -> {
                if (SortQueryBo.SortTypeEnum.ASC.getCode().equals(sortReq.
getSortTypeEnum().getCode())) {
                query.orderByAsc(sortReq.getFieldName());
            } else {  query.orderByDesc(sortReq.getFieldName()); }
        });
    }
    /** 是否包含删除的数据 */
    protected void addIncludeDeleted(QueryWrapper<?> queryWrapper) {
        if (!includeDeleted) {
            queryWrapper.eq(BaseEntity.DELETE_TIME, 0);
        }
    }
    /** 根据id查询 */
    protected void addId(QueryWrapper<?> queryWrapper) {
        if (id!=null) {
   queryWrapper.eq(BaseEntity.ID, id);
        }
    }
}
```

通过以上几个类，可以很方便地构建出一些查询条件，快速完成业务开发。

20.4.3 用户收藏功能

此处的用户收藏是指用户收藏资源。这里增加一个中间表 r_users_collections 记录用户和资源的收藏关系，通过 MyBatis-Plus 生成相关代码。

用户收藏资源的请求方法如代码清单 20-71 所示。

代码清单 20-71 用户收藏资源的请求方法

```
/** 用户收藏资源  */
@RequestMapping(value="/favoriteResources", method=RequestMethod.POST )
public ResultModel favoriteResources(Integer resourceId,HttpServletRequest
request, HttpServletResponse response) {
    UsersEntity usersEntity=checkResourcesParam(resourceId, request);
    RUsersCollectionsEntity rUsersCollectionsEntity=new RUsersCollectionsEntity();
    rUsersCollectionsEntity.setUserId(usersEntity.getId());
    rUsersCollectionsEntity.setResourceId(resourceId);
    rUsersCollectionsService.save(rUsersCollectionsEntity);
    return ResultModel.success();
}
```

其中 checkResourcesParam 方法如代码清单 20-72 所示。

代码清单 20-72 checkResourcesParam 方法

```
/**
 * 校验收藏资源参数的方法
 * @param resourceId 资源id
 * @param request 请求
 * @return 当前登录的用户信息
 */
private UsersEntity checkResourcesParam(Integer resourceId, HttpServletRequest
request) {
    if (resourceId=null || resourceId < 1) {
        throw new CheckoutException(ResultCodeEnum.PARAMETER_ERROR);
    }
    // 查询资源是否存在
    ResourceEntity resourceEntity=resourceService.getById(resourceId);
    if (resourceEntity==null) {
        throw new CheckoutException(ResultCodeEnum.PARAMETER_ERROR);
    }
    UsersEntity usersEntity=getLoginInfo(request);
    if (usersEntity==null) {
        throw new CheckoutException(ResultCodeEnum.USER_NOT_LOGGED);
    }
    return usersEntity;
}
```

同用户关注一样，也有取消收藏资源方法，如代码清单 20-73 所示。

代码清单 20-73　取消收藏资源方法

```
/** 取消收藏资源方法 */
@RequestMapping(value="/cancelResources", method=RequestMethod.POST )
public ResultModel cancelResources(Integer resourceId,HttpServletRequest request,
HttpServletResponse response) {
    UsersEntity usersEntity=checkResourcesParam(resourceId, request);
      rUsersCollectionsService.removeByUserIdAndResourceId(usersEntity.
getId(),resourceId);
    return ResultModel.success();
}
```

这两个链接也需要添加到登录拦截器中，只有在用户登录后方可访问。用户收藏资源功能与用户关注功能类似，所以该功能也有一个分页列表方法，如代码清单 20-74 所示。

代码清单 20-74　用户收藏资源的分页列表方法

```
/** 用户收藏资源的分页列表方法 */
@RequestMapping(value="/pageResourceList", method=RequestMethod.POST )
public ResultModel pageResourceList(Integer userId,Integer pageNum, Integer
pageSize, HttpServletRequest request, HttpServletResponse response) {
    if (pageNum==null || pageNum<1) { pageNum=1; }
    if (pageSize==null || pageSize<1) { pageSize=20;  }
    if (userId==null || userId<0) {
        throw new CheckoutException(ResultCodeEnum.USER_NOT_LOGGED);
    }
    RUsersCollectionsQueryBo rUsersCollectionsQueryBo=new RUsersCollectionsQueryBo();
    rUsersCollectionsQueryBo.setUserId(userId);
    rUsersCollectionsQueryBo.buildQuery();
      IPage<RUsersCollectionsEntity> entityIPage=rUsersCollectionsService.
getPage(pageNum, pageSize, rUsersCollectionsQueryBo);
    Page<RUsersCollectionsPageDTO> rUsersCollectionsPageDTOPage=new Page<>();
    rUsersCollectionsPageDTOPage.setPageSize((int) entityIPage.getSize());
    rUsersCollectionsPageDTOPage.setCurrentIndex((int) entityIPage.getCurrent());
    rUsersCollectionsPageDTOPage.setTotalNumber((int) entityIPage.getTotal());
    List<RUsersCollectionsPageDTO> rUsersCollectionsPageDTOS=new ArrayList<>();
    // 批量查询用户信息
    List<Integer> resourceIds=new ArrayList<>();
    for (RUsersCollectionsEntity record : entityIPage.getRecords()) {
        resourceIds.add(record.getResourceId());
    }
    if (!CollectionUtils.isEmpty(resourceIds)) {
            Collection<ResourceEntity> resourceEntities=resourceService.
listByIds(resourceIds);
        rUsersCollectionsPageDTOS= ResourceConvert.INSTANCE.entityToRUsersCollectio
nsPageDto(resourceEntities);
    }
    rUsersCollectionsPageDTOPage.setItems(rUsersCollectionsPageDTOS);
    return ResultModel.success(rUsersCollectionsPageDTOPage);
}
```

20.5 用户评论功能

如果仅是实现一个用户的评论功能，那非常简单。为了系统的健壮性，要对一些敏感词进行过滤处理。如果全部使用人工审核，则在用户量不大的情况下是可行的，但是用户多了之后，人工审核就行不通了，所以就需要实现自动过滤敏感词功能。另外，还需要防止不法用户进行 XSS（Cross Site Seripting，跨站脚本）攻击。

20.5.1 简单的评论功能

增加一个用户评论表 users_comment，利用 MyBatis-Plus 生成代码。用户评论请求接口如代码清单 20-75 所示。

代码清单 20-75 用户评论请求接口

```
/** 用户评论 */
@RequestMapping(value="/comment", method=RequestMethod.POST )
public ResultModel comment(UsersCommentReq usersCommentReq,HttpServletRequest
request, HttpServletResponse response) {
    // 参数校验
    ValidateUtils.validate(usersCommentReq);
    // 判断资源是否存在
     ResourceEntity resourceEntity=resourceService.getById(usersCommentReq.
getResourceId());
    if(resourceEntity==null){
        throw new CheckoutException(ResultCodeEnum.PARAMETER_ERROR);
    }
    UsersEntity usersEntity=getLoginInfo(request,true);
     UsersCommentEntity usersCommentEntity =UsersCommentConvert.INSTANCE.
reqToEntity(usersCommentReq);
    usersCommentEntity.setUserId(usersEntity.getId());
    // 需要过滤敏感词
    usersCommentEntity.setDetails(usersCommentReq.getRealDetails());
    usersCommentEntity.setState(UsersCommentEnum.NORMAL.getValue());
    usersCommentService.save(usersCommentEntity);
    return ResultModel.success();
}
```

注意：将链接添加到登录拦截器中。在代码清单 20-75 中，只是简单地将评论进行了保存，后面将对评论进行敏感词的过滤。接下来继续扩展与评论相关的功能，为用户提供删除评论的功能，如代码清单 20-76 所示。

代码清单 20-76 用户删除评论请求接口

```
/** 用户删除评论 */
@RequestMapping(value="/deleteComment", method=RequestMethod.POST )
```

```
public ResultModel deleteComment(Integer commentId,HttpServletRequest request,
HttpServletResponse response) {
    if(commentId==null || commentId<1){
        throw new CheckoutException(ResultCodeEnum.PARAMETER_ERROR);
    }
    // 获取登录用户
    UsersEntity usersEntity=getLoginInfo(request,true);
    // 获取评论
    UsersCommentEntity usersCommentEntity=usersCommentService.getById(commentId);
    if(usersCommentEntity==null){
        throw new CheckoutException(ResultCodeEnum.PARAMETER_ERROR);
    }
    if(!usersCommentEntity.getUserId().equals(usersEntity.getId())){
        // 无权限删除
        throw new CheckoutException(ResultCodeEnum.NO_PRIVILEGE);
    }
    // 软删评论
    usersCommentService.updateDeleteTimeById(commentId);
    return ResultModel.success();
}
```

其中，updateDeleteTimeById 方法如代码清单 20-77 所示。

代码清单 20-77　updateDeleteTimeById 方法

```
@Override
public void updateDeleteTimeById(Integer commentId) {
    QueryWrapper<UsersCommentEntity> queryWrapper=new QueryWrapper<>();
    queryWrapper.eq(UsersCommentEntity.ID, commentId);
    UsersCommentEntity usersCommentEntity=new UsersCommentEntity();
    usersCommentEntity.setDeleteTime(DateUtils.getIntDateTimeS());
    this.update(usersCommentEntity,queryWrapper);
}
```

其中，DateUtils.getIntDateTimeS 方法的实现就是 (int) System.currentTimeMillis()/1000。用户既然能添加、删除评论，自然也能编辑评论。用户编辑评论请求接口如代码清单 20-78 所示。

代码清单 20-78　用户编辑评论请求接口

```
/** 用户编辑评论 */
@RequestMapping(value="/updateComment", method=RequestMethod.POST )
public ResultModel updateComment(UsersCommentReq usersCommentReq,HttpServletReque
st request, HttpServletResponse response) {
    // 参数校验
    ValidateUtils.validate(usersCommentReq);
    // 判断资源是否存在
     ResourceEntity resourceEntity=resourceService.getById(usersCommentReq.
getResourceId());
    if (resourceEntity==null) {
        throw new CheckoutException(ResultCodeEnum.PARAMETER_ERROR);
    }
```

```
    // 判断评论是否存在且未被删除
    UsersCommentEntity usersCommentEntity= usersCommentService.
getById(usersCommentReq.getCommentId());
    if(usersCommentEntity==null){
        throw new CheckoutException(ResultCodeEnum.PARAMETER_ERROR);
    }
    if(UsersCommentEnum.FORBIDDEN.getValue().equals(usersCommentEntity.getState())
|| !DeleteEnum.NO_DELETE.getValue().equals(usersCommentEntity.getDeleteTime())){
        throw new CheckoutException(ResultCodeEnum.STATUS_EXCEPTION);
    }
    UsersEntity usersEntity=getLoginInfo(request, true);
    if(!usersEntity.getId().equals(usersCommentEntity.getUserId())){
        throw new CheckoutException(ResultCodeEnum.NO_PRIVILEGE);
    }
    UsersCommentEntity newUsersCommentEntity=new UsersCommentEntity();
    newUsersCommentEntity.setId(usersCommentEntity.getId());
    newUsersCommentEntity.setRealDetails(usersCommentEntity.getRealDetails());
    // 敏感词需要过滤
    usersCommentEntity.setDetails(usersCommentEntity.getRealDetails());
    usersCommentService.updateById(usersCommentEntity);
    return ResultModel.success();
}
```


20.5.2 评论分页功能

评论分页功能类似前面的用户关注功能和用户收藏资源功能，需要获取有用户的评论列表和资源的评论列表。接下来针对这两个场景，分别开发两个接口。首先是获取用户的评论列表，也就是获取单个用户的评论列表，如代码清单 20-79 所示。

代码清单 20-79 获取用户的评论列表

```
/** 获取用户的评论列表 */
@RequestMapping(value="/pageUsersList", method=RequestMethod.POST)
public ResultModel pageUsersList(Integer userId, Integer pageNum, Integer
pageSize, HttpServletRequest request, HttpServletResponse response) {
    if (pageNum==null || pageNum<1) {
        pageNum=1;
    }
    if (pageSize==null || pageSize<1) {
        pageSize=20;
    }
    if (userId==null || userId<0) {
        throw new CheckoutException(ResultCodeEnum.PARAMETER_ERROR);
    }
    // 获取用户信息
    UsersEntity usersEntity=usersServices.getById(userId);
    if(usersEntity==null){
```

```
        throw new CheckoutException(ResultCodeEnum.PARAMETER_ERROR);
    }
    UsersCommentQueryBo usersCommentQueryBo=new UsersCommentQueryBo();
    usersCommentQueryBo.setUserId(userId);
    usersCommentQueryBo.buildQuery();
    IPage<UsersCommentEntity> entityIPage=usersCommentService.getPage(pageNum,
pageSize, usersCommentQueryBo);
    Page<UsersCommentPageDTO> usersCommentDTOPage=new Page<>();
    usersCommentDTOPage.setPageSize((int) entityIPage.getSize());
    usersCommentDTOPage.setCurrentIndex((int) entityIPage.getCurrent());
    usersCommentDTOPage.setTotalNumber((int) entityIPage.getTotal());
    List<UsersCommentPageDTO> usersFocusPageDTOS= UsersCommentConvert.INSTANCE.
entityToPageList(entityIPage.getRecords());
    usersCommentDTOPage.setItems(usersFocusPageDTOS);
    return ResultModel.success(usersCommentDTOPage);
}
```

其中，有查询类及分页的 DTO 类，请前往 GitHub 进行查看。由于源码太多，请读者务必结合项目源码对照操作。核心方法 usersCommentService.getPage 如代码清单 20-80 所示。

代码清单 20-80　usersCommentService.getPage 方法

```
@Override
public IPage<UsersCommentEntity> getPage(Integer pageNum, Integer pageSize,
UsersCommentQueryBo usersCommentQueryBo) {
    Page<UsersCommentEntity> page=new Page<>();
    page.setSize(pageSize);
    page.setCurrent(pageNum);
    QueryWrapper<UsersCommentEntity> queryWrapper=usersCommentQueryBo.
buildQuery();
    return this.page(page, queryWrapper);
}
```

接下来介绍资源评论的分页列表接口。由于是资源的评论，每个评论的用户都不相同，所以每个评论的数据至少都需要带上用户昵称，如代码清单 20-81 所示。

代码清单 20-81　资源评论的分页列表接口

```
/** 获取资源评论的分页列表 */
@RequestMapping(value="/pageResourceList", method=RequestMethod.POST)
public ResultModel pageResourceList(Integer resourceId, Integer pageNum, Integer
pageSize, HttpServletRequest request, HttpServletResponse response) {
    if (pageNum==null || pageNum<1) {
        pageNum=1;
    }
    if (pageSize==null || pageSize<1) {
        pageSize=20;
    }
    if (resourceId==null || resourceId<0) {
        throw new CheckoutException(ResultCodeEnum.PARAMETER_ERROR);
```

```
    }
    // 获取资源
    ResourceEntity resourceEntity=resourceService.getById(resourceId);
    if(resourceEntity==null){
        throw new CheckoutException(ResultCodeEnum.PARAMETER_ERROR);
    }
    UsersCommentQueryBo usersCommentQueryBo=new UsersCommentQueryBo();
    usersCommentQueryBo.setResourceId(resourceId);
    usersCommentQueryBo.buildQuery();
     IPage<UsersCommentEntity> entityIPage=usersCommentService.getPage(pageNum,
pageSize, usersCommentQueryBo);
    Page<UsersCommentPageDTO> usersCommentDTOPage=new Page<>();
    usersCommentDTOPage.setPageSize((int) entityIPage.getSize());
    usersCommentDTOPage.setCurrentIndex((int) entityIPage.getCurrent());
    usersCommentDTOPage.setTotalNumber((int) entityIPage.getTotal());
    List<UsersCommentEntity> usersCommentEntities=entityIPage.getRecords();
    List<UsersCommentPageDTO> usersFocusPageDTOS=new ArrayList<>();
    List<Integer> userIds=new ArrayList<>();
    if(!CollectionUtils.isEmpty(usersCommentEntities)){
        // 获取用户id
        for (UsersCommentEntity usersCommentEntity : usersCommentEntities) {
            userIds.add(usersCommentEntity.getUserId());
        }
    }
    // 查询用户信息
    Collection<UsersEntity> usersEntities=usersServices.listByIds(userIds);
    Map<Integer,UsersEntity> usersEntityMap0z=new HashMap<>();
    if(!CollectionUtils.isEmpty(usersEntities)){
        for (UsersEntity usersEntity : usersEntities) {
            usersEntityMap.put(usersEntity.getId(),usersEntity);
        }
    }
    for (UsersCommentEntity usersCommentEntity : usersCommentEntities) {
        UsersCommentPageDTO usersCommentPageDTO= UsersCommentConvert.INSTANCE.enti
tyToDto(usersCommentEntity);
           usersCommentPageDTO.setUsername(usersEntityMap.get(usersCommentEntity.
getUserId()).
getUsername());
        usersFocusPageDTOS.add(usersCommentPageDTO);
    }
    usersCommentDTOPage.setItems(usersFocusPageDTOS);
    return ResultModel.success(usersCommentDTOPage);
}
```

通过前面几个接口的设置，与评论有关的功能就完成了。接下来实现敏感词过滤与防范 XSS 攻击的设置。

20.5.3 评论敏感词过滤

在实现文字过滤的算法中，DFA（Deterministic Finite Automaton，确定有穷自动机）是比较好的实现算法。它通过 event 和当前的 state 得到下一个 state，即 event+state=nextstate。其简单原理就是遍历需要过滤敏感词的字符串，判断是不是敏感词开头，如果是，则继续匹配敏感词的下一个字，否则继续遍历字符串。关于 DFA 算法，读者可自行查相关资料了解。接下来直接介绍代码。

首先，利用一个类存储敏感词的首字符，通过该类可以快速定位到包含敏感词的首字符。其次，增加停顿词的概念，假设“大家好”是敏感词，但是写成“大，家，好”，根据汉语语义来说，依然可以理解为“大家好”这三个字，在这里便可以将“,”标识为停顿词，如果敏感词之间是由停顿词隔开的，那么忽略停顿词。核心敏感词过滤类如代码清单 20-82 所示。

代码清单 20-82 核心敏感词过滤类

```
public class WordFilterUtils {
    /** 存储首字符 */
    private static final FilterSet FILTER_SET=new FilterSet();
    /** 存储节点 */
    private static final Map<Integer, WordNode> NODES=new HashMap<>(1024, 1);
    /** 停顿词 */
    private static final Set<Integer> STOPWD_SET=new HashSet<>();
    /** 敏感词过滤替换 */
    private static final char SIGN = '*';
    /** 增加停顿词、遍历停顿词，将停顿词添加到STOPWD_SET中 */
    public static void addStopWord(final List<String> words) {
        if (!isEmpty(words)) {
            char[] chs;
            for (String curr : words) {
                chs=curr.toCharArray();
                for (char c : chs) {
                    STOPWD_SET.add(charConvert(c));
                }
            }
        }
    }
    /** 添加DFA节点、添加敏感词开头的字符，以及增加敏感词节点  */
    public static void addSensitiveWord(final List<String> words) {
        if (!isEmpty(words)) {
            char[] chs;
            int fchar;
            int lastIndex;
            WordNode fnode; // 首字符节点
            for (String curr : words) {
                chs=curr.toCharArray();
                fchar=charConvert(chs[0]);
                //没有首字符定义
                if (!FILTER_SET.contains(fchar)) {
                    // 首字符标志位可重复使用add，已经判断，不再重复
```

```
                FILTER_SET.add(fchar);
                fnode=new WordNode(fchar, chs.length==1);
                NODES.put(fchar, fnode);
            } else {
                fnode=NODES.get(fchar);
                if (!fnode.isLast() && chs.length==1) {
                    fnode.setLast(true);
                }
            }
            lastIndex=chs.length - 1;
            for (int i=1; i<chs.length; i++) {
                fnode=fnode.addIfNoExist(charConvert(chs[i]), i==lastIndex);
            }
        }
    }
}
/** 过滤判断，将敏感词转换为屏蔽词 */
public static String doFilter(final String src) {
    char[] chs=src.toCharArray();
    int length=chs.length;
    // 当前检查的字符
    int currc;
    // 当前检查字符的备份
    int cpcurrc;
    int k;
    WordNode node;
    for (int i=0; i<length; i++) {
        currc=charConvert(chs[i]);
        if (!FILTER_SET.contains(currc)) {
            continue;
        }
        node=NODES.get(currc);
        if (node==null) {
            continue;
        }
        boolean couldMark=false;
        int markNum=-1;
        if (node.isLast()) {
            // 单字匹配
            couldMark=true;
            markNum=0;
        }
        // 继续匹配，长单字优先
        k=i;
        // 当前字符的复制
        cpcurrc=currc;
        for (; ++k<length; ) {
            int temp=charConvert(chs[k]);
            if (temp==cpcurrc) {
```

20

项目功能实现

```
                    continue;
                }
                if (STOPWD_SET.contains(temp)) {
                    continue;
                }
                node=node.querySub(temp);
                if (node==null) {
                    break;
                }
                if (node.isLast()) {
                    couldMark=true;
                    markNum=k-i;
                }
                cpcurrc=temp;
            }
            if (couldMark) {
                for (k = 0; k<=markNum; k++) {
                    chs[k+i]=SIGN;
                }
                i=i+markNum;
            }
        }
        return new String(chs);
    }
    /**
     * 是否包含敏感词
     * @return true表示包含
     */
    public static boolean isContains(final String src) {
        if (FILTER_SET.size()>0 && NODES.size() > 0) {
            char[] chs=src.toCharArray();
            int length=chs.length;
            // 当前检查的字符
            int currc;
            // 当前检查字符的备份
            int cpcurrc;
            int k;
            WordNode node;
            for (int i=0; i<length; i++) {
                currc=charConvert(chs[i]);
                if (!FILTER_SET.contains(currc)) {
                    continue;
                }
                node=NODES.get(currc);
                if (node==null) {
                    // 不会发生
                    continue;
                }
                boolean couldMark=false;
```

```
                if (node.isLast()) {
                    // 单字匹配
                    couldMark=true;
                }
                // 继续匹配，长单字优先
                k=i;
                cpcurrc=currc;
                for (; ++k < length; ) {
                    int temp=charConvert(chs[k]);
                    if (temp==cpcurrc) {
                        continue;
                    }
                    if (STOPWD_SET.contains(temp)) {
                        continue;
                    }
                    node=node.querySub(temp);
                    if (node==null) {
                        break;
                    }
                    if (node.isLast()) {
                        couldMark=true;
                    }
                    cpcurrc=temp;
                }
                if (couldMark) {
                    return true;
                }
            }
        }
        return false;
    }
    /** 大写转换为小写，全角转换为半角 */
    private static int charConvert(char src) {
        int r=BCConvert.qj2bj(src);
        return (r>= 'A' && r<= 'Z') ? r + 32 : r;
    }
    /** 判断一个集合是否为空 */
    public static <T> boolean isEmpty(final Collection<T> col) {
        return col==null || col.isEmpty();
    }
}
```

在这里，选择使用数据库存储敏感词和停顿词，在 Spring 初始化时进行加载。首先创建两个表，并导入相关敏感词和停顿词。

新建敏感词表与停顿词表的 SQL 语句如代码清单 20-83 所示。

代码清单 20-83　新建敏感词表与停顿词表的 SQL 语句

```
CREATE TABLE `stop_word` (
  `id` int(11) NOT NULL AUTO_INCREMENT,
```

```
  `word` varchar(32) NOT NULL COMMENT '停顿词',
  PRIMARY KEY (`id`) USING BTREE,
  UNIQUE KEY `uniq_word` (`word`) USING BTREE
) ENGINE=InnoDB AUTO_INCREMENT=1 DEFAULT CHARSET=utf8mb4 COMMENT='敏感词过滤之停顿词';
CREATE TABLE `sensitive_word` (
  `id` int(11) NOT NULL AUTO_INCREMENT,
  `word` varchar(128) NOT NULL COMMENT '敏感词',
  PRIMARY KEY (`id`) USING BTREE,
  UNIQUE KEY `uniq_word` (`word`) USING BTREE
) ENGINE=InnoDB AUTO_INCREMENT=1 DEFAULT CHARSET=utf8mb4 COMMENT='敏感词';
```

接下来使用 MyBatis-Plus 生成代码。

在 Spring 初始化后，进行敏感词和停顿词加载，如代码清单 20-84 所示。

代码清单 20-84　加载敏感词和停顿词

```
/** Spring容器启动后，初始化数据 */
@Component
@Slf4j
public class InitCommentWordConfig implements InitializingBean {
    @Autowired
    private StopWordService stopWordService;
    @Autowired
    private SensitiveWordService sensitiveWordService;

    @Override
    public void afterPropertiesSet() throws Exception {
        long s=System.currentTimeMillis();
        log.info("初始化数据开始");
        // 初始化评论过滤词，获取敏感词
        List<SensitiveWordEntity> sensitiveWordList=sensitiveWordService.list();
        // 获取停顿词
        List<StopWordEntity> stopWordList=stopWordService.list();
           log.info("初始化评论过滤词-敏感词数量:{},停顿词数量:{}", sensitiveWordList.
size(), stopWordList.size());
        List<String> wordStr=new ArrayList<>();
        for (SensitiveWordEntity word : sensitiveWordList) {
            wordStr.add(word.getWord());
        }
        // 加载敏感词
        WordFilterUtils.addSensitiveWord(wordStr);
        List<String> stopStr=new ArrayList<>();
        for (StopWordEntity word : stopWordList) {
            stopStr.add(word.getWord());
        }
        // 加载停顿词
        WordFilterUtils.addStopWord(stopStr);
        long e=System.currentTimeMillis();
        log.info("初始化数据结束，一共消耗时间:{}毫秒 相当于 {}秒", (e - s), (e - s) / 1000.00);
    }
}
```

Spring 启动后初始化敏感词和停顿词，系统在调用新增评论和编辑评论时，将用户的评论数据进行过滤，即可完成评论内容脱敏任务。

20.5.4 防范 XSS 攻击

评论区可能会出现 XSS 攻击，接下来讲解如何防范 XSS 攻击。

XSS 攻击是 Web 程序中常见的攻击手段，其原理是攻击者往 Web 页面里插入恶意的脚本代码（CSS 代码、JavaScript 代码等），当用户浏览该页面时，嵌入其中的脚本代码会被执行，从而达到恶意攻击用户的目的，如盗取用户 Cookie、破坏页面结构、重定向到其他网站等。

理论上来讲，Web 页面中所有可以由用户输入的地方，如果没有对输入的数据进行过滤处理，都会存在 XSS 攻击漏洞。当然，也可以选择对模板视图中的输出数据进行过滤。简单来说，就是阻止用户输入一些 JS 代码。接下来增加 XSS 攻击过滤工具类，如代码清单 20-85 所示。

代码清单 20-85　XSS 攻击过滤工具类

```
public class XssUtils {
    private static Pattern scriptPattern=Pattern.compile("<[\r\n| | ]*script[\r\n|
| ]*>(.*?)</[\r\n| | ]*script[\r\n| | ]*>", Pattern.CASE_INSENSITIVE);
    private static Pattern scriptPattern2=Pattern.compile("src[\r\n| | ]*=[\r\n|
| ]*[\\\"|\\\'](.*?)[\\\"|\\\']", Pattern.CASE_INSENSITIVE | Pattern.MULTILINE |
Pattern.DOTALL);
    private static Pattern scriptPattern3=Pattern.compile("</[\r\n| | ]*script[\
r\n| | ]*>", Pattern.CASE_INSENSITIVE);
    private static Pattern scriptPattern4=Pattern.compile("<[\r\n| |
]*script(.*?)>", Pattern.CASE_INSENSITIVE | Pattern.MULTILINE | Pattern.DOTALL);
    private static Pattern scriptPattern5=Pattern.compile("eval\\((.*?)\\)",
Pattern.CASE_INSENSITIVE | Pattern.MULTILINE | Pattern.DOTALL);
    private static Pattern scriptPattern6=Pattern.compile("e-xpression\\((.*?)\\)",
Pattern.CASE_INSENSITIVE | Pattern.MULTILINE | Pattern.DOTALL);
    private static Pattern scriptPattern7 = Pattern.compile("javascript[\r\n| |
]*:[\r\n| | ]*", Pattern.CASE_INSENSITIVE);
    private static Pattern scriptPattern8=Pattern.compile("vbscript[\r\n| |
]*:[\r\n| | ]*", Pattern.CASE_INSENSITIVE);
    private static Pattern scriptPattern9=Pattern.compile("onload(.*?)=", Pattern.
CASE_INSENSITIVE | Pattern.MULTILINE | Pattern.DOTALL);
    /** 特殊符号处理 */
    public static String getXssFilter(String summary, int maxLength) {
        return getXssFilter(summary, maxLength, "");
    }
    public static String getXssFilter(String text, int maxLength, String
replacement) {
        if (StringUtils.isEmpty(text)) {
            return "";
        }
        // 处理
        text=text.replace("<", "&lt;");
        text=text.replace(">", "&gt;");
```

```
        text=text.replace("\"", "'");
        text=text.replace("\n", replacement);
        text=xssHand(text);
        if (text.length()>maxLength) {
            text=text.substring(0, maxLength-1);
        }
        return text;
    }
    /** 将容易引起XSS、SQL漏洞的半角字符直接替换成全角字符 */
    public static String xssEncode(String s) {
        if (s==null || s.isEmpty()) {
            return s;
        } else {
            s=stripXSSAndSql(s);
        }
        return xssHand(s);
    }
    /** 半角转换为全角 */
    public static String xssHand(String s) {
        StringBuilder sb=new StringBuilder();
        for (int i=0; i<s.length(); i++) {
            char c=s.charAt(i);
            switch (c) {
                case ‘>':
                    // 转义大于号
                    sb.append("＞");
                    break;
                case '<':
                    // 转义小于号
                    sb.append("＜");
                    break;
                case '&':
                    // 转义“&”
                    sb.append("＆");
                    break;
                case '#':
                    // 转义“#”
                    sb.append("＃");
                    break;
                default: sb.append(c); break;
            }
        }
        return sb.toString();
    }
    /** 防范XSS攻击(根据实际情况调整、替换) */
    private static String stripXSSAndSql(String value) {
        if (value!=null) {
            value=scriptPattern.matcher(value).replaceAll("");
            value=scriptPattern2.matcher(value).replaceAll("");
```

```java
            // Remove any lonesome </script> tag
            value=scriptPattern3.matcher(value).replaceAll("");
            // Remove any lonesome <script ...> tag
            value=scriptPattern4.matcher(value).replaceAll("");
            // Avoid eval(...) expressions
            value=scriptPattern5.matcher(value).replaceAll("");
            // Avoid e-xpression(...) expressions
            value=scriptPattern6.matcher(value).replaceAll("");
            // Avoid javascript:... expressions
            value=scriptPattern7.matcher(value).replaceAll("");
            // Avoid vbscript:... expressions
            value=scriptPattern8.matcher(value).replaceAll("");
            // Avoid onload=expressions
            value=scriptPattern9.matcher(value).replaceAll("");
        }
        return value;
    }
}
```

通过调用 xssEncode 方法，便可将字符串中容易引起 XSS 攻击的文本过滤掉。在有用户输入功能的情况下，永远不要轻信用户输入的内容，一定要对输入的内容做过滤处理。

SQL 注入是一种注入攻击，通过将 SQL 语句插入数据库执行，使攻击者能够完全控制 Web 应用程序后面的数据库服务器。攻击者可以使用 SQL 注入漏洞绕过应用程序的安全措施，绕过网页或 Web 应用程序的身份验证和授权，检索整个 SQL 数据库的内容，还可以使用 SQL 注入来添加、修改和删除数据库中的记录。当使用 MyBatis 时，掌握好“#{}”和“${}”的使用时机，便可以充分避免 SQL 注入。

20.6 资源数据分页功能

本节讲解资源数据的分页功能，如专题资源分页数据、分类资源分类数据、标签资源分页数据等。

20.6.1 专题资源分页数据

由于使用了 MyBatis-Plus 工具，并且在用户评论分页数据查询中使用了查询条件构造器，所以分页查询代码编写起来非常简单。首先创建专题资源分页查询条件构造器，如代码清单 20-86 所示。

代码清单 20-86 专题资源分页查询条件构造器

```java
/** 专题资源分页查询条件构造器 */
@Data
public class RResourceSubjectQueryBo extends BaseQueryBo {
    /** 资源表id */
    private Integer resourceId;
```

```
        /** 专题表id */
        private Integer subjectId;
        public QueryWrapper<RResourceSubjectEntity> buildQuery() {
            QueryWrapper<RResourceSubjectEntity> queryWrapper = new QueryWrapper<>();
            if (resourceId!=null) {
                // 构造资源id的查询条件
                queryWrapper.eq(RResourceSubjectEntity.RESOURCE_ID, resourceId);
            }
            if (subjectId!=null) {
                // 构造专题id的查询条件
                queryWrapper.eq(RResourceSubjectEntity.SUBJECT_ID, subjectId);
            }
            //  添加排序条件
            addSortQuery(queryWrapper);
            // 是否包含删除的数据
            addIncludeDeleted(queryWrapper);
            return queryWrapper;
        }
    }
```

通过该类，可以非常方便地根据资源 id 与专题 id 创建出 MyBatis-Plus 的查询封装器。接下来构建 Service 层的分页查询方法，其参数非常简单，就是当前页、一页的数据量和查询构造器，如代码清单 20-87 所示。

代码清单 20-87　Service 层的分页查询方法

```
    @Override
    public IPage<RResourceSubjectEntity> getPage(Integer pageNum, Integer pageSize,
    RResourceSubjectQueryBo queryBo) {
        Page<RResourceSubjectEntity> page=new Page<>();
        page.setSize(pageSize);
        page.setCurrent(pageNum);
        QueryWrapper<RResourceSubjectEntity> queryWrapper = queryBo.buildQuery();
        return this.page(page, queryWrapper);
    }
```

然后分析前端控制器的入口方法，如代码清单 20-88 所示。

代码清单 20-88　前端控制器的入口方法

```
    /** 获取专题资源，包括列表、分页  */
    @RequestMapping(value="/pageList", method=RequestMethod.POST )
    public ResultModel pageList(Integer subjectId, Integer pageNum, Integer pageSize,
    HttpServletRequest request, HttpServletResponse response) {
        if (pageNum==null || pageNum<1) {
            pageNum=1;
        }
        if (pageSize==null || pageSize<1) {
            pageSize=20;
```

```
    }
    if (subjectId==null || subjectId<0) {
        throw new CheckoutException(ResultCodeEnum.PARAMETER_ERROR);
    }
    // 获取专题
    ResourceSubjectEntity resourceSubjectEntity=resourceSubjectService.
getById(subjectId);
    if (resourceSubjectEntity==null) {
        throw new CheckoutException(ResultCodeEnum.PARAMETER_ERROR);
    }
    if (!DeleteEnum.NO_DELETE.getValue().equals(resourceSubjectEntity.
getDeleteTime())) {
        throw new CheckoutException(ResultCodeEnum.DATA_DOES_NOT_EXIST);
    }
    RResourceSubjectQueryBo rResourceSubjectQueryBo=new RResourceSubjectQueryBo();
    rResourceSubjectQueryBo.setSubjectId(subjectId);
    rResourceSubjectQueryBo.buildQuery();
    IPage<RResourceSubjectEntity> entityIPage=rResourceSubjectService.
getPage(pageNum, pageSize, rResourceSubjectQueryBo);

    List<RResourceSubjectEntity> rResourceSubjectEntities=entityIPage.
getRecords();
    List<Integer> resourceIds=new ArrayList<>();
    for (RResourceSubjectEntity rResourceSubjectEntity : rResourceSubjectEntities) {
        // 资源id
        resourceIds.add(rResourceSubjectEntity.getResourceId());
    }
    Collection<ResourceEntity> resourceEntities=new ArrayList<>();
    if (!CollectionUtils.isEmpty(resourceIds)) {
        resourceEntities = resourceService.listByIds(resourceIds);
    }

    Page<ResourcePageDTO> resourcePageDTOPage=new Page<>();
    resourcePageDTOPage.setPageSize((int) entityIPage.getSize());
    resourcePageDTOPage.setCurrentIndex((int) entityIPage.getCurrent());
    resourcePageDTOPage.setTotalNumber((int) entityIPage.getTotal());

    List<ResourcePageDTO> resourcePageDTOS= ResourceConvert.INSTANCE.entityToPageL
ist(resourceEntities);
    resourcePageDTOPage.setItems(resourcePageDTOS);
    return ResultModel.success(resourcePageDTOPage);
}
```

该查询方法一共分为以下 3 步：

①参数校验，判断资源或专题是否已被删除。

②到关系表中查询专题的所有资源。

③根据查询的资源组装资源 id 集合，一次查询分页数据的所有信息。

20.6.2 分类资源分页数据

在页面展示资源的分类数据时，常常需要分页查询资源数据。与前面的专题资源分页数据的构建一样，先创建分页资源分页查询条件构造器，如代码清单 20-89 所示。

代码清单 20-89 分页资源分页查询条件构造器

```
@Data
public class RResourceTypeQueryBo extends BaseQueryBo<RResourceTypeEntity> {
    /** 资源id */
    private Integer resourceId;
    /** 资源分类id */
    private Integer resourceTypeId;

    @Override
    public QueryWrapper<RResourceTypeEntity> buildQuery() {
        QueryWrapper<RResourceTypeEntity> queryWrapper=new QueryWrapper<>();
        if (resourceId!=null) {
            queryWrapper.eq(RResourceTypeEntity.RESOURCE_ID, resourceId);

        }
        if (resourceTypeId!=null) {
            queryWrapper.eq(RResourceTypeEntity.RESOURCE_TYPE_ID, resourceTypeId);
        }
        addSortQuery(queryWrapper);
        addIncludeDeleted(queryWrapper);
        return queryWrapper;
    }
}
```

其 Service 层的分页查询方法与代码清单 20-87 的方法类似，具体实现代码不再粘贴出来。下面直接看其前端控制器的分类资源分页数据查询接口的方法，如代码清单 20-90 所示。

代码清单 20-90 分类资源分页数据查询接口的方法

```
/** 获取分类资源，包括列表、分页 */
@RequestMapping(value="/pageList", method=RequestMethod.POST)
public ResultModel pageList(Integer typetId, Integer pageNum, Integer pageSize,
HttpServletRequest
        request, HttpServletResponse response) {
    if (pageNum==null || pageNum<1) {
        pageNum=1;
    }
    if (pageSize==null || pageSize<1) {
        pageSize = 20;
    }
    if (typetId==null || typetId<0) {
        throw new CheckoutException(ResultCodeEnum.PARAMETER_ERROR);
    }
```

```
    // 获取分类
    ResourceTypeEntity resourceTypeEntity=resourceTypeService.getById(typetId);
    if (resourceTypeEntity==null) {
        throw new CheckoutException(ResultCodeEnum.PARAMETER_ERROR);
    }
      if (!DeleteEnum.NO_DELETE.getValue().equals(resourceTypeEntity.
getDeleteTime())) {
        throw new CheckoutException(ResultCodeEnum.DATA_DOES_NOT_EXIST);
    }

    RResourceTypeQueryBo rResourceTypeQueryBo=new RResourceTypeQueryBo();
    rResourceTypeQueryBo.setResourceTypeId(typetId);
    rResourceTypeQueryBo.buildQuery();
     IPage<RResourceTypeEntity> entityIPage=rResourceTypeService.getPage(pageNum,
pageSize, rResourceTypeQueryBo);

    List<RResourceTypeEntity> rResourceSubjectEntities=entityIPage.getRecords();
    List<Integer> resourceIds=new ArrayList<>();
    for (RResourceTypeEntity rResourceSubjectEntity : rResourceSubjectEntities) {
        // 资源id
        resourceIds.add(rResourceSubjectEntity.getResourceId());
    }
    Collection<ResourceEntity> resourceEntities=new ArrayList<>();
    if (!CollectionUtils.isEmpty(resourceIds)) {
        resourceEntities=resourceService.listByIds(resourceIds);
    }

    Page<ResourcePageDTO> resourcePageDTOPage=new Page<>();
    resourcePageDTOPage.setPageSize((int) entityIPage.getSize());
    resourcePageDTOPage.setCurrentIndex((int) entityIPage.getCurrent());
    resourcePageDTOPage.setTotalNumber((int) entityIPage.getTotal());

    List<ResourcePageDTO> resourcePageDTOS= ResourceConvert.INSTANCE.entityToPageL
ist(resourceEntities);
    resourcePageDTOPage.setItems(resourcePageDTOS);
    return ResultModel.success(resourcePageDTOPage);
}
```

该查询方法也是分为 3 个步骤，即先进行参数校验，再查询分类中的资源数据，然后通过资源 id 的集合查询资源数据。

需要注意的是，这里不建议使用级联查询，单表查询完全够用。虽然使用单表查询的步骤会多出一步，但是会对系统以后的水平扩展及垂直扩展方便很多。

20.6.3 标签资源分页数据

标签资源分页数据与前面的专题资源分页数据、分类资源分页数据类似，由于也是多对多关系，所以也要先创建一个中间表，开发时需要先根据标签在关联表中查询出该标签的分页资源 id。首先创建标签资源分页查询条件构造器，如代码清单 20-91 所示。

代码清单 20-91　标签资源分页查询条件构造器

```
@Data
public class RResourceTagsQueryBo extends BaseQueryBo<RResourceTagsEntity> {
    /** 资源表id */
    private Integer resourceId;
    /** 标签表id */
    private Integer tagsId;
    @Override
    public QueryWrapper<RResourceTagsEntity> buildQuery() {
        QueryWrapper<RResourceTagsEntity> queryWrapper=new QueryWrapper<>();
        if (resourceId!=null) {
            // 资源id条件
            queryWrapper.eq(RResourceTagsEntity.RESOURCE_ID, resourceId);
        }
        if (tagsId!=null) {
            // 标签id条件
            queryWrapper.eq(RResourceTagsEntity.TAGS_ID, tagsId);
        }
        addSortQuery(queryWrapper);
        addIncludeDeleted(queryWrapper);
        return queryWrapper;
    }
}
```

Service 层的分页查询方法如代码清单 20-92 所示。

代码清单 20-92　Service 层的分页查询方法

```
@Override
public IPage<RResourceTagsEntity> getPage(Integer pageNum, Integer pageSize,
RResourceTagsQueryBo queryBo) {
    Page<RResourceTagsEntity> page=new Page<>();
    page.setSize(pageSize);
    page.setCurrent(pageNum);
    QueryWrapper<RResourceTagsEntity> queryWrapper=queryBo.buildQuery();
    return this.page(page, queryWrapper);
}
```

其前端控制器的实现方法与前面的也非常类似，如代码清单 20-93 所示。

代码清单 20-93　前端控制器标签资源分页数据查询接口方法

```
/** 获取标签资源，包括列表、分页 */
@RequestMapping(value="/pageList", method=RequestMethod.POST)
public ResultModel pageList(Integer tagsId, Integer pageNum, Integer pageSize,
HttpServletRequest
        request, HttpServletResponse response) {
    if (pageNum==null || pageNum<1) {
        pageNum = 1;
    }
```

```java
    if (pageSize==null || pageSize<1) {
        pageSize = 20;
    }
    if (tagsId==null || tagsId<0) {
        throw new CheckoutException(ResultCodeEnum.PARAMETER_ERROR);
    }
    // 获取分类
    TagsEntity tagsEntity=tagsService.getById(tagsId);
    if (tagsEntity==null) {
        throw new CheckoutException(ResultCodeEnum.PARAMETER_ERROR);
    }
    if (!DeleteEnum.NO_DELETE.getValue().equals(tagsEntity.getDeleteTime())) {
        throw new CheckoutException(ResultCodeEnum.DATA_DOES_NOT_EXIST);
    }

    RResourceTagsQueryBo rResourceTagsQueryBo=new RResourceTagsQueryBo();
    rResourceTagsQueryBo.setTagsId(tagsId);
    rResourceTagsQueryBo.buildQuery();
    IPage<RResourceTagsEntity> entityIPage=rResourceTagsService.getPage(pageNum, pageSize,
rResourceTagsQueryBo);

    List<RResourceTagsEntity> rResourceSubjectEntities=entityIPage.getRecords();
    List<Integer> resourceIds=new ArrayList<>();
    for (RResourceTagsEntity rResourceSubjectEntity : rResourceSubjectEntities) {
        // 资源id
        resourceIds.add(rResourceSubjectEntity.getResourceId());
    }
    Collection<ResourceEntity> resourceEntities=new ArrayList<>();
    if (!CollectionUtils.isEmpty(resourceIds)) {
        resourceEntities = resourceService.listByIds(resourceIds);
    }

    Page<ResourcePageDTO> resourcePageDTOPage=new Page<>();
    resourcePageDTOPage.setPageSize((int) entityIPage.getSize());
    resourcePageDTOPage.setCurrentIndex((int) entityIPage.getCurrent());
    resourcePageDTOPage.setTotalNumber((int) entityIPage.getTotal());

    List<ResourcePageDTO> resourcePageDTOS= ResourceConvert.INSTANCE.entityToPageL
ist(resourceEntities);
    resourcePageDTOPage.setItems(resourcePageDTOS);
    return ResultModel.success(resourcePageDTOPage);
}
```

开发到这里，相信读者已经发现有非常多类似的代码。那么，是否可以通过优化代码来减少重复性工作呢？答案是肯定的。

下面介绍两种处理方式对类似的步骤进行统一处理。

- 第 1 种：开发自动代码生成工具，类似 MyBatis-Plus。通过插件或者代码形式，基于模板引擎与配置文件生成固定步骤、风格类似的代码。
- 第 2 种：改造代码，通过设计模式、泛型、多态与反射机制构建动态代码传参，从而通过一个方法完成多种数据的分页查询。

20.6.4 优化程序分类、专题、标签数据

在菜单页面中，还需要展示一些如分类菜单、专题菜单、标签云等的功能，所以还需要有查询这些数据的接口。首先设置分类，由于分类不多，可以一次查询全部，然后放入 Redis 缓存中，时间可以设置为一天。如果更新分类，就必须删除缓存或者等待缓存失效，切记：只更新数据库的数据是无法生效的！

查询所有分类菜单，如代码清单 20-94 所示。

代码清单 20-94　查询所有分类菜单

```
/** 获取分类的所有数据 */
@RequestMapping(value="/all", method=RequestMethod.POST )
public ResultModel all(HttpServletRequest request, HttpServletResponse response) {
    List<ResourceTypeDTO> resourceTypeDTOList=new ArrayList<>();
Collection<ResourceTypeEntity> resourceTypeEntities=resourceTypeService.
listNoDelete();
    Map<Integer, List<ResourceTypeEntity>> resourceTypeListMap=new HashMap<>();
    for (ResourceTypeEntity resourceTypeEntity : resourceTypeEntities) {
        if (resourceTypeListMap.containsKey(resourceTypeEntity.getPid())) {
             List<ResourceTypeEntity> resourceTypeEntityList= resourceTypeListMap.
get(resourceTypeEntity.getPid());
            // 由于是引用，不用再重新“put”集合到Map中去
            resourceTypeEntityList.add(resourceTypeEntity);
        } else {
            List<ResourceTypeEntity> resourceTypeEntityList = new ArrayList<>();
            resourceTypeEntityList.add(resourceTypeEntity);
                resourceTypeListMap.put(resourceTypeEntity.getPid(),
resourceTypeEntityList);
        }
    }
    // 按照结构分层，首先获取最上层的数据
    Integer pid = 0;
    List<ResourceTypeEntity> resourceTypeEntityList=resourceTypeListMap.get(pid);
    if (!CollectionUtils.isEmpty(resourceTypeEntityList)) {
        for (ResourceTypeEntity resourceTypeEntity : resourceTypeEntityList) {
            ResourceTypeDTO resourceTypeDTO=new ResourceTypeDTO();
            resourceTypeDTO.setId(resourceTypeEntity.getId());
            resourceTypeDTO.setName(resourceTypeEntity.getName());
            resourceTypeDTOList.add(resourceTypeDTO);
        }
    }
    // 遍历第1层的节点
    for (ResourceTypeDTO resourceTypeDTO : resourceTypeDTOList) {
        List<ResourceTypeDTO> cResourceTypeDTOList=addChildNode(resourceTypeDTO,
resourceTypeListMap);
        resourceTypeDTO.setResourceTypeDTOS(cResourceTypeDTOList);
```

```
    }
    return ResultModel.success(resourceTypeDTOList);
}
/** 递归无限的分类层级 */
private List<ResourceTypeDTO> addChildNode(ResourceTypeDTO resourceTypeDTO,
Map<Integer, List<ResourceTypeEntity>> resourceTypeListMap) {
    List<ResourceTypeDTO> cResourceTypeDTOList=new ArrayList<>();
      List<ResourceTypeEntity> resourceTypeEntityList= resourceTypeListMap.
get(resourceTypeDTO.getId());
    if (!CollectionUtils.isEmpty(resourceTypeEntityList)) {
        for (ResourceTypeEntity resourceTypeEntity : resourceTypeEntityList) {
            ResourceTypeDTO resourceTypeDTO1 = new ResourceTypeDTO();
            resourceTypeDTO1.setId(resourceTypeEntity.getId());
        resourceTypeDTO1.setName(resourceTypeEntity.getName());
              List<ResourceTypeDTO> resourceTypeDTOList=addChildNode(resourceTypeD
TO1, resourceTypeListMap);
            resourceTypeDTO1.setResourceTypeDTOS(resourceTypeDTOList);
            cResourceTypeDTOList.add(resourceTypeDTO1);
        }
    }
    return cResourceTypeDTOList;
}
```

在 addChildNode 方法中，通过递归的形式可以将分类的任意层级递归出来，并添加到集合中，然后通过访问接口直接进行测试。

由于分类很少变动，以及为了获取更快的返回速度，可以给该方法的返回数据增加 Redis 缓存。改动后的代码如代码清单 20-95 所示。

代码清单 20-95　增加 Redis 缓存

```
@RequestMapping(value="/all", method=RequestMethod.POST)
public ResultModel all(HttpServletRequest request, HttpServletResponse response) {
      List<ResourceTypeDTO> resourceTypeDTOList= redisClient.get(RedisConstants.
getAllResourceTypeKey()).getList(ResourceTypeDTO.class);
    if(!CollectionUtils.isEmpty(resourceTypeDTOList)){
        return ResultModel.success(resourceTypeDTOList);
    } ...//省略代码
    // 增加缓存
redisClient.set(RedisConstants.getAllResourceTypeKey(),resourceTypeDTOList,RedisCons
tants.
REG_MAX_TIME_1_DAY);
    return ResultModel.success(resourceTypeDTOList);
}
```

接下来分析获取所有专题信息的接口。专题没有分层级，比较简单，如代码清单 20-96 所示。

代码清单 20-96　获取专题信息接口

```
/** 获取专题 */
@RequestMapping(value="/all", method=RequestMethod.POST)
```

```
public ResultModel all() {
    Collection<ResourceSubjectEntity> resourceSubjectEntities= resourceSubjectService.
listNoDelete();
    List<ResourceSubjectDTO> resourceSubjectDTOS= ResourceSubjectConvert.INSTANCE.
entityTo(resourceSubjectEntities);
    return ResultModel.success(resourceSubjectDTOS);
}
```

其中，listNoDelete 方法为获取未删除数据的方法，如代码清单 20-97 所示。

代码清单 20-97　获取未删除数据的方法

```
@Override
public Collection<ResourceSubjectEntity> listNoDelete() {
    QueryWrapper<ResourceSubjectEntity> queryWrapper=new QueryWrapper<>();
    queryWrapper.eq(BaseEntity.DELETE_TIME, 0);
    return this.list(queryWrapper);
}
```

由于接口是可以自定义的,在这里选择标签分页获取。标签分页获取接口如代码清单20-98所示。

代码清单 20-98　获取标签分页接口

```
/** 获取标签分页数据 */
@RequestMapping(value="/pageTagsList", method=RequestMethod.POST )
public ResultModel pageTagsList(Integer pageNum, Integer pageSize, HttpServletRequest
        request, HttpServletResponse response) {
    if (pageNum==null || pageNum<1) {
        pageNum=1;
    }
    if (pageSize==null || pageSize<1) {
        pageSize=20;
    }
    TagsQueryBo tagsQueryBo=new TagsQueryBo();
    // 查询分页数据
     IPage<TagsEntity> tagsEntityIPage=tagsService.getPage(pageNum, pageSize,
tagsQueryBo);
    // 设置分页数据
    Page<TagsDTO> tagsDTOPage=new Page<>();
    tagsDTOPage.setPageSize((int) tagsEntityIPage.getSize());
    tagsDTOPage.setCurrentIndex((int) tagsEntityIPage.getCurrent());
    tagsDTOPage.setTotalNumber((int) tagsEntityIPage.getTotal());
    // 复制集合数据
     List<TagsDTO> resourcePageDTOS= TagsConvert.INSTANCE.entityToDTOList(tagsEntityIPage.
getRecords());
    tagsDTOPage.setItems(resourcePageDTOS);
    return ResultModel.success(tagsDTOPage);
}
```

其中，getPage 方法的实现相信读者已经非常了解了，这里不再粘贴出具体代码，读者请以 GitHub 仓库中的代码为准进行学习。

20.7 交易功能

在本项目中定义的虚拟货币为U币。接下来的章节将分析用户如何购买U币、如何使用U币获取资源的下载，以及分享者如何获取U币。

本项目中采用支付宝支付方式，需要读者到支付宝开发者中心开通“当面付”功能，才可以通过API动态生成收款码进行收款。接下来主要讲解交易结构设计、支付功能实现、管理员登录拦截器实现、日志拦截器。

20.7.1 设计交易结构

首先从数据存储方面进行讲解，现在已经有了users表，users表中有一个ub列存储用户可用的U币，相当于统计用户U币的总数；另外需要一个充值记录表。

新增users_recharge_ub表，用来记录用户的充值信息，仅仅记录比较简略的、实际支付成功的信息。users_recharge_ub表的SQL语句如代码清单20-99所示。

代码清单 20-99　users_recharge_ub 表的 SQL 语句

```
CREATE TABLE `users_recharge_ub` (
  `id` int(11) NOT NULL AUTO_INCREMENT COMMENT '唯一标识码',
  `user_id` int(11) NOT NULL COMMENT '用户id',
  `money` decimal(11,2) NOT NULL COMMENT '充值金额-单位元',
  `ub_number` int(11) NOT NULL COMMENT '兑换的U币数量',
  `order_number` varchar(32) CHARACTER SET utf8 COLLATE utf8_general_ci NOT NULL COMMENT
'充值的订单编号,格式为当前年月日时分秒毫秒+随机生成10位数字，如:"20170228155316339*****"',
  `pay_type_en_name` varchar(64) NOT NULL COMMENT '支付类型名称',
  `create_time` datetime NOT NULL DEFAULT CURRENT_TIMESTAMP,
  `update_time` datetime NOT NULL DEFAULT CURRENT_TIMESTAMP ON UPDATE CURRENT_
TIMESTAMP,
  `delete_time` int(11) NOT NULL DEFAULT '0',
  `users_pay_id` int(11) NOT NULL COMMENT '用户支付详细信息id',
  PRIMARY KEY (`id`),
  UNIQUE KEY `uniq_order_number` (`order_number`) USING BTREE,
  KEY `fk_user_recharge_ub_user_id` (`user_id`)
) ENGINE=InnoDB AUTO_INCREMENT=2 DEFAULT CHARSET=utf8 COMMENT='用户充值U币表，存放用户
充值U币信息。';
```

为了查看更加详细的信息，再新增一个users_pay表，用来记录用户支付的详细信息。users_pay表的创建语句如代码清单20-100所示。

代码清单 20-100　users_pay 表的 SQL 语句

```
CREATE TABLE `users_pay` (
  `id` int(11) NOT NULL AUTO_INCREMENT,
  `money` decimal(11,2) NOT NULL COMMENT '金额',
```

```
  `info` varchar(200) CHARACTER SET utf8mb4 COLLATE utf8mb4_0900_ai_ci NOT NULL
DEFAULT '' COMMENT '留言',
  `create_time` datetime NOT NULL DEFAULT CURRENT_TIMESTAMP COMMENT '创建时间',
  `update_time` datetime NOT NULL DEFAULT CURRENT_TIMESTAMP ON UPDATE CURRENT_
TIMESTAMP COMMENT '修改时间',
  `email` varchar(100) CHARACTER SET utf8mb4 COLLATE utf8mb4_0900_ai_ci NOT NULL
DEFAULT '' COMMENT '通知邮箱',
  `state` tinyint(4) NOT NULL DEFAULT '0' COMMENT '显示状态 0待审核 1通过 2驳回 3已扫码',
  `pay_type_en_name` varchar(64) NOT NULL COMMENT 'pay_type的en_name，支付类型',
  `pay_num` varchar(128) NOT NULL DEFAULT '' COMMENT '支付标识',
  `custom` tinyint(4) NOT NULL DEFAULT '0' COMMENT '是否自定义输入，0否',
  `mobile` tinyint(4) NOT NULL DEFAULT '0' COMMENT '是否移动端，0否',
  `device` varchar(1000) CHARACTER SET utf8mb4 COLLATE utf8mb4_0900_ai_ci NOT NULL
DEFAULT '' COMMENT '用户支付设备信息',
  `token_num` varchar(64) CHARACTER SET utf8mb4 COLLATE utf8mb4_0900_ai_ci NOT NULL
DEFAULT '' COMMENT '生成二维码编号标识token',
  `pass_url` varchar(200) CHARACTER SET utf8mb4 COLLATE utf8mb4_0900_ai_ci NOT NULL
DEFAULT '' COMMENT '通过审核的URL',
  `back_url` varchar(200) CHARACTER SET utf8mb4 COLLATE utf8mb4_0900_ai_ci NOT NULL
DEFAULT '' COMMENT '操作后的URL',
  `close_url` varchar(200) CHARACTER SET utf8mb4 COLLATE utf8mb4_0900_ai_ci NOT
NULL DEFAULT '' COMMENT '关闭交易的URL',
  `delete_time` int(11) NOT NULL DEFAULT '0',
  `order_number` varchar(32) NOT NULL COMMENT '订单号',
  `users_id` int(11) NOT NULL COMMENT '用户id',
  PRIMARY KEY (`id`),
  UNIQUE KEY `uniq_order_number` (`order_number`) USING BTREE,
  KEY `fk_users_id` (`users_id`) USING BTREE
) ENGINE=InnoDB AUTO_INCREMENT=6 DEFAULT CHARSET=utf8mb4 COLLATE=utf8mb4_0900_ai_
ci COMMENT='用户支付信息详情表';
```

以上是与支付相关的主要表结构，用于记录用户充值信息和用户支付的详细信息。用户支付的详细信息用于记录交易过程，与实际账单是一致的。如果需要对账，则通过对比该表与平台方的支付成功账单，二者完全一致才可以。

本系统中的所有表结构，在 ssm-uifuture 项目 resource 目录的 .sql 文件下都可以找到。

20.7.2 实现支付功能

关于支付功能，读者可以先构思一下。本项目使用的是支付宝支付，"当面付"功能是指商家给用户展示二维码，二维码固定了金额，用户通过支付宝扫码进行支付。为了能收到用户支付成功的消息，开发者需要在开放平台配置支付回调通知接口。在该接口中可以获取平台方给开发者的支付通知（成功或失败）。这就是实现网站支付功能的完整流程。

要使用阿里的支付宝开放平台，首先需要在 pom 文件中引入阿里的 SDK 包，如代码清单 20-101 所示。

代码清单 20-101　引入阿里的 SDK 包

```
<!--阿里支付-->
<!-- https://mvnrepository.com/artifact/com.alipay.sdk/alipay-sdk-java -->
<dependency>
    <groupId>com.alipay.sdk</groupId>
    <artifactId>alipay-sdk-java</artifactId>
    <version>4.8.10.ALL</version>
</dependency>
```

既然要生成二维码，那么还需要一个生成支付宝收款码的接口，如代码清单 20-102 所示。

代码清单 20-102　生成支付宝收款码的接口

```
/** 生成支付二维码 */
 @RequestMapping(value="/alipay/createQrCode", method=RequestMethod.GET)
 public ResultModel alipayCreateQrCode(UsersPayReq usersPayReq, HttpServletRequest
request) throws AlipayApiException {
// 参数校验
ValidateUtils.validate(usersPayReq);
// 获取登录的用户
UsersEntity usersEntity=getLoginInfo(request, true);
if (usersPayReq.getMoney().compareTo(MIN_PAY)<0 || usersPayReq.getMoney().
compareTo(MAX_PAY)>0) {
return ResultModel.errorNoData("请填写正确的金额，当面付单笔金额不得大于1000");
}
String email=usersPayReq.getEmail();
if (StringUtils.isEmpty(email)) {
email=usersEntity.getEmail();
if (StringUtils.isEmpty(email)) {
return ResultModel.errorNoData("请填写正确的邮箱，或者使用账号绑定邮箱，以便接收通知邮件");
            }
        }

// 防止频繁提交
String key=RedisConstants.getAlipayCreateQrCode(usersEntity.getUsername());
String isOne redisClient.get(key).getObject(String.class);
if (StringUtils.isNotEmpty(isOne)) {
// 获取过期时间
Long expire=redisClient.getRedisTemplate().getExpire(key, TimeUnit.SECONDS);
return ResultModel.errorNoData("提交过于频繁请" + expire + "秒后再试");
        }
// 添加支付信息
UsersPayEntityusersPayEntity=UsersPayConvert.INSTANCE.usersPayReqToUsersPayEntity(u
sersPayReq);
usersPayEntity.setEmail(email);
usersPayEntity.setPayTypeEnName(PayTypeEnNameEnum.ALI_APY.getValue());
String orderToken=PasswordUtils.getToken();
usersPayEntity.setOrderNumber(orderToken);
usersPayService.save(usersPayEntity);
```

```
    // 记录缓存
    redisClient.set(key, "1", 30);

    AlipayClient alipayClient=new DefaultAlipayClient("https://openapi.alipay.com/
gateway.do",
    appId, sysConfig.getAliPayPrivateKey(), "json", "GBK", sysConfig.
getAliPayPublicKey(), "RSA");
    AlipayTradePrecreateRequest r=new AlipayTradePrecreateRequest();
    r.setBizContent("{" +"\"out_trade_no\":\"" + usersPayEntity.getOrderNumber() +
"\"," + "\"total_amount\":" + usersPayEntity.getMoney() + "," + "\"subject\":\"充值
U币\"" + "  }");
    // 设置通知回调链接
     r.setNotifyUrl(notifyUrl);
     AlipayTradePrecreateResponse response = alipayClient.execute(r);
    if (!response.isSuccess()) {
     return ResultModel.errorNoData("调用支付宝接口生成二维码失败，请联系客服");
    }
            Map<String, Object> result=new HashMap<>(16);
            result.put("order_number", usersPayEntity.getOrderNumber());
            result.put("qr_code", response.getQrCode());
            return ResultModel.success(result);
    }
```

该方法中比较重要的是如何接入阿里开放平台，另外就是防止用户频繁刷单。

先介绍一下幂等性，其约定大致为：包括网络错误或请求超时在内，请求同一个接口多次或单次，对资源所造成的影响是一致的。幂等性关注的是对资源造成的影响，而非请求返回的结果。在这里，为了防止用户频繁刷单，进行了条件设置，即一个用户在 30s 内允许下单一次。

用户支付是否成功要依赖支付方的通知才能知晓，新增回调通知接口，如代码清单 20-103 所示。

代码清单 20-103　回调通知接口

```
    /** 支付二维码通知接口 */
    @RequestMapping(value="/alipay/notify", method=RequestMethod.POST)
    public ResultModel alipayNotify(@RequestParam(required=false, name="out_trade_
no") String outTradeNo, @RequestParam(required=false, name="trade_status") String
tradeStatus) {
        if ("TRADE_SUCCESS".equals(tradeStatus)) {
            sendActiveEmail(outTradeNo);
        }
        return ResultModel.success("success");
    }
    /** 异步发送通知邮件 */
    private void sendActiveEmail(String outTradeNo) {
        if (StringUtils.isEmpty(outTradeNo)) {
            return;
        }
        UsersPayEntity usersPayEntity=usersPayService.getByOrderNumber(outTradeNo);
        if (usersPayEntity==null) {
```

```
            return;
        }
          if (!UsersPayStateEnum.TO_BE_AUDITED.getValue().equals(usersPayEntity.
getState())) {
            log.warn("[UsersPayController->sendActiveEmail]接收的订单支付信息状态不对,订单
号={},数据库订单信息={}", outTradeNo, JSON.toJSONString(usersPayEntity));
            return;
        }
        // 通过token控制，一天内有效
        String token=PasswordUtils.getSalt();
        // 放入Redis
        redisClient.set(token, "1", RedisConstants.REG_MAX_TIME_1_DAY);
        // 查询用户信息
        UsersEntity usersEntity=usersService.getById(usersPayEntity.getUsersId());
        if (usersEntity==null) {
            log.warn("[UsersPayController->sendActiveEmail]非法用户进行支付,订单号={},数据
库订单信息={},usersEntity={}", outTradeNo, JSON.toJSONString(usersPayEntity), JSON.
toJSONString(usersEntity));
            return;
        }
        // 发送邮件，附带审核的链接即可，需拼接订单号与token
        SendPayCheckEmail sendPayCheckEmail = new SendPayCheckEmail() {
            @Override
            public String getEmail() {
                // 管理员邮箱，用于审核支付订单
                return "code@uifuture.com";
            }
            @Override
            public String getUrl() {
                return passUrl + "?orderNumber=" + outTradeNo + "&token=" + token;
            }
            @Override
            public String getUserInfo() {
                return JSON.toJSONString(usersEntity);
            }
            @Override
            public String getPay() {
                return JSON.toJSONString(usersPayEntity);
            }
        };
        SendPayCheckEmailCallable sendEmailCallable=new SendPayCheckEmailCallable(email
Config, sendPayCheckEmail);
        ExecutorService executorService=Executors.newSingleThreadExecutor();
        executorService.submit(sendEmailCallable);
    }
```

支付方通知用户支付成功的信息后，可以给管理员发送一封邮件。管理员可以通过邮件中的提示信息进行审核，如果收到对应的金额，则可以单击审核通过的链接，给用户增加对应的 U 币。在这里使用了 UUID 作为令牌，防止被人攻击；另外，设置了审核链接的有效期为一天。

其实也可以将功能调整为自动充值，在回调通知中是可以得知实际的订单支付状态的，还可以给用户发送一封充值成功的邮件。

接下来实现用户支付审核通过接口，管理员可以通过支付通知邮件，确认用户支付成功，给用户增加对应的 U 币。用户支付审核通过接口如代码清单 20-104 所示。

代码清单 20-104　用户支付审核通过接口

```
/** 订单支付审核通过 */
@RequestMapping(value="/payPass", method=RequestMethod.POST)
public ResultModel pass(String orderNumber, String token) {
    if (StringUtils.isEmpty(orderNumber)) {
        return ResultModel.errorNoData("参数错误");
    }
    if (StringUtils.isEmpty(token)) {
        return ResultModel.errorNoData("参数错误");
    }
    // 校验token未超时
    String value=redisClient.get(token).getString();
    if (StringUtils.isEmpty(value)) {
        return ResultModel.errorNoData("该订单已经无法再进行审核");
    }
    UsersPayEntity usersPayEntity=usersPayService.getByOrderNumber(orderNumber);
    if (usersPayEntity==null) {
        return ResultModel.errorNoData("参数错误");
    }
    if (!UsersPayStateEnum.TO_BE_AUDITED.getValue().equals(usersPayEntity.getState())) {
        return ResultModel.errorNoData("订单状态错误");
    }
    //用户增加记录和U币
    usersService.addUB(usersPayEntity);
    return ResultModel.success("success");
}
```

该 token 就是前面存储在 Redis 中利用 UUID 生成的令牌，有效期为一天。核心是 usersService 中的 addUB 方法，如代码清单 20-105 所示。

代码清单 20-105　addUB 方法

```
@Override
public void addUB(UsersPayEntity usersPayEntity) {
    if (usersPayEntity==null) {
        throw new ParameterException(ResultCodeEnum.PARAMETER_ERROR);
    }
    // 修改支付信息的状态
    UsersPayEntity updateUsersPay=new UsersPayEntity();
    updateUsersPay.setId(usersPayEntity.getId());
    updateUsersPay.setState(UsersPayStateEnum.ADOPT.getValue());
    // 记录用户的增加U币信息
    UsersRechargeUbEntity usersRechargeUbEntity=new UsersRechargeUbEntity();
    usersRechargeUbEntity.setUserId(usersPayEntity.getUsersId());
```

```
    usersRechargeUbEntity.setMoney(usersPayEntity.getMoney());
// 在这里U币的最小整数单位为分(人民币)，确保数字一定为整数可以使用Long
    Integer ub=usersPayEntity.getMoney().multiply(new BigDecimal("100")).
intValue();
    usersRechargeUbEntity.setUbNumber(ub);
    usersRechargeUbEntity.setOrderNumber(usersPayEntity.getOrderNumber());
    usersRechargeUbEntity.setUsersPayId(usersPayEntity.getId());
    usersRechargeUbEntity.setPayTypeEnName(usersPayEntity.getPayTypeEnName());
    // 用户增加U币
    UsersEntity usersEntity=this.getById(usersPayEntity.getUsersId());
    if (usersEntity==null) {
        throw new ParameterException(ResultCodeEnum.USERS_DOES_NOT_EXIST);
    }
    // 开启事务
    TransactionStatus transaction=null;
    try {
        transaction=transactionManager.getTransaction(new DefaultTransactionDefinit
ion());
        usersPayService.updateById(updateUsersPay);
        usersRechargeUbService.save(usersRechargeUbEntity);
        usersMapper.operateUB(usersPayEntity.getUsersId(), ub);
        // 事务提交
        transactionManager.commit(transaction);
    } catch (Exception e) {
        log.error("[UsersServiceImpl->addUB], 事务回滚", e);
        if (transaction!=null) {
            // 回滚
            transactionManager.rollback(transaction);
        }
    }
}
```

从这里可以看出，金额与U币的充值比例为1∶100，并使用了事务保证充值的原子性。需要注意，在这里并没有对事务回滚应用补偿机制，限于篇幅，此处不进行具体实现了，解决方法是使用延时队列，一旦该事务回滚便发送一个消息到消息队列中，在另外一个类中对该消息队列进行消费，如果消费失败，则延时进行重试；当失败多次时用邮件或短信通知，然后人工介入处理。

接下来展示订单处理接口，如代码清单20-106所示。

代码清单 20-106　订单处理接口

```
/** 根据订单号获取支付信息 */
@RequestMapping(value="/getPayInfo", method=RequestMethod.GET)
public ResultModel getPayInfo(String orderNumber) {
    if (StringUtils.isEmpty(orderNumber)) {
        return ResultModel.errorNoData("参数错误");
    }
    UsersPayEntity usersPayEntity=usersPayService.getByOrderNumber(orderNumber);
    UsersPayDTO usersPayDTO= UsersPayConvert.INSTANCE.usersPayEntityToUsersPayDTO(u
sersPayEntity);
```

```
        return ResultModel.success(usersPayDTO);
    }
    /** 获取待审核的订单列表 */
    @RequestMapping(value="/getCheckList", method=RequestMethod.GET)
    public ResultModel getCheckList() {
        List<UsersPayEntity> usersPayEntityS= usersPayService.selectAllByState(UsersPay
StateEnum.TO_BE_AUDITED.getValue());
        List<UsersPayDTO> usersPayDTOS= UsersPayConvert.INSTANCE.usersPayEntityToUsersP
ayDTOList(usersPayEntityS);
        return ResultModel.success(usersPayDTOS);
    }
    /** 获取未进行支付的订单列表 */
    @RequestMapping(value="/getNoPayList", method=RequestMethod.GET)
    public ResultModel getNoPayList() {
        List<UsersPayEntity> usersPayEntityS= usersPayService.selectAllByState(UsersPay
StateEnum.SWEEP_CODE.getValue());
        List<UsersPayDTO> usersPayDTOS= UsersPayConvert.INSTANCE.usersPayEntityToUsersP
ayDTOList(usersPayEntityS);
        return ResultModel.success(usersPayDTOS);
    }
    /** 获取未通过审核的订单列表 */
    @RequestMapping(value="/getRejectList", method=RequestMethod.GET)
    public ResultModel getRejectList() {
        List<UsersPayEntity> usersPayEntityS= usersPayService.selectAllByState(UsersPay
StateEnum.REJECT.getValue());
        List<UsersPayDTO> usersPayDTOS = UsersPayConvert.INSTANCE.usersPayEntityToUsers
PayDTOList(usersPayEntityS);
        return ResultModel.success(usersPayDTOS);
    }
    /** 获取订单支付状态 */
    @RequestMapping(value="/getPayState", method=RequestMethod.GET)
    public ResultModel getPayState(String orderNumber) {
        if (StringUtils.isEmpty(orderNumber)) {
            return ResultModel.errorNoData("参数错误");
        }
        UsersPayEntity usersPayEntity=usersPayService.getStateByOrderNumber(orderNumber);
        UsersPayDTO usersPayDTO= UsersPayConvert.INSTANCE.usersPayEntityToUsersPayDTO(u
sersPayEntity);
         usersPayDTO.setStateStr(UsersPayStateEnum.getByValue(usersPayEntity.
getState()).getName());
        return ResultModel.success(usersPayDTO);
    }
    /** 驳回订单 */
    @RequestMapping(value="/payBack", method=RequestMethod.POST)
    public ResultModel payBack(String orderNumber) {
        if (StringUtils.isEmpty(orderNumber)) {
            return ResultModel.errorNoData("参数错误");
        }
```

```
    UsersPayEntity usersPayEntity=usersPayService.getByOrderNumber(orderNumber);
    if (usersPayEntity==null) {
        return ResultModel.errorNoData("参数错误");
    }
    if (!UsersPayStateEnum.TO_BE_AUDITED.getValue().equals(usersPayEntity.getState())) {
        return ResultModel.errorNoData("订单状态错误");
    }
    // 修改支付信息的状态
    UsersPayEntity updateUsersPay=new UsersPayEntity();
    updateUsersPay.setId(usersPayEntity.getId());
    updateUsersPay.setState(UsersPayStateEnum.REJECT.getValue());
    usersPayService.updateById(updateUsersPay);
    return ResultModel.success("success");
}

/** 分页获取订单 */
@RequestMapping(value="/page", method=RequestMethod.GET)
public ResultModel page(PageDTO pageDTO) {
    if (pageDTO == null) {
        return ResultModel.errorNoData("参数错误");
    }
    UsersPayQueryBo usersPayQueryBo=new UsersPayQueryBo();
    usersPayQueryBo.setUsersId(pageDTO.getUsersId());
     IPage<UsersPayEntity> entityIPage=usersPayService.getPage(pageDTO.
getCurrentIndex(), pageDTO.getPageSize(), usersPayQueryBo);
    List<UsersPayDTO> usersPayDTOS= UsersPayConvert.INSTANCE.usersPayEntityToUsersP
ayDTOList(entityIPage.getRecords());
     Page<UsersPayDTO> usersPayDTOPage=BeanConvertUtils.pageConvert(entityIPage,
usersPayDTOS);
    return ResultModel.success(usersPayDTOPage);
}
```

以上接口实现的都是一些简单的增、删、改、查操作，不再一一阐述。这些接口显然不能直接对外暴露，需要用户登录，且是管理员登录才能操作。为方便处理，需实现管理员登录拦截器。

20.7.3 实现管理员登录拦截器

本小节讲解如何实现管理员登录拦截器，以及 Controller 层和 Service 层的接口入参、出参日志拦截器。实现 Spring MVC 的拦截器有多种方式，此处使用 HandlerInterceptor 接口进行拦截器的实现，如代码清单 20-107 所示。

代码清单 20-107 管理员登录拦截器

```
/** 管理员登录拦截器，必须管理员登录才能访问 */
@Slf4j
@Configuration
public class AdminLoginInterceptor extends BaseController implements
HandlerInterceptor {
```

```
    @Override
      public boolean preHandle(HttpServletRequest request, HttpServletResponse
response, Object handler) throws Exception {
        // 判断是否登录
        UsersEntity users=getLoginInfo(request);
        // 已经登录
        if (users==null) {
            // 写入提醒
            ResultModel resultModel=new ResultModel(500, "登录后方可操作");
            responseResult(response, resultModel);
            return false;
        }
        // 非管理员不能访问
        return UsersTypeEnum.ADMIN.getValue().equals(users.getType());
    }
}
```

另外，需要在 Spring MVC 的 XML 配置文件中配置需要拦截的链接，如代码清单 20-108 所示。

代码清单 20-108　配置需要拦截的链接

```
<!--配置需要拦截的链接-->
<mvc:interceptor>
    <mvc:mapping path="/users-pay-entity/getPayInfo"/>
    <mvc:mapping path="/users-pay-entity/getCheckList"/>
    <mvc:mapping path="/users-pay-entity/getNoPayList"/>
    <mvc:mapping path="/users-pay-entity/getPayState"/>
    <mvc:mapping path="/users-pay-entity/getRejectList"/>
    <mvc:mapping path="/users-pay-entity/payBack"/>
    <mvc:mapping path="/users-pay-entity/page"/>
    <bean class="com.uifuture.ssm.interceptors.AdminLoginInterceptor"/>
</mvc:interceptor>
```

为了更好地定位线上问题，还需要统一的日志拦截器来输出日志。

20.7.4　实现日志拦截器

配置一层的拦截器是比较简单的，先讲解如何实现 Service 层的拦截器配置。新建一个类，如代码清单 20-109 所示。

代码清单 20-109　Service 层日志拦截器

```
@Aspect
@Component
@Slf4j
public class ServiceLogAspect {
    /** 通过@Around声明一个建言，此建言直接使用拦截规则作为参数，拦截接口实现 */
    @Around("execution(public * com.uifuture.ssm.service.impl.*.*(..))")
    public Object around(ProceedingJoinPoint joinPoint) throws Throwable {
            MethodSignature methodSignature=(MethodSignature) joinPoint.
getSignature();
```

```
            String methodName=methodSignature.getName();
            String className=methodSignature.getDeclaringTypeName();
            Object[] args=joinPoint.getArgs();
            Object result=null;
            try {
                    log.info("方  法  名:{}#{},  入  参:{}", className, methodName, JSON.
    toJSONString(args));
                long s=System.currentTimeMillis();
                result=joinPoint.proceed();
                long e=System.currentTimeMillis();
                if (e-s>1000) {
                    log.warn("方法名:{}#{}, 消耗时间:{}ms", className, methodName, (e - s));
                }
            } catch (Exception e) {
                  log.error("方法名:" + className + "#" + methodName + ", 入参:" + JSON.
    toJSONString(args) + ",出现异常", e);
                throw e;
            }
            return result;
        }
    }
```

要使 AOP 生效，需要在 Spring 配置文件中配置 AOP，如代码清单 20-110 所示。

代码清单 20-110　在 Spring 配置文件中配置 AOP

```
<?xml version="1.0" encoding="UTF-8"?>
<beans xmlns:xsi="http://www.w3.org/2001/XMLSchema-instance" xmlns:aop="http://
www.springframework.org/schema/aop"
       xmlns="http://www.springframework.org/schema/beans"
       xsi:schemaLocation="http://www.springframework.org/schema/beans
http://www.springframework.org/schema/beans/spring-beans.xsd
http://www.springframework.org/schema/aop
http://www.springframework.org/schema/aop/spring-aop.xsd">
    <!--基于AspectJ的自动代理-->
    <aop:aspectj-autoproxy proxy-target-class="false"/>
    <!--proxy-target-class="true",Spring建议通过 <tx:annotation-driven/> 标记的 proxy
-target-class属性值来控制是基于接口还是基于类的代理被创建。如果 proxy-target-class属性值被设
置为true，那么基于类的代理将起作用(此时CGLIB库cglib.jar要在CLASSPATH中);
如果proxy-target-class属性值被设置为false，或者这个属性被省略，那么标准的JDK基于接口的代理将
起作用。接口上的注解不能继承 -->
</beans>
```

proxy-target-class 属性的默认值是 false，当该值为 true 时，Spring 将使用 CGLIB 动态代理，而不再是默认的 JDK 动态代理。

Controller 层的拦截器，如果仅仅增加拦截器类，那么是无法实现拦截的。

因为 Spring 的 Bean 扫描和 Spring MVC 的 Bean 扫描是分开的，两者的 Bean 位于两个不同的 Application，而且 Spring MVC 的 Bean 扫描要早于 Spring 的 Bean 扫描，所以当 Controller 层的 Bean 生成后，再执行 Spring 的 Bean 扫描，则 Spring 要被 AOP 代理的 Controller 层的 Bean 已经

在容器中存在，配置 AOP 就无效了。同样的情况也存在于数据库事务中，如果 Service 层的 Bean 扫描配置在 Spring MVC 的配置文件中，而数据库事务管理器配置在 Spring 的配置文件中，也会导致数据库事务失效，原理是一样的。

在这里选择在 Spring MVC 配置中再增加 AOP 注解的扫描，也就是增加配置：<aop:aspectj-autoproxy proxy-target-class="false"/>，这样 Spring MVC 也会扫描拦截器，但是注意在 Spring 中不扫描 Controller 注解和 RestController 注解，而 Spring MVC 配置文件中只扫描 Controller 层的包。Spring 与 Spring MVC 的配置如代码清单 20-111 所示。

代码清单 20-111　Spring 与 Spring MVC 的配置

```
// Spring的包扫描配置
<context:component-scan base-package="com.uifuture.ssm">
    <!-- 使用annotation自动注册Bean,并保证@Required、@Autowired的属性被注入 -->
    <!-- 包自动扫描，不扫描Controller注解-->
     <context:exclude-filter type="annotation" expression="org.springframework.
stereotype.Controller"/>
    <context:exclude-filter type="annotation" expression="org.springframework.web.
bind.annotation.RestController"/>
</context:component-scan>

// Spring MVC的包扫描配置
<!-- 配置自定义扫描的包 -->
<context:component-scan base-package="com.uifuture.ssm.controller"/>
```

Controller 层的拦截器类如代码清单 20-112 所示。

代码清单 20-112　Controller 层的拦截器类

```
@Aspect
@Component
@Slf4j
public class ControllerLogAspect {
    /** 通过@Around声明一个建言,此建言直接使用拦截规则作为参数
     * 拦截接口实现 */
    @Around("execution(public * com.uifuture.ssm.controller.*.*(..))")
    public Object aroundController(ProceedingJoinPoint joinPoint) throws Throwable {
            MethodSignature methodSignature = (MethodSignature) joinPoint.
getSignature();
        String methodName=methodSignature.getName();
        String className=methodSignature.getDeclaringTypeName();
        Object[] args=joinPoint.getArgs();
        Object result=null;
        try {
                log.info("方 法 名:{}#{}, 入 参:{}", className, methodName, JSON.
toJSONString(args));
            long s=System.currentTimeMillis();
            result=joinPoint.proceed();
            long e=System.currentTimeMillis();
            if (e-s>1000) {
```

```
                log.warn("方法名:{}#{}, 消耗时间:{}ms", className, methodName, (e -
s));
            }
        } catch (Exception e) {
              log.error("方法名:" + className + "#" + methodName + ", 入参:" + JSON.
toJSONString(args) + ",出现异常", e);
            throw e;
        }
        return result;
    }
}
```

至此，该项目的核心功能就都实现完了。更多的功能和代码可以访问本书配套的 GitHub 项目资源链接。

最后，送读者一条建议，作为本书的结尾：相较于学习技术，更重要的是学习开发思维。因为技术会更替，而思维可以长存。

附　　录

类型处理器	Java 类型	JDBC 类型
BooleanTypeHandler	java.lang.Boolean,boolean	数据库兼容的 BOOLEAN
ByteTypeHandler	java.lang.Byte,byte	数据库兼容的 NUMERIC 或 BYTE
ShortTypeHandler	java.lang.Short,short	数据库兼容的 NUMERIC 或 SHORTINTEGER
IntegerTypeHandler	java.lang.Integer,int	数据库兼容的 NUMERIC 或 INTEGER
LongTypeHandler	java.lang.Long,long	数据库兼容的 NUMERIC 或 LONGINTEGER
FloatTypeHandler	java.lang.Float,float	数据库兼容的 NUMERIC 或 FLOAT
DoubleTypeHandler	java.lang.Double,double	数据库兼容的 NUMERIC 或 DOUBLE
BigDecimalTypeHandler	java.math.BigDecimal	数据库兼容的 NUMERIC 或 DECIMAL
StringTypeHandler	java.lang.String	CHAR、VARCHAR
ClobReaderTypeHandler	java.io.Reader	—
ClobTypeHandler	java.lang.String	CLOB、LONGVARCHAR
NStringTypeHandler	java.lang.String	NVARCHAR、NCHAR
NClobTypeHandler	java.lang.String	NCLOB
BlobInputStreamTypeHandler	java.io.InputStream	—

续表

类型处理器	Java 类型	JDBC 类型
ByteArrayTypeHandler	byte[]	数据库兼容的字节流类型
BlobTypeHandler	byte[]	BLOB、LONGVARBINARY
DateTypeHandler	java.util.Date	TIMESTAMP
DateOnlyTypeHandler	java.util.Date	DATE
TimeOnlyTypeHandler	java.util.Date	TIME
SqlTimestampTypeHandler	java.sql.Timestamp	TIMESTAMP
SqlDateTypeHandler	java.sql.Date	DATE
SqlTimeTypeHandler	java.sql.Time	TIME
ObjectTypeHandler	Any	OTHER 或未指定类型
EnumTypeHandler	Enumeration Type	VARCHAR，任何兼容的字符串类型，存储枚举的名称（而不是索引）
EnumOrdinalTypeHandler	Enumeration Type	任何兼容的 NUMERIC 或 DOUBLE 类型，存储枚举的索引（而不是名称）
InstantTypeHandler	java.time.Instant	TIMESTAMP
LocalDateTimeTypeHandler	java.time.LocalDateTime	TIMESTAMP
LocalDateTypeHandler	java.time.LocalDate	DATE
LocalTimeTypeHandler	java.time.LocalTime	TIME
OffsetDateTimeTypeHandler	java.time.OffsetDateTime	TIMESTAMP
OffsetTimeTypeHandler	java.time.OffsetTime	TIME
ZonedDateTimeTypeHandler	java.time.ZonedDateTime	TIMESTAMP
YearTypeHandler	java.time.Year	INTEGER

续表

类型处理器	Java 类型	JDBC 类型
MonthTypeHandler	java.time.Month	INTEGER
YearMonthTypeHandler	java.time.YearMonth	VARCHARorLONGVARCHAR